FIFTY YEARS OF PEELING AWAY THE LEAD PAINT PROBLEM

Saving Our Children's Future With Healthy Housing

FIFTY YEARS OF PEELING AWAY THE LEAD PAINT PROBLEM

Saving Our Children's Future With Healthy Housing

DAVID E. JACOBS

ELSEVIER

ACADEMIC PRESS
An imprint of Elsevier

Academic Press is an imprint of Elsevier
125 London Wall, London EC2Y 5AS, United Kingdom
525 B Street, Suite 1650, San Diego, CA 92101, United States
50 Hampshire Street, 5th Floor, Cambridge, MA 02139, United States
The Boulevard, Langford Lane, Kidlington, Oxford OX5 1GB, United Kingdom

Notices
Knowledge and best practice in this field are constantly changing. As new research and experience broaden our understanding, changes in research methods, professional practices, or medical treatment may become necessary.

Practitioners and researchers must always rely on their own experience and knowledge in evaluating and using any information, methods, compounds, or experiments described herein. In using such information or methods they should be mindful of their own safety and the safety of others, including parties for whom they have a professional responsibility.

To the fullest extent of the law, neither the Publisher nor the authors, contributors, or editors, assume any liability for any injury and/or damage to persons or property as a matter of products liability, negligence or otherwise, or from any use or operation of any methods, products, instructions, or ideas contained in the material herein.

ISBN: 978-0-443-18736-0

For Information on all Academic Press publications
visit our website at https://www.elsevier.com/books-and-journals

Publisher: Stacy Masucci
Acquisitions Editor: Elizabeth A. Brown
Editorial Project Manager: Pat Gonzalez
Production Project Manager: Kiruthika Govindaraju
Cover Designer: Vicky Pearson Esser

Typeset by MPS Limited, Chennai, India

Working together
to grow libraries in
developing countries

www.elsevier.com • www.bookaid.org

Dedication

For the children and the many who struggled tirelessly to protect them.

Contents

About the author   xi
Foreword   xiii
Preface   xxv
Acknowledgments   xxix
A short summary: lead is a long-lasting
insidious poison   xxxi

Part 1

Paralysis and the abject failure to address lead paint before 1985

Chapter 1 Banning lead paint:
the missed opportunity   3

1.1   The beginning   3
1.2   Finger pointing—food versus gasoline versus paint
versus bad landlords versus bad parents   4
1.3   "It's right in front of you—it's the paint!"   11
1.4   Inadequate measurement methods obscure the
lead paint problem   13
1.5   Where lead was used   17
1.6   The first international lead paint ban   18
1.7   How the US government changed from
promoting lead paint to banning it   21
1.8   Lead-free versus lead-safe   25
1.9   Toxicity research and intervention solution
research   27
References   28

Chapter 2 Early failures and the seeds of
success   35

2.1   Limitations of the Medical Model: The 1971
Lead Based Paint Poisoning Prevention Act   36
2.2   Treatment versus prevention   36
2.3   Surveillance and population surveys   41
2.4   Reagan's "New Federalism": lead poisoning
disappears from the policy agenda   43

2.5   Where was the housing profession?   44
2.5.1   Decent safe and sanitary housing   44
2.6   Early lead paint removal efforts backfire   45
2.7   Housing codes fail to regulate lead paint   49
2.8   Housing law and public housing   50
2.9   Lawsuits and affordable housing   52
2.10 Seeds of success: new pathway studies reveal
the importance of lead dust from paint   53
2.11 Time for change   55
References   55

Part 2

Breaking the Barriers to Progress (1986–2001)

Chapter 3 Solutions take shape: the lead
paint Title X law   61

3.1   Science, policy, and practice meet—an
unlikely venue   62
3.2   Bombshell report to Congress: "corrective
actions have been a clear failure"   67
3.3   Congress tries again: the 1987 Housing Act
and the 1988 Stewart McKinney
Amendments   69
3.4   Leaders in public housing take action: the
birth of lead paint risk assessments   71
3.5   A scandal prompts Congress to create the
HUD lead paint office   75
3.6   Moving remediation from the Centers for
Disease Control and Prevention to the
Department of Housing and Urban
Development: The 1990 Public Housing
Guidelines   79
3.7   Sticker shock: The Department of Housing
and Urban Development's comprehensive and
workable plan   84
3.8   A new Alliance to End Childhood Lead
Poisoning marshals political will   85

3.9   The nation's health secretary declares lead poisoning the number one childhood environmental disease over White House objections 88

3.10 The Centers for Disease Control and Prevention issues public health strategic plan and new medical guidance 88

3.11 Congress acts: Title X of the 1992 Housing and Community Development Act 91
    3.11.1 Moving from reaction to prevention 94
    3.11.2 Creating the workforce 96
    3.11.3 Seven principal purposes 97
    3.11.4 Using science to define "lead paint hazard" 97
    3.11.5 Show me the money: new Congressional appropriations for private housing 98
    3.11.6 Reforming housing regulations 98
    3.11.7 The right to know 99
    3.11.8 Renovation, repair, and painting 101

3.12 Bringing science to bear: a new National Center for Lead-Safe Housing brings health and housing together 102

3.13 Confidence emerges 105

References 105

Chapter 4 Growing pains—new regulations, enforcement, capacity, and proof emerge 113

4.1 The 1995 rescission and bringing science to the Department of Housing and Urban Development 114

4.2 The 1995 Department of Housing and Urban Development lead paint guidelines 122

4.3 The Title X task force fills in the gaps 126

4.4 The fight over lead dust standards 129
    4.4.1 "Model" wars 135

4.5 Do the new remediation methods work? 140

4.6 The struggle to reform all federal housing lead paint regulations 147
    4.6.1 Cold feet 159

4.7 Improved lead paint testing—how government stimulated private innovation 162

4.8 National lead laboratory accreditation program 165

4.9 First enforcement actions 166
4.10 Mustering the proof 174
References 175

Chapter 5 The Nation Acts: community organizing, a 10-year solution from the President's Cabinet, and political sabotage 187

5.1 Parents and communities 188

5.2 The Campaign for a Lead-Safe America 194

5.3 The Community Environmental Health Resource Center 200

5.4 Community groups and the press 203
    5.4.1 Advocates and scientists 204

5.5 The President's Cabinet approves a 10-year strategy, 2000–10 206
    5.5.1 Lead paint industry interference 215
    5.5.2 Why the 2010 goal was not achieved 218

5.6 Political sabotage 219
    5.6.1 Attack and counterattack at HUD, 2004 219
    5.6.2 Attempted elimination of the CDC lead program, 2012 227

References 231

Chapter 6 Research ethics and the *Grimes* court case 243

6.1 The context: lead poisoning and the courts 243
    6.1.1 The Kennedy Krieger Institute and health research 245
    6.1.2 City Homes and affordable housing 247

6.2 Legal and scientific evidence 248

6.3 Research ethics and protection of research study participants 252

6.4 The Baltimore lead paint abatement and repair and maintenance study 254

6.5 The *Grimes* decision 260
    6.5.1 Therapeutic and nontherapeutic research 261
    6.5.2 Protecting children or putting them in harm's way? 263
    6.5.3 New ground 264
    6.5.4 Parents versus the court of appeals 264
    6.5.5 Institutional review boards 265
    6.5.6 Special duty 265

6.5.7   Informed consent   266
6.5.8   The facts in the case   267
6.5.9   Confusion on lead dust testing methods   267
6.5.10   Federal oversight   268
6.5.11   The Appeals Court comparison to Nazi and Tuskegee research   269
6.6   Ethics in housing intervention research   270
6.7   The best of intentions or the best of community-based science?   276
6.8   Environmental justice and community participation in research   280
6.9   The legacy of the Maryland Court of Appeals *Grimes* decision   281
References   283

Part 3

The new consensus (2001–22)

Chapter 7 If "you make a mess, you have to clean it up"- the Rhode Island and California court decisions   291

7.1   Local jurisdiction lawsuits against the lead paint industry   291
7.2   The Rhode Island court decision, 1999–2008   295
       7.2.1   Public nuisance law   297
7.3   The California court decision, 2000–2022   299
7.4   The industry fights back   306
7.5   The new consensus   308
References   309

Chapter 8 The US and international healthy homes movement   313

8.1   The detective scientists who solved the Cleveland mold mystery   313
8.2   The Department of Housing and Urban Development healthy homes report to Congress   319
8.3   The Surgeon General's Call to Action   325
8.4   Assembling the evidence   326
       8.4.1   Does it work? A systematic review of housing interventions and health   329

8.5   The formation of the National Safe and Healthy Housing Coalition   330
8.6   The World Health Organization Healthy Homes Movement   331
8.7   "A Kid Who Grew Up in Public Housing"   339
Appendix: Vilnius Declaration   341
References   345

Chapter 9 Reframing health, environment, and housing   351

9.1   Health: reframing communicable and noncommunicable disease   351
9.2   Environment: reframing the shared commons   354
9.3   Housing: reframing wealth, affordability, and equity   359
9.4   Toward a healthy housing consensus   366
References   368

Chapter 10 Conclusion: the triumph of science and citizen action over policy paralysis   373

10.1   Knowing and doing   373
10.2   Two steps forward, one step back   377
10.3   The Find It, Fix It, Fund It campaign   379
10.4   Getting the housing market to work   383
10.5   Getting government to work   384
10.6   Getting the procedures right and recruiting the necessary expertise   385
10.7   Strategic plans   386
        10.7.1   A new forecast to eliminate lead paint poisoning by 2027   388
10.8   The influence of industry   388
10.9   Nine lessons from lead paint poisoning prevention   389
10.10   Ending a policy paralysis paradox   391
References   392

Appendix 1: US government agencies involved in lead paint   397
Appendix 2: Honor role-leaders in lead paint poisoning prevention and healthy housing   399
Glossary   417
Index   421

About the author

David E. Jacobs led the nation's childhood lead poisoning prevention efforts from 1995 to 2004, when he directed the lead paint and healthy homes office at the US Department of Housing and Urban Development. The rare scientist to work across the housing, health, and environmental fields, Dr. Jacobs helped design, reform, and implement policies, laws, and regulations, which together with mobilized citizens and parents protected millions of children from lead poisoning.

Currently the chief scientist at the National Center for Healthy Housing, he holds appointments at the Schools of Public Health at the University of Illinois Chicago and Johns Hopkins University. He is the director of the US Collaborating Center for Healthy Housing Research and Training for the World Health Organization. For more than a decade, he has served as the board president of Lincoln Westmoreland  Housing, a unique interracial nonprofit organization providing low-income housing and community services. He holds degrees in technology and science policy, environmental health, engineering, and political science and is board certified in occupational health (industrial hygiene).

Foreword

Don Ryan

Alliance to End Childhood Lead Poisoning, Served as Executive Director from 1990 to 2005

This book is the inside story about why it took the United States more than 50 years to come to grips with lead paint, which has poisoned millions of children in their homes. This story has villains, vested interests, victims, and heroes, but at the core it is about why public policy must be based on science. This is a cautionary tale for legislators, judges, agency staff, advocates—indeed, all of us—about the pitfalls of ignoring scientific evidence. But ultimately, this is a success story about how researchers, public health and housing experts, government agency and congressional staff, parents of poisoned children, and advocates for children's health overcame daunting obstacles to put the nation on the path to real solutions.

No one is better suited to tell this story than Dave Jacobs who was "in the room when it happened" at many key junctures. I served as executive director of the Alliance to End Childhood Lead Poisoning from 1990 to 2005 and as congressional staff before that. To set the stage for this book, I map the policy and political landscape that existed as advances in science challenged old assumptions and beliefs and demanded new approaches to protect children from poisoning.

1 A social crime

Rene' Dubos, the microbiologist and philosopher who made famous the maxim "Think globally, act locally," was a keen observer of science and society. In 1969 Dubos reflected on childhood lead poisoning: "... the problem is so well-defined, so neatly packaged, with both causes and cures known, that if we don't eliminate this social crime, our society deserves all the disasters that have been forecast for it." Dubos was justified in calling childhood lead poisoning a social crime, and he was correct in highlighting lead paint, which eclipsed exposures from other sources at the time. But he sorely underestimated what it would take to bring about the "cure."

The tragedy of childhood lead poisoning rightfully outraged Dubos. Although millions of children across all income strata were at risk, national health survey data would later show striking inequalities by race and income: Black children were at eight times higher risk for poisoning, and children of low-income families were at five times higher risk. Viewed through the lens of Black Lives Matter, the fact that lead poisoning was widely dismissed before the 1950s is appalling today, just as it should have been then.

In the 1940s and 1950s many children died from lead poisoning as extremely high blood lead levels, typically in the range of 100–200 µg/dL (micrograms of lead per tenth of a liter of blood), caused coma and convulsions due to brain encephalopathy

xiii

and other organ damage. To save the lives of severely poisoned children, doctors administered drugs called chelating agents to leach some of the lead out of their bodies through urine. Although chelation therapy did save lives, it was a clumsy and limited tool at best: children had to be hospitalized for painful injections, and there were often harsh side effects. This disease has no medical cure.

2 Recognition of lead's extreme toxicity to children

Doctors and researchers then discovered that children who survived severe lead poisoning still suffered lasting neurological damage, including mental retardation. This realization led to federally funded research, sustained over several decades, that confirmed lead's neurotoxicity at lower and lower blood lead levels.

The dangers of adults' occupational exposures to lead had been known for thousands of years, starting with Nikander (the Greek physician and poet) in the second century BC. In 1786 Benjamin Franklin rued the failure to protect printers from "the dangles," wrist drop due to nerve damage long known to be caused by lead. But preschool children's developing brain and nervous system put them at highest risk for lead's neurotoxic effects: reduced intelligence and attention span, learning disabilities, and behavior problems—among many other ailments. The effects of lead poisoning in childhood persist for years and dim lifetime potential.

Based on the steadily growing evidence of lead's neurotoxicity, the Centers for Disease Control and Prevention (CDC) progressively lowered the blood lead level that triggered action from 80 μg/dL in the 1950s to 60 (1960), 30 (1970), 25 (1985), 10 (1991), 5 (2012), and 3.5 (2021).

In the 1970s growing recognition that children were suffering adverse effects far below levels that present clinically observable symptoms led state and local health departments to concentrate on "screening" to identify children with elevated blood lead levels and "case management" to investigate lead hazards in the child's home and track changes in their status over time. In public health parlance, screening and case management constitute "secondary prevention," trying to detect a disease early and prevent it from getting worse (but in the case of lead, only after toxic exposure had already occurred). Yet most children and virtually all houses remained untested.

3 The lead pigment, and paint industries' fierce opposition

The lead, pigment, and paint industries' relentless efforts to challenge regulation of lead have been well documented (see for example Markowitz and Rosner's aptly titled book *Deceit and Denial*). Similar to the tobacco industry's attempts to cloud the link between smoking and cancer, the Lead Industries Association and its allies conducted a relentless campaign to stall regulations, challenge the findings of damning studies, sponsor shoddy research to muddy the waters, and use charges of scientific misconduct to intimidate researchers.

The lead, pigment, paint, and petroleum industries also engaged in a kind of "toxic shell game," arguing that the ubiquity of lead in the environment and exposures from a multitude of uses (e.g., gasoline vs paint vs drinking water vs industrial emissions) offered an excuse not to regulate lead at all. Finally, the lead and paint industries tried to shirk responsibility by blaming slumlords for poor maintenance and blaming parents for poor housekeeping and bad childcare. The quest to hold the lead paint manufacturers accountable for the problem they created is one thread in this story.

4 Progress in regulating new uses of lead

In 1971 the tide began to turn in regulating ongoing uses of lead when the US Congress passed the first Lead-Based Paint Poisoning Prevention Act and directed the Consumer Product Safety Commission (CPSC) to limit the maximum amount of lead in new paint to 1%, and 2 years later reduced this threshold to 0.06%. (Today the limit for lead in new residential paint is 0.009%). Over the next 15 years the Environmental Protection Agency (EPA) effectively banned lead from gasoline, plumbing fixtures, pesticides, fertilizers, and printing ink. The Food and Drug Administration eliminated the use of lead solder in food and baby formula cans, and CPSC began regulating lead in many other consumer products, such as makeup, jewelry, and toys. The Occupational Safety and Health Administration regulated lead exposures in industrial workplaces in 1978, partly due to concern for workers' children being exposed to lead brought home from the workplace on clothing.

Childhood lead poisoning prevention was justifiably heralded as an environmental health success story as national health surveys documented steady declines in children's average blood lead levels. Indeed, from the 1970s to the late 2010s lead levels in children's blood declined by a remarkable 93.6%, a ringing public health success.

5 Grappling with the legacy of past uses of lead

Despite these population-wide gains, lead poisoning remained widespread in the 1980s among preschool children in cities across the United States—disproportionately children in low-income families and children of color. Lead paint in millions of homes and apartments was the foremost source, followed by soil contaminated by both exterior paint and past emissions from leaded gasoline and in some locales from industrial emissions. In the 1920's most industrialized countries endorsed an international convention to eliminate lead in residential paint, but the lead and paint industries successfully lobbied to delay US regulation for decades, burdening tens of millions more homes and apartments with lead paint.

Local health department staff who screened children for lead poisoning and investigated their homes for hazards had long recognized lead paint as the primary culprit. If lead paint is the primary cause of poisoning, the solution must be to remove all the leaded paint from US housing, right?

If only the cure was so neat and simple.

6 "Traditional" practices were aggravating the problem

Baltimore's older, dilapidated housing made it one of the epicenters of lead poisoning, with elevated lead levels in nearly 40% of children in some older, low-income neighborhoods (for context, these blood lead levels were orders of magnitude higher than exposures due to the outrageous drinking water recent debacle in Flint, Michigan). Like most cities, the Baltimore Health Department struggled to enforce clean-up measures after a home had poisoned a child. Disturbed by seeing generation after generation of children poisoned by lead in the same neighborhoods and houses, J. Julian Chisolm and Mark Farfel, two scientists from the Kennedy Krieger Institute (KKI), a community health nonprofit associated with Johns Hopkins University, took a hard look at the effectiveness of prevailing paint removal measures.

Their before-and-after comparisons of lead dust levels in children's homes led to the shocking realization that practices widely used to remove lead paint were *increasing* children's exposure to lead. Their findings, first published in 1983 and punctuated by the reports of children who had to be hospitalized or rehospitalized following paint removal, sent a wake-up call to practitioners and policymakers: "traditional" lead paint removal methods (e.g., open flame burning, power sanding, dry scraping, and abrasive blasting) generated dangerous levels of lead dust. These widely used lead paint removal methods, which some cities mandated in regulation, were making the problem *worse*.

7 A widening gap between science and policy

KKI's ground-breaking study spurred other researchers to delve more deeply into how children were actually exposed to lead from paint and to evaluate a range of hazard control measures. By 1988 this led Chisolm and Farfel and other researchers to the realization that lead-based paint is the primary *source* of children's high-dose exposures but settled lead dust is the primary *pathway* of poisoning. Lead dust particles, which are usually not visible to the naked eye, settle quickly and remain on windowsills, floors, and children's toys and hands. Ingestion of lead dust due to hand-to-mouth transmission is the primary pathway of exposure for most young children. Researchers found paint condition and lead dust levels on surfaces such as floors and windowsills to be much stronger predictors of risk than the paint's lead content.

This discovery demanded attention to lead dust and development of a range of hazard control strategies beyond simply removing lead paint. But policy lagged behind science in two important respects. First, as researchers worked to define lead *hazards*, policymakers remained preoccupied by the *presence* of lead paint and the concentration of lead in the paint. Second, as researchers began evaluating a range of promising strategies to control these hazards in both the short and long run, policymakers (and some advocates) held to the simplistic belief that removing all lead paint was the only solution, despite its inherent risks. Translating researchers' seismic shift in understanding into policy and practice posed an enormous challenge.

8 The Department of Housing and Urban Development's Foot-Dragging outrages Congress

In 1973 Congress gave the Department of Housing and Urban Development (HUD) primary authority "to eliminate as far as practicable the hazards of lead-based paint poisoning" in federally assisted or insured properties. With no grounding in science, HUD was ill equipped to carry out its duties under this vague directive. Moreover, the added cost associated with lead paint stood at odds with HUD's mission of expanding affordable housing. HUD dragged its feet for more than two decades, its only notable contribution being support for research to advance X-ray fluorescence technology to measure the amount of lead in painted surfaces, which served to reinforce preoccupation with the presence of lead paint and measuring its lead content.

In a landmark class action lawsuit brought by public housing tenants in Washington, DC, a US district court in 1982 criticized HUD's failure to include intact leaded surfaces as an immediate hazard. After HUD changed its hazard definition to comply, Congress passed a major lead poisoning law in 1988 codifying in statute that "intact and nonintact paint on both interior and exterior surfaces" constituted an immediate hazard. Congress's expansive hazard definition stood sharply at odds with the growing body of scientific evidence. The 1988 law was silent on lead dust.

9 Congress dictates "all or nothing" for public housing

This 1988 law also directed public housing authorities to inspect for lead paint and prescribed full abatement as the only response, but the timing of abatement was problematic because it depended on the availability of funding for "comprehensive modernization." Congress placed public housing authorities in an untenable position: they had to inspect for lead paint, identify every leaded surface, inform their tenants, and then typically wait for years, sometimes decades, to conduct abatement. This "all or nothing" dilemma ignored the urgent question facing public housing authorities—and the owners of millions of other properties—"What should we be doing *right now* to protect children whose homes contain lead paint?" The result was predictable: more lawsuits and more poisoning.

This clumsy 1988 law pushed public housing authorities to the forefront in figuring out how to manage lead paint in US housing. Some public housing authorities responded by working to develop a common set of interim safety measures and creating their own liability insurance pool for authorities that followed these measures. They turned to Jacobs, an occupational health scientist from the Georgia Institute of Technology (and the author of this book), to develop a prototype "risk assessment" protocol as well as a prescribed set of measures to manage lead paint until it could be abated. Jacobs went on to serve as deputy director of the National Center for Lead-Safe Housing and then led HUD's Office of Lead-Based Paint and Healthy Homes from 1995 to 2004. As this book details, Jacobs worked directly with four HUD Secretaries whose regard for science and interest in protecting children from lead poisoning varied widely.

10 Striking fear in the hearts of owners, insurers, and lenders

Although the 1988 law only impacted public housing (less than 1% of the US housing stock), it signaled Congress' view that intact lead paint constituted an immediate hazard. This cast a pall over millions of homes and apartments with lead paint and sent a chill through the housing, real estate, liability insurance, and lending industries. Property and casualty insurers responded by writing "lead exclusions" into their liability policies, and lenders began to question financing any property with lead paint. Responsible property owners were confused about what steps they should take to provide due care. Anxiety and paralysis ruled the day, what Jacobs calls "policy paralysis."

11 The Helter-Skelter of lead poisoning lawsuits

Tort lawsuits (and fear of them) also directed national attention to childhood lead poisoning. Parents of lead-poisoned children brought hundreds of suits against rental property owners in Baltimore, New York, Philadelphia, and other lead poisoning hot spots. The vast majority of lead-poisoned children never received any compensation at all, but the occasional "jackpot" award captured headlines and struck fear into the hearts of rental property owners and their insurers. Many "slumlords" simply created shell corporations to insulate themselves from liability.

Uncertainty over what constituted a lead hazard and the lack of clear lead safety standards made lead poisoning lawsuits extremely difficult for plaintiffs to prove—and, paradoxically, equally difficult for landlords to defend against. As a result, the tort system operated so randomly that no clear "standards of care" emerged, leaving many yearning for clarity about how to deal with lead paint in homes and apartments.

12 Bombshell report to Congress confronts policymakers

In my 10 years as professional staff to the House Appropriations Committee, the most compelling report I read was "The Nature and Extent of Lead Poisoning in Children in the United States," which the Agency for Toxic Substances and Disease Registry issued in 1988. Paul Mushak, one of the nation's leading toxicologists and the report's primary author, resigned from the Public Health Service in protest after stiff internal opposition within the executive branch threatened to water it down. The report pulled no punches, confronting policymakers in both Congress and executive agencies with deeply disturbing facts:

– The scientific evidence is overwhelming that, even at very low levels, lead's neurotoxic effects hurt young children's developing brains and reduced their intelligence, learning, attention span, and behavior, causing lifelong damage.
– More than one million US children have elevated lead levels in their blood that are causing such adverse health effects.
– Lead poisoning disproportionately affects low-income children living in older housing especially in communities of color; remodeling and renovation projects that inadvertently create lead dust hazards also threatened children from higher income families.
– Lead-based paint is the primary unattended source of US children's exposure to lead, but lead-contaminated dust is the primary pathway of children's exposure.
– And the bottom line: "Corrective actions on lead paint have been a complete failure."

13 Stuck in the rut of reacting to poisoned children

Thirty-seven large cities and states operated Childhood Lead Poisoning Prevention Programs funded by CDC grants during the 1980s. In truth, most of these programs were prevention in name only, as they struggled first to identify poisoned children by testing their blood and then to force corrective action. They failed to live up to their name for multiple reasons:

1. A poisoned child was the only trigger for action—Most state and local laws

had no general requirements for lead safety, the only trigger for action being identification of an already-poisoned child. Even when the system worked, the response came only after the child's neurotoxic exposure had already occurred, often with irreversible effects.

2. Most lead-poisoned children were missed—The vast majority of children with elevated lead levels were never identified. Federal funding for screening programs was drastically reduced in 1980, and many jurisdictions performed no blood lead screening. Many pediatricians discounted the significance of a disease that presented no clinical symptoms and wrongly believed that children from higher income families were not at risk. And physicians were not trained to recognize and provide advice on what was fundamentally a housing problem.

3. Lead hazards in most poisoned children's homes were never controlled. With only a few exceptions, most state and local laws were inadequate and lax enforcement frustrated efforts to correct lead hazards. For example, Baltimore based its lead abatement orders on the presence of a poisoned child, which meant a landlord could moot the order by evicting the family. As public health staff followed the child to their new address for ongoing screening, the landlord would rent the home to another family, repeating the cycle. *The Baltimore Sun* later identified one rental property that poisoned children from seven different families. At their peak, state and local lead poisoning prevention programs secured corrective measures in only about 25,000 homes per year nationwide, a generous estimate since programs failed to consistently track and report this embarrassing statistic. At this rate, it would take 2000 years to solve the problem in US housing.

4. When abatement did occur, blood lead levels often remained elevated or got even worse. In the rare instances when health departments succeeded in securing corrective action, many children's blood lead levels remained stubbornly elevated. In many cases, simplistic paint removal created lead dust and made the problem even worse. One researcher called this "urban lead mining."

5. Parent education was not effective—Most health departments sought to educate parents about the importance of high-calcium and iron diets and housekeeping, but parents did not have it in their power to protect their children from poisoning. Study after study showed education alone just did not work. At best, such education sought to teach parents to raise their children in a hazardous environment. At worst, it shifted responsibility from rental property owners and set up parents for failure and blame.

14 Protecting children required a whole new approach

A growing number of experts and advocates for children's health recognized the inherent limitations and multiple failures of relying on identification of a poisoned child as the sole trigger for action. In 1991 CDC's Strategic Plan to Eliminate Childhood Lead Poisoning called for a shift in approach to "primary prevention," making housing safe for children in the first place. While this public health agency's call for fundamentally shifting the national approach was bold, its plan offered only a fuzzy vision for how to make US housing safe for children.

At one level, the promise of primary prevention had appeal as wide as "motherhood and apple pie:" Who could oppose taking preventive action before a child is poisoned? But realizing this vision required answering technically complex and politically loaded questions: What conditions constitute a hazard? What measures effectively control hazards? What conditions or events should trigger action by private property owners? How could priority be directed to children at highest risk? How could the reliability of hazard identification and control be assured? And who should pay: property owners, taxpayers, or the lead paint manufacturers?

Federal agencies were clearly not up to the challenge of answering these vexing questions. CDC supported local lead poisoning prevention programs but had little grounding in housing, as its strategic plan reflected. HUD had no grounding in health or science and its regulatory reach extended only to federally assisted housing, which generally posed lower risks than purely private low-income housing. EPA regulated many sources of lead in the environment but claimed its catch-all authority under the Toxic Substances Control Act did not encompass lead paint, which its staff viewed as "a HUD program."

Incredibly, no one in the key offices at EPA or HUD could name a single person at the other agency when I asked them in 1988. Lead paint had fallen through the cracks.

15 A big surprise: 57 million American homes contain lead paint

Congress' 1988 law also directed HUD to submit a "comprehensive and workable plan" for inspecting and abating lead paint in privately-owned housing. Although HUD's December 1990 report proved to be anything but workable, it included the first national survey of lead paint and lead hazards in US housing. The findings shocked everyone:

- 57 million homes, fully three-fourths of all US housing built before 1980, contain some lead-based paint (a later reanalysis estimated the real number was closer to 64 million).
- Lead paint is prevalent in homes across the income spectrum.
- 20 million homes—more than one in five of all US housing units—contain a lead hazard, either nonintact lead paint or elevated lead dust levels, or both.
- 3.6 million homes with immediate lead hazards are occupied by a young child, an arresting statistic that highlighted the urgent need for action to protect millions of preschool children at immediate risk.
- Fully abating all lead paint in US housing would cost $500 billion dollars.

For the first time, the enormity of lead paint's diverse challenges confronted legislators, policymakers, property owners, public health and housing professionals, researchers, and advocates for children's health and affordable housing.

16 Wildly differing reactions to our lead-contaminated housing stock

The discovery that more than half of all US housing was burdened by lead paint drew a wide range of reactions. Paralyzed by the price tag, some property owners found the problem "too big to solve" and used the high cost of removing lead paint as an excuse for inaction. Some small landlords engaged in scare tactics by threatening to abandon rental properties if governments mandated any hazard controls. Some large rental property owners demanded "safe harbor" (immunity from lawsuits) for taking unspecified measures to manage lead paint safely in place.

At the same time, advocates held conflicting views based on their perspectives and priorities. Some health advocates equated primary prevention to full paint removal, viewing short-term measures as a "risky gamble." Some legal services attorneys feared that federal recognition of short-term hazard control strategies would undercut their leverage in seeking full abatement for their lead-poisoned clients.

On the other hand, advocates for affordable housing and homelessness viewed lead paint as a direct threat to affordable housing, as well as children's health. They believed calls for removing all lead paint to be an unreasonable squandering of scarce resources which could be better used to meet other critical housing needs. And they feared that, absent deep public subsidies for abatement, owners of very low-income housing would abandon their properties, exacerbating the shortage of affordable housing.

Researchers and some public health experts emphasized the need for the emerging science to guide policy. They noted the reality that even if full removal of lead paint in US housing were somehow to become a national priority, it would take decades to complete—and that urgent action was needed in any case to protect children in the meantime. Some researchers also noted the inherent risk that widescale efforts to remove intact lead paint could inadvertently increase children's lead dust exposures.

17 A nonprofit organization steps into the breach

Convinced that Congress and federal agencies could not chart a path to primary prevention and that solutions would require action by state and local governments as well as the private sector, experts and advocates created a new national nonprofit policy and advocacy organization called the Alliance to End Childhood Lead Poisoning, which opened its offices in October 1990.

Lead poisoning's grip on me (Don Ryan) prompted a radical change in my career, as I was grabbed by the scope of the problem, the stark disparities in risk, the "all or nothing" irrationality of Congress' 1988 law, and the lack of coordination among federal agencies. So, I left the professional staff of the House Appropriations Committee, where I had overseen budgets for EPA, CPSC, and some HUD programs, to launch the Alliance and serve as its first executive director.

Experts and advocates from a range of fields and disciplines stepped forward to serve on the Alliance's board, including Herb Needleman, a leading researcher on lead's toxicity; Stephanie Pollack, an attorney who had worked on Massachusetts' landmark lead law; Phil Landrigan, a pediatrician and strong advocate for lead poisoning prevention; Cushing Dolbeare, founder of the National Low-Income Housing Coalition; Bailus Walker, former Washington, DC health director; and Nick Farr, with the Enterprise Foundation, a nonprofit that expanded financing for affordable housing.

Intently focused on primary prevention, the Alliance tackled "hot potato" issues such as what constitutes a hazard and how to balance short-term and permanent hazard controls. To maintain its independence and credibility, the Alliance refused funding from any organizations with an economic stake in the issue. The Alliance sought grants from private foundations with mixed success, but the House and Senate Appropriations Committees provided funds for the Alliance, and EPA, HUD, and CDC welcomed its assistance in tackling tough questions and building consensus.

The Alliance sought to educate policy-makers by putting childhood lead poisoning on the national agenda, inspiring feature stories in the national press and media, capped by *Newsweek's* July 14, 1991 cover story. The Alliance organized a coalition of national groups to advocate for prevention, tapping the political muscle of the education, labor, and environmental movements and bringing affordable housing advocates "into the tent." And the Alliance anchored a network of over 300 grassroots organizations committed to tenant rights, affordable housing, environmental justice, and lead poisoning prevention, which added political clout. The Alliance brought parents of poisoned children together, who themselves formed United Parents Against Lead and several other groups. This book tells their story for the first time, and how important parents were in forging progress and building political will.

The Alliance's major contribution was advancing national prevention policy. Alliance staff tapped experts, researchers, and practitioners from across the country on its Technical Advisory Committee to identify key issues and weigh alternative approaches. In its first year, the Alliance developed a policy framework for primary prevention, which was vetted at its October 1991 national conference, the first to focus on lead paint. Over 800 experts, advocates, and parents attended this conference, which provided the platform for major announcements by then HHS Secretary Louis Sullivan. Against the White House Office of Management and Budget's direct orders, Sullivan declared, "Lead poisoning is the No. 1 environmental threat to the health of children in the United States" and announced that CDC was lowering the "level of concern" in blood lead levels from 25 to 10 μg/dL, which recognized a 10-fold increase in the number of children affected. For good measure Sullivan added, "Let me state unequivocally that President Bush and I are committed to ending this senseless, totally preventable tragedy."

18 National Center for Lead-Safe Housing (later National Center for Healthy Housing)

As a policy and advocacy organization, the Alliance itself lacked scientific expertise. All sides recognized the need for a trusted science-based organization to render objective judgments about key technical questions. Concerned that lead paint threatened affordable housing as well as children's health, the Fannie Mae Foundation made a $5.2 million grant in 1992 to create the nonprofit National Center for Lead-Safe Housing, which was jointly sponsored by the Alliance and the Enterprise Foundation. Nick Farr left the Enterprise Foundation to lead NCHH, and Jacobs left his post at Georgia Tech to become its first deputy director, before moving on to lead national prevention efforts at HUD.

NCHH would develop the national technical guidelines for lead paint as well as training for risk assessors and hazard control professionals. Its research supported lowering the hazard standard for lead dust, and its evaluation of HUD's lead hazard control grants, which tracked the impact of lead safety treatments in 3000 homes in 14 jurisdictions over 3 years, provided authoritative evidence that the new methods reduced lead in both house dust and children's blood.

19 A landmark federal law

In 1992 an unlikely legislative break-through shifted the national approach to primary prevention. The housing, real estate, insurance, and lending industries' extreme discomfort with the status quo set the stage for enactment of a comprehensive national lead poisoning prevention law, commonly known as Title X of the Housing and Community Development Act of 1992. Bruce Katz, who later championed progress on lead poisoning as chief of staff to HUD Secretary Cisneros, was the top staff person on the Senate Banking Committee's Housing Subcommittee. After hearings made clear the urgent need for comprehensive legislation, Katz and his staff assistant Cheryl Fox relied heavily on Cushing Dolbeare and me to shape the bill, draft many of its key sections, and rebut criticism from housing and real estate interests.

Reflecting the Alliance's principles for prevention, this landmark law defined lead hazards based on science, mandated both immediate and long-term hazard controls in federally assisted housing, expanded research, established a lead hazard control grant program for highest risk, privately-owned housing, created a national system for training and certifying contractors, and assigned responsibilities to federal agencies and set deadlines.

This book tells the inside story of how this legislation was passed into law, and the decades-long fight to get it implemented.

20 The critical question facing the nation: what makes homes safe for children?

Title X recognized the need for both short-term strategies and permanent solutions to deal with lead paint in US housing. But the evidence was unclear about "what works" to prevent lead exposure in a child's home and what combinations of strategies offered the best "bang for the buck:" replacing windows, making floors "smooth and cleanable," cladding window sills, enclosing lead paint, stabilizing peeling paint, intensive cleaning, covering bare soil, and dozens of other measures.

Congress mandated research to ensure that cities and states receiving lead hazard control grants from HUD used the best mix of strategies and funded a multiyear evaluation to confirm the results. Mark Farfel, the researcher with Johns Hopkins' KKI in Baltimore who had highlighted the dangers of traditional paint removal practices, was among the first to step up to tackle the challenge of "what works." KKI's Abatement, Repair and Maintenance Study validated the effectiveness of new hazard control strategies by examining both lead dust levels and children's blood lead levels over time.

In a chilling example of judicial over-reach, ignoring both the underlying science and the facts of the case, Maryland's highest court leapt to unfounded conclusions, condemning KKI's evaluation as racist research and impugning Farfel's reputation. This book sets the record straight for the first time. In fact, once juries in Baltimore had a chance to consider the facts, they consistently exonerated the researchers.

21 Federal agencies struggle with their assignments

Title X gave marching orders and strict deadlines to federal agencies, primarily HUD and EPA, but both agencies struggled to meet their responsibilities. HUD turned the corner in 1995, when Bruce Katz (then chief of staff at HUD) persuaded Secretary Cisneros to hire Jacobs to lead the Department's efforts. Jacobs hired other scientists and talented staff who overhauled HUD's lead safety regulations and turned around HUD's lead hazard control grants, which were later recognized as one of HUD's top performing programs. HUD and EPA promulgated and enforced lead hazard disclosure regulations (with the help of the Department of Justice and local prosecutors) that gave people the right to know about lead paint in their homes, which resulted in millions of abated homes. HUD funded a number of parent and community groups to enable them to tell their stories and press for protective action. HUD worked with CDC and local health departments to create a system to match lead-poisoned children with safe federally assisted housing and passed new standards for lead dust, which EPA later adopted. And HUD led development of a multi-agency, Cabinet-level strategic plan, which helped secure record funding from Congress.

The inside story of how HUD came to lead national efforts on lead paint is described for the first time in these pages— along with how Secretary Alphonso Jackson later attempted to diminish the program and shuffled Jacobs aside.

22 Lead poisoning prevention and healthy homes

Progress over the next decade led to realization that most homes that were poisoning children with lead also had moisture, mold, pest, and other problems, which caused childhood asthma, injuries, and many other maladies. Many other countries had already embraced the concept of healthy homes, but in the United States, similar to lead poisoning, it took the tragedy of multiple infant deaths in Cleveland from mold exposure due to defective ventilation systems to direct attention and galvanize political will.

A courageous pediatric pulmonologist, Dorr Dearborn, a local health commissioner, Terry Allan, and government scientists started the healthy homes movement in the United States, with the help of Louis Stokes, who represented Cleveland in Congress. In 1999 Representative Stokes launched the Healthy Homes Initiative by appropriating funds and directing HUD to develop the seminal plan for healthy homes, an effort which Jacobs led.

By 2020 healthy homes became a truly global movement with the publication of the first international Housing and Health Guidelines by the World Health Organization. Healthy homes bridged the divide of many nations with very different housing challenges that shared recognition of the importance of housing quality to health and the benefits of holistic approaches.

A new consensus emerged. Lead and other health hazards in housing are solvable problems at reasonable cost and large benefits. The cure that Dubos could only imagine became increasingly realized. As the Covid-19 pandemic required sheltering at home, having a healthy home has never been more important. This book explains why policy and practice must be guided by science.

December 2020

Preface

If told there is something dangerous like lead paint in your home, what is your first response? Likely, it is "Get it out as fast as possible!" But what if that turned out to be a dangerous and wrong strategy? What should be done instead? How can you know your children are safe? Will corrective measures work and not backfire, making the problem worse? Who caused this headache in the first place, and shouldn't they help fix it? Is it up to each of us alone to figure out what to do in each home? How can we be sure other children are not also harmed? How can you pay for it? Most importantly, where is the evidence on what should be done? These are all questions this book examines.

Many believe the problem was fixed when the United States finally banned the use of lead in new residential paint in 1978, something that many industrialized countries had already done a half century earlier in the 1920s. But the dilemma of what to do about the lead paint already coating tens of millions of older homes across the country remained largely untouched in the 1970s and 80s.

This is the first book to show how we as a country found answers and implemented them. It traces the history from 1970 to 2022, illuminating the scientific discoveries, enlightened policies, and the power of citizens acting on them both to bring about change. Knowledgeable and compassionate people with the right skills and power in the right place doing the right thing at the right time backed by political will created by engaged citizens were all key

ingredients to overcoming the many wrong directions. The book charts the disasters that happened when the facts were unknown, ignored, or obscured, and how many obstacles to recognize, measure, define, remediate, and fund the solutions to address lead paint threats in housing were addressed, ending a decades-long policy paralysis by providing a path forward to create solutions. The science drove policy, and those policies were put into practice to ensure they worked in the real world.

I first became involved with lead when I was asked to inspect lead battery plants and other factories in the 1980s, where I found many workers were overexposed. The neurological problems that can come with lead exposure resonated with me. I have a neurological condition, epilepsy (thankfully now well controlled after a childhood with seizures). I remember my mother wondering whether my condition somehow might have been related to my chewing on our painted dining room table legs when I was a child growing up in inner city Philadelphia during the 1950s.

In the 1980s I measured worker's exposure to lead paint as they removed it in public housing using torches and power sanding, methods that are now banned because they produce enormous amounts of dust and fume. As is often the case, the high exposures among workers also indicated high exposures to children who moved into those supposedly safe homes. Earlier researchers found the same thing.

The history on lead paint between 1970 and 2020 resonates with important lessons for solving other large problems for which solutions seem elusive or even hopeless. The story cuts across many walks of life. Readers and students interested in public health, housing, history, policy, environment, ethics, law, urban planning, lead poisoning prevention, and science will find many surprises in these pages.

I chose not to go into detail on lead toxicity, medical treatment, adult exposures, other lead sources such as water, food and air, and the role of the lead paint industry before 1970 because other books have documented the massive evidence. Instead, I offer a short summary of lead toxicity to underscore the importance of addressing exposures (see Figure 1). Other books have focused on outrage, or cover-ups, or what new source of lead exposure was identified. Still other accounts focus on blood lead testing, or why people in charge failed to do what they were supposed to do, or why the industry and others continued to sell and promote lead paint after they knew better. Instead, this book endeavors to show the progress made and what remains to be done.

There are three parts of this book. The first documents the scant progress made before 1985, including useless finger-pointing, endless lawsuits, the limitations of identifying and treating children only after their blood lead levels rose, thorny scientific dilemmas, and the abdication of the housing world. This first part closes with the remarkable advances in scientific knowledge of how exposures truly occurred.

The second part describes initial progress. It shows how Congress reacted to the new science and new calls for action (including attempts to conceal the evidence) by ultimately passing the country's main lead paint law, Title X of the 1992 Housing and Community Development Act. This second part documents the tortuous path of how that law was passed and then (mostly) implemented throughout the 1990s, showing how scientists and citizens inspired political leadership and overcame political interference to end the paralysis wrought by the 1970s and 1980s. The second part closes with an examination of research ethics and a controversial Appeals Court decision in Maryland regarding a major lead paint study in 2001, ultimately rejected by juries who considered the facts.

The third part shows how a new consensus was achieved. It opens with other precedent-setting court decisions in Rhode Island and California requiring the lead, pigment, and paint industries to help pay for remediation for the first time. The story of the birth of the healthy housing movement in the United States and around the world follows. Although the book is intended primarily for a US audience, the global nature of today's healthy homes efforts is a remarkable development. Together, these marked a still emerging consensus on the connections among housing, health, environment, and other sectors between 2000 and 2020. The book concludes with the lessons from the lead paint experience, more recent developments, and what remains to be done, outlining the elements of a new strategic plan with both benefits and costs.

Most chapters are accompanied by a timeline that summarizes the key scientific policy and other developments for each period. References are provided to document the history, some revealed for the first time.

The book has an "honor roll" of many who played key roles in this time period. I was privileged to know and learn from them—parents, scientists, low-income housing providers, policy analysts, lawyers, advocates, doctors, and many others (my apologies to anyone I may have overlooked).

In various parts of the book, I struggled with how to tell the story and my role in it, from a perspective of how science informs both policy and practice in the real world. I document how political interference and attacks on the nation's lead prevention efforts were overcome at HUD, CDC and other agencies, which included an ultimately unsuccessful attempt to fire me by HUD Secretary, Alphonso Jackson in 2004. Many others paid a much higher price. I will forever be perplexed at why the lead poisoning prevention and healthy homes field has had so many attacks where "no good deed shall go unpunished," that sadly stalked the many who fought to protect children.

I hope this book will underscore the challenges in getting political leaders to pay attention to the science and the importance of taking action.

As a part of this book's development, I have created an archive of lead paint historical documents that I hope future historians will analyze (available at: the University of Illinois Chicago Special Collections & University Archives, School of Public Health, "David E. Jacobs papers"). More importantly, I hope that future leaders in the lead, healthy homes, and allied fields will learn from both our failures and our triumphs, making a future filled with homes that do not fail our children but instead create that special place enabling them to grow and be the best and brightest to make our future full of hope.

Over the years, thousands dedicated themselves to conquering childhood poisoning from lead paint and other housing-related diseases and injuries. This is really their story, and that is why I have dedicated the book to them and our children.

The story of lead paint and healthy housing between 1970 and 2022 remains largely untold. I truly hope this book will set the record straight. Most of all, I hope that readers will share my optimism, rooted in the idea that obstacles can be overcome with evidence, but only if the evidence is applied in both enlightened policy and practice.

Dave Jacobs

Acknowledgments

The Henry Halloran Trust at the University of Sydney in Australia provided early support to make this book possible. The Australians not only were the first to recognize the dangers of lead paint before the 1900s, they were also the first to recognize that what was done to address it between 1970 and 2020 had not yet been told. Professor Peter Phibbs there provided valuable insights through the lens of urban planning and healthy homes. I will be forever grateful to him, the Trust, and the university for the encouragement, friendship, and opportunity to write this book.

Many others also provided key insights and new materials. I am indebted to Don Ryan for writing the preface, outlining the policy dilemmas that needed to be overcome and for his friendship and encouragement. I am also deeply indebted to many others for their thoughtful reviews of earlier drafts, including Jon and Marsha Baker, John Bartlett, Whitlynn Battle, Miranda Brazeal, Mary Jean Brown, Susan Buchannan, Salvatore Cali, Karen Dannemiller, Joan Davis, Dorr Dearborn, Mark Farfel, Neal Freuden, Barry Goldstein, Jerry Hershovitz, Leann Howell, Maura Jackson, Katrina Korfmacher, Thomas Matte, Rebecca Morley, David Ormandy, Janet Phoenix, Amanda Reddy, Steve Schwartzberg, Claudia Thurber, Dahn Warner, Lee Wasserman, Steve Weil, Jonathan Wilson, and Alan Woolf. All these (as well as other anonymous reviewers) provided important insights on earlier versions. Chris Bloom and Sarah Keeley provided assistance with a few graphics inside the book. I am also indebted to the publisher Elsevier, and to their capable and helpful editors, Elizabeth Brown, Pat Gonzalez, Stacy Masucci, and Kirthika Govindaraju. Any errors are solely mine.

Most of all, I want to thank my two children, Paul and Robin, my partner Joan Davis, and my extended family for all they did to support me over many decades. I could not have done this without their love and deep and abiding commitment to doing the right thing, justice, science, the law, and people. They inspired me to tell this story so that all children are recognized for the precious gift they are.

David E. Jacobs, 2022

A short summary: lead is a long-lasting insidious poison

Lead is a metal we do not need in our bodies, unlike iron or zinc. It performs no useful biological function and attacks many organs and bodily systems, most importantly the brain.

Some of the poisonous effects are lifelong and may be irreversible, for two main reasons.

First, lead inhibits normal brain development, both the branching of the neural network during infancy and the later pruning that are both parts of healthy development. We need the neural pathways to enable our brains to receive the signals from our senses, and we also need the pruning to be able to interpret them. Lead diminishes both branching and pruning.

Second, the body mostly treats lead like calcium, with much of it ending up in our bones, where it can remain for decades. Over time, it leaches out to blood, where it is distributed internally, causing lifelong problems. In pregnant women, lead can cross the placental barrier and affect the developing fetus. Calcium also plays an important role in our nerve cells as a transmitter of electrical signals. Lead slows down and interferes with those electrical signals.

Lead is perhaps the best studied of all toxic substances; one recent estimate is that there are more than 28,900 studies documenting its many adverse health effects in both children and adults. Yet only a few have focused on exposure prevention and remediation. The main health effects of lead in children are shown in Fig. 1.

It does not take much lead exposure to cause harm. In the United States, blood lead levels are measured in micrograms of lead per deciliter (10th of a liter) of blood (μg/dL). A microgram is a millionth of a gram. If a teaspoon of sugar weighs 1 g, then only a few crystals weigh a microgram. A single lead paint chip can weigh many thousands of micrograms and can contaminate house dust and soil. The best way to diagnose lead poisoning is with a blood test.

There is no medical cure for lead poisoning. Exposure prevention is the solution.

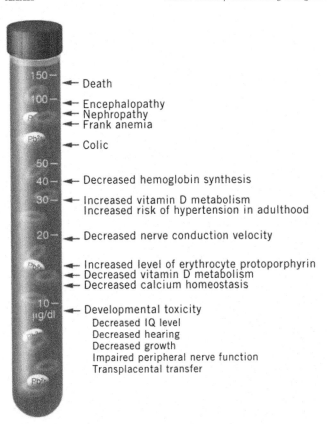

150 — ← Death

100 — ← Encephalopathy
        ← Nephropathy
        ← Frank anemia

        ← Colic

50 —
40 — ← Decreased hemoglobin synthesis

30 — ← Increased vitamin D metabolism
        Increased risk of hypertension in adulthood

20 — ← Decreased nerve conduction velocity

        ← Increased level of erythrocyte protoporphyrin
        ← Decreased vitamin D metabolism
        ← Decreased calcium homeostasis

10 — ← Developmental toxicity
μg/dl   Decreased IQ level
          Decreased hearing
          Decreased growth
          Impaired peripheral nerve function
          Transplacental transfer

FIGURE 1   Some of the toxic lead effects in children (micrograms of lead per deciliter of blood, μg/dL). Source: *With permission from Bellinger and Bellinger. Childhood lead poisoning: the torturous path from science to policy.* J Clin Invest. *2006;116:853—857.*

Paralysis and the abject failure to address lead paint before 1985

Banning lead paint: the missed opportunity

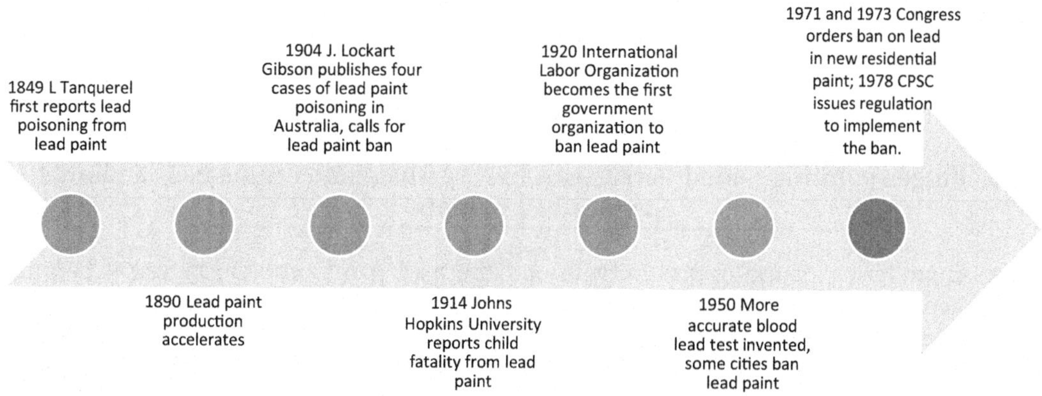

1849 L Tanquerel first reports lead poisoning from lead paint

1904 J. Lockart Gibson publishes four cases of lead paint poisoning in Australia, calls for lead paint ban

1920 International Labor Organization becomes the first government organization to ban lead paint

1971 and 1973 Congress orders ban on lead in new residential paint; 1978 CPSC issues regulation to implement the ban.

1890 Lead paint production accelerates

1914 Johns Hopkins University reports child fatality from lead paint

1950 More accurate blood lead test invented, some cities ban lead paint

FIGURE 1.0 Timeline.

1.1 The beginning

As soon as lead was put into many house paints in the mid- to late 1800s, public health professionals and authorities almost immediately called for its prohibition.[1-3] They pointed out that once it became dispersed throughout millions of homes, controlling exposures would be costly, complex, and less effective. Prohibiting lead gasoline and lead in food canning would also prove to be important, but these were easier to address because they could be eliminated through central sources (retooling gasoline refineries and canning factories). But lead paint production and sales were more widespread and diffuse, making them harder to eliminate.

The lead paint industries argued that exposures would not be a problem if their product was used "correctly"; or could be controlled by routine maintenance; or even by somehow changing children's behavior. They lobbied hard against a ban. They even asked, "Does your

Fifty Years of Peeling Away the Lead Paint Problem
DOI: https://doi.org/10.1016/B978-0-443-18736-0.00015-7

paint have enough lead in it?" From the mid-late 1800s to 1978 in the United States, the industry won and effectively prevented a ban on lead in new residential paint.[4]

The debate over banning versus other solutions still rages on today. Some suggest that a so-called "permanent" solution (removing all existing lead paint in all housing—a kind of "retroactive ban") is the only truly effective option. Others suggest that short-term solutions are more effective because they produce less dust and are more feasible. The historical record shows limiting actions to only "permanent" or "short term" ones was a false choice.

Combining short- and long-term solutions is not fundamentally different than most other housing repairs. Maintenance and capital improvements are short- and long-term solutions, respectively. For example, patching a roof with a leak is a short-term fix (maintenance), but eventually the roof will require replacement (capital improvement). Patching the roof or replacing it will both be effective in preventing leaks. The short-term solution requires continual inspections and higher operating costs, and the long-term solution has fewer ongoing inspection needs and lower operating expenses, but higher up-front costs.

In the years immediately following the introduction of lead into paint in the mid-1800s, harm and both long- and short-term costs might have been avoided entirely had the initial calls to proactively ban the product been heeded. But retroactive bans attempted in the 1970s and 1980s only produced paralysis and worsened exposures.

1.2 Finger pointing—food versus gasoline versus paint versus bad landlords versus bad parents

The squandering of the opportunity to end the lead paint problem before it became widespread was aided and abetted by early industry attempts to confuse the public by saying other sources of lead were more important.

Lead paint was initially marketed as a "sanitizing" agent that would also promote durability and cleanliness of housing, a history documented elsewhere.[4] The industry even produced a children's coloring book that promoted lead paint's brightness: In this telling, Dutch Boy paint would be a "party" and create "harmony in the home" (Fig. 1.1).[5]

FIGURE 1.1 Dutch boy coloring book for children. Source: *National Lead Company. The Dutch Boy Conquers Old Man Gloom: A Paint Book for Boys and Girls; 1929, 14 pages.*

Lead was put into paint for primarily three purposes: corrosion control, "hiding" ability (coverage of existing painted surfaces), and as a drying agent.[6] A 1938 publication from the National Lead Company stated: "White lead paint wears by slow and gradual chalking. This advantageous quality permits the white lead film to remain unbroken, smooth and even...."[7] This ability of lead paint to enter house dust would prove to be important in establishing lead paint policies in later years. "White lead" refers to lead carbonate, which was the main lead ingredient in housepaint.

One account suggests that some lead was added to house paint for 400–500 years,[8] but other accounts found that white lead was first manufactured in significant quantities in the United States in 1804.[6] Following the publication of an 1839 investigation of severe lead poisoning in 863 painters and lead paint production workers,[9] Europeans started to end the use of residential lead paint. Much safer substitutes were readily available, such as zinc-based paint. But the lead, pigment and paint industries were intent on expanding their markets, despite the knowledge that lead was toxic. One advertisement appeared in 1897, stating its paint "is nontoxic and is nonpoisonous" because it "is not made with lead" (Fig. 1.2).

The evidence that lead in paint was a problem had emerged.

The lead industry adopted the "Dutch Boy" logo, based on the process of making white lead (residential lead carbonate). The "Dutch process" changed metallic lead to the basic carbonate form by subjecting it to the vapors of acetic acid, water, and carbon dioxide, in the presence of heat, for a period of about 90 days.

The appearance of titanium dioxide in the 1950s began the slow decline in the use of residential lead paint, although its widespread production persisted in the United States until it was banned in 1978, coating millions of homes over the years. It is still being

FIGURE 1.2 Advertisement for Aspinall's enamel, which appeared in the Diamond Jubilee issue of the Illustrated London News, 1897. Source: *Illustration reproduced from Measuring lead exposure in infants, children, and other sensitive populations.* National Research Council (US) Committee on Measuring Lead in Critical Populations. *Washington (DC): National Academies Press (US); 1993.*

manufactured in many other countries, sometimes by United States corporations such as Sherwin-Williams.[10] In 2014 the world's biggest paint company, Akzol Nobel, admitted, "There is no good reason to put lead into paint today"[11] and stopped making lead paint in 2011.[12]

In the late 1800s and throughout the first half of the 20th century, the industry succeeded in putting lead into many consumer products, not only paint. Whether the introduction of lead into such products would produce dangerous exposures was left to scientific studies to determine after the exposure had already occurred. The industry did fund some initial studies to determine what an "allowable" lead exposure might be, but these were widely regarded as biased and unethical. One 1953 study included feeding lead to humans to determine how much was absorbed, concluding that food was the main exposure source.[13,14]

Internal documents show that the industry knew its product was harmful well before it was finally eliminated in new US residential paint by the government in 1978. To the extent that lead in paint was regarded to be a problem, it was depicted by the Sherwin Williams paint company as an issue "only" in dilapidated inner-city housing that would be difficult to solve. There was a continuing focus on "flakes" of lead paint, not lead dust.

An internal memorandum from Sherwin Williams reported a meeting in 1969 highlighting the company's need for secrecy:

> A memo of the meeting will be prepared by the University, but no individual comments will be quoted.... When acute cases are detected, they nearly always can show by x-ray that the child has ingested flakes of paint. The child is hospitalized and treated ... for about 10–20 days at $100 per day.... However, this de-leading of the blood is ... temporary.... The child may be in trouble again, particularly if he is returned to a home that is not properly repaired and ingests more lead materials. If these individuals in the adult age, because of the mental impairment, become wards of the State, it is estimated their total cost to the State during their lifetime could range between $200–300,000.... On a national basis the total number of children afflicted is very high indeed.... As to a solution to the problem, a very simple statement, but very difficult to carry out, would be to remove the source of lead or put it behind barriers so that the children could not get to it.... The entire population is becoming more and more exposed to lead.... The entire problem is certainly depressing and the outlook for an economical practical solution is not too optimistic.[15]

Yet Sherwin Williams continued to resist calls to stop making lead paint internationally as recently as 2014, when a group of investors requested the Securities and Exchange Commission (SEC) to require Sherwin-Williams to include a proposal that the company "establish a policy and eliminate the use of all lead compounds in its products" in its annual meeting proxy materials for its shareholders. But Sherwin-Williams successfully argued that the proposal be excluded from its shareholder meeting, because it thought that new lead paint manufacturing was part of its "ordinary business operations" and thus not amenable to shareholder action. The SEC caved and took no action.[16] Another group of shareholders (accompanied by a few nuns from a nearby church) attended the company's annual meeting making the same request in 2016, but they were rebuffed.[17]

Because the industry had been so successful in spreading the use of lead into many products, the routes of exposure became quite complex. Lead became a "multimedia" pollutant, meaning that it had contaminated many parts of the environment. The importance of paint became obscured, particularly in the environmental community.

Although lead is a naturally occurring element in the earth's crust, it was not until Clair Patterson's work in the 1960s demonstrated that lead contamination had become widespread. This was an accidental discovery associated with the difficulty in establishing laboratories that did not suffer from high background lead contamination.[18] He believed that one consequence of the focus on protecting workers in lead industries resulted in a widespread belief that everyone else had little or no lead in their bodies and that what they did have was "normal"[19] (Fig. 1.3).

Patterson is best known for establishing the age of the earth, which involved the analysis of lead stable isotopes that needed laboratories without background lead contamination. In 1980 he wrote a scathing letter to the FDA Commissioner that highlighted deficiencies in measurement technologies and the concept of "normal" lead levels, stating that lead from industry:

> is responsible for rendering the older urban areas in the US more or less uninhabitable to children for generations to come.... The most influential force has been the widespread generation of erroneous and misleading lead analytical data in thousands of laboratories....[20]

Fig. 1.4 shows some of the significant pathways and sources of exposure, but EPA's view minimized the importance of paint. "Housing" is not shown as either a source or a pathway of exposure. Air, water, soil, and drinking water are all prominent in Fig. 1.4, but paint is barely shown at all, and even then, only as a small subset of indoor and outdoor dust and soil. The transmission pathways of exposure are limited to "air, soil, water, and

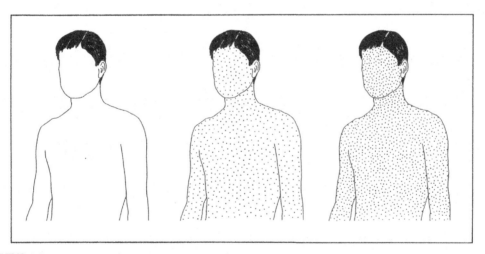

FIGURE 1.3   Prehistoric, average, and poisoning lead levels in the body. The left figure shows the body burden of lead in ancient people uncontaminated by industrial lead; the middle figure shows typical Americans in the mid- to late 1990s; and the right figure shows people with severe clinical lead poisoning. Each dot represents 40 micrograms of lead. Source: *Patterson CC. Natural Levels of Lead in Humans. Carolina Environmental Essay Series III, University of North Carolina at Chapel Hill, March 22; 1982. The University of Illinois Chicago Special Collections & University Archives, School of Public Health, "David E. Jacobs papers" the University of Illinois Chicago School of Public Health Library.*

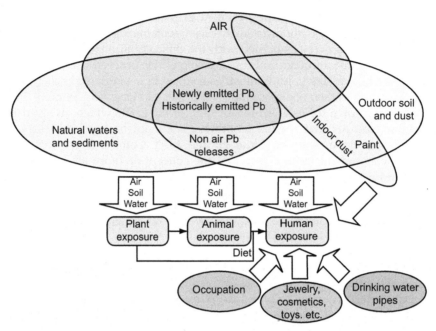

FIGURE 1.4 EPA graphic showing their assessment of sources and pathways of lead exposure, 2013. Source: *EPA*. Integrated Science Assessment for Lead; *June 2013. EPA/600/R-10/075F.*

indoor dust," but not paint. This reflects a narrow "shared commons" concept, which undergirds most environmental regulation and policy (Chapter 9).

Still others suggested that lead in food and food canning was the primary source. In the 1970s, the Food and Drug Administration stated that lead from solder used to seal the seams of cans accounted for as much as one-third of the lead ingested from all food.[21] In 1973 EPA stated, "Food is the largest contributor of lead to the general population."[22] In 1975 the Bureau of Foods published its findings, showing that the highest levels were found in canned foods. Canned tomatoes had the highest levels in adult food and orange and apple juice had the highest levels in baby foods.[23] The canning industry falsely claimed that the lead solder used in food cans manufacturing was only on the exterior and could not get into the food product, most importantly canned baby formula.[24] One local health department in California later found cans of drinking water provided by the US Coast Guard for emergency use that had been sealed with lead.[25]

It was not by accident that it was a nutrition expert, Kathryn Mahaffey, then at the FDA, who first requested (and achieved) the first measurement of lead in children's blood in a nationally representative sample of the US population in the late 1970s (Chapter 2).[26,27] The author of this book (Jacobs) and Mahaffey would later marry, finding "there is romance in heavy metal." They enjoyed many happy years together until her untimely passing in 2009. The initial impetus for undertaking such a survey was concern about lead in diet and baby formula cans, but the data later proved to be essential in properly identifying the key sources and pathways, including paint and the settled contaminated house dust and soil that it generates.

These national surveys led to decreases in the lead content of both infant and adult foods (Fig. 1.5). Between 1980 and 1985, lead in food declined by 77% according to FDA.[28] By the mid-1980s, the FDA required that no more than 0.3 ppm of lead be allowed in evaporated milk. FDA also required a 99% conversion to nonlead-soldered cans for infant formula, and a conversion to glass jars for infant juices, all of which led to an 85% decrease in lead levels in infant foods and juices. Lead in all canned foods dropped 40% by 1985.[29] But the food industry initially resisted; for example, the Evaporated Milk Association proposed an allowable limit of lead in evaporated milk, but others called for a phase-out of lead-soldered cans.[30]

Mahaffey emerged as one of the nation's foremost lead toxicologists, helping to write a major report from the National Academy of Sciences on the topic in 1993.[31] She served as chair of the federal government's Committee on Lead in Foods.[32] She was among the first in the post-1970 era to recognize the importance of housing and paint as a source of lead poisoning and also played a leading role in phasing out the use of lead in gasoline.[33] Visiting one of the world's biggest lead mines in Broken Hill Australia as part of an international lead conference,[34] she worked with Brian Gulson, an Australian researcher whose

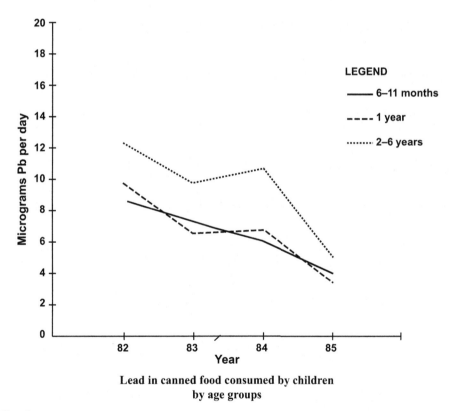

**Lead in canned food consumed by children
by age groups**

FIGURE 1.5   Children's lead intake from canned food, 1982–1985. Source: Task Force Report to the Board of Directors. *National Institute of Building Sciences, February 20; 1988 (the University of Illinois Chicago Special Collections & University Archives, School of Public Health, "David E. Jacobs papers" the University of Illinois Chicago School of Public Health library).*

stable isotope research would point to lead paint and lead-contaminated house dust as major contributors of lead in children's blood.[35] She also worked on an Institute of Medicine report on Lead in the Americas highlighting the importance of both population health and housing surveys.[36] In addition to her work on lead, she was the primary author of EPA's 8-volume Mercury Report to Congress.[37]

Lead in gasoline was also a source of exposure as well as food canning, but its successful phaseout in the late 1970s and 1980s was used to delay action on lead paint. In 1973 EPA stated: "In EPA's opinion, lead in gasoline is *the most important remaining source* of controllable lead entering the environment" (emphasis in original).[38] Some in the environmental and public health community thought that the phaseout of lead in gasoline was the end of the story and some within both government and industry used it as an excuse to avoid action on lead paint.

There were two competing views of lead exposure within the environmental community, depicted below (Figs. 1.6 and 1.7). The first was that most of the lead problem had been solved by the phaseout of lead in gasoline (Fig. 1.6), a commonly held view that blood lead levels tracked lead in gasoline and with its phaseout, the lead problem had been solved.

The second view held that gasoline lead was indeed an important source, but its phaseout eliminated only a small fraction of overall lead exposure. This view meant that in fact exposures remained very large. The Centers for Disease Control and Prevention (CDC) blood lead level of concern in 1992 was 10 µg/dL, a *thousand* times higher than the "natural" background blood lead level, which was estimated to be 0.016 µg/dL (Fig. 1.7). This "natural" background estimate was determined by measuring lead in prehistoric human bones and then deriving the associated blood lead level.[39] Fig. 1.7 shows the CDC blood lead level of concern and the much, much lower "natural" background blood lead level. It meant that lead in paint was still a huge problem, despite the progress in eliminating it from food canning and gasoline.

These two competing views (gasoline and leaded food canning phaseouts solved the problem versus the problem remained very large) led to the view that paint was relatively unimportant.

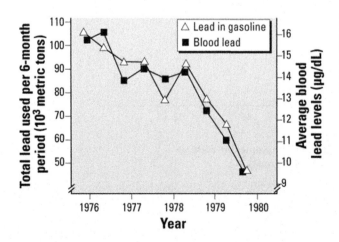

FIGURE 1.6 EPA graphic showing blood lead and gasoline lead 1976–1980. Source: *Bridbord K, Hanson D. A personal perspective on the initial federal health-based regulation to remove lead from gasoline. Environ Health Perspect 2009;117(8):1195–1201.*

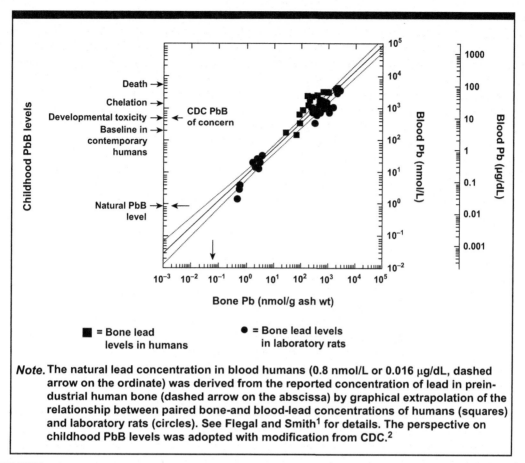

FIGURE 1.7 Blood lead and "natural" background lead toxicity. Source: *Smith DR, Flegal AR. The public health implications of humans' natural levels of lead. Am J Public Health 1992;82(11):1565–1566.*

The lead paint industry actively promoted this concept. A paint chemist with the Sherwin-Williams paint company stated: "the excessive lead in the blood of small children derives at least 90% from vapor, dust and soil spewed out from gasoline combustion and less than 10% from historic lead in old paint."[40]

In later chapters, we shall see how this odd convergence between the lead paint companies and the environmental community led to policy paralysis when it came to paint.

1.3 "It's right in front of you—it's the paint!"

EPA launched a major lead study that focused on lead in soil in the 1980s in three cities: Baltimore, Boston, and Cincinnati. Examined in greater detail in Chapter 4, that study focused on soil removal and excavation because it was thought to have been contaminated by gasoline fallout in earlier years. The Baltimore part was led by J. Julian Chisolm, who recounted an encounter with an inner-city resident where the study was taking place. After being told that the soil was going to be removed and why, the resident took on a bemused look. Pointing to the dirt on the ground and the air above and then the painted wall, he said, "It's not down there and it's not up there—it's right in front of you! It's the paint!" Having devoted decades to drawing attention to the paint problem, Chisolm replied, "You're right."[41] The study results bore that out: There was no improvement in blood lead levels following soil excavation in Cincinnati and Baltimore and only a very small improvement in Boston.[42]

By the late 1980s, other new research had demonstrated the importance of paint and the contaminated dust and soil it produced (Chapters 2 and 3). But throughout the 1970s and early 1980s, both industry and the US Department of Housing and Urban Development (HUD) seized on the gasoline lead phaseout, holding that nothing more needed to be done on paint. Decades later, Donna Shala, who served as an Assistant Secretary at HUD in the late 1970s and then as Secretary of the Department of Health and Human Services in the 1990s admitted, "We were clearly wrong about that. The scientists told us it was gasoline."[43]

Before the 1980s, the main route of exposure to lead paint was thought to be the relatively rare ingestion of lead paint chips.[44] Fig. 1.8 is an X-ray of a child's abdomen, showing internal lead paint chips that had been ingested (because lead blocks X-rays, the paint chips show as white specs, denoted by the arrows). This seemed to support industry attempts to blame bad landlords or bad parents for supposedly inadequate maintenance or poor supervision of children that had allowed them to eat paint chips. The industry even tried to blame child behavior, such as chewing on painted surfaces,[45] supposedly because lead tastes "sweet." However, almost all lead used in paint was white lead carbonate, which has a chalky (not sweet) taste.

Of course, inadequate maintenance is associated with other housing problems. But if there is no lead paint present, then lead poisoning would be far less likely. Nevertheless, the industry succeeded in shifting the focus from lead paint. The deferral of regulation of lead paint before the 1970s proved to be one of the industry's major successes and internal documents from them said so.[46]

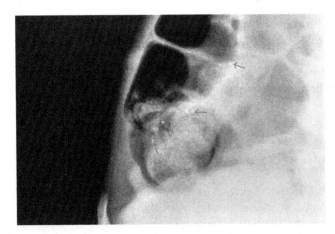

FIGURE 1.8 Radiograph showing ingested lead paint chips. Source: *Lorne Garrettson, MD, reproduced with permission.*

In 1938 the Lead Industries Association stated, "In many cities, we have successfully opposed ordinance or regulation revisions which would have reduced or eliminated the use of lead."[47] In 1958 the association stated, "Every effort is being made to confine ... regulatory measures ... to warning labels on paint cans ... which are less detrimental to our interest than would be any legislation of a prohibitory nature."[48]

The industry argued that lead in paint was fully contained in the paint binder and therefore could not be available to young children. But intact paint does not stay that way. There are many deteriorated paint conditions formally used by the painting and decorating industries to describe routine paint failures. These include "alligatoring, blistering, checking, cracking, flaking, chalking and peeling."[49] Even in 1938, the lead paint industry recognized these forms of deterioration, and added a few others, including, "scaling, peeling, spotting, wrinkling, discoloration, bleeding, sagging, early loss of gloss and running."[7] The idea that old lead-based paint would remain intact and not become available for ingestion and dust and soil contamination became widely discredited.

Although lead in canning, gasoline, water, and other media was clearly important, the evidence that lead paint and the contaminated dust and soil it generated was the main exposure source and pathway finally began to emerge. Chapter 2 describes the Lead-Based Paint Poisoning Prevention Act and the Congressional hearings leading up its passage in 1971. In 1988 a major report to Congress ranked lead paint exposures as the number one problem (Chapter 3).[50] The National Academy of Sciences also listed lead paint first in its major report of 1993.[31] Approximately 70% of elevated blood lead levels in children were estimated to be from lead paint.[51]

There were three types of evidence that all converged to show how lead in paint was highly correlated with children's blood lead: case studies, environmental correlate studies, and stable isotope ratio studies.[45] One such isotope study in 1983 concluded, "lead in the soil was derived mainly from weathering of lead-based exterior paints and the lead-contaminated soil was a proximate source of lead in the blood of the children."[52]

Fig. 1.9 demonstrates how blood lead levels were overwhelmingly associated with older homes where lead paint was both more prevalent and more concentrated. Importantly, Fig. 1.6 on gasoline presented above only showed average blood lead levels. But when it came to poisoned children with an elevated blood lead level, the importance of lead paint was inescapable. For all poisoned children, the percentage doubled in housing built before 1946 (9% vs 4%); for low-income children in such housing, it was 16%; and for African American children in such housing, it was an astonishing 22%, more than five times the national average in 1991−1994.

The large number of poisoned children was clearly being driven by lead paint. The idea that the lead problem had been solved by eliminating it from food canning and gasoline and new paint gave way to the recognition that existing lead paint in housing remained an enormous problem.

1.4 Inadequate measurement methods obscure the lead paint problem

The technical challenge of measuring where the lead is, the pathways through which it enters the human body, and what happens once it has been absorbed also contributed to the failure to recognize lead paint. If the problem could not be defined, then it could not be

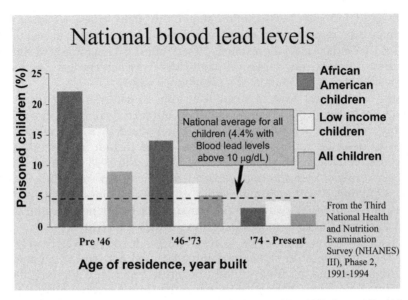

FIGURE 1.9   Blood lead levels and housing age by race and income, 1991–1994. Source: *David Jacobs.*

solved. The effort to measure lead in paint reliably did not begin until the 1930s and for existing paint, such measurements would not be fully validated until the mid-1990s.[31,53]

Ultraclean techniques in laboratories in the 1970s showed that reported concentrations of lead in some environmental media had been off by orders of magnitude.[54] The flawed nature of some reported lead data due to this contamination was initially documented in oceanographic research, largely due to background contamination.[55] In 1993 the National Academy of Sciences reported, "Problems of lead contamination [measurement] are pronounced because of the ubiquity of lead."[31]

Not until the 1950s did the measurement of lead in blood begin in earnest, and its invention and standardization are usually credited primarily to the work of J. Julian Chisolm at Johns Hopkins University and Sergio Piomelli at Columbia University.[56] Hopkins was involved in both the early recognition of the problem in the 1920s and later it (together with its sister nonprofit Kennedy Krieger Institute) became a center of research on successful lead paint mitigation methods in the 1980s and 1990s, led by Mark Farfel and Chisolm (covered later).

Before the invention and validation of blood lead measurements, clinicians were forced to diagnose lead poisoning primarily through either observation of misshapen (stipling) red blood cells viewed under a microscope, X-rays of growing long bones, and/or by observation of acute clinical symptoms such as problems in kidney excretion, blindness, convulsions, encephalopathy (severe changes to the brain structure and functioning) and in the worst cases, death. Until the early 1990s, most blood lead screening tests were based on the method pioneered by Piomelli, which used zinc or free erythrocyte protoporphyrin as a surrogate.[57] These are found in red blood cells and are elevated by lead exposure, but they could also be caused by iron deficiency or other conditions and were not capable of reliably measuring blood lead levels less than 15 μg/dL.[29,58]

Chisolm noted the historic challenge this way:

> ...when widespread screening of children was first proposed in [the US] in 1970, there were no suitable screening methods available. In my opinion, it took about five years then to establish fairly widespread screening [using erythrocyte protoporphyrin]. ... The 1991 CDC statement, which recommend blood lead testing [to detect blood lead levels of concern of 10 µg/dL] presents a similar enormous challenge.[59]

Although blood lead level measurements emerged as a key tool in lead poisoning prevention, they did suffer from an important limitation—blood lead measurements integrate *all* sources of exposure, which means that by itself, a blood lead measurement cannot conclusively determine the source or pathway of the exposure. That was done by inspectors and risk assessors and was not something for which clinicians were trained to do. Blood lead measurements were also influenced by other factors, such as season, age of child, previous exposures, lead in bone stores, nutritional status, and other factors.[60]

Reliable measurement technology for lead in existing paint, first mandated by Congress in the 1971 Lead-Based Paint Poisoning Prevention Act, would not be in place until the 1990s, when HUD and EPA finally published instrument performance data for X-ray fluorescence (XRF) portable lead paint analyzers. Initially used in the mining industry, a prototype for applying the technology to lead paint was reported in the journal *Science* in 1971.[61] An experimental model was used in New York City low-income housing and another experimental model appeared in 1976 but both required the use of liquid nitrogen and were not truly portable.[62,63] HUD and the National Bureau of Standards funded further instrument development between 1972 and 1977, but they were widely recognized as both inaccurate (inability to measure the true level of lead in existing paint) and imprecise (inability to get the same result repeatedly).[64] These early studies tended to focus on where lead paint was located, not its condition.[65] They were initially regarded as a qualitative measurement method until later research in the late 1980s and early 1990s produced more reliable quantitative measurements.

In 1977 HUD contracted with the National Bureau of Standards to produce lead paint films with known amounts of lead,[66] which could be used to initially check the calibration of XRF instruments. But the calibration was set at the factory, not in the field, introducing another potential source of measurement error. The history of how the problem of in situ lead paint measurements using XRF was solved has been described elsewhere, but it would not have happened if government had not created the necessary standards and published data on how well each instrument performed in the 1990s. The standards and data publication created a level playing field for instrument manufacturers that resulted in more reliable, faster and less expensive instruments.[67]

Effective lead paint mitigation relied on proper identification of exactly where lead hazards were located within a home. Initial attempts to develop validated methods to detect the presence of lead paint fell short in the 1970s and 1980s.[67] HUD's initial research to develop portable XRF lead-based paint analyzers was hampered by the fact that HUD was not a scientific agency, with little expertise in scientific instrument development and validation.

Several studies on XRF reliability to detect lead in paint were conducted in the 1970s at the National Bureau of Standards, which reported the instruments had errors on the order of 30%–50% of the true value.[68] The poor performance of XRFs seemed to be due to interference from the material under the paint as well as controversy over which type of X-ray frequency was best to use. The poor performance of XRFs was not overcome until the 1990s through HUD and EPA standardization (Chapter 4). But during the 1970s through the 1980s, few trusted the results of lead inspections, including housing professionals.

One alternative to XRF analysis for existing paint was to scrape it off from painted surfaces and submit the paint chip samples to laboratories for better analysis, a costly destructive process that was understandably resisted by property owners and residents alike. Paint chip laboratory analysis also suffered from important sources of error, such as failure to remove all paint layers (potentially excluding the lead paint). Another was the inclusion of substrate material (the surface underneath the paint) in the sample, which could dilute the amount of lead found in paint films. Removing all layers without including the substrate proved to be difficult.

Another alternative was to use a chemical that when applied to lead paint would turn black (lead sulfide),[69] which was popular in Massachusetts in the 1970s and 1980s, or pink (lead rhodizonate), but these methods suffered from both subjective assessment of color change and an inability to penetrate layers of paint that did not contain lead to reach the underlying layers where lead paint was more likely to exist. These so-called "spot test" kits were widely available in hardware stores but were not validated even as late as 2020. They were also plagued by high measurement error rates.[70,71]

Similarly, the ability to measure lead in settled house dust did not become standardized until the 1990s and was not widely practiced in the 1970s and 1980s. Although the Australians who first called for a ban of lead in paint in 1895 recognized the importance of both paint chips and dust, and in fact did some rudimentary dust analyses discussed later in this chapter, their findings often depended only on observation of acute symptoms, such as seizures or blindness.

The first studies to measure lead in settled house dust and correlate it with blood lead were completed in the 1970s and early 1980s.[72,73] There were subsequent attempts to measure so-called "bioavailable" lead in dust, and to use various vacuum and wipe sampling methods (Chapters 3 and 4), and even whether the amount of lead present should be expressed in parts per million (concentration) or weight of lead per surface area (loading). But a standardized lead dust sampling method that was conclusively correlated with blood lead and validated with essential laboratory quality control procedures did not emerge until HUD and EPA implemented them in the 1990s (Chapter 5). This paved the way for lead dust standards promulgated in 1999 for federally assisted housing and in 2001 for other housing.

Why did this historical inability before the 1990s to accurately and precisely measure lead in paint, dust, soil, and blood matter? Without such measurements, it was not possible to fully understand both the sources and pathways of exposure, and more importantly how to eliminate or interrupt them, respectively. It made countering the industry claim that the problem was not their lead paint but some other source of lead more difficult and debatable. The use of clinically observable acute symptoms and mere visual examination of paint condition, instead of its lead content, meant that a full understanding of how lead in the paint became available to children would not occur until the late 1980s.

This inability to measure reliably meant that before the 1990s, many policies held that a "lead-based paint hazard" was simply the presence of old deteriorated paint, which may or may not actually contain lead. Or, it meant any lead paint, regardless of its condition, was an immediate hazard. Even today, most housing codes regard any deteriorated paint to be a minor code violation, regardless of whether it contains lead.[74] Worse, such codes often assume deteriorated paint is simply a cosmetic defect—a relatively unimportant problem, instead of a very real health hazard.

This is why the 1971 Lead-Based Paint Poisoning Prevention Act first defined lead paint to be merely a concentration of lead in paint, with no recognition of the importance of dust and soil contaminated by the lead paint or of the degree of paint film deterioration. In short, the focus on the amount of lead in paint and not exposure, caused in part by measurement deficiencies, led to policy paralysis between 1971 and 1992.

1.5 Where lead was used

The first factory to produce white-lead pigments for paint in the United States was established in 1804 in Philadelphia, but thousands more did so in the decades to follow and production accelerated starting around 1880.[75] The total amount of lead used in gasoline and paint was about the same from 1880 to 1980,[76] and production of lead paint occurred over a longer time period. Lead paint was far more concentrated yet also dispersed throughout millions of homes.

Lead in water pipes was also used extensively. Lead solder was used for electronics, food canning, and sealing plumbing connections. Lead was also used to make lead service lines that connect household water supply to street drinking water main lines.

One way to estimate the historical lead use in water pipes is by using the number of lead water pipes still in use in 2016 (6.1 million),[77] the average length of the pipe (55 ft),[78] and the weight of lead per linear foot of pipe:[79–81]

6.1 million lead pipes $\times$ 55 linear ft/pipe $\times$ 6 pounds/linear ft/2000 pounds/ton = 1 million tons.

This estimate of 1 million tons of lead water pipes is likely too large because some lead pipes have been removed since the last national survey, which used data from 2011. Lead in drinking water is discussed in more detail elsewhere[82] and in Chapter 5.

By comparison, hundreds of millions of tons of lead were used in paint. Nevertheless, lead in water pipes remains a concern in communities where corrosion control was not used or where pipes were deteriorating or not being phased out. Corrosion control involves the addition of certain chemicals to the water supply to help create and preserve an internal lining in lead water pipes to minimize its leaching into the water. These chemicals and the acidity of the water supply can change the concentration of lead in water. Failure to implement corrosion control and phase out lead pipes led to the disaster in Flint, MI.[82]

Lead is not only in water service lines, however. Private wells, which are prevalent in rural areas can also be a problem (testing such wells is not required by EPA).[83]

The history of lead in water pipes and how the lead industry promoted its use, despite early attempts to stop it in the 1920s has been described in detail elsewhere.[84] Lead use in water pipes was not banned until 1986 in the United States. Lead was used in water pipes because, although it was more expensive to use than iron, lead pipes lasted about 35 years longer and was more malleable and more easily bent around existing structures.

Clearly, it was a bad idea to put lead into paint, gasoline, canning, or water pipes or the thousands of other unnecessary uses. The myriad uses, sources, and pathways of exposure were dizzying and daunting. Fig. 1.10 suggests the magnitude of the scientific puzzle. The most important pathway (paint to house dust) for children would not be grasped until the mid-1980s, which had dramatic consequences for policy, and how to define, identify and remediate lead paint hazards.

1.6 The first international lead paint ban

Lead paint was first recognized as a problem for children in the late 1800s.[45,85] Although it was first reported in 1848,[86] the Australian ophthalmologist J. Lockhart Gibson is credited with the first significant finding of four cases in 1904 where ocular degeneration (decreased eye vision) and severely reduced kidney excretion were identified as signs of lead poisoning. His initial findings showed how difficult it was to identify lead paint as the source, and even then, he recognized how important lead in dust was:

> Shortly, I am able, I believe, to advance a very strong plea for painted walls and railings as the source of the lead, and for the biting of fingernails or sucking of fingers, as in a majority of cases, the means of conveyance of the lead to the patient.... A majority, probably all the children we have observed affected with lead poisoning, have lived in houses whose rooms have been painted and had either sticky or deteriorated surfaces.... Analysis of calico cloths [unprocessed cotton cloths] and of the dust samples gave the following result:

> 1st—From the cloth rubbed on the ceiling whose paint was very glossy and showed no signs of wear and tear, no lead was obtained.

> 2nd—From the cloths rubbed on the walls where the painted surface showed signs of wear and tear, 2 milligrams of lead carbonate were obtained.

> 3rd—A similar quantity of lead carbonate was obtained from the cloth rubbed on the painted floor margin.

> 4th—From the cloth rubbed on the verandah floor 12 milligrammes of lead were obtained.

> I have until now been in the habit of saying to patients' parents, "You had better avoid tank water for drinking purposes;" but I do not know the source of the lead—I only know that lead is the cause of the affliction. I shall henceforth, unless I am offered a better explanation, *blame paint*. (emphasis added).[2]

Despite this account, many physicians and others did not believe it, because the industry claimed lead was contained in the paint binder and could not be released into the home environment where children would be exposed, a claim that persisted for decades.[87−89]

But evidence to the contrary accumulated. By 1907, American physicians learned of the Australian studies from a chapter in a medical textbook[90] and the Australians continued to publish.[91] This literature was used by US researchers when they began to document cases of lead poisoning in the 1910s and 1920s.

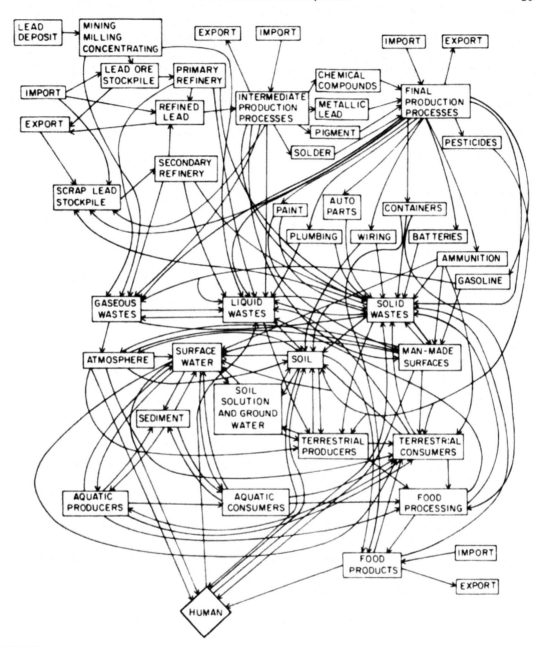

FIGURE 1.10 Sources and pathways of lead exposure, 1980. Source: *National Research Council. 1980. Lead in the Human Environment: A Report. Reproduced with permission from the National Academy of Sciences, Courtesy of the National Academies Press, Washington, D.C.*

The first US documented case of childhood lead poisoning from paint came in 1914 when physicians at Johns Hopkins University reported a Baltimore boy who died of lead poisoning from paint on the railing of his crib.[92] In that case, the boy had seizures and was treated in the hospital, only to be released back into his home environment, where he was re-exposed and died.

By 1926, there were at least 15 separate medical publications showing the link between lead paint and childhood lead poisoning.[93] A compelling case study published in 1943 was the first indication that the adverse health effects of lead poisoning were persistent and affected children's academic success and behavior, even when earlier acute symptoms were described as "mild."[94]

The evidence became so compelling that an arm of the League of Nations (the predecessor to the United Nations) acted. Under the authority of protecting *occupational* health, the International Labour Organization was the first government agency to finally ban the use of the lead in paint in 1920.[95] Although focused on adults, it also recognized the increased risk faced by children (or younger employees). It stated:

> Each Member of the International Labour Organisation ratifying the present Convention undertakes to prohibit, with the exceptions provided for in Article 2 [artistic paint], the use of white lead and sulphate of lead and of all products containing these pigments, in the internal painting of buildings…. The employment of males under eighteen years of age and of all females shall be prohibited in any painting work of an industrial character involving the use of white lead or sulphate of lead or other products containing these pigments.
>
> (a) Adequate facilities shall be provided to enable working painters to wash during and on cessation of work;
>
> (b) overalls shall be worn by working painters during the whole of the working period;
>
> (c) suitable arrangements shall be made to prevent clothing put off during working hours being soiled by painting material.

Although the convention made no explicit mention of protecting children, controlling worker exposures was regulated in detail; the initial focus on solutions to address both existing paint and new paint was important. This leadership from the occupational health profession would be repeated in the US in later years.

Alice Hamilton, widely credited with founding the occupational health profession in the United States, investigated 23 of the 25 US factories known to manufacture white lead for paint and discovered 358 specific cases of lead poisoning, 16 of them fatal, occurring between January 1910 and April 1911.[96] Hamilton's many contributions are summarized later in this book.

Many countries adopted this convention, mostly in the 1920s (although one country ratified it as recently as 2006). By 2022 there were 63 countries ratifying the ban (see https://www.ilo.org/dyn/normlex/en/f?p = 1000:11300:0::NO:11300:P11300_INSTRUMENT_ID:312158).

But the United States did not ban the use of lead in residential paint until 1978. It never signed onto the International Labour Organization (ILO) convention due to lobbying from the National Paint, Oil and Varnish Association and its industry successors.

Efforts to ban the use of new lead paint continued through 2022 by the United Nations/World Health Organization's Global Alliance to Eliminate Lead Paint.[97] According to the United Nations, as of September 30, 2018, only 36% of countries had confirmed that they had legally binding controls on the production, import, sale, and use of new lead paints.[98] By December 31, 2021, 84 countries had banned the use of lead in residential paint (out of a total of 195 countries).[99] But exposures to lead paint continued to be a problem in other countries and in the US. For example, immigrant children arriving in the United States were 10 times more likely to have higher blood lead levels compared to children born in the United States.[100] CDC reported a fatality from lead paint in US housing in an immigrant child in 2000, who was healthy before coming to the United States.[101]

Some local governments in the United States moved to ban new lead paint in their housing before Congress finally did so in 1978. In 1951, Baltimore prohibited the use of lead paint on interior surfaces of dwellings and in 1958 the city required warning labels on lead paint cans.[102] But this was more the exception than the rule. The only other cities with pre-1978 regulations banning the use of lead-based paint on interior surfaces were Chicago, IL; Cincinnati, OH; Jersey City, NJ; New Haven, CT; New York; Philadelphia, PA; St. Louis, MO; Washington, DC; Wilmington, and DC.[103] The vast majority of cities and states did not ban lead paint.

There are no data on how well the 1920 ILO ban was enforced by the countries signing it. For example, in France lead poisoning was "rediscovered" in 1985.[104,105] It is also not known how well local government bans in the United States were actually enforced. The sobering reality was that lead paint production still remained extensive around the globe in 2022. Some of these paints had extraordinarily high lead concentrations.[106]

1.7 How the US government changed from promoting lead paint to banning it

Instead of signing onto the ILO ban, US government agencies actually *recommended* lead-based paint for residential purposes during the first half of the 20th century. This included the National Bureau of Standards, Federal Security Agency, US Housing Authority, the Public Works Administration and the US Department of Commerce, all due to the extensive lobbying of the lead industry.[107]

But by the late 1920s the large number of lead poisoning cases in the United States received the attention of at least one insurance company[108] and the US Bureau of Labor Statistics began to track the incidence of lead poisoning in both children and adult workers.[109] Studies of numerous cases of lead paint poisoning were conducted in Baltimore,[56,110] Boston,[111] New York City,[112] and Chicago.[113] By the end of the 1950s, over 6000 cases had been reported, including hundreds of child fatalities.

The American National Standards Institute (ANSI) adopted a voluntary industry consensus standard limiting the lead content in surface coatings to 1% in 1955, which was due to a change in paint chemistry (i.e., the industry's change to titanium-based paints), not public health concerns.[114] However, the ANSI standard was not enforceable and no federal laws addressing existing lead-based paint already applied to dwellings were passed until 2 decades later.

In 1970 Congress finally held hearings on the problem. It found that about 200 children died each year from lead poisoning, and of the 12,000—16,000 children who did not, half were left severely mentally retarded or developmentally delayed.[115,116] Between 6% and 28% of urban children had blood lead levels greater than 50 μg/dL,[117] which is slightly less than half of the fatal dose and today would likely require hospitalization, according to CDC guidelines. In 1971 HUD funded the Department of Commerce to draft guidelines on lead paint removal, but they were primitive and not based on any evidence of effectiveness.[118] More robust evidence-based HUD guidelines on remediation would not appear until the 1990s.

But the industry continued to claim that paint was not the problem. Well into the 1980s, it continued the finger pointing at anyone other than themselves, such as housewives who did not clean enough. A paint chemist associated with Sherwin Williams argued that:

> Simple, vigorous, periodic scrubbing of floors, sills, walls of inner home surfaces ... can reduce dramatically and sufficiently the perceived and persistent lead now detected in homes of children.... "Cleanliness is next to Godliness" was practiced by those legendary Dutch housewives who vigorously scrubbed their homes [and is] now needed above all other aspects of the lead-in-child[ren] problem...[40]

In 1971 Congress finally reversed the trend of government requiring or advocating *for* the use of lead paint and instead moved to finally ban it in new residential paint through the Lead-Based Paint Poisoning Prevention Act.[119] Congress defined lead paint as meaning

> Any paint containing more than 1 per centum lead by weight (calculated as lead metal) in the total non-volatile content of liquid paints or in the dried film of paint already applied.[120]

The 1% level was consistent with the ANSI voluntary standard at the time and was perceived to be the lowest level that could be detected reliably and make production of non-lead paint feasible due to background contamination, such as that caused by lead gasoline and industrial emissions. Again, measurement ability proved to be an important limiting factor in establishing lead paint policy.

After the 1971 Act was passed, the Department of Health, Education, and Welfare (which later became the Department of Health and Human Services) focused on the early detection of children with toxicity and their medical care. Neither DHHS nor HUD ever requested appropriations authorized under Title II of the 1971 Act for grants to local agencies to eliminate lead paint hazards in dwellings, although Congress had authorized such grants.[121] Instead, the funding provided under the 1971 Act and 1973 amendments largely went to fund local health departments for blood lead screening programs to identify children whose blood lead levels had increased to potentially fatal levels. The hearings leading to the 1971 Act were driven largely by citizen movements focused on poverty.[122]

In 1973 Congress amended the 1971 Act to further lower the allowable lead content in paint applied to federally assisted housing to 0.5% until December 31, 1974 and to 0.06% thereafter. Although the language of the statute was ambiguous about whether it was intended to apply to existing residential paint or only new paint, it was in practice applied to only new paint. The 0.06% level was based on a recommendation from the American Academy of Pediatrics and a report from the National Academy of Sciences. That report focused on preventing lead in blood from rising above 60 μg/dL,[123] which was well above the CDC reference value in 2021 of 3.5 μg/dL.

The 1973 Act also ordered the Consumer Product Safety Commission (CPSC) to:

> conduct appropriate research on multiple layers of dried paint film, containing the various lead compounds commonly used, in order to ascertain the safe level of lead in residential paint products. No later than December 31, 1974, the Chairman [of CPSC] shall submit to Congress a full and complete report of his findings and recommendations.[124]

It took CPSC another 3 years to finally reduce the allowable level to 0.06% and it clearly applied to only new paint, not existing paint.[125,126] This level was based mostly on the ability to measure it reliably and the feasibility of producing new paint. A small study in 1992 showed that lead levels in new paint were all less than 0.01% and most were below 0.0025%.[127]

Importantly, neither the 1971 nor the 1973 Acts said anything at all about the *condition* of existing paint already in homes. These first Acts only regulated the *amount* of lead in paint, not exposure pathways.

The 0.06% level was based on a health risk assessment that made a series of assumptions about the frequency, amount, and bodily absorption of lead paint. These assumptions were that lead paint was six coats thick, that a typical child ate the equivalent of about 1 square inch of paint per day, and that absorption of lead from that paint was about 10% (the same as lead from food).[128] But none of these were validated, with the result that lead content in new paint remained largely based on the ability to measure it reliably. (In 2008 Congress reduced the allowable level of lead in new residential paint to 0.009%, again based largely on limitations of measurement reliability.)[129]

The 1971 Act and its 1973 Amendments provided quite limited funding (Table 1.1). Significant funding to remediate existing lead paint in housing would not appear until the early 1990s when hundreds of millions annually would be appropriated (described in this book's concluding chapter).

After the 1971 and 1973 lead paint acts were passed, HUD did not act on lead paint beyond small-scale contracting for an instrument to measure lead in existing paint, some research on covering materials to create a barrier between lead paint and children and an "Experimental Hazard Elimination Program," which was piloted in three cities. The latter examined costs, but not the effectiveness of various abatement methods.[130] The person who led the Department's lead poisoning efforts in those years, Irving Billick (director of the Environmental Research Group within HUD's Office of Policy Development and Research) was adamant that it was gasoline, not paint, that was the main source of exposure and he effectively prevented HUD from taking action.[131]

In 1978 HUD reviewed its activities on lead paint in a Report to Congress. Donna Shalala, then the HUD Assistant Secretary for Policy Development and Research, wrote in the forward to that report:

> the report raises many questions about the lead problem and its solution for which we do not have satisfactory answers.

Written by Billick, the report repeated the industry's position, stating:

> Lead-based paint is only one of many sources of environmental lead available to children ... too little is known and too little data are available on the relative contribution of the different sources of lead....[132]

TABLE 1.1   Federal lead paint funding, 1971–1982 (Appropriation is the amount actually provided to federal agencies).

Fiscal year	Authorization (millions of dollars)	Appropriation (millions of dollars)
1971	10.0	-0-
1972	20.0	6.5
1973	*	6.5*,**
1974	63.0	11.0**
1975	63.0	9.0
1976	10.0	3.5
1977	12.0	8.5
1978	14.0	8.5
1979	14.0	10.3
1980	14.0	11.3
1981	15.0	10.1
1982	**	**

Data sources: S. Bailey. Legislative history of the lead-based paint poisoning prevention program. In *House Committee on Energy and Commerce, Lead Poisoning and Children, 97th Congress*, ad session, December 2; 1982:3–12.
*The program was not reauthorized until fiscal year (FY) 1974; funding for FY 1973 came from a continuing resolution.
**An additional $4.6 million was appropriated for FY 1973, but the administration impounded it. The impounded funds were added to the official appropriations for FY 1974.
Based on Table 4 in Warren, C Brush with Death: A Social History of Lead Poisoning, Johns Hopkins U. Pr.

Throughout the 1970s and most of the 1980s, Billick and other HUD officials argued specifically against using a more scientific pathway analysis that later proved to be so important:

> Strong arguments can be made for using statistical analysis [of gasoline lead for] policy decisions rather than waiting for more detailed experimental data on pathways.[131]

Pathway studies finally appeared in the mid-1980s (Chapter 2), leading to a major report to Congress and Title X of the 1992 Housing and Community Development Act. These studies paved the way to a more scientifically based definition of what a lead paint hazard meant.

This history demonstrates that the call to ban new lead paint has always been accompanied by a designation of what level of lead should be present in new paint. The level of 0.009% became broadly accepted around the world for new paint and is used by the United Nations and the World Health Organization in its Global Alliance to Eliminate Lead Paint in 2022.

It is ironic that in Australia, where the calls to ban new lead paint first began more than a century ago, the allowable level remains considerably greater. The recommended concentration of lead in Australian domestic paint has declined from 50% before 1965, to 1% in 1965. In 1992, it was reduced to 0.25%, and in 1997 it was further reduced to 0.1%. But the current allowable lead paint level there in 2022 is still 0.1%, more than a hundred times above 0.009%,[133] although citizens there are advocating for an improvement (Chapter 8).

1.8 Lead-free versus lead-safe

Had the Australian physicians' early calls to proactively ban new lead paint been heeded, millions of homes would not have been contaminated and millions of children would not have suffered lead poisoning. Since then, many public health entities have called for proactive bans of all nonessential uses of lead, such as the American Public Health Association in 2008[134] and the International Society for Environmental Epidemiology in 2015.[135]

A *proactive* ban of lead paint that has not yet been released into the home environment can be distinguished from a ban of existing lead paint, that is, its complete removal from all existing housing. In later chapters, we will examine how little was achieved when ill-informed but well-meaning policies were advanced to attempt to remove all lead paint from US housing.

In subsequent years, the term "lead free" and "lead safe" came to describe housing that either had all lead paint removed and those that still had lead paint present but not immediately hazardous, respectively.

This nuance emerged as a key policy solution to making progress in the years following 1990. Yet as we shall see, making the distinction proved difficult for both professionals and the public alike. After all, if no "safe" amount of lead in the body had yet been found, then how could anything other than a "lead free" house be acceptable? And because some amount of lead is always present in dust, soil, and as we have just seen, even in nonlead new paint, is "lead-free" merely some ideal that cannot be truly achieved in the real world? Does the fact that no safe level of lead in blood has yet been established mean that all hazard and cleanup standards for lead exposure in dust, soil, paint, and other media must be zero? How can a zero standard possibly be obtained and indeed enforced? And how is zero different from some changing analytical detection limit? The solution to these dilemmas was to be found in exposure pathway studies (Chapter 2).

Consider an analogy: To minimize automobile accidents, there are certain standards and laws. You must not go through a red light and you must obey the speed limit. When that happens, we conclude that we are generally "safe," even though traffic accidents can still occur.

The realization that lead safety could be achieved with the development of scientifically based exposure limits was key to overcoming the paralysis inadvertently caused by the simplistic (indeed dangerous) "removal" mindset.

The concept of standards for exposure limits and both immediate and long-term action was borrowed from occupational health and applied to the lead paint field. In occupational health, the "hierarchy of controls" was established and widely practiced and promoted by numerous safety organizations (Fig. 1.11). In brief, the principle holds that elimination is the preferred option, followed by substituting a less hazardous product, followed by isolating people from hazards, followed by administrative methods and lastly by personal protective equipment, such as masks and protective clothing.[136] At its root, the hierarchy approach was based on the idea that there was not a single solution and that all of them need to act in concert.

As applied to lead paint (Fig. 1.12), the early calls to eliminate its use before release into homes was the most preferred option (the ILO ban in 1920). Next, substitutes like titanium and zinc paints were implemented, even as the lead paint industry opposed elimination

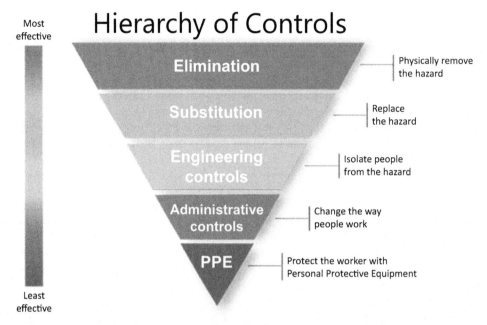

FIGURE 1.11   Hierarchy of controls by effectiveness. Source: *National Institute for Occupational Safety and Health. https://www.cdc.gov/niosh/topics/hierarchy/default.html.*

FIGURE 1.12   Hierarchy of lead paint hazard control. Source: *David Jacobs.*

bans. Remediation of lead paint was a form of engineering control. Cleaning and placing temporary barriers between the child and lead paint were later shown to be ineffective (Chapter 2).

But once it had already been applied to millions of homes, some persisted in believing that only complete removal of lead paint from existing housing could be effective, that is, a retroactive "ban."[137] Anything else they argued was only a partial solution and ineffective, and some even went so far as to call anything less than complete removal a "sellout"

(Chapters 5 and 6). At the other extreme, the lead industry suggested only routine maintenance or cleaning was needed and that bans should never be entertained.

The history shows that both positions (complete removal or only routine maintenance/
cleaning) produced the same result—inaction and continued poisoning.

1.9 Toxicity research and intervention solution research

Standards and methods of detection and mitigation of lead paint could not be well-
informed ones if the science had not been undertaken and validated. That research was
needed to determine exactly which lead paint hazard identification and control methods
worked and which ones did not.

This underscored a tremendous and persistent imbalance in research. In 2012 the
National Toxicology Program stated that lead is one of the most "extensively studied environmental toxicants, with more than 28,900 publications on health effects and exposure in
the peer-reviewed literature." This was based on a PubMed literature search at the time,
so the total is probably even higher today.[138] Yet only a fraction, perhaps only 34 studies
were identified by EPA that focused on determining the effectiveness of the various lead
paint interventions, or "solutions" research.[139] Of course, both types of research were
needed and interdependent, because no one will do interventions on something that has
not been shown to harmful.

This imbalance between toxicity research and solution research persisted. It was a
reflection of part of a broader gap between basic versus applied research or biomedical
versus public health investment. It is also reflected in policy research, which often neglects
to examine if the policy under investigation was implemented and put into practice.

Where would support for the needed intervention research come from? In the early
1990s, the National Institute for Occupational Safety and Health carried out construction
worker exposure studies, but these were not focused on children. As we shall see, it
improbably fell to a housing agency (HUD), not a health or environmental research agency
to develop, implement and fund the needed research to inform which hazard identification and remediation methods worked and were actually implemented (and which ones
should be abandoned).

Intervention research proved to be far more difficult. Although determining the many
ways in which lead produces harm was critically important, it was in some ways far sim-
pler to carry out: Assess exposures and measure health outcomes. Of course, it was often
difficult to measure exposures, although the presence of a biomarker (blood lead and less
frequently bone lead) made that easier. Similarly, measuring the health outcomes
attributable to lead exposure and not to some other confounding variable was also challenging and the industry repeatedly challenged findings that even so-called "low-level
lead exposures" could cause harm, a story described elsewhere.[140]

But those issues pale in comparison to the complexity of carrying out real-world studies
in homes where participants go about their normal day-to-day lives. It involves changing
the home environment to eliminate hazards and determining if the effects were due to the
intervention instead of other factors or mere chance. It also involved working with not just
a researcher and a human subject, but also with the parents, community, building owners,

housing and health departments, building code officials, private construction and inspection firms, financial institutions, and many others. And without proven solutions, policymakers were far less likely to act and fund them to be broadly implemented.

This complexity was overcome, in part, by assembling truly multidisciplinary research teams from housing, exposure assessment, occupational health, construction, medicine, epidemiology, public health, environmental sampling, social sciences and economics, and most important the communities affected. The National Center for Lead Safe Housing, Johns Hopkins University and the University of Cincinnati and many others played important roles in conducting this intervention research from 1985 to 2020. All had occupational health experts among their leaders.

This imbalance in toxicity versus intervention research has broader implications. If we are to solve other seemingly intractable problems such as climate change, or the continuing entry of new untested chemicals and materials into commerce, we cannot afford to restrict our research to only documenting the health and many other adverse effects caused by exposure—there must be much more research on solutions and how to implement them (and perhaps most important whether they really work).

The failure to proactively ban lead paint in the 1800s was a missed opportunity. The attempt to ban it retroactively was a disaster in the few homes where it was attempted. Solutions to the ongoing exposures caused by lead paint that had already been used in homes were eventually found, but it would take more than a century to find and implement them. To do that, better collaboration between housing and health professionals would be required. Yet there were important political and professional barriers standing in the way.

References

1. Gibson L, Love W, Hardie D, Bancroft P, Turner AJ. Notes on lead poisoning as observed among children in Brisbane. *Intercolonial Med Congress Aust*. 1892;78–83.
2. Lockhart Gibson J. A plea for painted railings and painted walls of rooms as the source of lead poisoning amongst queensland children. *Australasian Med Gaz*. 1904.
3. Turner AJ. Lead poisoning among queensland children. *Australas Med Gaz*. 1897;16:475–479.
4. Markowitz G, Rosner D. *Deceit and Denial*. 1st ed. University of California Press; 2013.
5. National Lead Company. *The Dutch Boy Conquers Old Man Gloom: A Paint Book for Boys and Girls*; 1929, 14 pages.
6. Gooch JW. *Lead-Based Paint Handbook*. New York: Springer Science + Business Media; 2002. Available from: 10.1007/b113550. eBook ISBN 978-0-306-46905-3, Hardcover ISBN, 978-0-306-44448-7.
7. The Handbook on Painting. National Lead Company, New York; 1938:6. The University of Illinois Chicago Special Collections & University Archives, School of Public Health, "David E. Jacobs papers" the University of Illinois Chicago School of Public Health Library.
8. Stevenson LG. *A History of Lead Poisoning* [Ph.D. dissertation]. Baltimore, MD: Johns Hopkins University; 1949.
9. Tanquerel des Planches L. *Trait des maladies de plomb o saturines. 1839. (Lead diseases: a treatise)*. Lowell: Daniel Bixby and Company; 1848.
10. Jacobs DE. Lead poisoning: focusing on the fix. Invited editorial. *J Public Health Manage Pract*. 2016;22 (4):326–330.
11. Time for Action on Lead Compounds in Paint. By Julian Hunter, AkzoNobel's Director of Sustainable Value; 2018. https://www.akzonobel.com/en/for-media/media-releases-and-features/time-action-lead-compounds-paint.
12. Pianoforte K. AkzoNobel works with United Nations to eliminate the use of lead paint. *Coatings World*; February 26, 2014.
13. Kehoe RA. Experimental studies on the inhalation of lead by human subjects. *Occup Health*. 1953;12(10):161. PMID 14957352.

14. Kehoe RA. The metabolism of lead in man in health and disease. The harben lectures, 1960. *J R Inst Public Health Hyg*. 1961;. Reprint by McCorquodale and Co., Ltd, London.

15. Baldwin EC. *Building Research Advisory Board Meeting*, New York City. Inter-Office Letter for Cleveland Executive Office of Sherwin Williams Paint Company; June 2, 1969. The University of Illinois Chicago Special Collections & University Archives, School of Public Health, "David E. Jacobs papers" the University of Illinois Chicago School of Public Health Library.

16. Letter from Jones-Day on behalf of Sherwin-Williams to the US Securities and Exchange Commission, December 11; 2014. <https://www.sec.gov/divisions/corpfin/cf-noaction/14a-8/2015/chetrinityhealthsw-co012115-14a8.pdf>. Also see Letter from US Securities Exchange Commission to Ann Foulkes, February 26; 2015. <https://www.sec.gov/divisions/corpfin/cf-noaction/14a-8/2015/chetrinityhealth022615-14a8.pdf>.

17. Dissell R, Zeitner B. Protesters Bring Lead Paint Fight to Sherwin-Williams Shareholder Meeting. Cleveland Plain Dealer, April 20; 2016. <https://www.cleveland.com/healthfit/2016/04/protesters_bring_lead_paint_fi.html>.

18. Patterson CC. Contaminated and natural lead environments of man. *Arch Environ Health*. 1965;11(3):344−360. Available from: https://doi.org/10.1080/00039896.1965.10664229.

19. Patterson CC. *Natural Levels of Lead in Humans*. Carolina Environmental Essay Series III, University of North Carolina at Chapel Hill, March 22; 1982. The University of Illinois Chicago Special Collections & University Archives, School of Public Health, "David E. Jacobs papers" the University of Illinois Chicago School of Public Health Library. Figure reproduced from p. 3.

20. Letter from Claire Patterson, California Institute of Technology to Jere Goyan, FDA Commissioner, April 23, 1980 and FDA response (the University of Illinois Chicago Special Collections & University Archives, School of Public Health, "David E. Jacobs papers" the University of Illinois Chicago School of Public Health Library.).

21. Food and Drug Administration. Lead in food. Advance notice of proposed rulemaking. *Fed Register*. 1979;44:51233−51242.

22. EPA. *EPA's Position on the Health Implications of Airborne Lead*, EPA; November 28, 1973. P. VIII-6 (the University of Illinois Chicago Special Collections & University Archives, School of Public Health, "David E. Jacobs papers" the University of Illinois Chicago School of Public Health Library).

23. Bureau of Foods, Compliance Program Evaluation, FY 1974 Heavy Metals in Foods Survey (7320.13C); June 19, 1975. The University of Illinois Chicago Special Collections & University Archives, School of Public Health, "David E. Jacobs papers" the University of Illinois Chicago School of Public Health Library.

24. Capar SG. Survey of lead and cadmium in adult canned foods eaten by young children. *J Assoc Anal Chem*. 1990;73(3):357−364.

25. Personal Communication. Steve Schwartzberg (previous director of the Alameda County lead program) and David Jacobs; January 2021.

26. Mahaffey KR, Annest JR, Roberts J, Murphy RS. National estimates of blood lead levels, US 1976−1980. *N Eng J Med*. 1982;307(10):573−579.

27. In Memoriam. Kathryn R. Mahaffey: 1943−2009. *Env Health Perspect*. 2009;117(8):A336. PMCID: PMC2721882.

28. FDA data cited in: National Institute of Building Sciences. Report to the Board of Directors, "Lead Based Paint in Housing"; February 20. 1988:13. The University of Illinois Chicago Special Collections & University Archives, School of Public Health, "David E. Jacobs papers" the University of Illinois Chicago School of Public Health library.

29. Farfel MR. Reducing lead exposure in children. *Ann Rev Public Health*. 1985;6:333−360.

30. Memorandum of Department of Health, Education and Welfare Meeting; July 14, 1980 (the University of Illinois Chicago Special Collections & University Archives, School of Public Health, "David E. Jacobs papers" the University of Illinois Chicago School of Public Health library). Also see letter from Health Associates (New York City) to the U. S. Bureau of Foods, May 26, 1980 (the University of Illinois Chicago Special Collections & University Archives, School of Public Health, "David E. Jacobs papers" the University of Illinois Chicago School of Public Health library).

31. National Academy of Science. *Measuring lead exposure in infants, children, and other sensitive populations. National Research Council (US) Committee on Measuring Lead in Critical Populations*. Washington, DC: National Academies Press (US); 1993. ISBN-10: 0-309-04927-X.

32. Memorandum to Richard Ronk from Kathryn Mahaffey; July 24, 1980 (The University of Illinois Chicago Special Collections & University Archives, School of Public Health, "David E. Jacobs papers" the University of Illinois Chicago School of Public Health library).

33. Bridbord K, Hanson D. A personal perspective on the initial federal health-based regulation to remove lead from gasoline. *Env Health Perspect*. 2009;117:1195−1201.

34. International Lead Abatement and Remediation Conference, New South Wales Health Department, Commonwealth EPA; June 1–3, 1994. NSW Government Lead Issues Paper. The University of Illinois Chicago Special Collections & University Archives, School of Public Health, "David E. Jacobs papers" the University of Illinois Chicago School of Public Health library.

35. Gulson B. Stable lead isotopes in environmental health with emphasis on human investigations. *Sci Total Environ*. 2008;75–92. Available from: http://doi.org/10.1016/j.scitotenv.2008.06.059.

36. Mahaffey K, Jacobs D.E. *Lead in the Americas: A Call to Action. Working Group Two: Paint*. Edited by Christopher Howson, Mauricio Hernandez-Avila and David P. Rall. Committee to Reduce Lead Exposure in the Americas. Board on International Health. Institute of Medicine, Washington, DC, USA, In Collaboration with The National Institute of Public Health, Cuernavaca, Morelos, Mexico; 1996.

37. United States Environmental Protection Agency. *Mercury Study Report to Congress* EPA-452/R-97-003; December 1997.

38. EPA's Position on the Health Implications of Airborne Lead. Prepared by EPA; November 28, 1973. P. VIII-7 (the University of Illinois Chicago Special Collections & University Archives, School of Public Health). "David E. Jacobs papers" the University of Illinois Chicago School of Public Health library.

39. Flegal AR, Smith DR. Lead levels in preindustrial humans. *N Eng J MedL*. 1992;326:1293–1294.

40. Weaver JC. A white paper on white lead. *ASTM Stand News*. 1989;34–38.

41. Personal Communication. J Julian Chisolm to David Jacobs; 1989.

42. EPA. *Three City Urban Soil Lead Abatement Demonstration Project*; October 1992. https://nepis.epa.gov/Exe/ZyNET.exe/9100UHKD.TXT?ZyActionD = ZyDocument&Client = EPA&Index = 1991 + Thru + 1994&Docs = &Query = &Time = &EndTime = &SearchMethod = 1&TocRestrict = n&Toc = &TocEntry = &QField = &QFieldYear = &QFieldMonth = &QFieldDay = &IntQFieldOp = 0&ExtQFieldOp = 0&XmlQuery = &File = D%3A%5Czyfiles%5CIndex%20Data%5C91thru94%5CTxt%5C00000026%5C9100UHKD.txt&User = ANONYMOUS&Password = anonymous&SortMethod = h%7C-&MaximumDocuments = 1&FuzzyDegree = 0&ImageQuality = r75g8/r75g8/x150y150g16/i425&Display = hpfr&DefSeekPage = x&SearchBack = ZyActionL&Back = ZyActionS&BackDesc = Results%20page&MaximumPages = 1&ZyEntry = 1&SeekPage = x&ZyPURL.

43. Personal Communication. Donna Shalala to David Jacobs; 1995.

44. McElvaine MD, DeUngria EG, Matte TD, Copley CC, Binder S. Prevalence of radiographic evidence of paint chip ingestion among children with moderate to severe lead poisoning, St Louis, Missouri, 1989 Through 1990. *Pediatrics*. 1992;89(4).

45. Jacobs D. Lead-based paint as a major source of childhood lead poisoning: A review of the evidence. In: Beard M, Ikse A, eds. *Lead in Paint, Soil and Dust: Health Risks, Exposure Studies, Control Measures and Quality Assurance*. Philadelphia: ASTM; 1995:175–187.

46. Felix Wormser, deferral of lead regulation has been our biggest success. *Quoted in Markowitz and Rosner, Deceit and Denial*. 1st ed. University of California Press; 2013.

47. Secretary's Report. Report presented at: Lead Industries Association; November 28, 1938; New York, NY. Quoted in: Rabin R. The lead industry and lead water pipes, a modest campaign. *Am J Public Health*. 2008;98:1584–1592.

48. Lead Industries Association. Quarterly report of the secretary, October 1 to December 31, 1957. Quoted in In: Markowitz G, Rosner D, eds. *Deceit and Denial*. Los Angeles, CA: University of California Press; 2002.

49. National Paint and Decorating Council. *Paint Problem Solver*. 4th ed. St. Louis: National Decorating Products Association; 1990.

50. The Nature and Extent of Childhood Lead Poisoning: A Report to Congress. Agency for Toxic Substances and Disease Registry. US Public Health Service; July 1988.

51. Levin R, Brown MJ, Kashtock ME, et al. Lead exposure in US children, 2008: Implications for prevention. *Env Health Perspect*. 2008;116:1285–1293.

52. Yaffe Y, Flessel CP, Wesolowski JJ, et al. Identification of lead sources in California children using the stable isotope ratio technique. *Arch Env Health*. 1983;38(4):237–245. Available from: https://doi.org/10.1080/00039896.1983.10545809. PMID: 6615005.

53. HUD Guidelines for the Evaluation and Control of Lead Based Paint Hazards in Housing. *1995*. Chapter 7. XRF Performance Characteristics Sheets.

54. Patterson CC, Settle DM. The reduction of orders of magnitude errors in lead analyses of biological materials and natural waters by evaluating and controlling the extent and sources of industrial lead contamination

introduction during sample collecting, handling and analysis. In: LaFluer P, ed. *Accuracy in Trace Analysis: Sampling, Sampling Handling and Analysis.* Vol. 1. Washington, DC: US Department of Commerce; 1976:321–351. National Bureau of Standards Special Publ 422.

55. Bruland KW. Trace elements in sea water. In: Riley, Chester R, eds. *Chemical Oceanography.* Vol. 8. London: Academic Press; 1983:157–220.

56. Chisolm JJ, Harrison HE. The exposure of children to lead. *Pediatrics.* 1956;18(6):943–958.

57. Piomelli S, Davidow B, Guinee VF, Young P, Gay G. The FEP (free erythrocyte porphyrins) test: a screening micromethod for lead poisoning. *Pediatrics.* 1973;51:254–259.

58. CDC Preventing Lead Poisoning in Children; October 1991. <https://wonder.cdc.gov/wonder/prevguid/p0000029/p0000029.asp#head001000000000000>.

59. Chisolm JJ. *Overview and Historical Considerations of Childhood Lead Poisoning.* Undated. The University of Illinois Chicago Special Collections & University Archives, School of Public Health, "David E. Jacobs papers" the University of Illinois Chicago School of Public Health Library.

60. Agency for Toxic Substances and Disease Registry. *Toxicological Profile for Lead;* 2020. <https://www.atsdr.cdc.gov/toxprofiles/tp13-p.pdf>.

61. Laurer G, Kniep TJ, Albert RE, Kent FS. X-ray fluorescence: detection of lead in wall paint. *Science.* 1971;172:466–468.

62. Campbell JL, Crosse LA. Non-destructive analysis of lead on painted apartment walls. *Can J Public Health.* 1976;67:506–510.

63. Laurer, et al. The distribution of lead paint in New York City Tenement Buildings. *Am J Public Health.* 1973;63(2):163.

64. Lead-Based Paint Poisoning Research, Review and Evaluation, 1971–1977. US Department of Housing and Urban Development, LBP-00049; July 1978.

65. National Bureau of Standards. *Analysis of Housing Data Collected in a Lead-Based Paint Survey in Pittsburgh PA. PB-268 150;* March 1977. The University of Illinois Chicago Special Collections & University Archives, School of Public Health, "David E. Jacobs papers" the University of Illinois Chicago School of Public Health Library.

66. Lead Paint Reference Materials. National Bureau of Standards, HUD Contract IAA-H-38-76TQ; July 1977.

67. DeKosky R. Developing chemical instrumentation for environmental use in the late twentieth century: detecting lead in paint using portable X-ray fluorescence spectrometry. *Ambix.* 2009;56(2):138–162.

68. Rasberry SD. Investigation of portable X-ray fluorescence analyzers for determining lead on painted surfaces. *Appl Spectrosc.* 1973;27:102–108. on 103.

69. Sayre JW, Wilson DJ. A spot test for detection of lead in paint. *Pediatrics.* 1970;46(5):783–785.

70. Jacobs DE. A*n Evaluation of Five Commercially Available Lead-Based Paint Field Test Kits,* EPA Contract No. OD-4913NAEX, Georgia Tech Research Project A-8830; July 1991.

71. Rossiter Jr. WJ, Vangel MG, McKnight ME, Dewalt G. *NIST Spot Test Kits for Detecting Lead in Household Paint: A Laboratory Evaluation NISTIR 6398.* Gaithersburg, MD: National Institute of Standards and Technology; 2000.

72. Vostal L, Taves E, Sayre JW, Charney E. Lead analysis of house dust: a method for the detection of another source of lead exposure in inner city children. *Environ Health Perspect.* 1974;91–97.

73. Charney E, Sayre JW, Coulter M. Increased lead absorption in inner city children: where does the lead come from? *Pediatrics* 1980;65:226–231. 19. Charney E, Kessler B, Farfel M, Jackson D. Childhood lead poisoning: a controlled trial of the effect of dust-control measures on blood lead levels. *N Engl J Med.* 1983;309:1089–1093.

74. International Property Maintenance Code. International Code Council; 2018. https://codes.iccsafe.org/content/IPMC2018/preface.

75. McKnight ME, Eric BW, Roberts WE, Lagergren ES. (December 1989), Methods for measuring lead concentrations in paint films (NISTIR 89–4209). US Department of commerce, National Institute of Standards and Technology. In: Mattiello JJ, ed. *Protective and Decorative Coatings.* Vol. II. New York: John Wiley and Sons, Inc; 1942.

76. Clark S, Bornschein R, Succop P, Roda S, Peace B. Urban lead exposures of children in Cincinnati, Ohio. *Chem Speciat Bioavailab.* 1991;3(3–4):163–171.

77. National Survey of Lead Service Line Occurrence by Cornwell DA, Brown RA, Steve H. via, *J Am Water Work Assoc* 108(4):e182–191.

78. EPA. *Contribution of Service Line and Plumbing Fixtures to Lead and Copper Rule Compliance Issues*; 2008:17. <https://archive.epa.gov/region03/dclead/web/pdf/91229.pdf>.

79. A one-inch diameter pipe weighs 6 pounds per linear foot.

80. Steve Via and Tom Neltner. *Personal Communications with David Jacobs.*

81. *National Survey of Lead Service Line Occurrence Correction(s)* for this article David A. Cornwell, Richard A. Brown, Steve H. Via First published: 01 April 2016. <https://doi.org/10.5942/jawwa.2016.108.0086>.

82. What the Eyes Don't See: A Story of Crisis, Resistance, and Hope in an American City. By Mona Hanna-Attisha. Penguin Random House; 2009. ISBN 9780399590856.

83. Geiger SD, Bressler J, Kelly W, et al. Predictors of water lead levels in drinking water of homes with domestic wells in three Illinois counties. *J Public Health Manage Pract*. 2020. Accepted July 24.

84. Rabin R. The Lead Industry and Lead Water Pipes. A modest campaign. *Am J Public Health*. 2008;98:1584–1592. Available from: https://doi.org/10.2105/AJPH.2007.113555.

85. Rosner D, Markowitz G, Lanphear B, Lockhart J. Gibson and the discovery of the impact of lead pigments on children's health: a review of a century of knowledge. *Public Health Rep*. 2005;120:296–300.

86. Dana SL. *Lead Diseases: A Treatise from the French of L. Tanquerel des Planches*. Lowell, MA: D. Bixby; 1848.

87. McKhann CF. Lead poisoning in children. *Am J Dis Child*. 1926;32:386–392.

88. McKhann CF. Lead poisoning in children. *Arch Neurol Psychiat*. 1932;27:294–295.

89. McKhann C, Vogt EC. Lead poisoning in children. *JAMA*. 1933;149:1135.

90. Edsall DL. Chronic lead poisoning. In: Osler W, ed. *Modern medicine*. Philadelphia and New York: Lea Brothers & Co.; 1907:87. 92, 108.

91. Gibson JL. Plumbic ocular neuritis in Queensland children. *BMJ*. 1908;2:1488–1490.

92. Thomas H, Blackfan K. Recurrent meningitis due to lead in a child of five years. *Am J Dis Child*. 1914;8:377–380.

93. Reich P. *The Hour of Lead*; June 1992. Env Defense Fund (the University of Illinois Chicago Special Collections & University Archives, School of Public Health, "David E. Jacobs papers" the University of Illinois Chicago School of Public Health library).

94. Byers R, Lord E. Late effects of lead poisoning on later mental development. *Arch Dis Child*. 1943;66:471–494.

95. *White Lead Convention – International Labour Organization Convention Ban*; 1920. https://www.ilo.org/dyn/normlex/en/f?p = NORMLEXPUB:12100:0::NO::P12100_ILO_CODE:C013.

96. Hamilton A. The White-Lead Industry in the United States, with an Appendix on the Lead-Oxide Industry, *Bulletin of the Bureau of Labor*; July 1911:189–259. https://fraser.stlouisfed.org/title/bulletin-united-states-bureau-labor-3943/july-1911-477666.

97. World Health Organization. *Global Alliance to Eliminate Lead Paint*; 2011. https://www.who.int/initiatives/global-alliance-to-eliminate-lead-paint.

98. Global Health Observatory. https://www.who.int/gho/phe/chemical_safety/lead_paint_regulations/en/.

99. United Nations Environment Program and World Health Organization. *Eliminating Lead Paint Matters! Global Alliance to End Lead Paint*; March 2022. https://mail.google.com/mail/u/0/#all/FMfcgzGmvLPZHbMVq-ZrBZrwzDpDDZcZB.

100. Tehranifar P, Leighton J, Auchincloss AH, et al. Immigration and risk of childhood lead poisoning: findings from a case–control study of New York City children. *Am J Public Health*. 2008;98:92–97. Available from: https://doi.org/10.2105/AJPH.2006.093229.

101. Fatal Pediatric Lead Poisoning—New Hampshire, 2000. *Morbidity and Mortality Weekly Report*. June 8;2001:50 (2). Reported by RM Caron, R DiPentima, C Alvarado, P Alexakos, J Filiano, T Gilson, J Greenblatt, G Robinson, N Twitchell, L Speikers, MA Abdel-Nasser, HA ElHenawy, M Markowitz, P Ashley. https://www.cdc.gov/mmwr/PDF/wk/mm5022.pdf.

102. Baltimore City Health Dept. Chronology of lead poisoning control: Baltimore, 1931–1971. Baltimore Health News 1971;48:34–40. Cited in: Farfel MR. Reducing lead exposure in children. *Ann Rev Public Health*. 1985;6:333–360.

103. Gilsinn JF. *Estimates of the Nature and Extent of Lead Paint Poisoning in the United States*. U.S. Department of Commerce, National Bureau of Standards, Table 1;1972:11.

104. Fassin D, Naudé A-J. Plumbism reinvented. *Am J Public Health*. 2004;94:1854–1863. Available from: https://doi.org/10.2105/AJPH.94.11.1854.

105. Etchevers A, Bretin P, Lecoffre C, et al. Blood lead levels and risk factors in young children in France, 2008−2009. *Int J Hyg Environ Health*. 2014;217(4−5):528−537. Available from: https://doi.org/10.1016/j.ijheh.2013.10.002.

106. IPEN. Lead In Solvent-Based Paints For Home Use Global Report; October 2017. https://ipen.org/sites/default/files/documents/ipen-global-lead-report-2017-v1_2-en.pdf.

107. Walker P, Hickson E. Paint Manual, With Particular Reference to Federal Specifications. Washington, DC: U.S. Department of Commerce; 1945.

108. Metropolitan Life Insurance Company. Chronic lead poisoning in infancy and early childhood. *Stat Bull*. 1930;11(10):4.

109. Hoffman FL. *Deaths from lead poisoning. Bulletin No. 426, Bureau of Labor Statistics*. Washington, DC: Government Printing Office; 1927.

110. Williams H, Kaplan E, Couchman CE, Sayres RR. Lead poisoning in young children. *Public Health Rep*. 1952;67:230−236.

111. Byers RK. Urinary excretion of lead in children. *Am J Dis Child*. 1954;87:548−558.

112. McLaughlin MC. Lead poisoning in children in New York City, 1940−1954. *NY State Med J*. 1956;56:3711−3714.

113. Jenkins CD, Mellins RB. Lead poisoning in children. *Am Med Assoc Arch Neurol Psychol*. 1957;77:70−78.

114. ANSI Z66.1-1955. *Specifications to Minimize Hazards to Children from Surface Coating Materials*. New York: American National Standards Institute, Inc.; 1955.

115. Statement of Senator Edward Kennedy. *Hearings on the Lead Paint Poisoning Prevention Act Before the Subcommittee on Health of the Senate Comm. on Labor and Public Welfare*, 91st Cong., 2d Sess. 20; 1970.

116. Testimony of Richard Robert, National Bureau of Standards. *Lead Based Paint Hearing*. The University of Illinois Chicago Special Collections & University Archives, School of Public Health, "David E. Jacobs papers" the University of Illinois Chicago School of Public Health Library; May 20, 1974.

117. Hearings Before the Subcommittee on Health of the Committee on Labor and Public Welfare. Government Printing Office, Washington, DC; November 23, 1970.

118. Hattis D et al. *Procedures for Lead Paint Removal and Detoxification: Guidelines and Attributes*. National Bureau of Standards Report 10-658; December 10, 1971.

119. Lead-Based Paint Poisoning Prevention Act. 42 USC Ch. 63; 1971.

120. Public Law 91-695-Jan. 13; 1971. Fed Reg p. 2080.

121. Billick IH, Gray YE. Lead-based paint poisoning research: review and evaluation, 1971−1977. Washington DC: US Dept. Housing and Urban Development cited in Farfel MR. Reducing lead exposure in children. *Ann Rev Public Health 1985*. 1978;6:333−360.

122. Warren C. *Brush with Death: A Social History of Lead Poisoning. E-book*. Baltimore: Johns Hopkins University Press; 2001. Available from: https://hdl-handle-net.proxy.cc.uic.edu/2027/heb.30909. Accessed 8 February 2021.

123. National Academy of Sciences. *Report of the Ad Hoc Committee to Evaluate the Hazard of Lead in Paint*. Prepared for CPSC, contract #FDA 70-22, task order 6; 1973.

124. Public Law 93.151-NOV. 9; 1973.

125. U.S. Consumer Product Safety Commission. Ban of lead-containing paint and certain consumer products bearing lead-containing paint. 16 CFR 1303. *Fed Reg*. 1977;42:44199.

126. *Final Environmental Impact Statement on Lead Content in Paint*. Washington, DC: U.S. Consumer Product Safety Commission; 1977.

127. McKnight M, Roberts W. *Lead Concentration in Consumer Paints: A Pilot Study*. US Department of Commerce and US Department of Housing and Urban Development. NISTIR; March 1992 (the University of Illinois Chicago Special Collections & University Archives, School of Public Health, "David E. Jacobs papers" the University of Illinois Chicago School of Public Health library).

128. Brian Lee memo. Consumer Product Safety Commission, Washington, DC, June 22; 1990 (the University of Illinois Chicago Special Collections & University Archives, School of Public Health, "David E. Jacobs papers" the University of Illinois Chicago School of Public Health library).

129. Code of Federal Regulations. *Title 16: Commercial Practices PART 1303—Ban of Lead-Containing Paint and Certain Consumer Products Bearing Lead-Containing Paint*. Effective August 14; 2009. <https://www.ecfr.gov/cgi-bin/text-idx?SID = 97c5d853226258f4cf8412312c3baff6&mc = true&node = se16.2.1303_11&rgn = div8>.

130. Farfel M. *Evaluation of Health and Environmental Effects of Two Methods for Residential Lead Paint Removal*. Thesis. Johns Hopkins University; 1987:35.

131. Billick IH, Curran AS, Shrier DR. Relation of pediatric blood lead levels to lead in gasoline. *Env Health Persp.* 1980;34:213–217.

132. Lead Based Paint Poisoning Research: Review and Evaluation, 1971–1977. By Irwin Harold Billick, V. Eugene Gray. US Department of Housing and Urban Development. US Government Printing Office. Stock N. 023-000-00480-1; 1978. <https://books.google.com/books?id = DR4-AAAAMAAJ&printsec = frontcover&source = gbs_ge_summary_r&cad = 0#v = onepage&q&f = false>.

133. Australian Department of Energy and Environment. *Lead Paint.* Last updated Oct 10, 2021 <http://www.environment.gov.au/protection/chemicals-management/lead/lead-in-house-paint>.

134. APHA. *Calling for a Global Ban on Lead Use in Residential Indoor and Outdoor Paints, Children's Products, and All Nonessential Uses in Consumer Products.* Policy Date: 10/28/2008. Policy Number: 20084.

135. ISEE. Call for action for global control of lead exposure to eliminate lead poisoning. 23.09.2015. *Epidemiology.* 2015;26(5):774–777. Available from: https://doi.org/10.1097/EDE.0000000000000352.

136. National Institute for Occupational Safety and Health. *Hierarchy of Controls.* <https://www.cdc.gov/niosh/topics/hierarchy/>.

137. Rosner and Markowitz. *Lead Wars: The Politics of Science and the Fate of America's Children.* University of California Press; 2014.

138. NTP Monograph. *Health Effects of Low-Level Lead.* US Department of Health and Human Services, June 13; 2012. <https://ntp.niehs.nih.gov/ntp/ohat/lead/final/monographhealtheffectslowlevellead_newissn_508.pdf>.

139. EPA. *Review of Studies Addressing Lead Abatement Effectiveness: Updated Edition.* EPA 747-B-98-001; December 1998.

140. Denworth L. *Toxic Truth: a Scientist, a Doctor, and the Battle over Lead.* Beacon Press; 2008.

CHAPTER

2

Early failures and the seeds of success

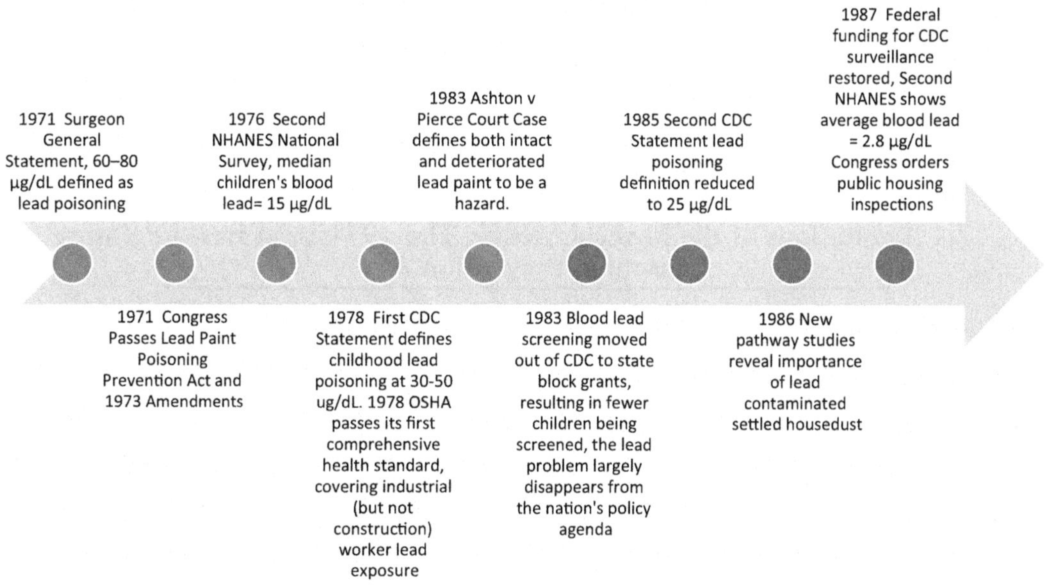
1971 Surgeon General Statement, 60–80 µg/dL defined as lead poisoning
1971 Congress Passes Lead Paint Poisoning Prevention Act and 1973 Amendments
1976 Second NHANES National Survey, median children's blood lead= 15 µg/dL
1978 First CDC Statement defines childhood lead poisoning at 30-50 ug/dL. 1978 OSHA passes its first comprehensive health standard, covering industrial (but not construction) worker lead exposure
1983 Ashton v Pierce Court Case defines both intact and deteriorated lead paint to be a hazard.
1983 Blood lead screening moved out of CDC to state block grants, resulting in fewer children being screened, the lead problem largely disappears from the nation's policy agenda
1985 Second CDC Statement lead poisoning definition reduced to 25 µg/dL
1986 New pathway studies reveal importance of lead contaminated settled housedust
1987 Federal funding for CDC surveillance restored, Second NHANES shows average blood lead = 2.8 µg/dL Congress orders public housing inspections

FIGURE 2.0 Timeline.

If failure offers the best way to learn and then find solutions, then the lead paint story is surely a seminal example. From the failure to fully adopt and enforce the early ban that originated with the Australians and passed by the International Labor Organization in 1920 to the failure to implement effective solutions across the housing stock, the landscape of lead paint poisoning prevention before the 1990s was marked by policy paralysis. But some seeds of success began to emerge during the late 1980s. Before those seeds could sprout, the failures in both medicine and housing came to be recognized. That is the story of this chapter.

Occupational and Child Health

During the late 1970s and early 1980s, Jacobs first entered the occupational safety and health field because of his own personal experience working in an automobile factory, where he was a press operator in a stamping plant. He met and befriended workers who

DOI: https://doi.org/10.1016/B978-0-443-18736-0.00006-6

(cont'd)

---

had missing fingers, hearing loss, and other health problems from workplace hazards. Jacobs ran the health and safety committee for the union local, and after being laid off, he obtained a degree in occupational health. He briefly taught a chemistry laboratory course, where he acquired skills that would be essential in the subsequent development of the lead paint risk assessment process. Jacobs also learned how to help people organize to voice their concerns and create solutions. From there, he went to Georgia Tech, where he conducted over 300 factory and workplace inspections (including some lead battery plants). Those inspections included measuring exposures to various hazards and designing control systems to eliminate or minimize them, the same approach he applied later in the lead paint field. Jacobs' first contact with the lead paint problem occurred in the occupational setting, but it was clear that children were also at risk. "Occupational health *beyond* the workplace" is a regular lecture he still provides. Occupational health informed public health for children. Georgia Tech became one of the nation's largest lead paint training centers.

2.1 Limitations of the Medical Model: The 1971 Lead Based Paint Poisoning Prevention Act

Because physicians were among the first to recognize lead poisoning as a disease, it was only natural that the medical approach would be employed first. That approach emphasized the identification of children with very severe lead poisoning through blood testing (a medical procedure) or observation of clinically observable symptoms and then treatment in a doctor's office or hospital, often followed by release back into the same home. Because lead poisoning is a largely asymptomatic disease, most symptoms are not observable until blood lead levels are quite high.

Most of the US lead poisoning prevention policies in the 1970s and 1980s originated with the Centers for Disease Control and Prevention (CDC) in the form of guidance to physicians. Both the 1971 Lead-Based Paint Poisoning Prevention Act[1] and its amendments in 1973[2] placed the main authority to address the problem in a health agency, the US Department of Health and Human Services, where CDC is located.

2.2 Treatment versus prevention

The first official national medical guidance on lead appeared in 1971 from the US Surgeon General and recommended "screening" (blood testing) for at-risk children. It stated, "The prime goal of screening programs is the *prevention* of lead poisoning" (emphasis added).[3] But if elevations in blood lead were found, then the exposure had already occurred, making "prevention" a hollow claim.

Within the public health world, there are three forms of "prevention." In the lead paint context, primary prevention means taking action *before* children are exposed. Secondary prevention typically means blood lead screening followed by case management only *after* an elevated blood lead level is found and is meant to stop exposures from getting worse. Finally, tertiary prevention typically means hospitalization and often drug treatment known as chelation to reduce lead in the blood through increased excretion, often in a desperate attempt to avoid the child's death. (Tertiary prevention can also include providing educational and other resources to the child's caregivers.)

Although the Surgeon General did mention the need for correction of housing with lead paint problems, the 1971 statement was really aimed at medical management strategies based on different blood lead levels. Importantly, the 1971 Surgeon General's Statement said, "*sources* of lead must be removed from the environment" (emphasis added). In other words, there was no recognition of key exposure *pathways* like settled house dust (and the need to eliminate them).

The Surgeon General called on physicians to report cases of elevations in blood lead, the beginnings of the nation's surveillance and case management efforts, which did not begin in earnest until the 1990s. (Surveillance and case management mean tracking children with already elevated blood lead levels. There was no analogous surveillance system to track homes with lead paint hazards at the time.)

In 1971 the Surgeon General identified a blood lead level of 80 µg/dL to be an "unequivocal case of lead poisoning." This is not that far removed from a fatal dose (which is between 100 and 150 µg/dL). CDC defined lead poisoning as being at or above 80 µg/dL.[4] Other medical interventions were recommended at 60 µg/dL. By 2021, CDC used a blood lead reference value of 3.5 µg/dL to define an "elevated blood lead level."[5]

Some of the 1971 Surgeon General's medical guidance numbers were based on limitations in the ability to measure lead in blood reliably, as well as a continued reliance on clinically observable symptoms.

The medical focus remained largely unchanged a few years later, when in 1978 the first comprehensive statement on lead poisoning from CDC was issued. It featured a recognition that screening was only necessary until "removal" of lead from the environment could occur. The recommendations on screening, diagnosis, treatment, and follow-up of children with undue lead absorption and lead poisoning were all described in detail. But CDC also stated:

> The ultimate preventive goal is identification and *removal* of lead in the environment before it enters the child. Until this occurs, screening, diagnosis, treatment, and environmental management will continue to be necessary public health activities. (emphasis added)[6]

Like the earlier Surgeon General's statement, the focus from CDC remained on medical treatment; a simplistic "removal" of lead without a recognition of exposure pathways; and reporting. In the 1978 CDC statement, the definition of lead poisoning was lowered to 50 µg/dL if symptoms were observed, and other interventions were recommended at 30 µg/dL due to the growing research showing adverse health effects at ever-lower blood lead levels. There was also discussion of accuracy problems in blood lead measurements and use of free erythrocyte protoporphyrin as a surrogate measure until blood lead measurement technology could be improved to detect lower blood lead levels reliably. Erythrocyte protoporphyrin is a precursor to heme and was used to mainly assess anemia (iron deficiency), but also had value in detecting high blood lead levels.

The 1971 Surgeon General and the 1978 CDC statements generated important activity. By 1978, the recently formed national Lead Paint Poisoning Prevention Program had approximately

60 state and local health department programs. An evaluation showed that, using a sample of children with very high blood lead levels, over 90% of children in the program had begun medical pediatric management, and about 75% of the homes were reportedly investigated. Over 75% of these children showed some improvement in their blood lead levels, but the average was reduced by approximately 30%. This meant that blood lead levels remained quite high. Furthermore, 58% and 85% of Class III (20–44 μg/dL) and Class IV (45–69 μg/dL) children, respectively, were lost to follow-up by 1 year (classes of lead poisoning in use at the time are defined in Table 2.1). About one-fifth to one-quarter of the children did not improve or did not have their homes inspected or remediated.[7]

The limitations of medical management were becoming clearer. At higher blood lead levels above 45 μg/dL, chelation therapy was recommended. But for lower levels, chelation did not appear to be effective and many children who lived in homes where the paint was removed using dangerous methods had to be readmitted to the hospital because the dust was not controlled (discussed later in this chapter).

Perhaps the most definitive study of chelation therapy appeared in 2001, which was led by Walter Rogan from NIEHS. The study looked at children with lower blood lead levels between 20 and 44 μg/dL. It found that:

> The mean IQ score of children given succimer [a chelating medication] was 1 point lower than that of children given placebo, and the behavior of children given succimer was slightly worse as rated by a parent. However, the children given succimer scored slightly better on the Developmental Neuropsychological Assessment, a battery of tests designed to measure neuropsychological deficits thought to interfere with learning. All these differences were small, and none were statistically significant...chelation therapy is not indicated for children with these blood lead levels.[8]

TABLE 2.1   Medical interpretation of child blood lead classes, 1978.

Class	Blood lead concentration (μg/dL)	Comment
I	≤ 9	A child in Class I is not considered to be lead-poisoned
IIA	10–14	Many children (or a large proportion of children) with blood lead levels in this range should trigger communitywide childhood lead poisoning prevention activities. Children in this range may need to be rescreened more frequently
IIB	15–19	A child in class IIB should receive nutritional and educational interventions and more frequent screening. If the blood lead level persists in this range, environmental investigation and intervention should be done
III	20–44	A child in class III should receive environmental evaluation and remediation and a medical evaluation. Such a child may need pharmacologic treatment for lead poisoning
IV	45–69	A child in class IV will need both medical and environmental interventions, including chelation therapy
V	≥ 70	A child in class V lead poisoning is a medical emergency. Medical and environmental management must begin immediately

From CDC Lead Poisoning Prevention Statement; 1978.

This study, perhaps more than any other in this time period, showed the limitations of the medical approach; there simply was no medical "cure," other than addressing exposures. It was primarily a drug trial, but this study also included limited cleaning. Although the cleaning did reduce dust lead levels temporarily, the study concluded:

> Despite these substantial reductions in dust lead loadings, a single professional cleaning did not reduce the lead loadings of all dust samples to levels below current federal standards for lead in residential dust. Attainment of dust levels below current standards will require more intensive cleaning and lead hazard reduction strategies.[9]

The limitations of chelation therapy (instead of exposure reduction) were summarized in 1984 by doctors in four of the nation's largest lead treatment clinics in New York City, Baltimore, and Boston:

> Medical treatment with chelating agents must not be considered a substitute for dedicated preventive efforts to eradicate controllable sources of lead (e.g., substandard housing . . .)[10]

This limitation was reported most recently in 2021 in a major review of the medical aspects of lead poisoning.[11]

Perhaps one of the more bizarre proposals in the 1970s came from the Grow Chemical Corporation, which conducted animal tests on adding chelating drugs to paint. The company proposed painting the walls of homes with children in a two square block area in Newark with such a paint. When asked in 1974 by Housing and Urban Development (HUD) officials in a public hearing what he thought of this idea, J. Julian Chisolm from Johns Hopkins University responded, "The idea of putting chelating agents into paint on the walls horrifies me."[12] Among the many problems with such an approach was the fact that the dose a child might receive would be entirely uncontrolled, that other essential metals like zinc or iron might be removed from the child's body from chelation, and that chelation needed to be done only under close medical supervision. The proposal went nowhere but underscored the fact that medical treatment options were limited.

Another proposal came from the Peerless Paint and Varnish Company, which wanted to add chemicals to paint that would make it taste bitter, supposedly to prevent children from eating paint chips. These so-called "repulsive tasting paints" contained denatonium benzoate; upon further questioning, the company admitted that "pure denatonium benzoate might be toxic," but also stated it had done some tests with the health departments in New York City, Hartford, New Haven, and Newark.[13] Interventions to add drugs or other chemicals to paint ultimately were never implemented because of the obvious dangers.

However, there was some progress at the Occupational Safety and Health Administration (OSHA) for adult workers in this time period. In 1978 OSHA passed its first comprehensive health standard after President Carter appointed Eula Bingham to lead the agency. This standard covered industrial workers from lead exposure. Yet the workers doing lead paint mitigation were construction (not industrial) workers and where not covered by the 1978 OSHA regulation. This loophole would not be corrected until 1996 as a result of Congressional action (Chapter 3). The story of how the OSHA standard on lead was passed, despite fierce industry opposition, is described in Chapter 5.

By 1985, the second CDC childhood lead poisoning statement marked a clearer move away from relying only on clinical symptoms. It provided more detail on biomarkers in blood and related measurement difficulties. CDC stated for the first time that a "precise threshold for the harmful effects of lead" was not known. It also reduced the definition of elevated blood lead level to 25 µg/dL due to the emerging research documenting more subtle subclinical symptoms.[14]

Importantly, that 1985 CDC statement also mentioned a lead industry member: "Dr. Jerome F. Cole of the International Lead Zinc Research Organization did not support the (CDC) recommendations." Earlier, the Lead Industries Association sued CDC to prevent the 1985 statement from being issued, alleging that the CDC committee was not "balanced."[15] The director of the CDC Center for Environmental Health at the time, Vernon Houk, stated:

> To attempt to prohibit a responsible health agency from making recommendations regarding the health of children is unique. The data on lead is so voluminous and so strong that if we can't handle lead, we can't handle any of these things we have much less strong data on.[16]

For the first time, CDC included a major section on lead paint remediation in the 1985 Statement. It noted the importance of dust control, which coincided with the appearance of the pathway studies discussed later in this chapter. But only three of the Statement's 22 pages were devoted to the detection and remediation of lead paint, and it defined remediation as including just four steps:

1. Removing lead paint from wood trim or walls;
2. Thorough vacuuming to clean up the debris;
3. Wet scrubbing for maximum elimination of fine lead-bearing particles; and
4. Repainting the area with lead-free paint (i.e., paint containing less than 0.06% of lead in the final dried solid).

Although dust was recognized, the focus was still on the removal of lead paint and only rudimentary cleanup procedures. Of the statement's 13 authors, none had any background in housing.

CDC had provided some limited funding in the 1970s to local jurisdictions for blood lead surveillance, but the CDC grant funds could not be used for remediation of privately owned low-income housing where lead paint hazards were most severe. No federal funding for remediation of lead paint in privately owned homes would become available until 1990 when Congress provided the first such funds to HUD.

In short, CDC could only advocate for housing remediation, not actually implement it because it had no authority over housing and because it was not a regulatory agency.

CDC issued important subsequent lead poisoning prevention statements, most importantly in 1991 and again in 2012, which are discussed in later chapters. As more information became known about adverse health effects, the blood lead levels that triggered action were also lowered (Fig. 2.1).[17]

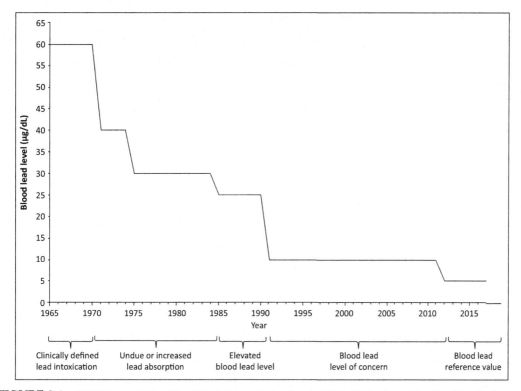

FIGURE 2.1 CDC historical blood lead levels triggering action (in 2021 CDC reduced the blood lead reference value from 5 µg/dL to 3.5 µg/dL). Reproduced with permission: *Ettinger et al. CDC lead poisoning prevention program.* J Public Health Manage Protect *2019; Jan/Feb (suppl 1):S5–S12.*

2.3 Surveillance and population surveys

The 1971 Lead Paint Poisoning Prevention Act mandated that the nation begin counting childhood lead poisoning cases, although only $3 million was authorized to local cities and states to do so at the time. The very first 1971 Surgeon General report also recommended reporting such cases but failed to provide resources to do so. By 1991, such local programs were only in 13 states and 2 large cities.[18] The inability to construct a truly national surveillance system in the 1970s and 1980s was a clear failure. But by 2019, CDC funded surveillance programs in all but five states.[19]

On the other hand, instead of local surveillance programs, population-wide surveys of the prevalence of children with elevated blood lead levels, beginning in the 1970s, were enormously important in understanding the magnitude of the problem, as well as informing both sources and pathways of exposure. The National Health and Nutrition Examination Survey (NHANES) did not include blood lead levels in its first wave in the early 1970s, but it did so in the second wave (1976–1980).

The findings showed that blood lead levels were much higher in the late 1970s compared to 2022. An astonishing 50% of children under 6 years old had a blood lead level greater than 15 μg/dL (i.e., the median blood lead level at the time), about 15 times higher than in 2020. Even more astonishing, among Black children under 6 years old (who were more likely to reside in older and poorer-quality homes), between 96% and 99% had a blood lead level greater than 10 μg/dL (the level of concern established by CDC in 1991). The NHANES findings discussed large disparities by degree of urbanization, income, and race. This powerfully demonstrated the relationship between environmental justice and childhood lead poisoning. But importantly, this first national blood lead survey did not report anything on housing age.[20]

This masked the influence of lead paint, which was more highly concentrated in older housing. Because the focus in this era was on lead in gasoline and lead in food from lead soldered canning, it is perhaps unsurprising that the authors chose not to report the age of housing. NHANES was led by scientists from the Food and Drug Administration and the National Center for Health Statistics at CDC, neither of which had housing expertise.

The housing connection to lead paint was missing. Indeed, a 1973 study from the National Bureau of Standards and funded by HUD woefully underestimated the true extent of the problem. The report relied on estimates of dilapidated housing and child occupancy and concluded that 500,000 children had elevated blood lead levels and that 7 million homes likely had deteriorated lead paint.[21] Although the report admitted these were likely to be underestimates, later surveys with actual measurements put the number closer to 4 million children and 64 million homes with lead paint in the 1990s.

This lack of understanding on housing and lead paint in population surveys would not be corrected until the third NHANES survey in 1988–1991.[22] By then the average blood lead level had improved to 2.8 μg/dL (down from 15 μg/dL), but once again race, income, and urban status were all correlated with blood lead.

For the first time, the association with housing age (by residence in the northeast US with its older housing) was also reported. Despite the improvement in average blood lead levels, 1.7 million children were reported to have blood lead levels above the CDC level of concern at the time (10 μg/dL) and 21% of Black children still had blood lead levels greater than that. As bad as this was, it marked an overall improvement of 78% in reduced blood lead levels compared to the late 1970s.

Both the second and third NHANES surveys were published in medical journals, reflecting the emphasis on clinical care. In later years, results would be published in journals with a much wider audience.

The effect of housing age and its association with lead paint finally became even clearer in the next NHANES survey, covering 1991–94. Despite continued improvement, 890,000 children were still reported to have blood lead levels greater than 10 μg/dL. The authors reported that "sociodemographic factors associated with higher blood lead levels in children were non-Hispanic black race/ethnicity, low income, and *residence in older housing*" (emphasis added).[23]

Older housing had finally emerged as a significant risk factor. Importantly, these data were published in an environmental, not medical journal for the first time, reflecting the broader focus lead poisoning was finally attaining beyond the medical profession.

The initial failures to recognize the importance of lead paint in older housing in these early NHANES surveys marked an important gap in understanding sources and pathways of exposure during the 1970s and 1980s. Although there were earlier estimates of the number of houses contaminated with lead paint during the 1980s, they were not reliable. The first nationally representative and reliable housing survey on lead paint prevalence did not occur until 1990 (Chapter 3).

The failure to combine nationally representative housing and health surveys into a single comprehensive one remained in 2022. CDC continued to conduct its NHANES surveys (which included blood lead data) and HUD conducted its American Healthy Housing survey (which included paint, soil and dust lead data). The two remained separate, making clear links to housing and health needlessly difficult. The challenge of disconnected housing and health surveys, starting in the late 1970s, remained a key knowledge gap in 2022 that contributed to the historic divide among the housing, health, and environmental communities (Chapter 9).

An important exception to this divide occurred between 1996 and 2000 when HUD entered into a unique agreement with CDC. HUD paid CDC to collect paint, dust, and soil data in the same NHANES homes where blood lead measurements also occurred. This move proved to be essential in developing dust and soil lead standards, eventually promulgated first by HUD in 1999 and EPA in 2001.[24,25] But after that, both HUD and CDC returned to doing their own separate surveys, despite clear evidence of the power of a combined housing and health nationally representative survey.[26] The World Health Organization recognized this same problem as well and conducted a massive combined housing and health survey in eight European cities (Chapter 8), with compelling and policy-relevant results.

A key lesson from the 1970s and 1980s was that the medical approach alone did not work for housing-related diseases and injuries such as lead poisoning. After all, remediation of dangerous housing conditions such as lead paint is a specialized form of housing rehabilitation and construction. Physicians are not trained to perform such a function, which perhaps explains why so little housing data were examined at the time. Physicians are well trained in the clinical setting to administer medical procedures such as a blood lead test and provide anticipatory guidance and education and medical treatment. But education alone was repeatedly shown to be ineffective on its own,[27] and medical treatment options were limited.

2.4 Reagan's "New Federalism": lead poisoning disappears from the policy agenda

The reliance on medical surveillance and population surveys made the lead paint problem particularly vulnerable if the surveys ended.

President Reagan assumed office in 1981. He campaigned and sought to govern on the platform that "the most important cause of our … problem has been the government itself." He thought responsibility for many domestic policies should be given to the states and he reduced government funding especially at HUD.[28]

During most of the 1970s, the federal lead program was administered primarily by CDC as "categorical" grants expressly for that purpose. From 1974 to 1978 about 400,000 children were screened.[29] But in 1981 the program was folded into a broad block grant program, the Maternal and Child Health Services Block Grant to the States, and lead poisoning prevention became only one of many eligible activities. Control over the funds shifted from city mayors to state governors. As a result, surveillance and population surveys largely ended, and with it the growing national recognition of the problem. In 1982 the budgets of blood lead screening programs were reduced by an average of 10% compared to 1981 and the number of children screened fell by the same amount,[30] with even further reductions throughout much of the 1980s.

Folding lead poisoning prevention into a block grant program proved to be a tragic mistake because most of the states simply stopped tracking lead poisoning and the problem disappeared from the nation's attention and policy agenda until the late 1980s (Chapter 3).

2.5 Where was the housing profession?

2.5.1 Decent safe and sanitary housing

Although the 1973 Amendments to the Lead Based Paint Poisoning Prevention Act stated that HUD was to be the agency leading the Federal effort regarding lead paint in housing, in fact, HUD primarily carried out only limited research from 1973 to 1987 and spent very little on mitigating lead paint in the nation's housing. The absence of leadership from the housing world in the 1970s and 1980s contributed to little progress on lead paint.

These first lead paint Congressional Acts mandated a HUD research and demonstration program to identify where lead paint existed and how to remediate it safely, as well as to improve lead paint X-ray fluorescence (XRF) analyzers. It was not until the late 1980s and early 1990s that Congress clearly put a housing agency in a leadership role with specific tasks and with significant funding to address the problem in low-income privately owned housing where the risks were greatest.

The Housing Act of 1937[31] created the US Housing Authority, a predecessor to HUD. That Act established "decent safe and sanitary" housing as a national goal. But lead paint did not get included in the definition of exactly what that meant until 2010:[32]

The term "decent, safe, and sanitary" in HUD regulations is defined as a dwelling that:

1. Meets applicable federal, state, and local housing and occupancy codes; including but not limited to the Uniform Building Code, National Electrical Code, ICBO Plumbing Code, the Uniform Mechanical Code, HUD Minimum Property Standards, and HUD Mobile Home Construction and Safety Standards (24 CFR part 4080);
2. Is structurally sound, clean, weathertight, and in good repair and has adequate living space and number of rooms;
3. Has an adequate and safe electrical wiring system for lighting and other electrical services where economically feasible;
4. *Meets the requirements of the HUD lead-based paint regulations (24 CFR part 35) issued under the Lead-Based Paint Poisoning Prevention Act (42 U.S.C. 4831et seq.) (this section 4 was added in 2010)*; (emphasis added)
5. In the case of a physically handicapped person, is free of any architectural barriers. To the extent that standards prescribed by the American National Standards Institute, Inc.,

in publication ANSI A117.1−1961 (R 1971), are pertinent, this provision will be considered met if it meets those standards;

6. Has heating as required by climatic conditions;
7. Has habitable sleeping area that is adequately ventilated and sufficient to accommodate the occupants;
8. Has a separate well-lighted and ventilated bathroom, affording privacy to the user, that contains a sink and bathtub or shower stall, properly connected to hot and cold water, and a flush toilet, all in good working order and properly connected to a sewage drainage system; and
9. In the case of new construction or modular housing, complies with the energy performance standards for new buildings set forth by the US Department of Energy.

Importantly, this definition placed compliance with housing codes first. Yet most housing codes failed to address lead paint, discussed later in this chapter.

The prevailing practice during the 1970s and 1980s on the part of housing professionals was to refer lead poisoning cases to health authorities for medical treatment, not to remediate housing. This abdication was driven in no small part by the absence of scientifically validated remediation methods.

2.6 Early lead paint removal efforts backfire

In the 1970s and early 1980s, there was little sustained effort to perform lead paint remediation. One press account in 1981 wrote:

> What was supposed to be an ambitious local and national fight against lead poisoning in toddlers has deteriorated into poorly financed and often-ignored sets of unrelated government programs and regulations.[33]

In 1982 the Wall Street Journal wrote:

> More than a decade after lead poisoning was first recognized as a major childhood health problem … community health officials seem more ill-equipped than ever to cope with the situation. They are tangled in red tape that slows enforcement actions…They are hurt by federal budget cuts for screening and treatment programs. They are overwhelmed by the sheer magnitude of the problem.[34]

One of the earliest proposals to study various methods of lead hazard control was presented to HUD by New York City in 1976,[35] but HUD never acted on it and performed very little research on lead paint interventions during the 1970s and early 1980s.

Studies in this time period showed that simplistic paint removal efforts often did more harm than good. For example, Mark Farfel at Johns Hopkins University in Baltimore reported that 40% of children in "traditionally" abated homes had to be hospitalized at least once within a year to undergo chelation therapy because their blood lead levels had "rebounded" to dangerous levels after the paint removal between 1978 and 1982.[29] There were also no reductions in average blood lead levels in both chelated and nonchelated children returning to homes that had undergone "traditional" lead paint removal methods (Fig. 2.2).

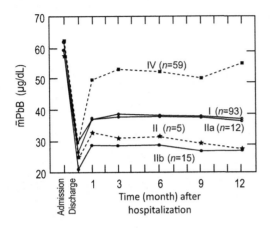

GEOMETRIC MEAN BLOOD LEAD CONCENTRATIONS (PbB) BEFORE
AND AFTER THERAPY ACCORDING TO HOUSING CLASSIFICATION

FIGURE 2.2 Blood lead levels following chelation therapy by housing category, 1985. Reproduced with permission: *Chisolm JJ et al. Relationship between level of lead absorption in children and type, age and condition of housing.* Environ Res *1985;38:31–45.*

Numbers in each housing group are shown in parentheses. Serial PbB data during the first 12 months of follow-up are shown above as their antilogarithms. Table 4 summarizes

Group

I   Old housing (50–100 years)—lead-in-paint hazards partially abated according to local ordinances.
II-a "Lead-free" public housing—child regularly *visits* Group I or Group IV housing
II-b "Lead-free" public housing—child *does not visit* Group I or Group IV housing
III   Old housing—completely gutted and renovated.
IV   Old housing—lead-in-paint hazards incompletely abated according to local ordinances

After inpatient chelation therapy, children discharged to and remaining exclusively in public "lead-free" housing (Group II-b) show significantly lower PbB than children discharged to or visiting old housing abated according to local ordinances (Groups I and II-a; *P* < 0.001). Among the 59 children (Group IV) discharged to old unabated housing, PbB ≥ 50 μg/dl recurred in 1 month in 29, in 3 months in 20, and in 6–12 months in 10.

High lead exposures were identified in workers removing lead paint in the 1980s.[36] This report drew on data from workers in broader construction trades as well as housing rehabilitation. It reported OSHA data that showed for workers with a potential for exposure to lead paint in construction, 49% of the air samples collected by OSHA were greater than its permissible exposure limit, which at the time was 200 μg/m^3 in construction. For workers removing lead paint by torching or dry power sanding, exposures were as high as 11,000 μg/m^3 (Fig. 2.3). More typically, workers had no protection at all (Fig. 2.4). These high occupational exposures were a sign that community exposures and exposure to children were also likely to be quite high.

Farfel and Chisolm found that improved abatement techniques resulted in lower blood lead levels in children compared to the dangerous "traditional" abatement measures.[37] In 1983 another study demonstrated that postabatement dust lead cleaning and dust testing were important in reducing blood lead levels, an early indication of the pathway studies to emerge a few years later.[38]

There were also case reports in the 1980s showing that when old lead-based paint was disturbed during ordinary housing rehabilitation, repainting, or traditional abatement

FIGURE 2.3 Power sanding and torching to remove lead paint (with some worker protection), 1986. Source: *David Jacobs*.

activities, large quantities of lead dust were generated and often resulted in elevated blood lead levels, for both the workers and the child occupants. A 1985 study reported that mean blood lead correlated significantly with the amount of lead in indoor paint and that refinishing activity in homes with lead paint was associated with an average 69% *increase* in blood lead level in the 249 infants studied.[39] Another study of 370 newly lead-poisoned children showed sources of lead poisoning included household renovation and paint chip ingestion.[40] Other researchers in this time period reported cases where remodeling or renovation activity resulted in elevated blood lead levels.[41,42] A study of children in Newark showed that the number of children being poisoned was *increasing* and was actually higher in 1980 than in 1972 when the local lead poisoning program began.[43]

One episode from the 1980s was particularly tragic and reported by Phillip Landrigan, who later emerged as a key researcher in the lead paint field:

> The dwelling involved was a 2-story, 19th century Victorian farmhouse with 10 rooms. Most of the wooden floors, moldings, walls, ceilings, and door frames had been painted with lead-based paint. The renovation work included restoration of surfaces by removing the paint down to the bare surface on floors and woodwork and recoating with new varnish. Ceilings were repaired, and wallpaper and paint were removed from a number of walls. Two workers used rotary power sanders, hand sanders, scrapers, torches, heat guns, and chemical paint strippers. The family left the house during most of the renovation work but returned after it was only partially completed. There was dust throughout the dwelling. After one of the family's dogs started to have seizures, a veterinarian determined that the dog was lead poisoned. The mother and two children were subsequently tested. The children had blood lead levels of $104\,\mu g/dL$ and $67\,\mu g/dL$, which is 6 to 10 times above the level of concern established by CDC in 1991 ($10\,\mu g/dL$). The mother had a blood lead level of $56\,\mu g/dL$. All three were admitted to a local hospital where they were treated for severe lead poisoning. The mother was 8 weeks pregnant and opted for a therapeutic abortion. A babysitter who had two children of her own sometimes cared for all four children in the home. The babysitter's two children were also tested and found to have blood lead levels of $80\,\mu g/dL$ and $68\,\mu g/dL$. These two children were also hospitalized and treated for severe lead poisoning.[44]

But dangerous removal still prevailed widely. During hearings conducted by HUD in 1974, George Gould (a lawyer with Legal Services from Philadelphia) testified:

> the best thing [HUD] can do is totally remove it [lead paint] ... I think [lead paint removal by scraping] is adequate to reduce the hazard. It is basically what is being done across the United States.[45]

Policies on lead paint at the local level, to the extent that there were any at all, tended to reflect the same mindset of simplistic removal. In 1971 Massachusetts was the first state to enact a lead paint statute.[46] The focus on removal was dominant. There, lead paint abatement came to be called "deleading," suggesting only removal was acceptable. At its beginning, the policy attempted to remove all lead paint on surfaces up to 5 ft, thinking that such surfaces were "accessible" to children. A 1984 "deleader's manual" makes no mention of testing dust for lead content at the conclusion of the job.[47] Initially, there was no dust testing after the work to ensure that cleanup was adequate, and the home was safe for children, although this was later included in the state's regulations.

Several studies found important problems in this "deleading" approach, most importantly that blood lead levels *increased* instead of improving.[48,49] One study that examined the Massachusetts practice of "deleading" found that abatement measures involving dry scraping of lead-based paint resulted in a statistically significant increase of blood lead levels from a mean of $36.4\,\mu g/dL$ to $42.1\,\mu g/dL$ ($P < .001$) in a cohort of 114 preschool children. However, when abatement was accomplished by covering or replacement of building components (i.e., minimizing the abrasion of the lead-based paint and consequent dust generation), blood lead levels significantly improved.[50]

Yet the Massachusetts requirements were quite advanced for their time because they included an important primary prevention element. Any home with a young child was required to undergo lead inspections and deleading, not just homes of children who already had high blood lead levels. This was an early marker of the shift away from the

FIGURE 2.4 Torching to remove lead paint, 1984 (with no worker protection). Reproduced with permission: *Mark Farfel.*

medical model. The state also provided important financial mechanisms to help property owners comply (Chapter 5).

In 1985 J. Julian Chisolm from Johns Hopkins University found that there were much lower blood lead levels for those children who had been treated for lead poisoning and released to public housing that had little or no lead paint (the changes were statistically significant). These children were compared to children who had been released to other housing where the lead paint had been removed using the traditional abatement methods at the time (open flame burning, sanding, and dry scraping).[51]

The early attempts to simply remove lead paint were failures, paving the way for a new effort focused on exposure reduction, not just paint removal. It also revealed that the focus on lead paint content alone was inconsistent with the emerging exposure science.

2.7 Housing codes fail to regulate lead paint

Most housing construction and rehabilitation is regulated by local housing codes and enforced by local housing agencies. The importance of codes is reflected in HUD's definition of a "decent safe and sanitary" house shown earlier. Although most jurisdictions have their own housing code, many in the US are derived from "model" codes developed by the so-called International Code Council (ICC) (the ICC is not a truly international body and is dominated by the United States).

In the 1980s there were attempts to include lead paint requirements in the model codes, but these were all rejected by the code-making bodies.[52] Attempts to train local code officials in lead paint issues met with failure in this time period. Instead, code officials who were not trained in lead paint detection or remediation referred such issues to local health departments. But local health departments had little or no experience in housing matters—the perfect storm that produced policy paralysis.

This abdication by code-making bodies and local housing code departments eventually led to the creation of three entirely new industries, mandated by Congress in the early 1990s. Lead paint risk assessment, inspection, and lead paint abatement employ licensed and specially trained professionals to identify and remediate lead paint. Starting in 1996, they were regulated by either EPA or delegated state government authorities.[53] Training was conducted by accredited providers. Most lead inspectors were not housing code compliance officers, with a few notable exceptions such as Rochester NY (Chapter 5).

The disconnection with housing codes continued into the 2010 decade. Even though there were by then legal lead requirements for housing renovation, repair, and painting, the model code-making bodies continued to refuse to include lead requirements.[54]

2.8 Housing law and public housing

To a large extent, the origin of all US housing laws, standards, and regulations was for health reasons. Epidemics of typhoid, tuberculosis, and cholera led to the "sanitation movement" around 1900, in which housing improvements in indoor plumbing and ventilation helped conquer such diseases, together with better medicine. Despite this successful collaboration, housing and health disciplines separated in the ensuing years. The housing sector became focused on housing as a way to promote economic development and wealth creation. Health was thought to be largely outside the housing mission, at least until lead paint poisoning came to their attention, beginning in earnest in the late 1980s. Improving housing, health and environmental collaboration became part of an emerging consensus by the 2010s (Chapter 9).

Much as medical doctors did not include housing professionals in their health policies on lead paint, housing professionals did not draw on medical expertise in developing housing policy. This also meant that lead paint hazards were seen differently from other housing deficiencies. From an owner's financial perspective, it made little sense to invest in fixing lead hazards in the home, because, unlike other home improvements, there was no resulting increase in housing value or price following lead abatement. The normal financing mechanisms, such as home improvement loans, mortgages, subsidies, and enforcement did not address lead paint during the 1970s and 1980s.

The absence of the housing industry in the lead poisoning prevention field made such housing finance mechanisms impossible during the 1970s and 1980s. This absence would lead to disastrous consequences for both the health care and housing markets. The costs of finding and treating lead-poisoned children were borne largely by the health care system, where costs were very high. Similarly, housing providers suffered from liabilities and lawsuits from lead-poisoned children, alleging that housing conditions and the presence of lead paint had caused harm, producing high housing costs.

Public housing authorities regulated by HUD mostly resisted lead remediation efforts in this time period, arguing that federal appropriations were too limited. If those resources were devoted to lead abatement, it would come at the expense of the already overwhelming need for more affordable housing and to repair other urgent housing problems. They had a point. By one estimate, HUD subsidies were able to meet the housing needs of only one-fourth of those low-income households that qualified in the 1980s,[55] which remained a problem in 2020.[56]

It would fall to leaders in the public housing world, notably Gordon Cavanaugh and Miles Mahoney in the late 1980s, to convince Congress to appropriate additional funds for lead abatement in public housing (Chapter 3).

Housing authorities were so resistant that they initially refused to attend training sessions on lead paint during the 1980s. It took a direct Congressional order to force housing authority executive directors (not just the staff) to attend such training. Senator Barbara Mikulski, a Democrat from Maryland, and Senator Kit Bond, a Republican from Missouri were both united in the belief that housing professionals and HUD should do more.

In a letter to all public housing authorities, Secretary Jack Kemp complied with the Senate's order and stated:

> Therefore, I want to make it clear that all housing authorities with developments constructed before 1980 are required to attend [lead paint training]. Travel funds and the registration fee ... are an allowable Comprehensive Improvement Assistance Program (CIAP) expense, or you may use operating reserves. The course is being offered in Atlanta, Chicago, San Francisco, Philadelphia, Providence, Seattle, Denver, Kansas City, and Dallas.[57]

Under threat of losing their subsidy, virtually all 3000 housing authorities reluctantly attended lead-based paint training in the late 1980s.

The training was offered by the Georgia Tech Research Institute, which at the time headed the EPA Southern Lead Paint Training Consortium. Previously, Georgia Tech had worked with many housing authorities through the Housing Authority Risk Retention Group (Chapter 3). In just 6 months, the course was offered 17 times and trained over 2200 public housing executive directors.[58]

The sessions included John Hiscox, then executive director of the Macon Housing Authority, where lead paint remediation research was done. Addressing his fellow housing authority directors, Hiscox reflected on their reluctance and the state of knowledge on lead inspection and abatement by saying,

> Lead paint abatement is like brain surgery with a new neurosurgeon: no one wants to be the first patient, but someone has to go first and in this case that's public housing authorities.[59]

Another executive director bemoaned the housing authority's untenable position this way:

> OK, here's your choice: Winter is coming on, and I have enough money in my operating budget to either abate our lead-based paint or to replace our old, inoperative furnaces. Which would you pick?[60]

The Macon housing authority and others mostly in New England threaded this needle. They and others led the way in providing real-world homes for studies to determine which methods of lead inspection and remediation could protect children and also be cost-effective.

Public housing became the first class of housing to undergo national lead inspection and abatement starting in the late 1980s (Chapter 3). As a result, by the late 2010s (with some notable exceptions), public housing had lower rates of lead poisoning compared to other low-income housing.[61,62]

But in the late 1980s and early 1990s most housing authorities were reluctant at best, as documented in a letter to HUD:

> Most housing authorities came to the training sessions with the idea that it would be better to hold off on completing [lead paint] testing until they had funds to do full-blown abatement, given the climate of litigation. In other words, why test, find a problem, and then expose oneself to lawsuits by previously undetected lead poisoned children if there was no money to correct the problem? In the training course, we encouraged a two-pronged strategy of pursuing both short-term and long-term solutions simultaneously. Our legal speaker made the point that if a housing authority was doing all it could be expected to do and was actively engaged in reducing immediately available sources of lead, the case against it is severely weakened.[63]

In 1982 under a consent decree in a class-action lawsuit (*Ashton* v. *Pierce*),[64,65] HUD agreed to conduct further rulemaking to define what a "lead-based paint hazard" meant. In that case, public housing tenants in the District of Columbia alleged that HUD's lead-based paint regulation was deficient for failing to define intact ("tight") lead-based paint surfaces as an "immediate hazard" requiring mitigation.

Because of the Court's ruling, HUD issued new regulations for all HUD housing programs that redefined "immediate hazard" to mean any and all lead paint. This definition would stand until Congress changed the definition of what a lead paint hazard meant in 1992, based on the new pathway science. But in 1987 Congress changed the definition of "hazard" to include "exterior as well as interior intact and nonintact painted surfaces."[66]

2.9 Lawsuits and affordable housing

During the 1970s and 1980s, there were an increasing number of lawsuits against private landlords and public housing authorities for lead paint, with some awards in the millions of dollars for an individual child. Yet most poisoned children received no compensation at all. This was exacerbated by the fact that housing owners did not have clear guidance as to their responsibilities; insurers were increasingly unwilling to provide lead liability coverage; the tort system was operating inefficiently and randomly for lead poisoning claims; and the vast majority of lead-poisoned children received nothing.[67] None of these suits resulted in remediating lead paint hazards.

This trend and occasional reports of multimillion-dollar lawsuits increased the concerns of rental property owners and their liability insurers. Average awards for tort cases, including but not limited to lead paint cases, rose by 10 times between 1960 and 1980.[68] So-called "jackpot" awards for lead paint cases occurred; one in Baltimore awarded $6 million to a single child.[69] One case against a landlord resulted in a $5,000,000 verdict.[70] In 1989 the Housing Authority of New Orleans had 87 lawsuits filed against it on behalf of 105 children and settled 1 case for $80,000, 1 for $75,000, 1 for $25,000, and 20 cases for $1500–$15,000 each.[71,72] The New York Times reported that over 1000 claims had been filed for lead poisoning and that the city could face payments as much as $500 million, with $230 million paid in 1994. A jury awarded $10 million to a single child who "suffered permanent retardation," which was later reduced to $4.4 million. Many cases dragged on for more than a decade.[73]

Recognizing this, many insurance policies began to adopt "pollution exclusion" clauses to avoid having to cover claims from lead-poisoned children. Such clauses started to appear in the 1980s, initially in response to litigation under EPA's Superfund cleanup program at hazardous waste sites, which did not involve lead paint. The insurance companies extended this to lead paint in housing. HUD stated that, "The risk of lead-based paint poisoning is noninsurable because a clear standard of care does not currently exist that could be used to consistently determine liability in negligence cases."[74] Public housing insurance would help fill this gap, but not until the late 1980s (Chapter 3).

Another case involved City Homes, a nonprofit housing provider in Baltimore, which eventually became bankrupt from lead paint lawsuits.[75] It provided hundreds of affordable housing units to low-income families since it was founded in 1986. In the late 1980s, one of the children in its housing units became lead poisoned.

Nick Farr, a board member and a vice president of the Enterprise Foundation (now Enterprise Community Partners), which had donated funds to City Homes, realized that there was an absence of standards, which threatened both children and low-income housing viability. He went on to serve on the board of the Alliance to End Childhood Lead Poisoning and helped to shape Title X of the 1992 Housing and Community Development Act. He helped persuade the Fannie Mae Foundation to create a new nonprofit research and policy organization, the National Center for Lead-Safe Housing in 1992, where he served as its first executive director. He said, "City Homes never intended to poison any children, but we did not know exactly what we should do to prevent it."[76] The Center would go on to develop such standards for key government agencies, hiring a unique combination of housing, health, and environmental professionals (Chapter 3).

Between 1970 and 1990, hundreds of lawsuits were brought. Most of them were against landlords, who argued in some cases that they had not put the lead paint in their housing units and therefore should not be held accountable. Other suits were brought against the paint, pigment and lead industries, arguing that they knew about the hazards of lead paint but continued to sell it anyway. Most of these suits against the lead paint industry failed for a variety of legal reasons, although a major one did succeed after a 20-year legal battle (Chapter 7).

But during the 1970s to 1990s, lawsuits failed to provide any significant funding or create any clear and accepted standards of care to remediate homes. The need for scientifically validated standards had become obvious.

2.10 Seeds of success: new pathway studies reveal the importance of lead dust from paint

One important example of pathway research from the mid-1980s that came to inform government policy based on science from the mid-1980s is shown in Fig. 2.5.[77] The solid and dotted lines show statistically significant and not significant exposure pathways, respectively. In this study, lead in paint was measured by XRF, lead in soil (PbS) was measured by chemical laboratory analysis, lead in settled house dust was measured by wipe

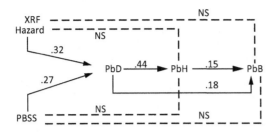

FIGURE 2.5 Lead exposure pathways in paint, soil, settled house dust, children's hands and blood, 1986. From: *Bornschein et al. Exterior surface dust lead, interior house dust lead and childhood lead exposure in an urban environment. In: Hemphill DD, ed.* Trace Substances in Environmental Health II, 1986 A Symposium (EPA). *University of Missouri, Columbia; 1986:327.*

Structural Equations: R^2

Ln(PbB) = 1.276 + .152 Ln (PbH) + .182 Ln (PbD)   .38
Ln(PbH) = -0.966 + .444 Ln (PbD)   .22
Ln(PbD) = 4.691 + .325 Ln (XRFHAZ) + .268 Ln (PbSS)   .52

All coefficients are significant at $P < .05$;NS-not significant

sampling (PbD), the amount of lead dust on children's hands was also measured by wipe sampling (PbH), and finally lead in children's blood (PbB) was measured by venipuncture and subsequent laboratory analysis.

This study was confirmed by many others[78] and showed that there was often no *direct* relationship between lead in the paint and children's blood lead. This was a dagger in the heart of the prevailing view that only the amount of lead in paint mattered.

Instead, the significant pathway of exposure was from paint and soil to interior settled lead house dust to lead dust on children's hands (from the inevitable contact with household surfaces), and then ingestion through normal hand-to-mouth contact and absorption into the bloodstream where it attacked the nervous system and other organs. This overturned earlier studies from the 1970s that thought both intact paint and nonintact paint were significant exposure pathways. For example, a 1979 study described its results this way:

> …intact painted surfaces as well as loose painted surfaces may be a lead hazard…elevated blood lead levels in the cases compared to controls most likely resulted from consumption of household paint (either loose, intact, or both)…. The results … lend support to childhood lead poisoning prevention programs which advocate the removal of lead paint from accessible surfaces. Removing only loose paint would protect some children, but it would provide little protection for children who chew surfaces.[79]

Importantly, the 1979 study was silent on the issue of lead-contaminated house dust and focused on chewing and eating paint chips.

The identification of lead dust by the newer studies in the 1980s could be described as a "Eureka!!" moment in the history of childhood lead paint poisoning prevention. Although later pathway studies in the 1990s and 2022 showed a more complex picture, the centrality of paint to settled dust to blood lead was validated repeatedly (Chapter 4).

These studies rebuffed industry claims that lead paint (and the dust and soil it contaminates) does not cause increases in blood lead. It was a revelation that changed federal policy in the ensuing decades.

The lead, pigment and paint industries attempted to use these studies to suggest that the presence of lead-based paint alone is not a problem if it is intact, an obvious attempt to shift the blame to maintenance, ignoring the emerging scientific evidence on pathways of exposure. But of course, all paint eventually deteriorates and if it is lead-based paint, such deterioration will inevitably cause exposure.

2.11 Time for change

The legacy of the 1970s until the late 1980s was that the lack of clear scientifically validated standards, inspection and mitigation methods, and a plan to implement them threatened children, the medical profession, and the affordable housing industry, all at the same time. The threat of increasing litigation was an important incentive driving the need for validated standards to be created. Another was the growing realization that even small amounts of lead paint were extremely damaging to children and medical treatment alone was mostly ineffective.

The shortcomings of the efforts to address lead paint in the 1970s and 1980s were substantial. The housing industry for the most part ignored the problem out of fear of high costs and/or confusion and the medical community attempted to treat it clinically. The courts ignored the emerging scientific evidence on the pathways of exposure. Scant federal funds were allocated to help address the problem during this time period.

By the end of the 1980s, it was increasingly clear that things had to change.

References

1. Pub. L. 91-695, Section 1, Jan. 13, 1971, 84 Stat. 2078. *Lead-Based Paint Poisoning Prevention Act.* <https://uscode.house.gov/view.xhtml?path = /prelim@title42/chapter63&edition = prelim>.
2. Public Law 93-151. *An Act to Amend the Lead Based Paint Poisoning Prevention Act, and for Other Purposes*; 11/09/1973.
3. Steinfeld JL. Medical aspects of childhood lead poisoning. *HSMHA Health Rep.* 1971;86(2):140−143.
4. Increased lead absorption and lead poisoning in children. A statement by the Center for Disease Control. *J Pediat.* 1975;87(5):824−830.
5. CDC. *Recommended Actions Based on Blood Lead Level*; 2019. <https://www.cdc.gov/nceh/lead/advisory/acclpp/actions-blls.htm>.
6. CDC. *Preventing Lead Poisoning in Young Children: A Statement by the Center for Disease Control: April 1978.* Atlanta, Georgia: Department of Health, Education, and Welfare; 1978 <https://stacks.cdc.gov/view/cdc/11184>.
7. Kennedy FD. *The Childhood Lead Poisoning Prevention Program: Evaluation.* Unpublished manuscript. Centers for Disease Control. US Public Health Service, Atlanta, GA; 1978. Cited in Farfel MR. Reducing lead exposure in children. *Annu Rev Public Health.* 1985:6:333−360.
8. Rogan WJ, Dietrich KN, Ware JH, et al. Treatment of Lead-Exposed Children Trial Group. The effect of chelation therapy with succimer on neuropsychological development in children exposed to lead. *N Engl J Med.* 2001;344:1421−1426. Available from: https://doi.org/10.1056/NEJM200105103441902.
9. Ettinger AS, Bornschein RL, Farfel M, et al. Assessment of cleaning to control lead dust in homes of children with moderate lead poisoning: treatment of lead-exposed children trial. *Env Health Perspect.* 2002;110: A773−A779.
10. Piomelli S, Rosen JF, Chisolm JJ. Graef. Management of childhood lead poisoning. *J Pediatrics.* 1984;105 (4):523−532.
11. Markowitz M. Lead poisoning: an update. *Pediatr Rev.* 2021;42(6):302−315.
12. HUD Lead Based Paint Hearings. 1974:117 and 125. (The University of Illinois Chicago Special Collections & University Archives, School of Public Health, "David E. Jacobs papers" the University of Illinois Chicago School of Public Health Library.)

13. HUD Lead Based Paint Hearings. 1974:166. (The University of Illinois Chicago Special Collections & University Archives, School of Public Health, "David E. Jacobs papers" the University of Illinois Chicago School of Public Health Library.)

14. CDC Preventing Lead Poisoning in Children; January 1985. <https://www.cdc.gov/nceh/lead/publications/plpyc1985.pdf>.

15. Comments of the Lead Industries Association, Inc. Before the Centers for Disease Control, May 31; 1984 (the University of Illinois Chicago Special Collections & University Archives, School of Public Health, "David E. Jacobs papers" the University of Illinois Chicago School of Public Health Library.). Also see Letter from Lead Industries Association attorneys to Vernon Houk, CDC, January 24; 1984 (the University of Illinois Chicago Special Collections & University Archives, School of Public Health, "David E. Jacobs papers" the University of Illinois Chicago School of Public Health Library.)

16. Cass Peterson. *Industry Battling CDC Employs Activist Tactic*, Washington Post; June 22, 1984:A17.

17. Ettinger AS, Leonard ML, Mason J. CDC's lead poisoning prevention program: a long-standing responsibility and commitment to protect children from lead exposure. *J Public Health Manage Pract.* 2019;25 (suppl 1): S5−S12. doi:10.1097/PHH.0000000000000868. PMID: 30507764; PMCID: PMC6320665.

18. CDC Preventing Lead Poisoning in Children; October 1991. <https://wonder.cdc.gov/wonder/prevguid/p0000029/p0000029.asp#head001000000000000>.

19. CDC National Childhood Blood Lead Surveillance Data. <https://www.cdc.gov/nceh/lead/data/national.htm>.

20. Mahaffey KR, Annest JR, Roberts J, Murphy RS. National estimates of blood lead levels, US 1976−1980. *New Eng J Med.* 1982;307(10):573−579.

21. Berger H. US Department of Commerce, National Bureau of Standards. *The NBS Lead Paint Poisoning Project: Housing and Other Aspects.* NBS Technical Note 759; February 1973.

22. Brody DJ1, Pirkle JL, Kramer RA, Flegal KM, Matte TD, Gunter EW. Paschal DC blood lead levels in the US population. Phase 1 of the third national health and nutrition examination survey (NHANES III, 1988 to 1991). *JAMA.* 1994;272(4):277−283.

23. Pirkle JL, Kaufmann RB, Brody DJ, Hickman T, Gunter EW, Paschal DC. Exposure of the U.S. population to lead, 1991−1994. *Environ Health Perspect.* 1998;106(11):745−750.

24. Dixon SL, Gaitens JM, Jacobs DE, et al. U.S. children's exposure to residential dust lead, 1999−2004: II. The contribution of lead-contaminated dust to children's blood lead levels. *Env Health Perspect.* 2009;117:468−474.

25. Gaitens JM, Dixon SL, Jacobs DE, et al. Children's exposure to residential dust lead, 1999−2004: I. Housing and demographic factors associated with lead-contaminated dust. *Environ Health Perspect.* 2009;117:461−467.

26. Jacobs DE, Dixon SL, Wilson JW, Smith J, Evens A. The relationship of housing and public health: a 30 year retrospective analysis in the US. *Environ Health Perspect.* 2009;117:597−604.

27. Yeoh B, Woolfenden S, Lanphear B, Ridley GF, Livingstone N. Household interventions for preventing domestic lead exposure in children. *Cochrane Database Syst Rev.* 2012;4(4). CD006047pmid:22513934.

28. Center for the Study of Federalism; New Federalism (Reagan). 2006. <https://encyclopedia.federalism.org/index.php/New_Federalism_(Reagan)>.

29. Evaluation of Health and Environmental Effects of Two Methods of Residential Lead Paint Removal. Mark Farfel Thesis, Johns Hopkins University, Baltimore; 1987:32. <https://catalyst.library.jhu.edu/catalog/bib_91837>.

30. Center for Science in the Public Interest. *Children, Lead Poisoning and Block Grants: A Year-End Review of How Block Grants Have Affected the Nation's Ten Most Crucial Lead Screening Programs.* Washington, DC: National Coalition for Lead Control; 1982. Cited in Farfel MR. Reducing lead exposure in children. *Ann Rev. Public Health.* 1985;6:333−360.

31. United States Housing Act Of 1937. Public Law 75−412. September;1:1937.

32. 25 CFR § 700. 55 − Decent, Safe, and Sanitary Dwelling.

33. Efforts to stamp out lead-paint poisoning are half-hearted failure, by Bill Prater. Kansas City Star; March 1981.

34. Lead poisoning takes a big continuing toll as cures prove elusive, by Michael Waldholz, Wall St Journal, Midwest edition; May 27, 1982.

35. Proposal for the evaluation of alternative intervention and hazard abatement techniques for the control of childhood lead based paint poisoning, To the HUD Assistant Secretary for Policy Development and Research, April 23, 1976 from the Medical and Health Research Association of New York and the New York City Department of Health (the University of Illinois Chicago Special Collections & University Archives, School of Public Health, "David E. Jacobs papers" the University of Illinois Chicago School of Public Health Library.)

36. Jacobs DE. Occupational exposures to lead-based paint in structural steel demolition and residential renovation. *Int J Environ Pollut.* 1998;9(1):126–139. Inderscience Enterprises, United Nations Educational, Scientific and Cultural Organization, Switzerland.

37. Farfel MR, Chisolm JJ. Health and environmental outcomes of traditional and modified practices for abatement of residential lead-based paint. *Am J Public Health.* 1990;80(10):1240–1245.

38. Charney E, Kessler B, Farfel M, Jackson D. A controlled trial of the effect of dust-control measures on blood lead levels. *N Engl J Med.* 1983;309(18):1089–1093.

39. Rabinowitz M, Leviton A, Bellinger D. Home refinishing, lead paint, and infant blood lead levels. *Am J Public Health.* 1985;75(4):403–404.

40. Shannon MW, Graef JW. Lead intoxication in infancy. *Pediatrics.* 1992;89(1):87–90.

41. Fischbein A, Anderson KE, Shigeru S, et al. Lead poisoning from "Do It Yourself" heat guns for removing lead-based paint: report of two cases. *Environ Res.* 1981;24:425–431.

42. Marino PE, Landrigan PJ, Graef J, et al. A case report of lead paint poisoning during renovation of a victorian farmhouse. *Am J Public Health.* 1990;80(10):1183–1185.

43. Jean Schneider D, MLavenhar. Lead poisoning: more than a medical problem. *Am J Public Health.* 1986;76:242–244.

44. HUD Guidelines for the Evaluation and Control of Lead Based Paint Hazards in Housing. Chapter 4, pp. 4–4, 1995 edition.

45. George Gould. Testimony at Lead-Based Paint Hearings; May 20–24, 1974:78 (the University of Illinois Chicago Special Collections & University Archives, School of Public Health, "David E. Jacobs papers" the University of Illinois Chicago School of Public Health Library.)

46. Lead Poisoning Prevention and Control Act. Mass. Gen. Laws Ann. ch. 111, Section 190-99.

47. Deleader's Manual: A Handbook for Safe Lead Paint Removal. Okonski L, ed. Massachusetts Department of Public Health, Childhood Lead Poisoning Prevention Program. May 1984. The University of Illinois Chicago Special Collections & University Archives, School of Public Health, "David E. Jacobs papers" the University of Illinois Chicago School of Public Health Library.

48. Landrigan PJ, Baker EL, Himmelstein JS, Stein GF, Weddig JP, Straub WE. Exposure to lead from the mystic river bridge: the dilemma of deleading. *N Engl J Med.* 1982;306:673–676.

49. Rey-Alvarez S, Menke-Hargrave T. Deleading dilemma: a pitfall in the management of childhood lead poisoning. *Pediatrics.* 1987;79:214–217.

50. Amitai Y, Brown MJ, Graef JW, Cosgrove E. Residential deleading: effects on the blood lead levels of lead-poisoned children. *Pediatrics.* 1991;88(5):893–897.

51. Chisolm JJ, Mellits ED, Quaskey SA. The relationship between the level of lead absorption in children and the age, type and condition of housing. *Environ Res.* 1985;38:31–45.

52. National Institute of Building Sciences, Draft HUD Guidelines; 1987 (the University of Illinois Chicago Special Collections & University Archives, School of Public Health, "David E. Jacobs papers" the University of Illinois Chicago School of Public Health Library.).

53. EPA 40 CFR Part 745 Subpart L. <https://www.ecfr.gov/cgi-bin/retrieveECFR?gp = 1&SID = 79d9123c529d8f8ca-792833afe6b664f&ty = HTML&h = L&n = pt40.31.745&r = PART#sp40.34.745.l>.

54. International Code Council International Property Maintenance Code; 2018. <https://codes.iccsafe.org/content/IPMC2018/preface>.

55. Friedman J, Weinberg DH. *The Economics of Housing Vouchers.* 1st ed. Academic Press; 1982. ISBN: 9781483260433.

56. It's a long wait for Section 8 housing in U.S. cities, by Aaron Schrank, January 3; 2018. <https://www.marketplace.org/2018/01/03/its-long-wait-section-8-housing-us-cities/>.

57. Notice from HUD Secretary Jack Kemp to Public Housing Authority Directors; June 14, 1991 (the University of Illinois Chicago Special Collections & University Archives, School of Public Health, "David E. Jacobs papers" the University of Illinois Chicago School of Public Health Library.)

58. Report to Michelle Price (EPA) from David Jacobs. EPA Contract No. CX-818586-01-0, Georgia Tech Project No. C-3000-100, Development and Delivery of Seventeen Lead-Based Paint Courses: Lead-Based Paint in Public and Indian Housing: How Housing Authorities Can Meet the Challenge.

59. Personal communication between John Hiscox, Executive Director of the Macon Georgia Housing Authority and David Jacobs. 1986

60. Personal communication reported by David Jacobs at the American Industrial Hygiene Association Conference. Lead Paint Roundtable; June 3, 1991.
61. Jacqueline MC, Maxine G, Casey C, Neil C. Pediatric blood lead levels within New York City public vs private housing, 2003−2017. *Am J Public Health.* 2019;109(6):906−911. Available from: https://doi.org/10.2105/AJPH0.2019.305021.
62. Ahrens KA, Haley BA, Rossen LM, Lloyd PC, Aoki Y. Housing assistance and blood lead levels: children in the United States, 2005−2012. *Am J Public Health.* 2016;106(11):2049−2056. Available from: https://doi.org/10.2105/AJPH.2016.303432.
63. Letter from David Jacobs to Ronald J. Morony, Director, Division of Innovative Technology, Office of Policy Development and Research, US Department of Housing and Urban Development. Comments on the draft document: "Reducing the Risks of Leading Poisoning in Public Housing: A Guide for In-Place Management". November 15, 1991. The University of Illinois Chicago Special Collections & University Archives, School of Public Health, "David E. Jacobs papers" the University of Illinois Chicago School of Public Health Library.
64. Ashton v. Pierce, 541 F. Supp. 635 (D.D.C. 1982). <https://law.justia.com/cases/federal/district-courts/FSupp/541/635/2288940/>.
65. Ashton v Pierce, amended. 716 Federal Reporter, 2nd Series; November 4, 1983:56−67.
66. Housing and Community Development Act of 1987. 101 STAT. 1946 Public Law 100−242—FEB. 5, 1988. Section 566c. <https://www.govinfo.gov/content/pkg/STATUTE-101/pdf/STATUTE-101-Pg1815.pdf>.
67. Putting the Pieces Together: Controlling Lead Hazards in the Nation's Housing − Report (HUD-1547-LBP). Washington, DC: U.S. Department of Housing and Urban Development; 1995.
68. Kagan RA. *Adversarial Legalism: The American Way of Law 2009.* Harvard University Press, June 30; 2009.
69. Doran C. Baltimore City Circuit Court Judge Refuses to Overturn Lead Paint Verdict, Daily Rec. (Baltimore, MD), March 31; 2008. <http://centerforamerica.org/speakers/lawsuit_abuse/ls_abuse2.html>.
70. Richwind Joint Venture v. Brunson, 335 Md. 661,645 A.2d 1147; 1994. <https://www.casemine.com/judgement/us/5914bdd6add7b049347a4bc2>.
71. Blum A. Suit Targets Paint Manufacturers; New Orleans Case a Message? National Law Journal, May 8, 1989, at 17.
72. Deborah R, Hensler Mark A. Peterson Brooklyn Law Review 59(3). Article 10 3-1-1993. Understanding Mass Personal Injury Litigation: A Socio-Legal Analysis.
73. New York Girding for Surge in Suits Over Lead Damage, By Matthew Purdy, New York Times; August 14, 1995.
74. Comprehensive and Workable Plan for the Abatement of Lead-Based Paint in Privately Owned Housing Report to Congress. US Department of Housing and Urban Development Washington, DC; December 7, 1990:5−15. <https://www.huduser.gov/portal/publications/affhsg/comp_work_plan_1990.html>.
75. City Homes files for bankruptcy protection by Luke Broadwater. Baltimore Sun SEP 12, 2013. <https://www.baltimoresun.com/maryland/bs-md-city-homes-20130912-story.html>.
76. Personal Communication between Nick Farr and David Jacobs; 1991.
77. Bornschein RL, Succop P, Kraft KM, Clark CS, Peace B, Hammond PB. Exterior surface dust lead, interior house dust lead and childhood lead exposure in an urban environment. In: Hemphill DD, ed. *Trace Substances in Environmental Health,* 20. *Proceedings of University of Missouri's 20th Annual Conference (EPA),* June 1986. University of Missouri, Columbia, Missouri; 1987.
78. Lanphear BP, Matte TD, Rogers J, et al. The contribution of lead-contaminated house dust and residential soil to children's blood lead levels. A pooled analysis of 12 epidemiologic studies. *Environ Res.* 1998;79(1):51−68. Available from: https://doi.org/10.1006/enrs.1998.3859. PubMed PMID: 9756680.
79. Gilbert C, et al. A comparison of lead hzards in the housing environment of lead poisoned children vs non-poisoned controls. *J Environ Sci Health.* 1979;A14(3):145−168.

Breaking the Barriers to Progress (1986–2001)

Solutions take shape: the lead paint Title X law

1985 Southern Lead Paint Training Consortium at Georgia Tech starts

1988. Public Housing Authority Risk Rentention Group and Lead Paint Risk Assessments Begin

1990 HUD issues interim public housing guidelines. Congress appropriates first lead paint grants for private housing

1991 Congress creates Lead Paint Office within HUD Secretary's Office. White paper on New Orleans Housing Authority leads to Congressional Hearings

1992 Congress passes Title X of the Housing and Community Development Act, redefines what a lead paint hazard means.

1988 Report To Congress Says Lead Paint a "Clear Failure"; Congress Requires All Public Housing to be inspected for lead paint, recognizes importance of lead dust and interim controls

1990 HUD issues first lead paint housing survey, 64 million houses with lead paint, cost for abatement estimated at $360 to $500 billion, leading to paralysis

1990 Alliance to End Lead Poisoning Founded.

1991 Third CDC Statement issued. CDC issues its Strategic Plan. Lead Poisoning declared to be the No. 1 children's environmental disease by Health and Human Services Secretary.

1992 National Center for Lead Safe Housing Founded

FIGURE 3.0 Timeline for Chapter 3.

The realization that childhood lead poisoning continued at epidemic rates during the 1970s and 1980s, and the abject failure of only chasing after children who had already been poisoned led to new attempts to restore the collaboration of the housing, health, and environmental communities. This chapter tells the story of how primary prevention (taking action *before* children are exposed) finally became the law of the land in 1992, but it would take new scientific and policy breakthroughs, scandal, courage, and new organizations to make it happen.

Fifty Years of Peeling Away the Lead Paint Problem
DOI: https://doi.org/10.1016/B978-0-443-18736-0.00001-7

3.1 Science, policy, and practice meet—an unlikely venue

The explosion of lead pathway and intervention-related research in the mid-1980s had to be translated so that those involved in lead inspection and remediation could put it into real-world practice. Although an unlikely venue for advancing knowledge, public policy, and scientific lead paint inspection, risk assessment, and remediation methods, new and larger lead paint training programs in the 1980s emerged. They provided a critical venue for sharing the emerging, safer, and more effective methods of lead remediation and lead detection beyond a research setting. Researchers, practitioners, housing, health and environmental professionals, parents, and policy experts all flocked to the courses to help translate the science to procedures that could be widely implemented.

These courses taught a far more detailed set of hazard identification and control methods than the 4 or 7 steps offered in the two Centers for Disease Control and Prevention (CDC) statements in 1978 and 1985. In the late 1980s, more private businesses sought to do the work of lead hazard detection and remediation in millions of homes. Some of the initial main training organizations included the Georgia Tech Research Institute in Atlanta, Lead Tech Services in Baltimore, and others largely centered in New England and the east. Many more would follow across the country in the years to come. By 2022, there were over a thousand training providers. Although CDC had been training local health department staff in lead issues since the 1970s (a popular course that continues to be offered in 2022), there were hardly any targeted to those who worked in housing until the 1980s.

In addition to lead paint training, the Georgia Institute of Technology also operated the Occupational Safety and Health Administration (OSHA) consultation program in that State and was staffed by occupational hygiene scientists and safety engineers, another example of the key role occupational health professionals came to play in the emerging housing-focused lead poisoning prevention field. The university measured worker exposures to lead and other hazardous agents in both housing and workplaces such as battery plants and helped design remediation procedures.

Georgia Tech had previously trained thousands of contractors in asbestos abatement and many of them signed up for the new lead paint courses, thinking that lead inspection and remediation might be the "new asbestos" and a business opportunity.

Indeed, there were early attempts to treat lead abatement exactly like asbestos abatement. However, asbestos fibers behave much differently than lead particles, which are denser and larger and therefore settle out the air more quickly. And much of the asbestos work had taken place in schools and commercial properties, not housing. In the end, the two methods for asbestos and lead remediation were quite different.

Part of the problem in these years was that it was not clear which profession would take on the task of overseeing lead paint inspection and remediation. In an address at a major conference of occupational health scientists and practitioners, the role the profession would need to play in the coming years was described:

> Let me share with you a dialogue that almost always occurs in every class I teach on lead-based paint at Georgia Tech. Our attendees include housing authority personnel, asbestos abatement contractors, architects, health care providers, housing providers, environmental consultants, and others. Typically, the asbestos abatement contractors jump up about halfway through the course, and state that "You're going about

this all wrong. We made the same mistakes in the early days of asbestos abatement. We must do lead paint abatement just like asbestos abatement to protect our workers and the public." Then, housing authorities and housing providers typically leap to their feet, arguing "There's no way we can afford another asbestos experience. If you insist on doing the work in the same way, we will be able to clean up only a very few properties. The vast majority will go unabated."

The challenge is to creatively apply the lessons learned from the asbestos experience – not just repeat it – to evaluate new technologies as they are developed. The nation needs a whole range of testing and abatement responses. We must learn to help housing authorities and other housing providers decide which properties pose immediate health threats to both their resident children and to their maintenance and construction workers. We must devise risk assessment and interim containment programs that protect health and make sense. We should fix those risk assessment methodologies that are simplistic and ineffective, and believe me, they exist. We must integrate abatement activities with on-going housing modernization and repair programs. And we must find a way to help the private homeowner, landlord and renter deal with lead paint hazards, given the reality that priority hazard sites are often dilapidated inner city housing. Clearly, we are talking about whole new programs, at every level of government and commerce.

Dr. Alice Hamilton, who pioneered much of the early work on occupational lead exposures and occupational health generally, said: "I am not one of those who believe that the use of lead ... can ever be made safe. You may control conditions within a factory, but how are you going to control the whole country? No lead industry, even under the strictest control, can lose all its dangers. Where there is lead, some cases of lead poisoning sooner or later develop, even under the strictest supervision."[1,2]

But it was a tough sell. Most occupational hygiene scientists were focused and trained on identifying problems in the workplace environment, not housing. Nevertheless, these scientists played a key role in developing the nation's capacity to inspect and remediate housing for lead paint hazards.

Furthermore, most housepainters were not unionized and were employed as sole proprietors, making surveillance of occupational lead exposure virtually impossible. However, unions were also interested in lead paint abatement. For example, the International Brotherhood of Painters and Allied Trades wrote in a letter to the National Institutes of Health:

For decades, lead-based paint was used in our industry. We mixed it, we applied it, and we removed it. We finalized a comprehensive lead-based paint abatement training course for our 150,000 members.[3]

Although lead paint training would not become regulated nationally until the mid-1990s by the Environmental Protection Agency (EPA), the Georgia Tech training course was unique in that it drew dozens of leaders in the lead paint field from across the country to help teach it, not just the usual one or two instructors. The week-long curriculum, which included sessions describing how large the problem was, clarified the potentially significant business opportunities for inspection and remediation contractors taking the course. The Georgia Tech course also taught policies, as well as inspection technologies, methods of solving hazards, and how to increase the public will to implement solutions, helping to create a new movement in the process.

For example, the National Lead Assessment and Abatement Council (later to become the Lead and Environmental Hazards Association) was created following one of the Georgia Tech courses in 1988. This was in response to a "Call to Form the American Association for Lead Paint Abatement" that was issued at the courses. This group later emerged as the main professional association of lead inspection and remediation companies. The initial activities of the new association included:

> ... a scientific forum for the exchange of new abatement and detection technologies through various symposia, meetings, and publications; activities to properly certify lead-based paint abatement and detection contractors; promote legislative, legal, and regulatory efforts to further enhance the professional nature of lead-based paint inspection, risk assessment and remediation, and work to ensure that remediation activities do not create new hazards; efforts to provide sufficient funding for lead-based paint abatement work in public and private homes, schools, and other buildings; petition the Occupational Safety and Health Administration to expand the scope of the General Industry Lead Standard to include construction activities, which will help ensure that workers exposed to lead during abatement activities receive the same level of protection accorded to those employed in industry; and develop a multi-year long-range plan to permit a rational, comprehensive approach to solving the complex lead-based paint problem.

> The lead-based paint issue has scientific, economic, social, political, and indeed moral dimensions. All will need careful consideration if we are to successfully address a problem that has already been responsible for huge losses in our children's intellectual prowess and a major cause of occupational disease.[4]

The course featured Lorne Garrettson, a pediatrician and mild-mannered Quaker, who dramatically displayed radiographs of a child's abdomen with lead paint chips that his patient had ingested. He would exclaim:

> I have treated thousands of children with lead poisoning in my many years of practice and I am still seeing them today![5]

It also featured lawyers such as Neil Leifer and Jennifer Willis, who would go on to sue lead paint companies, landlords, housing authorities, and others. Not generally part of technical training courses like this, they taught the legal basis for lead poisoning claims and how proper lead inspections and mitigation could reduce liability exposure for smart property owners.

X-ray fluorescence (XRF) instrument manufacturers showcased their instruments at the courses, describing how they worked and how to minimize measurement error. Because such instruments included radioactive sources, special training and licenses were needed, with physics expertise provided by Georgia Tech, where a small nuclear reactor was located. That provided a venue for research on XRF lead paint analyzers (led by Stan Lewis) that would figure prominently in Congressional action a few years later. The wisdom of locating a small research nuclear reactor in downtown Atlanta was questionable, but it did create an opportunity to learn more about X-ray fluorescence and how the technology could be improved to more reliably measure lead in paint (the reactor was decommissioned in 2002).

The course offered panel discussions of experienced contractors and others who had performed mitigation, with real-world solutions in a peer-to-peer format, including abatement and inspection contractors and trainers like John Zilka, Jim Keck, and Norman Fay. Because each had their own favorite methods of addressing lead paint problems, the interchange enabled students to understand the strengths and weaknesses of each. At the time, virtually nothing had been published comparing the various methods of controlling lead paint hazards, except for the early studies showing what *not* to do (open flame burning, power sanding, dry scraping, and abrasive blasting for paint removal).

Only construction professionals like these could speak the language of no-nonsense contractors and make it stick. John Zilka, Jim Keck, Jim McCabe, Susan Kleinhammer, and many others would go on to offer their own training courses for decades to come.

The Georgia Tech course also featured Steve Hayes (an occupational health professional), who described the importance of good specifications and standard operating procedures, rolling up his sleeves at the start of each lecture to contribute to the no-nonsense urgency of the matter.

Other occupational health professionals such as John Rekus showed how workers could be protected, even though at the time there was no comprehensive OSHA lead regulation for construction workers (this would not be promulgated until 1996). His sessions were among the liveliest, with a surprising opening question: "Does anyone in this room have a heart condition?" Usually, no one would raise their hand, so he would take his pointer which was fitted with a small noisemaker, slam it on the table setting it off, and declare to the startled audience, "Good! Now you will listen to me!" And they did. He went on to become one of the nation's top experts on hazards in confined spaces, but later remarked how rewarding he had found these attempts that bridged the medical, occupational health, public health, environmental, and housing fields.

The courses also featured speakers from the public housing world, such as Miles Mahoney, who had previously served as an appointed receiver to fix troubled housing authorities across the country over many years. Because funding was more available for lead paint inspection and remediation in public housing than in private housing, he described how both housing authority managers and contractors could find each other to produce safer housing for children and how such activities could be financed in light of all the nuances imposed by HUD regulations. Without the financing, little could happen. The influence of housing professionals would prove to be increasingly important.

The courses also included sessions on hazardous waste requirements, which at the time were unclear: Was it truly necessary to swamp the nation's capacity to handle as hazardous waste windows, doors, and other painted building components? It turned out that lead paint on building components did not leach into groundwater, so lined construction debris landfills were safe and feasible. However, there were special procedures taught in the course to ensure proper waste handling, including preventing so-called "recycling."

For example, old windows and doors coated with lead paint sometimes would be given to scavengers or antique stores or so-called "Re-Use Stores," even though they were still coated with lead paint. Surprisingly, some "green" building organizations still promoted recycling old building materials without testing them first in a laudable (if misguided) attempt to reduce waste. This problem still occurs today, with some still attempting to sell old windows and doors with lead paint that could contaminate new homes.

In 2020, one "green" website had no warning on the dangers of recycling old building materials with lead paint, promoting a "weathered appearance," presumably meaning deteriorated paint:

It's a staggering, shameful waste. Fortunately, reclaiming, reusing, and recycling is becoming more common in the construction and renovation industry. Reclaimed building materials like doors, windows, wood flooring, and much more are becoming increasingly easy to find. Not only is reuse much more eco-friendly, it's also incredibly budget-friendly: reclaimed materials can be 50 percent to 75 percent cheaper than their new counterparts. (Sometimes you can even find free materials! Hello dreamy, inexpensive sustainable home!) Plus, oftentimes your recycled building materials come with a colorful history or *weathered* appearance that adds a truly unique touch to your project. (emphasis added)[6]

Personnel from the EPA attended and helped teach the Georgia Tech courses as well. Rob Elias, an EPA researcher typically opened his talks with how important the science was. He recounted the story of how his driving buddy accused him of speeding while driving, to which he responded, "My speedometer says I'm obeying the law, and because I have conducted research and 'trust' my speedometer, I believe it is the truth!" He would go on to help develop the EPA's Integrated Exposure Biokinetic Uptake model that attempted to predict blood lead levels from environmental contamination that would become prominent in setting lead dust standards and performing cleanup at toxic waste sites,[7] many of which contained lead in nearby homes (discussed later in this book).

Although not normally considered part of a "technical" course, policy experts such as Stephanie Pollack and Susan Guyaux described how local leaders from Maryland, Massachusetts, New York City, and elsewhere had established some of the first local laws to address the problem, and what policy barriers had to be overcome. Pollack cited the importance of getting the support of city mayors who often had more direct control over housing and how the issue could be framed as a potential success story for them. Thomas Menino, mayor of Boston was one such example among many.

The course also included field trips so that students could apply the inspection and risk assessment procedures they had been taught in a hands-on fashion. These were sometimes to a local public housing development in Atlanta to provide real-world context. Students learned how to properly take lead paint measurements, collect dust and soil samples, and other technical procedures.

Thousands of students completed the course. Years later one student, who had started a successful lead abatement company, asked the course director "Do you remember me?" The man went on to say that he had been the very first student to complete the Georgia Tech course and had proudly framed his "number one" training certificate.

Beyond simply offering training courses, Georgia Tech created a forum to share what real-world lessons had been learned during the 1970s and 1980s. Distilling much of the knowledge presented and backed by many peer-reviewed studies, these lessons would later become embodied in the HUD Guidelines for the Evaluation and Control of Lead-Based Paint Hazards in 1995, a 800-page compendium of standardized practices and procedures that would come to define methods to be used in the field and the basis of federal state and local regulations (Chapter 4).[8]

In the opening speech to the inspection and risk assessment course, the purpose of the course was described as bringing to an end to

> the era of rank amateurism, outright fraud and data manufacturing that has dogged the lead paint inspection field for too many years ... that can only be stopped by providing a cadre of professional, licensed and regulated inspectors.[9]

But without the real-world knowledge from all the instructors and students in those courses, those standard methods may not have become so authoritative and widely accepted as being valid, feasible, and protective.

By 2022, lead paint training had become more specialized, with separate courses aimed at inspectors, risk assessors, abatement contractors, renovation, repair, and painting contractors. Policy-level courses became the exception. The technical courses are now required for state or EPA licensure and certification. This separation of technical and policy training in later years was unfortunate. The high demand for technical and policy training was also stimulated by new initiatives and legislation at the federal level, driven in part by a remarkable report to Congress in 1988, described next.

3.2 Bombshell report to Congress: "corrective actions have been a clear failure"

In 1986, Congress directed the Agency for Toxic Substances and Disease Registry (ATSDR) to examine methods and alternatives for reducing lead exposure in young children and issued unusually specific instructions. Through the Department of Commerce, HUD had issued an estimate of lead paint poisoning in 1972, but it was theoretical, and its authors admitted it was not validated.[10] The new 1986 directive called for:

(A) an estimate of the total number of children, arrayed according to Standard Metropolitan Statistical Area or other appropriate geographic unit, exposed to environmental sources of lead at concentrations sufficient to cause adverse health effects;

(B) an estimate of the total number of children exposed to environmental sources of lead arrayed according to source or source types;

(C) a statement of the long-term consequences for public health of unabated exposures to environmental sources of lead and including but not limited to, diminution in intelligence, increases in morbidity and mortality; and

(D) methods and alternatives available for reducing exposures of children to environmental sources of lead.[11]

It would take 2 years and a resignation to finally get it released. In 1988 two of the nation's foremost lead toxicologists, Paul Mushak and Annemarie Crocetti, completed this major report to Congress.[12] Mushak feared the bureaucracy would try to prevent its release or tone down its findings. He was right. After senior ATSDR officials attempted to send a shorter and distorted version to Congress, Mushak publicly resigned from the Public Health Service in protest, which was reported in the Washington Post. He said, "No way in hell can you comprehend the complexity of this problem in a boiled down, very misleading and essentially neutral document." Crocetti said, "It's one of the most subtle, nastiest rewrites I have ever seen."[13] ATSDR was forced to finally issue the unvarnished full hard-hitting report to Congress. Mushak would go on to conduct many other important studies, but he paid a steep price to get the truth out.

The report crystalized what was known about the number of children affected, ranked which sources of lead were most important, the consequences of failure to address the problem, and new methods of controlling hazards. The report was peer reviewed by public health, housing, environmental and other leaders, even including both the lead industry and environmental activist groups. The number of children at risk in both large and small cities was quantified, which was important in gaining Congressional support from many districts.

It also used Census data to estimate the number of older housing units with lead paint and found that even children in families *above* the poverty line were at risk. This changed the perception that the problem was a small one confined to a few low-income minority communities that the lead industry had been promoting for decades. Instead, it proved the problem was much larger, affecting most children, much like lead in food canning and gasoline.

Many of its findings were drawn from the early National Health and Nutrition Examination Survey (NHANES) survey discussed in the previous chapter, but this was the first time that the *numbers* of children (not just a percentage) were reported. It was also not confined to medical journals and carried more weight as a report to Congress.

For the first time, the 1988 ATSDR report to Congress described the pathways and sources of lead, ranking them by the number of children who could be affected. For lead paint, the report estimated 42 million homes contained lead paint (which was later shown to be an underestimate by HUD's later, more accurate 1990 survey—the real number was closer to 64 million) and 12 million children under 7 years old were at risk; for gasoline lead the estimate was 5.6 million children; for industrial point sources such as smelters, the estimate was 230,000 children; for dust and soil (from both paint and gasoline) the estimate was 5.9–11.7 million children; for lead in plumbing the estimate was 1.8 million children; and finally for lead in food, 1 million children were estimated to be exposed to enough lead in food to raise their blood lead to an "early toxicity risk level." (Table 3.1)

This was the first time that lead paint had been shown to affect the largest number of children compared to other sources. The report noted that the phase-out of lead in gasoline and improved regulation of smelters and food canning had succeeded in reducing the

TABLE 3.1 Number of children exposed to lead sources and pathways.

Source/pathway	Number of children at risk
Lead paint chip, dust, and soil ingestion	12 million
Lead gasoline inhalation	5.6 million
Industrial point sources	230,000
Lead plumbing and drinking water	1.8 million
Lead in food ingestion	1 million

Adapted from ATSDR. Nature and Extent of Childhood Lead Poisoning: A Report to Congress; *1988.*

importance of these other sources, but old existing lead paint and the contaminated dust and soil it generated remained a very large problem.

The ATSDR report to Congress highlighted the abject failure to address lead paint:

> Existing leaded paint in U.S. housing and public buildings remains an *untouched and enormously serious problem* despite some regulatory action in the 1970s to limit further input of new leaded paint to the environment. For this source, *corrective actions have been a clear failure* (emphasis added).

The report concluded:

> The "easiest" steps to lead abatement have already been taken or are being taken. These steps, not surprisingly, involved reducing lead in large-scale sources, such as gasoline and food, with more-or-less centralized distribution mechanisms [refineries and canning factories]. Enormous masses of lead remain in housing and public buildings, along with large amounts of lead in dust and soil. If these highly dispersed sources are to be abated, huge efforts will be required.

The report stated that childhood lead poisoning eclipsed "all other environmental health hazards found in the residential environment" and paved the way for new Congressional hearings, as well as a major declaration from the nation's leading health official, the Secretary of Health and Human Services only 4 years later.

For a government agency to declare in a report to Congress that previous efforts were a "clear failure" created a new sense of urgency and was a remarkable development. It marked a significant turning point.

3.3 Congress tries again: the 1987 Housing Act and the 1988 Stewart McKinney Amendments

In the wake of the ATSDR report to Congress, new amendments to the 1971 Lead-Based Paint Poisoning Prevention Act were passed in both 1987[14] and 1988.[15] The two laws were passed only 9 months apart, but the difference was remarkable.

The 1987 Act reflected the prevailing unscientific approach. For example, it was silent about the importance of lead dust, and it regarded *all* lead paint to be a immediate problem. The definition of lead paint hazard in the 1987 Act was revised to read: "accessible intact, intact, and nonintact interior and exterior painted surfaces that may contain lead in any such housing in which any child … resides or is expected to reside." The 1987 Act also called for an "abatement demonstration" program and ordered HUD to produce within 9 months a "comprehensive and workable plan" that would address all housing, not only public housing. There was no mention of interim controls. For the first time, the 1987 Act also required that essentially all family public housing be inspected for lead paint within 5 years and then abated whenever that housing development was to be "modernized" (a euphemism for substantial renovation).

But only a few months later in 1988 Congress reacted to the ATSDR report and began to recognize the science, the importance of lead dust, and interim controls. The 1988 law retained the focus on so-called "full abatement," but also recognized that immediate action was needed and called for an evaluation of:

> the merits of an interim containment protocol for public housing dwellings that are determined to have lead paint hazards but for which comprehensive improvement assistance under section 14 of the United States Housing Act of 1937 is not available.[15]

This was due to the experience of a lead paint public housing insurance effort described later in this chapter, as well as the pathway studies that had revealed the importance of lead dust.

In testimony from the Council of Large Public Housing Authorities before a Senate Subcommittee, Miles Mahoney (then the director of a public housing authority and who helped to teach the Georgia Tech lead paint courses) stated:

> We understand your subcommittee's frustration with HUD's failure to act expeditiously and responsibly in this area. As humane and responsible public officials, we share that frustration.... [There is] a dearth of knowledge and considerable misunderstanding in HUD and the Congress.... The major threat to children is the ingestion of dust or particles contaminated by lead. Neither the 1987 amendments nor the HUD regulations address that issue.... An interim containment strategy is needed.... There plainly is a need to fund the requirements for abatement of lead-based paint.[16]

Although Congress limited requirements to inspect for and abate lead paint to conventional public housing (less than 1% of the housing stock), it was widely recognized that such housing had much less lead paint than did low-income privately owned older housing because lead paint was more expensive and thus less likely to be used in public housing.[17] This was subsequently shown to be correct in a later HUD housing survey.[18] But because it was more regulated, public housing was addressed first, ultimately becoming a "proving ground" for lead paint remediation in private housing.

The 1988 Act also specifically recognized that occupational hygienists would need to become involved, reflecting the need for skills that combined both exposure assessment and engineering control, not only medicine. Reflecting continued problems with measurement, the Act also called for a demonstration to evaluate both XRF and paint chip testing technologies.

The 1988 Act eliminated requirements that had been shown over the years to simply not be feasible, including the requirement to test and abate public housing units only at turnover; testing all units within a housing development (instead of using a representative random sample of all housing units within a given housing development); and an extension of the period for completion of lead paint testing for all public housing units until 1994, due mainly to a shortage of trained inspectors across the country. The inability to complete these actions had been well documented by public housing authorities.[19]

Perhaps most importantly, the 1988 Act stated more clearly what Congress expected in a "Comprehensive and Workable Plan." It would need to include "an estimate of the total cost of abatement,"[20] described later in this chapter. This plan would turn out to be anything but "comprehensive and workable"; HUD would release it 2 years later in 1990 despite the Congressional order that it be produced in a few months.

3.4 Leaders in public housing take action: the birth of lead paint risk assessments

In truth, public housing authorities were in a difficult spot. On the one hand, they were required to identify where lead paint was located in each of their housing developments, but they had no way to deal with it immediately (and in some cases for years). Because modernization sometimes would not occur for decades in a housing development, abatement could not happen until then. Under HUD Congressional authority at the time, housing authorities could only fund abatement at the time of such rehabilitation, known as the Comprehensive Improvement Assistance Program.

Although the McKinney Act did not define exactly what "interim containment protocols" meant, Congress acknowledged for the first time that the interim options were clearly needed, because of the funding mechanism for abatement and modernization. If such modernization would not occur for years, what was a housing authority to do in the meantime?

Some housing authorities had already done lead paint inspections. Because they had knowledge of where the lead paint was located, but no ability to fix the problem until much later, lawsuits proliferated against housing authorities, involving millions of dollars spent on damage claims. Damage claims only compensated children who had already been harmed and did nothing to prevent even more children from becoming poisoned.

Gordon Cavanaugh, a lawyer who represented the Council of Large Public Housing Authorities and Miles Mahoney both became convinced that public housing should do more on lead paint. As leaders in the public housing world, they were positioned to help solve it.

Cavanaugh was a highly skilled lawyer who chose to devote his life to low-income families in public housing. Mahoney was a highly skilled manager who knew how housing authorities did (and did not) work and who also devoted his life to families in public housing.

Some of the larger housing authorities had banded together to self-insure against other risks, including general liability, errors, and omissions. The private insurance market generally did not cover risks associated with lead paint, using a "pollution exclusion" clause in most insurance policies, and also to avoid having to pay for ever-increasing litigation and damage claim costs from lead paint cases.

Created in 1987, the Housing Authority Risk Retention Group (HAARG) stepped into this gap to provide liability insurance coverage to public housing authorities.[21] It created a risk management system for lead paint that would control risks in the short term until a housing authority could permanently abate lead paint at the time of modernization. These short-term measures were not required by HUD at the time unless a child with an elevated blood lead level was identified. Interim control requirements were not formally promulgated until 1999 when HUD reformed virtually all its lead paint regulations for federally assisted housing, not only public housing. In a letter to HUD, HAARG stated it was the "only company that provides coverage to Public Housing Authorities for bodily injury caused by the ingestion of lead in housing developments not undergoing modernization."[22]

Mahoney and Cavanaugh asked the director of the Georgia Tech training courses to craft such a risk assessment/risk management system for the insurance company. The protocol involved an extensive analysis of a housing authority's history of maintenance (some would call this the "good, bad, and the ugly"), such as determining how many young children resided in a given housing development, age of housing, and the presence (or absence) of standard operating procedures, especially painting and maintenance procedures and other historical and demographic data (see box). But perhaps most importantly, it included a visual examination of painted surfaces, lead dust testing by wipe sampling, soil testing, and laboratory analysis to quantify the immediate risk. The result was a detailed "risk assessment" report. This focus on paint deterioration, dust, and soil reflected the emerging science of lead pathway exposures. None of this was required in HUD regulations at the time.

If deteriorated paint was identified and had not yet been tested for lead, the insurance company would do so. The risk assessment results led to specific hazard control actions short of abatement that would need to be taken to control immediate exposures to deteriorated lead paint, contaminated dust, and contaminated soil.

The only way a housing authority could get the insurance was to comply with the risk assessment interim control recommendations. The interim control costs could be absorbed by the housing authority's operating and maintenance budget because they were much less expensive than abatement. The degree of risk would also be factored into the premium that the authority would need to pay to the insurance company, which was also an eligible expense under HUD's regulations. The abatement requirements at the time of modernization still remained of course.

Elements of the HARRG Lead Paint Risk Assessment

Description of Housing Development; Reliability of Comprehensive Testing; Extent of Modernization Work Completed; Levels of Lead in Dust and Soil; Adequacy of Existing Management & Maintenance; Compliance with Local & State Regulations; Plans for implementation of abatement; Specific recommendation on interim containment; Review of Maintenance and Management; Frequency of inspections; Work Order System; Cleaning of Common Spaces; Maintenance Emergencies; Preventive Maintenance; Painting Cycles; Chronic Maintenance Problems; Maintenance Capacity; Existing Medical Surveillance for Workers Exposed to Lead; Respirator provision and training; HEPA Vacuum availability and training; Supervisory personnel training; Interviews with workers and supervisors; Worker and supervisor awareness of lead paint hazards and controls; Review of Elevated Blood Lead Level Emergency Response; Suitability of Temporary Relocation Housing; Emergency Abatement Program; Review of the Public Housing Authority's existing Tenant Selection and Assignment Plan; Blood lead screening of children in the housing development; Lead paint testing compliance with HUD Interim Guidelines; test surfaces and housing unit randomization; XRF calibration checks; Substrate effect correction; Review of Quality Control/Quality Assurance Plan; Qualifications of laboratory for confirmatory paint chip tests; paint chip sample collection procedures; data completeness and plausibility, data interpretation; abatement plan consistency with test results; General description of buildings, schematics, list of addresses to be covered by risk assessment; Reports associated with resident children with elevated blood lead levels; Copy of any building code inspection reports or code violations; Description of

---

(cont'd)

---

any previous remodeling, construction, and preventive maintenance work; An assessment from both the building owner and the risk assessor of the condition of all painted surfaces; Number of children present; Rate of housing unit turnover; Existing worker training program (if any); Dust and soil sampling according to a uniform design (usually 8–10 wipe and soil samples per unit); defective paint sampling (if previous testing has not been performed satisfactorily).

In short, the lead paint risk assessment protocols that came to be used across the nation were born from a public housing insurance need. They were adapted by HUD for public housing in 1992[23] and for other housing in 1995 when HUD released its new lead paint guidelines.

Because this was an entirely new approach, the insurance company had to set up new quality control procedures to ensure that accurate measurements were used to identify hazards. This included working with laboratories to develop procedures for lead dust wipe sample analysis, which at the time had not yet been standardized. Although Johns Hopkins University[24] and others[25] had done some earlier work in this area, there was no method to determine if laboratories analyzing dust, paint, and soil samples were doing the job adequately.

The insurance company decided to fund the creation of a national quality assurance system of dust wipe samples with known quantities of lead dust, prepared at Georgia Tech's laboratories. These "spike" samples were then included with normal field samples, with the laboratory blinded to which ones were real samples and which ones were spikes. The percent of the known lead dust that the laboratory recovered from the spikes was reported to help ensure laboratories were doing the job properly. If recoveries were inadequate (as they sometimes were in these years), the results would be suspect and discarded in favor of new validated results. Lead paint dust Standard Reference Materials (which have a known amount of lead), prepared by the National Institute for Standards and Technology at the Department of Commerce were used to ensure confidence in the results and overcome measurement deficiencies to make the risk quantifiable and insurable.

These early quality control efforts also had the benefit of the expertise of Terry Burke and others at the Wisconsin Occupational Health Laboratory, which was (and still is) responsible for analyzing much of the nation's OSHA consultation program samples, another example of the pivotal role played by the nation's occupational safety and health profession. Georgia Tech had a long-standing relationship with the lab, which helped to develop the analytical protocols to ensure adequate recoveries of lead in wipe and paint samples.

The public housing quality control system preceded the nationwide quality control procedures for all the nation's laboratories analyzing lead dust, paint and soil samples, which was not completed until the mid-1990s, when EPA and HUD launched the National Lead Laboratory Accreditation Program, described later in this book.

In the late 1980s, there were no standard wipe materials, so "baby wipes" were used instead. Hardly a laboratory grade material, they proved to be difficult for a laboratory to analyze reliably. Other analytical grade wipe materials, such as Whatman filters were available at the time, but they proved to be impractical because they were too fragile to withstand the rough and uneven surfaces in housing.

Therefore, the "baby wipe" method was used as an interim approach. This would involve more than a few trips to local department stores to buy all their supplies of certain brands of baby wipes that had proven to have low background lead levels and could be analyzed by laboratories for lead. The looks were incredulous when store clerks asked if all the wipes were being purchased for use at a large day care center. No, they were told, "they are being used to measure lead dust."

Years later, in 1996 baby wipes were replaced with manufactured analytical grade wipes according to an ASTM standard. These wipes were durable enough to withstand wiping rough household surfaces, but easier to analyze in a laboratory.[26] But at the beginning of the public housing risk assessment program, these and other scientific analytical challenges were overcome to ensure that the insurance company was correctly identifying hazards by making accurate measurements.

The public housing insurance program proved to be enormously successful and popular among housing authorities. Had it not been for the vision of Cavanaugh and Mahoney, their collaboration with occupational health scientists at Georgia Tech and various laboratories, support from HARRG, and earlier research at Johns Hopkins University, risk assessment methods and measurements would likely have been delayed or reduced to inaccurate subjective questionnaires. Teaming experienced housing professionals with health scientists was a key factor in creating policy solutions, in this case insurance policy that was implemented across the nation.

This paved the way for a better recognition that to ensure the availability of insurance, standards for risk assessments, inspections, and hazard control were needed, a finding contained in a Government Accountability Office report in 1994.[27]

The interim controls used by the insurance company were much more than simply repainting. They involved occupant and worker protection to remove the deteriorated lead paint and specialized use of a new two-coat paint system that was compatible with the existing painted surface. This was followed by special cleaning using filtered vacuums (high-efficiency particulate absorbing (HEPA) vacuums), wet cleaning, followed by repeated HEPA vacuuming and dust testing to ensure the cleanup had worked. This marked the first time that dust testing was implemented after interim controls, although in the early 1990s HUD did not require dust testing after abatement either.

A preliminary lead dust standard (an allowable amount of lead dust) following cleanup that was thought to be both feasible and protective had been developed at Johns Hopkins[28] and subsequently adopted in Maryland and Massachusetts.[24,29] These standards were used by the public housing insurance company to ensure cleaning was adequate and would protect children, even though formal federal government dust lead standards were still a decade away, finally being promulgated in 1999 and 2001.

The insurance company required that only after interim controls had been implemented and low dust levels achieved would children be allowed to reoccupy their home, a process that would come to be called "clearance testing." Lead paint hazards in the form of deteriorated lead paint, contaminated dust, and contaminated bare soil were all eliminated through these interim controls. Importantly, continued monitoring was required to ensure hazards did not re-emerge.

One result of all this is public housing is now generally safer for children and has much lower rates of lead poisoning compared to other low-income private housing.[30–32]

However, the requirement to fully abate lead paint in public housing at the time of modernization was not adequately enforced over the years. If it had, one would think that after 30 years since the abatement requirement was first put in place in 1988, all lead paint in public housing would have been abated by now. But, as seen in the New York City Housing Authority (NYCHA) and a few other housing authorities, some public housing lead paint remains unabated.[33]

The NYCHA case is perhaps the most egregious because the authority knowingly lied to HUD that it had complied with lead paint regulations, when in fact they had not. In 2018, NYCHA was the target of legal action from the US Department of Justice, which resulted in the appointment of a court-appointed monitor to ensure lead paint compliance.[34] Had NYCHA integrated lead abatement into its modernization and rehabilitation programs, it would likely have been far less expensive, compared to conducting it as a stand-alone activity under court supervision.

Importantly, NYCHA failed to use its lead paint inspection data to determine which surfaces did and did not have lead paint. It was as if the inspections were done to fulfill a regulation, but not put into actual practice to determine where abatement and interim controls were in fact most needed. In 1996, NYCHA asked Neal Freuden, who had been a housing authority director in Connecticut and later became a lead paint inspector, to review the NYCHA inspection reports. He found that many of them had been poorly done and would need to be repeated.[35] This mismanagement of public housing authorities is more the exception than the rule, but it is ironic that in one of the first cities to ban lead paint (New York City), it still remained such a large problem in 2022.

The problem also persisted in public housing in New Orleans, where in 1998 up to 29% of children continued to have high blood lead levels.[36]

The inaction at NYCHA can be contrasted with the many other housing authorities that did take it head-on in their public housing. Examples include Macon, GA, Baltimore, MD, Dover NH, and many others across New England and elsewhere.

3.5 A scandal prompts Congress to create the HUD lead paint office

The Housing Authority of New Orleans (HANO) had a long history of poor management that contributed to lead poisoning among hundreds of children in its public housing developments. HUD took over the authority on multiple occasions throughout the 1980s (and would do so again in later decades) to correct management problems.

In 1979, the New Orleans City Health Department sent a letter to HANO, stating:

> Although we have been repeatedly advised of the staff shortages, the number of children living in leaded public housing makes it necessary that we officially notify you that those conditions are

unjustifiable. To a degree, the Housing Authority of New Orleans and the New Orleans Health Department can be rightfully accused of not correcting an illegal and unsafe condition...the number of children that we see who have problems related to lead in those dwellings is dramatically escalating and we must require your utmost cooperation.[37]

But incredibly, in 1984 the Director of HANO wrote to HUD declining to apply for additional lead paint funds, saying, "There is neither need nor justification for extending our Lead-Based Paint Abatement Program beyond its present scope." By October 1985, so many lawsuits had been filed that HANO lost its insurance coverage. By 1988 over 100 damage suits were pending and in 1989 HANO sued pigment/paint producers.[37]

The New Orleans housing authority was the nation's 7th largest, with about 13,000 housing units for over 24,000 people. It was first designated as "troubled" by HUD in 1979. In 1984 HUD withheld approximately $10 million in modernization funding because its performance had not improved, although some thought that withholding funds only exacerbated the problem. In 1988, HUD's management review of HANO revealed 241 findings of deficiencies, many of them involving maintenance and painting problems.[38]

In the late 1980s, Cavanaugh and Mahoney commissioned a White Paper on the conditions in the New Orleans housing authority that had led to so many lead-poisoned children. They delivered it to Senator Barbara Mikulski, who chaired the HUD appropriations committee at the time.

The White Paper was blistering:

> The predominant attitude [among housing authorities] appears to be that lead-based paint is just another requirement that has been imposed with inadequate resources and non-existent technical assistance and aid. The results of this attitude are nothing short of disastrous. Children continue to be poisoned in the properties of several housing authorities, even in the face of repeated lawsuits.

> In New Orleans, for example, over 113 children have brought legal action against the housing authority as a result of lead poisoning. The [housing] development responsible for the majority of these cases remains in a severe state of disrepair, with lead-based paint still peeling from porch columns and other surfaces. Modernization activities that generate large amounts of dust are proceeding without proper testing to determine if a lead paint hazard exists before demolition and remodeling work, posing occupational health threats to construction workers in a wide variety of trades, as well as the children living there. Finally, there is virtually no attempt to perform risk assessments and identify those areas where the immediate hazards are greatest. As a result, there is no attempt to devise interim containment efforts, integrate interim cleanup and control measures into existing maintenance programs, or conduct resident educational campaigns....

> The source of most of these problems is HUD's failure to indicate how the attack on the subtle lead poisoning epidemic fits into its efforts to deal with other pressing problems, many of which are more obvious: homelessness and the lack of affordable housing, crime, drug abuse, etc. In short, HUD continues to deny in practice that lead poisoning is anywhere near the top in its ranking of priorities.

> The bottom line is that the federal government will need to step in and provide funding for abatement .... Until this happens on a large scale, testing and screening programs will do little but record the damage .... Preventive measures and primary prevention will remain the exception and the risk of injury will remain unacceptably high for the children being raised in our nation's public housing.[39]

At a Congressional hearing, Sen. Mikulski asked about the lead paint findings in the White Paper. There were two HUD witnesses: The Assistant Secretary for Public and Indian Housing (PIH) and the Assistant Secretary for Policy Development and Research

(PD&R). The HUD role since the 1973 Act had been confined primarily to research on the lead paint issue, and it was not until the 1988 McKinney amendments that HUD had been required to test lead paint in public housing, so it was unclear whether this was a public housing or policy issue.

Reflecting the agency's lack of focus, each witness pointed at the other as being responsible for the New Orleans situation. The public housing witness stated, "this is a policy issue" and referred to PD&R as the lead entity within HUD that was responsible. On the other hand, the PD&R witness stated that because the New Orleans situation was public housing, it was the responsibility of the PIH office, not the PD&R office.

Senators Mikulski, Bond, and others were outraged at the finger pointing and lack of accountability.

As a result, in 1989 the Senate report on appropriations, crafted principally by Congressional staffer Kevin Kelly, directed HUD to consolidate responsibility for lead safety into a single office within the Assistant Secretary for Public and Indian Housing, in an attempt to stop the bureaucratic "not my problem" finger pointing. The report also reflected the dilemma in public housing on timing of abatement:

> There is a compelling need for the immediate implementation of less costly measures of reducing or containing the risks from lead-paint poisoning in the interim period before full-scale abatement can be accomplished.[40]

In 1989 Congress included a statutory administrative provision restricting the use of appropriated public housing funds to temporarily prevent the implementation of regulations that required testing for and abatement of lead-based paint in public housing. Congress demanded that the HUD Secretary certify that it would be conducted in a cost-effective and safe manner. By freezing the expenditures of funds for lead abatement, Congress once again pointed to the lack of standard measurement procedures:[41]

> HUD's failure over the past several years to conduct the research and technical studies necessary for developing safe reliable and cost-effective methods for testing and abating lead-based paint remains a problem. Unfortunately, the vast majority of current abatement work relies on paint removal by sanding or heat guns, which usually spreads lead dust and increases exposures to workers and occupants. Despite these problems the Department proposed regulations … to require immediate widespread testing and abatement in public housing…. The Department is directed to submit a plan by December 1, 1988 for a comprehensive multi-year program to assure the quality and safety of lead paint abatement.[42]

Importantly, the 1989 appropriations Act provided $50 million for abating lead hazards in low-income *private* housing, the first time that dedicated federal funds had been appropriated. Congress doubled the $25 million requested by President George HW Bush. Plainly, the ATSDR report and the HARRG White Paper had prompted both Congress and the President to recognize the need for more action.

Responding to Cavanaugh and Mahoney's efforts, Congress also funded $25 million for lead risk assessments in public housing (not only inspections) for the first time, based on the successful public housing insurance program. Comments provided to Congress on the $25 million for public housing risk assessments outlined what they needed to include:

> HUD now has something they call a "risk assessment" in the [1990 Public Housing] Lead-Based Paint Guidelines. This protocol has been used in Kentucky, where it was found to be a poor predictor of the location

of poisoned children. The HUD protocol involves gathering some rudimentary demographic and housing data, and then arriving at a "number." In short, it is unlikely that this kind of assessment will yield useful information. It is vital that Congress direct HUD to re-examine their risk assessment methodology, if the $25 million is to be spent productively. Specifically, the risk assessment must measure immediately available sources of lead in dust and lead in soil if the interim containment measures are to be properly focused. Also, the risk assessment should provide for an independent third-party review by a trained risk assessment firm of the various management and maintenance practices at the individual housing development level. Ideally, the findings will provide the scientific basis for recommendations that alter the way in which training, maintenance, and other management practices are performed, so that lead-based paint hazards are adequately considered. Finally, there are a rather large number of environmental firms claiming that they can perform risk assessments. Hardly any of these understand how public and Indian housing authorities are organized.[43]

A year later, HUD had failed to create the new lead paint office that Congress had mandated and the money appropriated for lead abatement in private housing had not even begun to be spent. Predictably, Sen Mikulski and her colleagues on both sides of the aisle reacted in fury. She stated she did not want finger pointing and blame shifting; she wanted an authoritative office to be created within HUD. Together with Senator Bond and other Democratic and Republican leaders, she mandated the creation of a new office, not one buried within PIH or PD&R, because it was now clear that the problem extended far beyond public housing, was not only a research issue, and both PIH and PD&R were clearly incapable of responding. The lead issue had remained buried within HUD for more than 2 decades.

Instead, the Office of Lead Abatement and Poisoning Prevention was created within the Office of the Secretary of HUD at the highest level to ensure it had cross-cutting authority across all HUD programs, not only public housing or research.[44] The HUD Secretary at the time, Jack Kemp, initially resisted staffing the new office, but because he was entertaining a bid for President, he allowed it to go forward to some extent. This was due in no small part to Senator Mikulski's threat to restrict his travel budget. The Senate committee report stated:

> The new office of lead abatement and poisoning prevention ... has been given the ultimate responsibility, except for the Secretary himself, for control of all of HUD's lead-related activities.... Current HUD responsibility for the program now rests in separate program offices, and this effort is part of the Committee's desire for a one-stop shop.[45]

Importantly, none of the new office's initial staff included scientists, and most were drawn from PD&R (PD&R staff who had been working on lead paint issues were given the option to move to this new office or stay put). The new lead paint office was jokingly referred to as the "leper colony" within HUD. The HUD Memorandum creating the new office stated that there would be 20 staffers, "filled predominantly by lateral reassignment of staff principally from PD&R."[46]

A few competent PD&R staffers did decide to join the new office and two, in particular, went on to play leading roles. Ellis Goldman would direct the new lead hazard control grant program for private housing and Steve Weitz would direct the office's technical studies, policy, and regulatory reform. A new HUD Secretary, Henry Cisneros, and his chief of staff Bruce Katz (who had played a key role as Congressional staff in shaping Title X) would become the first to include scientists in the office, starting in 1995 (Chapter 4).

Congress also required the office to be headed by a member of the Senior Executive Service to help ensure that it was managed properly.[47] The office was to be created "in perpetuity" so that it could not later be moved or abolished or buried within the bureaucracy. This provision would later prove to be important when in 2005 another HUD Secretary, Alphonso Jackson, attempted to do just that in a concerted attack on HUD's lead program (Chapter 5).

3.6 Moving remediation from the Centers for Disease Control and Prevention to the Department of Housing and Urban Development: The 1990 Public Housing Guidelines

There had been earlier attempts to establish reliable procedures for lead hazard identification and remediation, but they were never broadly accepted. The first appeared in 1971 from the National Bureau of Standards under contract to HUD but was largely theoretical, not based on evidence.[48] Another appeared in 1986 from the director of Detroit's lead poisoning prevention program, funded by the Division of Maternal and Child Health in the Department of Health and Human Services, but this was focused on remediation of homes where children had already been exposed.[49] By the late 1980s it had become increasingly clear that health authorities such as CDC or environmental authorities like EPA did not have the expertise needed on lead paint detection and remediation, nor did they have the authority to make it happen. This responsibility increasingly shifted to HUD, initially as a result of the 1988 McKinney amendments to the Lead Paint Poisoning Prevention Act described earlier.

A stinging review of HUD's lead abatement requirements and practices in 1988 stated,

> There are no requirements for safe abatement procedures ... certifying or otherwise qualifying abatement personnel ... [and] determining whether lead dust in a housing unit either before or after abatement ... is at an acceptable level.[50]

In 1988, the House Appropriations Committee, where Don Ryan was a key staffer, directed HUD to develop comprehensive guidelines for testing, abatement, cleanup, and lead-based paint. To comply, HUD made a grant to the National Institute of Building Sciences (NIBS), which convened a broad-based committee to develop consensus guidelines. A year later it released a draft document titled "Guidelines for the Identification, Abatement, Clean-up and Disposal of Lead Paint Hazards," but the Office of Management and Budget at the White House blocked its release, and it was never officially published.[51]

The 1989 appropriations act also criticized HUD for delaying the publication of the Interim Public Housing Guidelines. It ordered HUD to issue technical guidelines on how to do lead abatement safely no later than 6 months, which would have been the Spring of 1989.[52] On September 15, 1989, HUD drafted a "limited edition" of public housing guidelines, but these also were never released in final form.[53]

The NIBs document was widely criticized and contained several minority reports from both health and housing professionals. Henry Falk from CDC wrote that the actual goal of the NIBS guidelines had never been defined and that "a more balanced presentation of optimal and acceptable approaches [is needed].... There is a real risk that we could end up worse than before." A group of housing professionals from the Council of Large Public Housing Authorities, the National Association of Housing and Redevelopment Officials, and several housing authorities wrote, "the [NIBS] guidelines are not ready to be released for implementation. It does not relate the detection and elimination of lead paint hazards directly to public housing management." They indicated 19 areas that were not addressed.

Five public health professionals and advocates wrote:

> As strong advocates of lead poisoning prevention, we cannot support the adoption of these [NIBS] guidelines ... they ignore entirely the issue of cost ... the failure to provide for external review of these guidelines is itself extraordinary ... If viewed as an authoritative demonstration that lead paint abatement is prohibitively expensive, the guidelines could serve as justification for Congress, HUD and private property owners to do nothing ... poor people will once again be shut out of a program initially designed for their benefit.[54]

Another reviewer agreed:

> In its present form, I doubt that the [NIBS] guidelines could be used by many (indeed, most) of the contractors who have attended my continuing education courses in lead paint abatement.... The more substantive problem appears to be ... the controversy ... with the failure to adopt a "cost-effective" approach to lead based paint abatement. The fear articulated in one of the minority [NIBS] reports is that this document would fail to reduce population blood lead levels in children significantly by specifying unnecessarily rigorous standards. The minority report states that this would result in a small number of very clean housing units, but much of the nation's leaded housing would remain unabated, leading to a poor allocation of limited resources.... In short, the key question facing us should not be how to make lead paint abatement more affordable by loosening various standards. Instead, we should be focusing on ways to accelerate lead-based paint activity and improve public awareness of this issue.

> Too many people believe this issue was settled with the ban of lead in gasoline [and] the ban of commercially available new leaded paint. Advocating further delays in the belief that new research will show how abatement can be made radically cheaper, or how the relative contributions from lead in air, water, soil, and paint stack up will simply perpetuate the way things are now, which I think we can all agree is inadequate and too slow. The bottom line is that the nation has not chosen to allocate sufficient resources to the lead paint problem—we must work to change that.[55]

Congress reacted in even more frustration over the continued delay and voted to require HUD to publish its public housing guidelines by April 1, 1990, in a vote by a joint House/Senate Appropriations Conference Committee.[56]

There was continuing press coverage of HUD's inaction. One New York Times article highlighted the delays and frustrations, quoting Sen Mikulski of Maryland, "I'd rather see HUD paying to clean up houses than Medicaid paying to clean up kids." The same article noted that even the best public housing could pose risks, noting that Millie (the dog at the White House) had become "acutely lead poisoned as a result of renovation work being done at the White House."[57]

Finally, in 1990, HUD released its new "interim" guidelines that implemented the widespread inspection and abatement efforts in public housing mandated by the 1988 McKinney amendments.

But there had been an important change: Of the 26 main contributors to the 1990 HUD Interim Public Housing Guidelines, only one was from CDC, one was from EPA, and only six were from health authorities or researchers; the remaining 18 authors were from the housing world. This marked an important shift from previous government efforts that had been led by CDC. Housing professionals had finally entered the fray.

The Guidelines stated they represented "the first national compilation of technical protocols, practices and procedures on testing, abatement, worker protection, clean-up and disposal."[58] Totaling 591 pages, they were far more detailed than the four or seven steps summarized earlier in short CDC documents. They discussed many additional methods that were essential if the new 1988 Congressional McKinney requirements were to be successfully implemented. They included operational details of how housing authorities actually worked and how lead paint work would be funded, something that health entities could not do and did not understand. There was a clear statement that although the guidelines were not legal regulatory requirements, they were "recommended," and thus defined standards of practice at the time.

Although national in scope, it was focused on local housing authorities. This HUD focus on cities was different than CDC's, which tended to rely on states. It also differed from EPA's, which tended to rely on its 10 regions for funding and implementation, not cities or states. This makes some sense, because many health programs, such as Medicaid and surveillance, are handled through states. EPA pollution control programs are addressed regionally because pollutants cross state boundaries and have impacts regionally.

In short, HUD (like housing generally) was more local. This difference in geographic emphasis became a bureaucratic and very real barrier to collaboration between HUD, CDC, and EPA in later years. How this disconnect played out and was overcome is described in later chapters. But because housing is regulated at the local level for the most part (such as through housing codes), HUD's focus on localities was essential to integrate lead paint hazard control into normal housing operations.

The 1990 HUD Interim Guidelines had a much more detailed focus on lead-contaminated dust and soil, reflecting the new research on pathways of exposure. For the first time, levels of lead in settled house dust and in soil that were deemed to be a "hazard" were defined numerically.

These 1990 guideline lead dust levels were far higher (less protective) than those used in 2022, but they established a clear yardstick to measure this important exposure pathway, reflecting the experience in the public housing insurance program and research. The lead dust level on floors was 200 $\mu g/ft^2$, far higher than the 10 $\mu g/ft^2$ codified in 2019 EPA regulations (matching earlier HUD action in 2017).[59,60]

However, the 1990 HUD interim guidelines continued to take the presence of lead paint to constitute a hazard, without regard to its condition. This was a reflection of the 1982 *Ashton v Pierce* court case (Chapter 6), which defined a lead paint hazard to be intact and nonintact lead paint. The Interim Guidelines simply stated, "The level of [lead paint] hazard is determined as 1.0 mg/cm^2 or 0.5% by weight." There was no mention of the paint's condition and lead in dust was not defined as a hazard.

Although the primary audience was housing authorities, the interim guidelines were also aimed at contractors who carried out the work; private inspectors; local health agencies and medical health care providers; and architects, engineers, and others who prepared work specifications. It discussed how blood lead screening was expected to interact with abatement operations, how housing authorities could prioritize their inspection and abatement activities, and how residents could best be educated. There were detailed recommendations on occupant and worker protection. In short, this was far more complicated and involved many more actors than blood lead screening, which typically involved a medical care provider, a public health authority and a laboratory.

The experiences from Georgia Tech and other training providers also helped to inform what to do about lead paint measurements by poorly performing XRF portable lead paint analyzers, which had been developed by funding from HUD in the late 1970s and 1980s. The XRF procedures in the interim guidelines included averaging multiple readings on the same testing location, attempts to correct the results for interference from the substrate (material underneath the paint), calibration checks, and laboratory analysis of paint chips to confirm certain inconclusive XRF results.

Although this greatly complicated lead paint measurements, it did increase their reliability and acceptance. Descriptions of how to obtain random samples of housing units within a housing development to help ensure that the results could be used to characterize an entire housing development also proved to be essential in improving confidence in the results and greatly expanding the number of housing units that were assessed for lead paint hazards.

These procedures were greatly aided by the involvement of the instrument manufacturers themselves and skilled statisticians like David Cox. Scientists who had characterized instrument error, such as Stan Lewis, a nuclear chemist from Georgia Tech also played a key role. Cox and his associates Kenn White and Gary DeWalt went on to establish the nation's first "lead paint archive." Developed with HUD funding, the archive contained real-world painted surface samples with known amounts of lead paint on different substrates that could be used to establish just how well the XRF instruments really worked.

This would ultimately lead to a whole new generation of XRF instruments that were more reliable, faster, and easier to use, but only after the government stepped in to create a level playing field in which instrument manufacturers could compete. But none of that would happen until the mid-1990s, so the 1990 Interim Public Housing guidelines proved to be an important step in increased measurement reliability and, with it, policymaker's confidence. HUD issued a notice to public housing authorities, requiring them to determine if the XRF testing they had already completed was in fact reliable. The notice specifically required a review of all written test reports, a time and method analysis (to help determine if the time inspectors had spent testing was sufficient to test all the surfaces they were supposed to test), and repeated random sampling; it stated HUD was "taking steps to follow up on reports of inadequate testing."[61]

Equally important, there was much more detail on exactly which methods of abating lead paint could be used and which ones should *not* be used, yet another example of how research succeeded in successfully informing policy. Specifically, using torches to melt off lead paint, power sanding, dry scraping, abrasive blasting, and a few other methods were prohibited because they produced very high lead dust and lead fumes that proved to be impossible to clean up adequately.

Recommended procedures were those that produced much less dust and would be easier to clean up, reducing the risks to both workers and children. Such abatement methods included replacing building components like windows and doors. This left much of the lead paint intact during building component removal, with little dust generation. On the other hand, such replacement also tended to be the most expensive and some historic preservation advocates objected, even if that meant leaving poisonous building materials in place. Other methods included using chemicals to strip off lead paint (both on-site and off-site), use of lower-temperature heat guns, and certain contained methods of abrasion, such as needle-guns equipped with shrouds and attached to special vacuums that would capture the dust at the point it was generated. Hazardous waste implications for each of the methods were also addressed in the interim guidelines, although formal EPA regulations would not appear until years later.

Detailed cleaning procedures were described, but the interim guidelines had only limited discussion of postabatement dust lead testing. Listings of laboratories were provided, but at the time there were no quality control procedures for labs tasked with analyzing lead samples in dust, soil, or paint, except for the comparatively small public housing insurance program.

The most *glaring omission* in the HUD interim guidelines was the absence of any discussion of options for how to address lead paint problems in the near term, that is, interim controls. To some extent, this was inevitable, because housing authorities did not have any formal requirements to address lead hazards immediately, only in the longer-term context of modernization.

Only two pages were devoted to what should be done in a home where a child with an elevated blood lead level was found. So-called "emergency" abatements were recommended in such cases, that is, outside of modernization. But the greater emphasis was placed on relocating the child to another home that had previously been abated or found to be free of lead paint. Such relocations, to the extent they were done at all, usually left the old housing unit with lead hazards available to be occupied by yet another child, who could then also be poisoned. Perhaps most important, the Guidelines identified no funding source for the emergency abatements.

This problem of so-called "repeat offender" houses would come up in future Congressional hearings. For example, Sen Jack Reed of Rhode Island once complained:

> I have a single federally assisted home in my state that has poisoned not one, not two, not three, but 4 different children. What is HUD doing to be sure that this home will not poison a 5th child?[62]

Of course, the answer involved not just moving the child but also moving away from the straight jacket of abatement only at some distant point in the future and supplementing that with short-term action.

Indeed, the HUD 1990 Interim Guidelines specifically warned *against* any immediate short-term measures, stating:

> The emergency intervention actions mentioned in the regulations should be taken only as a last resort, because removal of defective paint does not provide long-term safety and can increase lead dust levels in the air and on horizontal surfaces.[58]

It was a simplistic "all-or-nothing" approach, and "nothing" remained the norm.

The 1990 HUD interim guidelines did not directly address lead in soil, although it was recognized as a risk factor. They only referenced a yet-to-be-completed EPA study, known as the "3 cities study" for soil remediation then taking place in Cincinnati, Baltimore, and

Boston. That study ultimately showed that soil remediation by itself did not reduce children's blood lead levels significantly, except when the soil was initially very high, as was the case in Boston.[63]

3.7 Sticker shock: The Department of Housing and Urban Development's comprehensive and workable plan

As in the earlier NHANES health surveys that counted poisoned children, HUD also conducted an analogous national housing survey in 1989–90, the first truly robust estimate of how many houses contained lead paint and lead paint hazards. The estimates were released in a Report to Congress that the McKinney Act had mandated as part of HUD's Comprehensive and Workable Plan for the Abatement of Lead-Based Paint in Privately Owned Housing, along with new cost estimates.[64]

HUD's survey found that of the 77 million privately owned and occupied homes built before 1980, 57 million, or three-fourths, contained lead-based paint. It also stated an "unexpected" finding—that "lead-based paint is found as often in the homes of the well-to-do as the poor."

The *annual* costs were estimated by HUD to be $36 billion for encapsulation and $50 billion for lead paint removal over a 10-year period. Thus, the total costs were estimated at $360–$500 billion. This report compared those estimates for ordinary home repairs, which was estimated to be $101 billion per year. In other words, the report suggested that lead paint abatement might be half of what the nation spent on all home repairs each year. And there were no estimates of financial or other benefits. Later estimates in 2000 proved the costs to be far lower and were scientifically validated (Chapter 4). But in 1990, the sticker shock caused by HUD's estimates produced more policy paralysis.

Within HUD and the housing world, these 1990 numbers dampened any efforts to move forward on lead paint. Although the Plan stated that HUD's Secretary Jack Kemp would address the problem through various task forces, little was done until his successor, Henry Cisneros, was appointed by the new Clinton Administration as the HUD Secretary in 1991. Because HUD's plan was produced by HUD's Policy Development and Research office (the same office that thought lead in gasoline was the real problem), it focused on the need for more research and less on implementing an actual plan to fix houses.

Finally, the HUD 1990 Plan noted that there was no separate HUD program to fund lead abatement, although local jurisdictions could choose to spend certain HUD block grant funds if they chose to do so. In 1990, John Weicker (head of HUD's Policy Development and Research Office at the time) pushed through a $25 million request for lead abatement grants, which Congress doubled to $50 million. Weicker would also serve as an Assistant Secretary for HUD's FHA mortgage program from 2001 to 2005, when then HUD Secretary Mel Martinez made lead paint a priority (Chapter 5). But in the 1970s and 1980s, like the Maternal and Child Block Grants at the Department of Health and Human Services in the early 1980s, the HUD Community Development Block Grants failed to address lead hazards. Block grants at both health and housing agencies were a clear failure when it came to lead paint.

During the 1970s to early 1990s, HUD spent very little on the lead paint problem. Indeed, even the small HUD research budget for lead paint was eliminated in 1981[65] and

TABLE 3.2 Housing and Urban Development (HUD) budget and staff, 1966–2005.

	Budget outlays[a]		HUD staff	
	$	Percentage of Government	#	Percentage of Government
1966	$2.5 billion	1.8%	14,500	0.7%
1970	2.4 billion	1.2%	15,600	0.7%
1980	12.7 billion	2.2%	17,100	0.8%
1990	20.2 billion	1.6%	13,300	0.6%
2000	30.8 billion	1.7%	10,100	0.6%
2005	42.5 billion	1.8%	10,000	n/a

[a]*Outlays are one standardized measure of the federal budget. They count actual cash disbursements in a given fiscal year.*
HUD and OMB Historical Tables. Lawrence Thompson. A History of HUD; 2006.

was not restored until 1988. Out of a total HUD 1990 budget of $20 billion[66] (Table 3.2), the HUD lead paint plan stated:

> During FY 1990, approximately $11 million was obligated in support of lead-based paint activities. Of this amount, $160,000 is being spent on public information; $8.2 million on testing and abatement research; $1.6 million for research on health effects; and $770,000 on State and local capacity building.

The report identified no spending at all for widespread abatement. This paltry spending on lead could hardly be expected to make any dent at all in the 57 million homes with lead paint. HUD's estimate of $500 billion to address lead paint dwarfed the entire HUD budget and contributed to the policy paralysis in this era that suggested solving the lead problem was just not feasible.

3.8 A new Alliance to End Childhood Lead Poisoning marshals political will

With CDC focused on medical management, HUD preoccupied with high costs, and EPA only responding to nonhousing environmental lead sources, the paralysis that had defined much of the 1970s and 1980s threatened to continue, despite the outrage in Congress and elsewhere. Yet, the winds of change were gathering.

The Alliance to End Childhood Lead Poisoning was formed in April 1990 and played the key role in laying the groundwork for Title X, which was passed by Congress two and a half years later. Congress directed EPA to fund the Alliance's Primary Prevention Strategies project, convening working groups to tackle some of the thorniest issues.[67] The group played an instrumental role in defining and building consensus throughout the 1990s and early 2000s.

Deciding to leave his professional staff position at the Congressional House Appropriations Committee, Don Ryan became the founding executive director of the Alliance, which was a national, nonprofit, public interest, policy, and advocacy organization.

In an article on Ryan and the new Alliance, the Wall Street Journal reported:

> After eight years [as a staff member of the House Appropriations Committee apportioning budget dollars], the numbers took on faces and the clerk turned crusader. "The system was so badly botched," [Ryan] said

looking back. "I decided the system needed to be reached from outside".... The new venture [the Alliance] meant a big cut in pay.... The Alliance's survival owes much to Teresa Heinz, a respected advocate of children's issues and the widow of . . . Senator John Heinz.[68]

The Alliance also served as the main group to which local parent and advocacy groups turned for help. It played a major role in helping groups of parents of lead-poisoned children to establish themselves, telling their stories to decision-makers and pushing for policy change, including Congressional hearings.

Formed in 1990 by representatives of multiple fields and disciplines, including leaders of state and local programs, public health, environmental protection, affordable housing, education, civil rights, children's welfare, and parents of poisoned children, the Alliance was designed specifically to bring capacity and credibility to help achieve the large changes that Title X came to embody.

The Alliance had three major areas of activity: education, policy support, and advocacy.

Its education efforts informed policymakers, the press and the media, key private-sector players, health professionals, parents, property owners, and the general public of the need for prevention. In addition to directly benefiting at-risk children, these education efforts broadened the base of individuals and groups seeking solutions to childhood lead poisoning. It published two national directories of such local organizations. The Alliance inspired the July 15, 1991, Newsweek cover story that focused the nation's attention (Fig. 3.1).

FIGURE 3.1   National magazine: lead paint cover story. *Reproduced by permission: Newsweek Magazine, July 15, 1991.*

It kept the pressure on federal and local agencies and helped overcome private-sector obstacles. The Alliance's Model State Lead Poisoning Prevention Law assisted state and local advocates in pushing their legislatures toward primary prevention.

The Alliance offered a "Framework for Action for Lead-Safety in Private Housing,"[69] which focused attention on the best ways in which financing and capacity building could occur, laying the groundwork for deliberations of an important federal task force on financing options in 1994.

It also analyzed the role of Medicaid, finding that in 1992 only 21 state Medicaid agencies reimbursed for case management services for lead-poisoned children and only 5 states reimbursed for environmental investigations to find the source of a poisoned child's exposure.[70] The Alliance's Community Environmental Health Resource Center later helped community-based organizations master environmental sampling to leverage action by rental property owners and local governments.[71]

The Alliance was also active at the international level. It published an International Action Plan for Preventing Lead Poisoning and launched a "Global Lead Network," which predated the UN/WHO Global Alliance to Eliminate Lead Paint by about 20 years. Written primarily by Alliance staffers Maria Rapuano and Jim Rochow, it reflected input from over 250 people in 37 countries[72] who attended an international conference organized by the Alliance. The global campaign helped accelerate the phase-out of leaded gasoline internationally.[73,74]

Ryan and the Alliance were instrumental in shaping Title X, discussed later in this chapter. The Alliance insisted on relying on the pathway studies that had emerged in the 1980s and helped to advance the more scientific definition of what a lead paint "hazard" meant, set priorities, targeted subsidies, and developed private-sector incentives.

The Alliance's focus on primary prevention also benefited from the Massachusetts lead law, which was the first state to enact such a law in 1971. That law required remediation in any home where a child under 6 resided, regardless of their blood lead level. In this way, it departed from the more widespread secondary prevention. The Massachusetts law was revised in 1987 to require the use of trained and licensed contractors, relocation of occupants during abatement, and daily and final cleanup in units undergoing abatement. The revision also provided financial assistance to property owners in the form of a $1000 state income tax credit and a grant/loan program, as well as mandating that physicians screen all young children and that health insurers cover the costs. Further amendments were made in 1993 to allow owners to use interim controls for up to 2 years, a larger state income tax credit ($1500/unit), with a new state fund to help fund remediation and clearance dust testing.[75]

Patrick MacRoy succeeded Don Ryan as the Alliance's Director in 2006. The Alliance to End Childhood Lead Poisoning changed its name to the Alliance for Healthy Homes in 2002 and merged with the National Center for Lead Safe Housing (now the National Center for Healthy Housing) in 2010. Other key staffers at the Alliance included Anne Wengrovitch, Pierre Erville, Ralph Scott, Jim Rochow, Laura Fudala, Jane Malone, Eileen Quinn, Maria Rapuano, Julia Burgess, Janet Phoenix MD, Betsy Marzahn, Rachel Herzig, David Batts, Sylvia White, Bathsheba Philpott, and others.

3.9 The nation's health secretary declares lead poisoning the number one childhood environmental disease over White House objections

The Alliance convened the first national conference on childhood lead poisoning prevention in 1991, attended by over 800 leaders.[76] The keynote speaker was the Health and Human Services Secretary, Dr. Louis Sullivan. Courageously ignoring last-minute objections from the White House Office of Management and Budget, he used the occasion to declare: "Lead poisoning is the No. 1 environmental threat to the health of children in the United States."[77,78] He also used the occasion to announce CDC's new strategic plan and other CDC action.

3.10 The Centers for Disease Control and Prevention issues public health strategic plan and new medical guidance

In 1991 CDC issued its 3rd revision on medical management of blood lead levels,[79] which was released just after CDC also issued a public health strategic plan on lead poisoning prevention the same year.[80]

The 1991 CDC statement defined a new standard for blood lead that triggered an environmental inspection and intervention (15 µg/dL), as well as a "level of concern" (10 µg/dL) that would trigger community-wide action, although what "community action" meant was never really defined. The lowering of the blood lead level that triggered action substantially increased the number of children requiring case management and other interventions. It also necessitated a change in measurement technology from Free Erythrocyte Protoporphyrin to blood lead.

Congressman Henry Waxman, who later emerged as a key leader in lead poisoning prevention legislation stated in 1991:

> But the urgent question raised by the new [CDC blood lead] standard is: What are we going to do to help the millions of lead-poisoned children? Unfortunately, the answer from the [George Herbert Walker Bush] Administration is: not much.[81]

Waxman and a bipartisan group from Congress were expressing the widespread frustration that little beyond the medical approach was being done to address lead paint.

The 1991 CDC Strategic Plan contained a timeline (20 years to abate all pre-1950 housing) and for the first time a formal cost-benefit analysis (HUD's 1990 plan had only estimated costs, not benefits). That CDC analysis estimated the monetary benefits of addressing all pre-1950 housing with lead-based paint over a 20-year period to be $62 billion, with a cost for the first 5 years of $974 million. The benefits far outweighed the costs.

Yet the description of what to do to address lead paint hazards remained woefully shallow. For the first time, a few professionals from the housing world were included in the CDC statement. Nevertheless, much of the focus remained on clinical management, not housing. Nineteen of the 22 members of the committee who drafted the CDC 1991 statement were either medical doctors or toxicologists and only one had any experience in lead paint abatement.

In the 1991 CDC Statement's section on abatement, there was finally recognition that not all lead would need to be removed from all houses immediately and that other comprehensive measures should be employed as well. There was also discussion of both short- and long-term methods of controlling hazards for the first time. It mentioned the high error rates of XRF lead paint analyzers, and what an "immediate" and "potential" lead-based paint hazard meant (deteriorated lead paint and the presence of lead paint, respectively).

Abatement now consisted of seven steps (not just the simplistic four in the earlier 1985 statement) with a much more detailed discussion of each. The new seven steps included the following:

1. Proper training of abatement workers;
2. Worker protection;
3. Containment of dust and debris;
4. Replacing, encapsulating, or removing lead paint;
5. Cleanup;
6. Proper disposal of lead-contaminated waste; and
7. Final inspection for reoccupancy.

Despite the limited discussion of remediation methods, the 1991 CDC Strategic Plan constituted a major advance. It attracted significant press attention. In a front-page article in the Atlanta Georgia newspaper, Kathryn Mahaffey, who had begun to measure blood lead levels in the NHANES survey years earlier, said, "Numerous studies leave no doubt that low levels of lead have profound effects on children."[82]

The Alliance to End Childhood Lead Poisoning said that the CDC 1991 Strategic Plan was "the first time HHS has called concerted attention to childhood lead poisoning ... in contrast to the halting and ineffective prevention efforts of the past.... The document makes clear that lead-based paint, long dismissed as a nuisance housing issue, is the primary cause." But the Alliance also noted the President's 1992 budget had serious shortfalls, and that HUD and OMB had deleted lead paint abatement in the President's budget at the last minute. The Alliance also noted the plan did not state "which Federal agencies are responsible."[83]

The President finally requested funding for HUD's new lead grant program for private housing in 1991, although the FY 1991 President's budget request was only $25 million, far less than what CDC had estimated to be needed in its cost-benefit analysis. Nevertheless, this made it easier for Congress to appropriate funds for a program that had not yet been authorized. Congress doubled this to $50 million, marking the first time that federal dollars had been dedicated exclusively for remediating lead-based paint in privately owned housing.[47]

The 1991 CDC statement also called for universal screening of all children under 6 using a blood lead test, instead of the less sensitive erythrocyte protoporphyrin (EP) test that had been in wide use before that. The lower CDC blood lead "level of concern" ($10 \mu g/dL$) could not be detected reliably using the EP test. This guidance for universal screening created backlash among some in the medical community, with many doctors refusing to conduct the test. This is yet another example of the importance of scientifically valid measurement and the limitations of a medical only approach.

When asked by Sen Jack Reed why many physicians refused to conduct the blood lead test for their young patients at a Congressional hearing, Dick Jackson from CDC stated that he could not give a good answer, except that in the early years of life, physicians give about 20 injections for vaccines and other purposes and some doctors thought yet another blood test was too much. He said that physicians felt they had little to offer children with high blood lead levels given the limited resources available to remediate lead hazards in housing. He stated that CDC believed children must be tested and highlighted a CDC effort known as High Intensity Targeted Screening in Chicago.[84] With support from CDC's Director Vernon Houk, Jackson emerged as a key leader in getting CDC on board with true primary prevention (not only the secondary prevention marked by blood lead screening), along with key CDC scientists, such as Tom Matte, Sue Binder, Henry Falk, Jerry Hershovitz, and others.

Under pressure from the medical community, in 1997 CDC retracted the 1991 universal screening recommendation, leaving it to states to determine which areas were "high risk."[85] This marked a clear retreat by the medical profession, although universal screening was still urged in those states where data to identify high-risk children did not exist. The Government Accountability Office and other oversight agencies repeatedly criticized CDC and the nation's doctors for screening only a small fraction of the children who should be tested.

For example, all Medicaid children were required to be screened because they were three times as likely to have elevated blood lead levels. In 1998 Medicaid policy was revised to require all Medicaid-eligible children to undergo blood lead testing at 12- and 24-months old as well as for older children who had not already been screened.[86] But most states did not take this seriously. The Government Accountability Office reported that only 21% of Medicaid children had a blood lead test despite the policy.[87] The Alliance and others continued to pursue strategies to improve Medicaid screening which resulted in a new agency directive.[88]

Failure to conduct blood lead testing in most at-risk children was a constant problem in this era, stretching back to the 1970s, when local programs annually screened only 30% of targeted high-risk children under 6 years old living in inner city areas.[89] Follow up for children with lead poisoning was even worse. Very few children with a positive screening test ever received another blood lead test, housing remediation or other indicated treatments.[90] In Chicago with its "high intensity screening" the results were better, yet even there CDC found that only 61% of high-risk children had been tested.[91]

CDC thought it possible to identify high-risk children using easily accessible data such as the census, instead of screening. Starting in 1997, CDC recommended that state health departments use these data to ensure that children had blood lead tests. In the absence of such data, the recommendation to test all children 1 and 2 years old remained in place as did the requirement to screen all Medicaid-eligible children.[92] By 2008 66% of high-risk Medicaid children were tested as more state Medicaid agencies integrated blood lead testing into their reportable health indicators.[93,94] However, this rate remained stubbornly low; between 2008 and 2018, only 65% of high-risk Medicaid children were tested.[95]

Blood lead tests were used for both surveillance and population wide surveys. Surveillance systems were often operated at the local or state levels to identify community-wide risks and reduce the burden of lead exposure using community-based interventions. Biomonitoring of blood lead levels also was conducted at CDC through the population based NHANES survey.

The lack of universal screening led to an emphasis on only small so-called "hot spots" of lead exposure, instead of testing all children. This continued the medical profession's focus on high-risk children instead of all children, the traditional "triage" approach. The only estimates of the magnitude of the problem came from the NHANES, which involved a relatively small but more representative sample of the population. The small size led some to speculate that the confidentiality of individual child participants could be compromised in 2020, leading to calls that the size of the NHANES should be increased.[96]

In 1991, EPA released its own lead strategy, which also focused on developing methods to identify "hot spots" of lead poisoning, recycling, and public education, with little attention to housing.[97] The fact that CDC, HUD, and EPA each released their own separate plans underscored the lack of coordination. Furthermore, other health agencies, such as the National Institute for Environmental Health Sciences, the Surgeon General's Office, and the National Science Foundation did not contribute significantly to lead paint poisoning prevention during this time period. Clearly, assignments had to be made, and Congress finally did that the following year, described in the next section.

But other health entities outside government *did* take notice. For example, another major conference in 1991 attended by thousands of occupational health scientists and practitioners was held at the American Industrial Hygiene Conference and Exposition. For the first time, it featured a major session on new lead paint initiatives being implemented at the federal, state, and local level to control lead poisoning from lead paint, and how the occupational health profession could respond to new challenges in the construction and newly emerging lead abatement industries to protect both workers and children. Speakers included Don Ryan, Ellen Silbergeld, and other occupational and environmental health specialists who had become involved in housing lead hazard control, including Aaron Sussel, John Rekus, Greg Siwinski, and others.[98] The fact that occupational health professionals were increasingly involved at the intersection of public health and housing would prove to be important in the coming years.

3.11 Congress acts: Title X of the 1992 Housing and Community Development Act

Although enacted under the auspices of a housing law, Title X of the 1992 Housing and Community Development Act (also known as the Lead Hazard Reduction Act) was really a combined housing, public health, occupational health, and environmental law that still stands as the principal Congressional action on how to address lead paint in housing.[99] It has not been significantly amended since.

This is the first book to examine Title X in detail. One earlier book (discussed further in Chapter 6) stated that Title X marked a retreat because it required both permanent abatement and interim controls.[100] Another book thought that Title X only provided "incentives" for abatement, and "did not require abatement or provide for any government oversight."[101] As we shall see, these earlier accounts were both incorrect.

Unlike the 1971 Lead-Based Paint Poisoning Prevention Act, the 1992 law required the assessment and elimination of lead paint hazards for certain types of housing and it provided significant funds and support systems, not just "incentives". As we have seen, the 1971 Act failed, largely because it focused on medical blood lead screening, not housing remediation. Title X did not repeat that mistake.

Title X's importance in shaping the nation's strategy to address the problem cannot be overstated. It was an example of how science informed public policy, and how science helped overcome industry opposition to create consensus. Fundamentally, it shifted the national approach from reacting to already-poisoned children to making housing safe in the first place. Instead of only focusing on the presence of lead paint, as earlier Congressional legislation had done, it applied the new exposure science to redefine what a lead paint hazard meant. And it applied both short- and long-term methods to control such hazards. It created not only a new policy but was focused on the implementation of that policy.

From 1991 to 1992, the Senate Banking Committee and other Congressional Members convened unique "roundtable" hearings of various experts and parents to respond to the new awareness of the need to do more on the lead paint issue and to respond to the seeming inability of federal agencies to collaborate adequately to address the problem in a coordinated way. This was driven in part by press accounts of a 28-month-old boy who died with a blood lead level of 144 µg/dL in 1990 in Wisconsin. The Washington Post quoted Joe Schirmer, an epidemiologist for the state's health department, "This is a dramatic example of the worst-case scenario of what happens if paint is ignored as a source of lead exposure."[102]

The Congressional hearings included sessions before the:

US House of Representatives Subcommittee on Housing and Community Development, Washington, DC, April 29, 1992;
US Senate Subcommittee on Housing and Urban Affairs, Washington, DC, March 19, 1992; and
US Senate Subcommittee on Housing and Urban Affairs, Washington, DC, October 17, 1991.

The report from the Senate Banking Committee, quoted a scientist: "We now possess a good working knowledge of what constitutes safe planning, testing, and abatement." What is missing, he said, is a framework of workable, enforceable regulations to enable the lead hazard control industry to develop.[103]

Although the bill enjoyed broad support from the public and environmental health professionals and scientists, its passage was an uphill battle, because powerful private housing interests were opposed and because there were so many moving parts. At one point, Congressional staffer Bruce Katz told Don Ryan to lock himself into a room with Myron Ebell (from the National Association of Home Builders), Eileen Lee (from the National Multi-Housing Council), and a staffer from the National Association of Realtors. Katz told Ryan, "Don't come out until you hammer things out." Katz, Ryan, and another Congressional staffer, Cheryl Fox, deserve credit for the behind-the-scenes work to get the bill passed.

The housing industry lobbyists had been lodging general objections to the Title X bill, but they did not have specific amendments to propose. Ryan proceeded to walk them through Title X's new definition of a lead-based paint hazard based on the new science, not only the presence of lead paint. The National Multi-Housing Council eventually came to see that this new definition might reduce their uncertainty over the ever-mounting lawsuits their members faced. The Realtors came to see how the clear but limited duty of disclosure might reduce their members' legal liability.[104]

But in a sign of trouble to come, the National Association of Home Builders stated, "We are concerned that new regulations will be excessive and ineffective." It came from Dick

Morris, a technical advisor for the Homebuilders, which in general opposed regulation. He added that new laws would provide no more protection than if they simply washed and dusted their woodwork on a routine basis. As long as lead paint was bound to the surface, "it is hard to get into people's blood," he said. This repeated what the lead industry had been saying in its attempt to shift the blame to bad maintenance and poor parenting for decades.[105] But Ryan and others succeeded in getting them to stop blocking Title X and to instead focus on the new HUD lead paint guidelines that Title X mandated.

Morris would later serve on the 1995 HUD Guidelines drafting and review committee, where he continued the Homebuilder's opposition (Chapter 4) and attempted to delay its completion. He would be the only guidelines committee member to refuse to sign onto the final document.

The industry opposition also failed because of the widespread bipartisan support that Title X enjoyed, something that became increasingly rare in later years. For example, Democratic Senator Alan Cranston of California introduced an early version of the law (S. 2341) on March 11, 1992, with the support of Republican Senator Alfonse D'Amato from New York. Another version included S. 3031 on July 23, 1992, from the Banking Committee that involved financing.

The bipartisan support was fueled by HUD's paralysis and Congressional anger. When Senator Cranston introduced the "Urgent Lead Paint Hazard Prevention Act," he stated:

> This legislation will put an end to continued delays and hand wringing.... It is a national disgrace that little is being done to combat the No. 1 environmental problem facing America's children.... Although Congress has pressed for action for years, this [George Herbert Walker Bush] administration and the last [Ronald Reagan's presidency] have sat paralyzed before the lead paint problem like a mouse before a cobra ... Congress has long pushed for action on this problem.... After years of delay and litigation, a frustrated Congress moved in the 1987 Housing Act, which I coauthored, to give HUD strict timetables.... After 4 years of demonstrations and studies, the administration has not asked for any significant funding for effective solutions.... The administration has provided little more than token Federal support for testing and abatement in private and other federally assisted housing.[106]

Testimony on the need for the bill focused on lack of standards and cost:

> First, we must have better national standards that are uniformly enforced, not guidelines or advisories; and second we must provide resources that enable the average American homeowner, landlord, and renter to absorb the cost of proper professional testing and abatement services.... A lawyer explained to a crowd of contractors at a lead paint training course that it was his considered legal opinion that business people should not get into this field now, because of the absence of regulations and accepted practices, and their resulting excessive liability exposure.... The regulatory framework is virtually non-existent today ... HUD has no overall program policy statement describing the steps to be taken in dealing with lead based paint problems: testing, risk assessment, in-place management and abatement .... The situations at EPA and OSHA are essentially the same.[107]

In the House, Representative Henry Gonzales introduced H.R. 5334 following hearings on the integration of short- and long-term lead paint hazard control methods on June 5, 1992.

Called "in-place management" in the public housing context, interim controls became part of federal legislation and subsequent regulation. Congressional testimony highlighted the public housing dilemma this way:

> It has become apparent that there are certain deficiencies in the approach contained in the [1990] HUD Lead-Based Paint Interim Guidelines [for public housing].... That approach revolves around testing for the

concentration of lead in paint … and subsequent abatement in the context of modernization. The strength of this approach is that it should result in the complete abatement of lead paint in all public housing after they are cycled through comprehensive modernization.… The deficiency involves what to do in the near term to prevent lead poisoning, since many housing developments are not slated for modernization in the near future.[108]

An important question involved the balance between the short- and long-term approaches. The Title X Congressional testimony stated:

> The bill could be improved greatly by including a more balanced approach to long-term response measures. We have argued for some time that building owners and managers need to walk on two legs to advance the fight against lead poisoning: the short-term response and the long-term response. *Both must proceed at once if we are to make significant progress. This bill appears to compromise the long-term response and focuses nearly all efforts on the short-term response.* In the long run, failure to incorporate long-term responses now will result in a significant unnecessary increase in funds expended on lead paint (emphasis in original).[109]

On the other hand, a few health professionals also attacked Title X because they thought it only required a short-term approach. For example, John Rosen, a pediatrician at Montefiore Medical Center who had treated many lead-poisoned children wrote to the Alliance:

> The Alliance's support of the Cranston bill is totally inappropriate. This proposed law will make it impossible for low-income families to be protected from leaded paint.… This is a rare [President] Bush veto to support.[110]

The final version of the law passed by the Senate and the House included both risk assessment and inspection, as well as interim control and abatement requirements. It made clear assignments to the federal agencies, detailed below. And perhaps most importantly it authorized new federal money for lead paint hazard remediation in privately owned housing, where the risks were greatest. Congress stated the purpose of the newly authorized lead paint grant program was "to jump start the private market's response to lead paint hazards."[111]

Despite this progress, Title X punted on certain critical issues, such as standards of care; appraisals; liability and the concept of "safe harbor" (reduced liability for owners who proactively addressed lead hazards); alternative systems for victim compensation; underwriting standards; and private financing. The law directed HUD to create a Congressionally chartered task force to tackle these issues, which was completed 3 years later (Chapter 4).

Part of a much larger omnibus housing bill, Title X passed the Senate on October 8, 1992, the House on October 10, 1992, and was signed into law by President George Herbert Walker Bush on October 28, 1992.[99]

3.11.1 Moving from reaction to prevention

Although there had been lip service to the concept of not only reacting to poisoned children but preventing their exposure to lead in their homes in the first place, Title X was the first time that the nation's policy formally became primary prevention (acting before children are exposed). It implemented many priorities of federal policymakers that had been coalescing over the previous decades, and it made into law the prevention approaches that in some ways had already begun in public housing, where inspection and abatement were required regardless of whether children had yet been poisoned.

Title X took a highly prescriptive approach by specifying the definition of a lead hazard in statutory language. The groundwork for Title X's emphasis on primary prevention had been laid by the Alliance to End Childhood Lead Poisoning and others. In addition to conducting the first national conference on childhood lead paint poisoning described earlier, the Alliance also released a "Blueprint for Primary Prevention,"[67] which described the many benefits of the primary prevention approach, and how priorities could be established using the new science of pathway exposures. The document included a graphic depicting the presence of lead paint, housing age, presence of lead dust, deteriorated paint, and the presence of children to underscore how vital primary prevention was (Fig. 3.2).

Title X was a textbook example of Congress defining and assigning responsibilities to different agencies for which they were most suited. It essentially mandated a "three-legged stool,"

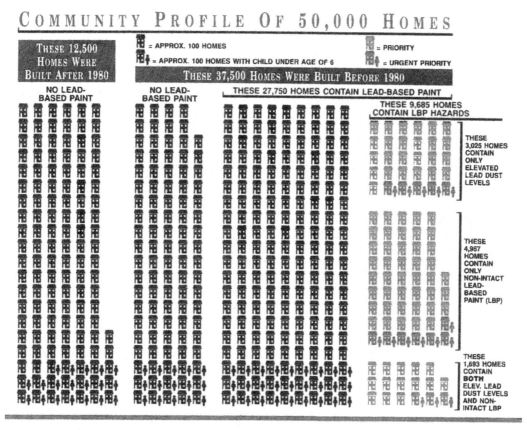

FIGURE 3.2 Lead paint priorities. Source. *Alliance to End Childhood Lead Poisoning, Childhood Lead Poisoning: Blueprint for Prevention (undated, ca 1991).*

where HUD would be responsible for inspecting and controlling housing lead paint hazards through its new authorized grants, federally assisted housing regulations and local government and capacity building. CDC would be responsible for blood lead surveillance and guidance to clinicians, national population surveys, and quality control for blood testing laboratories. EPA would be responsible for training and certifying workers, setting exposure standards, ensuring quality control for paint, dust, and soil sample laboratory and field analysis, and regulating renovation, repair, and painting activities. HUD and EPA were assigned joint responsibility for the disclosure of known lead paint at the time of sale or lease.

The law's initial findings underscored the importance of the issue, its scientific basis, and explication of both sources and pathways of exposure. Sec. 1002 (6) stated that both abatement and interim controls were effective:

> the danger posed by lead-based paint hazards can be reduced by abating lead-based paint or by taking interim measures to prevent paint deterioration and limit children's exposure to lead dust and chips.

This marked the first time that the federal government had required both immediate and long-term solutions.

3.11.2 Creating the workforce

Title X created a skilled trained workforce that was previously confined mostly to public housing and a few localities with lead laws, like Baltimore and Massachusetts. The new law said:

> the Federal Government must take a leadership role in building the infrastructure – including an informed public, State and local delivery systems, certified inspectors, contractors, laboratories, trained workers, and available financing and insurance – necessary to ensure that the national goal of eliminating lead-based paint hazards in housing can be achieved as expeditiously as possible.

The new law ended confusion over differing state and local laws and inconsistent use of terminology to provide a consistent national framework. Before Title X, states and local governments had been using different terms with different meanings to define lead hazards, lead paint, and various hazard control methods. Title X provided a consistent set of definitions and a common national vocabulary that standardized the evaluation and control of lead hazards across the country and supported consistent state contractor certification standards, including reciprocity among states.

Nevertheless, differing state laws and political realities sometimes made implementation difficult. For example, in Michigan, the governor at the time (John Engler) initially refused to allow state regulation of such workers, reflecting a knee-jerk opposition to all regulations that swept the nation in the mid to late 1990s. This meant that the state could not be eligible for the new HUD lead paint abatement grants, because Congress had said only states that ensured worker training were eligible. The result was a public fight between HUD and the Governor, who finally relented only after citizen groups demanded to know why the state would not accept federal funds to protect children (Chapter 4).

3.11.3 Seven principal purposes

The seven purposes articulated by Title X were as follows:

1. to develop a national strategy to build the infrastructure necessary to eliminate lead-based paint hazards in all housing as expeditiously as possible;
2. to reorient the national approach to the presence of lead-based paint in housing to implement, on a priority basis, a broad program to evaluate and reduce lead-based paint hazards in the Nation's housing stock;
3. to encourage effective action to prevent childhood lead poisoning by establishing a workable framework for lead-based paint hazard evaluation and reduction and by ending the current confusion over reasonable standards of care;
4. to ensure that the existence of lead-based paint hazards is taken into account in the development of Government housing policies and in the sale, rental, and renovation of homes and apartments;
5. to mobilize national resources expeditiously, through a partnership among all levels of government and the private sector, to develop the most promising, cost-effective methods for evaluating and reducing lead-based paint hazards;
6. to reduce the threat of childhood lead poisoning in housing owned, assisted, or transferred by the Federal Government; and
7. to educate the public concerning the hazards and sources of lead-based paint poisoning and steps to reduce and eliminate such hazards.

3.11.4 Using science to define "lead paint hazard"

Title X used the new pathway science to redefine what a "lead paint hazard" meant. No longer was it restricted to the content of lead in paint. Instead, Title X redefined a lead paint hazard to mean:

> The term "lead-based paint hazard" means any condition that causes exposure to lead from lead-contaminated dust, lead-contaminated soil, or lead-contaminated paint that is deteriorated or present in accessible surfaces, friction surfaces, or impact surfaces that would result in adverse human health effects....[112]

This marked a clear change from the *Ashton v Pierce* Court decision, the 1987 Housing Act, and the 1990 HUD Interim Public Housing Guidelines. In all these earlier cases a lead hazard was simply the presence of lead paint.

In 1971 Congress had defined lead paint to be "any paint containing more than 1 per centum lead by weight (calculated as lead metal) in the total non-volatile content of liquid paints or in the dried film of paint already applied."[113] This was reduced to 0.5% in 1973[114] and in 1978 it was reduced further to 0.06%,[115] applying to new paint only (in 2008 this new lead paint limit was reduced to 0.009%). As shown in the previous chapter, the 0.06% level was based on a report from the National Academy of Sciences that showed 0.06% was feasible and would not be exceeded due to background contamination.[116] Later in 1987, Congress redefined lead paint as greater than or equal to 1 mg/cm^2, but no mention of the weight percent standard.[117]

Title X provided a clear numerical definition of *existing* lead paint already applied to housing surfaces for the first time, distinguishing it from new paint and ending the confusion caused by the differing definitions of lead paint over the years, as well as providing better units of measurement.

Title X established a standard of $1 \, \text{mg/cm}^2$ *or* 0.5% for existing lead paint, overcoming a thorny technical problem. As was the case with dust, both loading (mg/cm^2) and concentration (weight percent) measurements were used to overcome a "dilemma of dilution." Why did Title X include both?

Consider this example: Adding a layer of paint with no lead onto a layer of lead paint would artificially reduce the lead concentration in the combined paint film by half. If the original layer of lead paint had 0.5%, then adding a new layer of non-lead paint would become 0.25% ppm in the total paint film, all things being equal (thickness, density, weight, etc.). Because older lead paint layers were often coated with many newer layers of non-lead paint, its presence could become so diluted that it would eventually "disappear," which of course made no sense because the lead paint layers were still there. The "solution to pollution" was *not* dilution, but using concentration alone obviously could not work.

On the other hand, loading (mg/cm^2) meant that the amount of lead within a square centimeter could not change, no matter how many non-lead paint layers had been added. This created a more stable measurement and increased confidence.

This problem did not exist for new paint, because there were no other existing paint layers, which explains why new paint definitions were expressed in weight percent.

Why did Congress prefer loading but also permit concentration in the definition of lead in existing paint? As a practical matter, loading required that the surface area be measured accurately, but in some cases, that could not be done reliably. For example, curved or ornate surfaces made accurate surface area measurements impossible. In short, this statutory definition provided a workable path, combining science, consistency, and practicality.

3.11.5 Show me the money: new Congressional appropriations for private housing

Although Congress had previously appropriated funds for identifying and remediating lead hazards in privately owned low-income housing a couple of years before Title X, it had never been formally authorized, making it vulnerable to future attacks. For the first time, Title X formally authorized competitive HUD grants to local governments to address lead-based paint hazards in low-income, highest risk privately owned housing, not only public or federally assisted housing. This grant program remains in place in 2022 (see Conclusion).

3.11.6 Reforming housing regulations

Virtually all of HUD's housing regulations were developed in the 1970s and 1980s and were out of date and inconsistent with the new pathway science by 1992. Some housing programs addressed only interior lead paint; others focused only on exterior paint; still others focused on so-called "chewable" surfaces; and some addressed "accessible" surfaces. None of these were based on the new science of sources and exposure pathways. More importantly, none explicitly prohibited dangerous methods of lead paint removal, such as power sanding, dry scraping, and torching. And none addressed lead dust and bare soil.

Title X directed HUD to reform virtually all its housing regulations pertaining to lead paint, which was completed in 1999 following significant opposition (Chapter 5). Although HUD had issued its Interim Public Housing Guidelines in 1990, the law ordered HUD to develop new permanent ones that covered more housing categories, which it did in 1995. It also directed EPA to establish a nationwide consistent system for certifying lead inspectors, risk assessors, and abatement contractors, recognizing that housing code-making bodies had failed to fill the void. EPA would later delegate certification responsibility to states, which mostly took on that role. For those that did not, EPA was ordered to certify such workers directly.

The reason that Title X mandated certification (licensing) was because of the proliferation of poorly done (in some cases fraudulent) lead-based paint inspections and abatement in public housing and other housing, some of which would be used as examples of how *not* to do lead inspections in the Georgia Tech classes. For example, one such inspection occurred in the Panama City housing authority at Massalina Memorial Homes in 1991. The inspector there submitted a report showing that not only was the number of XRF readings insufficient, but some readings were not actually taken.[118] The inspection reports from National Lead Detection Services were also used to demonstrate seriously deficient and fraudulent testing procedures.[119]

As part of the regulatory reform, Title X also directed EPA to set national standards for dangerous levels of lead in dust and bare soil (Chapter 4). Title X did this by amending the Toxic Substances Control Act.

3.11.7 The right to know

Although the realtors had backed down in their initial opposition to any form of disclosure due to their perception that disclosure could reduce their members' liability, they also won a compromise. The final version of the law only required "known" lead paint and/or lead paint hazards to be disclosed. If there had been no inspection, it was not "known" if lead was present or not, so there was nothing to disclose—and ignorance was rewarded. Most homes remained uninspected, with one important exception: the homes of lead-poisoned children. HUD sent a letter to all the nation's health departments, asking them to include standard language in the letters they sent, stating that the inspection results obtained by the health department would now need to be disclosed to future tenants or buyers, which proved to be important in later enforcement (Chapter 5).[120]

Congressman Henry Waxman first proposed lead paint disclosure in 1991 in one of the bills that eventually became part of the final Title X law. His bill would have made lead inspections mandatory before the sale or rental of pre-1978 housing, along with disclosure of the results. He later addressed the National Association of Realtors at their annual convention but knew that they were opposed to his bill,[121] which would have required a lead paint inspection, not just make it an option to be negotiated between buyer and seller. Before he spoke to the Realtors, he privately said "they will crucify me."[122]

In Congressional testimony, the Realtors opposed disclosure, stating it would have:

> a devastating effect on housing prices and home sales across the nation.... Why should the current homeowner bear the brunt of alleviating a condition he did not cause?... The most compelling single argument against mandatory lead-based paint testing is that lead poisoning is a public health issue, *not a housing issue*[123] (emphasis added).

The Realtors were joined by the National Association of Home Builders in calling for more testing of children, not homes. The Home Builders repeated the lead industry's claim that paint was not the problem. The New York Times put it this way:

> The National Association of Realtors and the National Association of Home Builders strongly advocated as much testing of children as possible to determine sources of lead hazards [and] questioned the need for widespread abatement ... [stating that] sources of lead included gasoline, soil, water, dental fillings and ceramics and crystal ... Karen Florini, senior attorney for the Environmental Defense Fund [said] "The bulk of the problem is from paint" ... Waxman criticized CDC for initially supporting disclosure, then opposing it, [saying] "all families need to know where this lead paint is located so they can take proper precautions."[124]

Another speech to the Realtors warned them of their liability:

> About six months ago, I remember discussing this issue with a realtor seated next to me on a flight. Her response to me was, "Oh please, don't give us something else to worry about, we already have our hands full with termites, pesticides, and radon." Because the problem is so widespread, lead paint liability should be a source of concern for realtors. In a recent settlement, rumored to be around $2 million, the court ruled that the property manager, who was also a realtor, could not be severed from the case. Acting prudently is therefore of paramount importance.[125]

But prudence did not prevail. The Realtors were indeed powerful, and they mostly won the argument in the 1990s. Senator Chafee proposed an alternative bill that did not require an inspection, which ultimately became part of Title X. The final version in Title X that passed Congress did not require an actual inspection or mitigation, hence nothing to disclose or act upon in most cases. But Title X did mark an important advance: before then, landlords and owners could legally conceal what they knew about lead paint in a property to be sold or rented. That now became a violation, with stiff penalties.

The disclosure law was meant to harness the housing market system, which often financed repairs at the time of sale. If both buyers and sellers knew where the hazards were, lead paint could become like any other housing deficiency. But because the final version of the law did not require a lead inspection, most houses remained untested. As a result, the housing finance system did not stimulate widespread privately financed lead paint remediation during this time. With millions of children occupying 37 million homes with lead paint, by 2022 some were calling for reform of the disclosure law so that it required a lead inspection, as Waxman had originally conceived (see Conclusion).

Although Title X only required disclosure of known hazards, it was creatively enforced in the late 1990s, making significant progress in the worst of the nation's housing. Millions of units that had not complied with disclosure were ultimately remediated using private funds from those who violated the law. In fact, one of the first enforcement cases involved a young child who died from lead paint exposure, and another resulted in a landlord being sentenced to 2 years in prison (Chapter 5).

3.11.8 Renovation, repair, and painting

The last major executive agency action to implement Title X occurred in 2010, 18 years after its enactment, when EPA finally issued and began enforcing regulations governing lead safety for the renovation, repair, and painting workforce.[126]

It was bitterly opposed by the National Association of Home Builders, explaining much of its delay. In 2004, then-Senators Hillary Clinton and Barack Obama questioned why even a proposed rule had not been issued. During an EPA Administrator's confirmation hearing, more than a decade after Title X had been passed (Title X had required that EPA promulgate a final regulation by 1996), they threatened to block his confirmation until such a rule had been proposed.[127] The threat worked. A proposed rule was released a short time later. But initially EPA had only proposed a voluntary program, a direct contradiction to the Congressional order for a regulation.

Despite (or perhaps because of) its breadth, the renovation, repair, and painting rule (RRP) was poorly enforced and flawed. It did not provide for clearance dust testing, unlike the requirement in federally assisted housing. This created a double standard: For children living in federally assisted housing, clearance testing was required after most repair activities that disturbed paint, but in all other housing it was not. Instead, EPA only required a "white glove test" comparing discoloration to a "cleaning verification card."

At a major lead poisoning prevention conference, EPA staffers were asked why the agency had chosen to ignore the science, use a test that had never been scientifically validated, that no one knew how to use, would depend on a subjective visual assessment of discoloration, and had never been correlated with children's blood lead. They had no answer.[128]

This departure from science was one of the few examples in Title X where the rule was widely ignored, largely due to the role of the National Association of Home Builders, which sued EPA repeatedly.[129] It also successfully lobbied behind the scenes to prevent EPA from adopting a true clearance test, the one that had been in place for more than a decade at HUD for federally assisted housing and had been fully validated. At least a few contractors opted to do a real clearance test, fearing liability from using the EPA RRP test that had never been validated.

Nevertheless, more than 100,000 remodeling contractors, painters, and other trades received training in lead-safe work practices. The National Center for Healthy Housing organized training for over 25,000 workers[130] as did many others. This rule covered perhaps more homes than any other, except the disclosure law.

The RRP rule continued to be the source of controversy and was amended at least seven times. Some have regarded it to be a paperwork regulation with little benefit. Others have found the disparity between RRP and the lead paint abatement workforce remarkable: On the one hand, abatement workers must be licensed and attend multiday training, with extensive oversight. On the other hand, the much larger renovation and painting workforce remained very poorly trained with little if any oversight. In recent years, a more uniform training system was advocated by some (see Conclusion).

In the waning days of the Trump Administration, EPA finally began serious enforcement of the RRP rule. It administered the biggest fine in the history of its Toxic Substances Control Act, where the RRP rule was located. The fine against Home Depot, a major hardware store chain that provides contractors to homeowners and property owners, totaled over $20 million.[131]

3.12 Bringing science to bear: a new National Center for Lead-Safe Housing brings health and housing together

Created on September 4, 1992, just before Title X became law, the National Center for Lead-Safe Housing (now the National Center for Healthy Housing) emerged as the authoritative technical, training, policy, and implementation organization for lead paint poisoning prevention and other housing-related diseases in this era. It produced evidence that the policies were not only scientifically valid but worked in practice. However, it almost never came to be.

Because HUD had so little scientific expertise, a more technically focused "think tank" was needed to help bring Title X's various mandates to reality. Numerous attempts to fund such an organization all initially met with failure.[132]

But the Alliance, the Enterprise Foundation and others persuaded an outgoing President of Fannie Mae (a "government-sponsored enterprise" that insures mortgages), David Maxwell, to make a $5.2 million donation to the Fannie Mae Foundation to fund the new Center. He made the donation after being criticized for taking a "golden parachute" retirement package.[132] The Alliance teamed with the Enterprise Foundation (now Enterprise Community Partners) to serve as founding parent organizations of the new Center, using the Fannie Mae donation to get started.[133] The proposal stated that its mission was to "develop, validate and promote the nationwide adoption of cost-effective strategies for sharply reducing childhood lead poisoning and preserve the nation's stock of affordable housing."[133] Health and housing affordability were its core principles.

The Center became the first national joint organization combining affordable housing and health scientists and professionals. Walter G. "Nick" Farr, a vice president with Enterprise, was asked to lead the new nonprofit. He and others had helped to frame Title X. He was joined by Jacobs as the deputy director at the fledgling organization, who left his post at Georgia Tech after running the Southern Lead Paint Training Consortium.

Farr was a former Yale Law School Professor, a former high-ranking HUD official during the Model Cities program of the 1970s, a former State department employee, and a former mortgage banker. His gregarious nature proved to be infectious. His decision to appoint a scientist as his deputy proved to be important in designing some of the most important lead studies in the early 1990s, with the help of a board of scientific advisors and the CDC.

Jacobs was also brought on board to assuage concerns by some health scientists that housing professionals like Farr would give in far too readily on protectiveness and would also not have the technical expertise needed. An internal memo stated Jacobs "...is extremely hard-nosed when it comes to full abatement vs in-place management ... [he] almost single handedly cleared up the confusion over lab analysis of dust wipe samples ... he is widely known and respected and seems to have no ardent enemies or detractors."[134]

Just before joining the Center, Jacobs was asked by New York City how best to address childhood lead poisoning based on risk, not only blood lead levels. He advanced three principles that would guide much of the Center's work in the decades to follow:

1. Resources should be allocated to measures that eliminate excessive exposure to both immediately available and potentially available lead-based paint, and the lead-contaminated dust and soil that come from it, while preserving affordable housing.
2. Resources should be allocated so that the greatest number of children receive the greatest possible benefit in the shortest period of time, while at the same time continuing to move toward the goal of permanent abatement of all lead and other housing-related health hazards. Movement along the short-term track and the long-term track should proceed simultaneously.
3. Primary prevention aimed at controlling the major sources and pathways of lead in a child's environment is far superior to identifying lead hazards principally through blood lead screening programs, which respond only to children who have already been affected.[135]

The Center played a key role in designing and helping to carry out some of the most important studies in this period. For example, the "Lead in Dust study," led by Bruce Lanphear and his mentor Michael Weitzman at the University of Rochester, laid the foundation to standardize wipe sampling to measure lead dust, correlating it with blood lead levels in hundreds of young children, carefully controlling for confounding variables.

The new Center also drafted the 1995 HUD Guidelines, which would go on to replace the earlier 1990 interim HUD public housing guidelines and ultimately become the basis for most state, federal and local regulations. The evidence base it assembled consisted of both science and practical experience from across the country. The Center also drafted EPA's first lead paint risk assessment training curriculum to accompany those guidelines, reflecting the previous Georgia Tech efforts. The Center's staff also played leading roles in the 1994 Title X Lead Paint Financing Task Force. It helped to implement some of the first state-based certification programs for inspectors, risk assessors, and abatement contractors in nine states, paving the way for other states to do so in later years. It produced key publications on case management of lead-poisoned children and conducted massive training across the country to implement new reforms of federal lead paint regulations in 1999, all described later in this book.

Sue Binder, who directed the CDC lead program in the early 1990s, realized the importance of housing studies to evaluate the different types of lead mitigation then in use. Binder and another CDC high-level manager, Henry Falk, authorized CDC scientist Thomas Matte to be stationed for a couple of days a week at the new National Center for Lead-Safe Housing. They also authorized another CDC employee, Stan Galik, to work at HUD to help develop the new Title X authorized private housing lead abatement grant program. The board of the new Center included both health and housing experts. David Rall, who founded the National Institute for Environmental Health Sciences, was joined by Ellen Silbergeld, a toxicologist. The housing experts on the board included Cushing Dolbeare, who founded the National Low Income Housing Coalition, Paul Brophy from the Enterprise Foundation, Larry Dale (from the Fannie Mae Foundation), and Don Ryan. Other key advisors to the new Center at its founding included Mark Farfel from Johns Hopkins University, Joel Schwartz from Harvard, John Graef from Boston Children's

Hospital, Scott Clark from the University of Cincinnati, and Miles Mahoney from the public housing world.[136]

Other staff included then-intern Jonathan Wilson who would later become the Center's research director. He later played a leading role in completing the Center's evaluation of the HUD lead hazard control grant program with the University of Cincinnati, the largest study of housing lead paint remediation ever completed (Chapter 4). The Center's evaluation of HUD's lead hazard control grant program, first chartered by Congress as part of Title X, resulted in periodic reports to Congress on initial findings. It emerged as the definitive lead paint remediation study for decades to come, enabling the HUD program to expand by creating confidence that it was indeed working.

A rare statistician gifted in translating complicated findings into English, Sherry Dixon joined the team. When questioned by Farr about statistics, she dutifully provided him with a textbook. He returned the following day joking, "I read this last night, but I still don't understand it."

The Center also brought on board people from the construction world, such as Bob Santucci and Armand Magnelli, who helped to standardize and create correct lead hazard control work specifications. Those specifications were used by HUD lead paint grantees across the country. Patricia McLean, a nurse by training, and Warren Galke, an epidemiologist, filled out the research team. Jill Breysse joined and led the Center's expansion into healthy housing targeted to seniors. Ron Jones, Heidi Most, Laura Fudala, Carol Kawecki, Jack Anderson, Michelle Harvey, Sarah Goodwin, Laura Titus, Tom Neltner, Susan Aceti, Chris Bloom, Judith Akoto, Phillip Dodge, Amy Murphy, Jo Miller, Tara Jordan-Radosevitch, Ruth Lindberg, Jane Malone, Peggy Hegarty-Stack, Ethel Bledsoe, Evelyne Bloomer, and many others also made significant contributions to the Center over the years.

The Center's deputy director (Jacobs) left in 1995 to direct HUD's Office of Lead Hazard Control before returning in 2006. Rebecca Morley, who had previously worked at HUD, became the Center's director in 2002 and renamed it the National Center for Healthy Housing. She and others founded the National Safe and Healthy Housing Coalition, which in 2022 had over 600 organizations (see Conclusion). Amanda Reddy became the Center's Director in 2017, after leading New York state asthma and healthy housing initiatives. She championed healthcare financing of healthy homes services, technical assistance to support the launch and growth of sustainable healthy homes programs, and the development of indicators for the HUD Healthy Communities Index.

In later years, the Center produced the first comparison of the nation's main green building standards to assess their impact on health. Following the aftermath of Hurricane Katrina, it launched a healthy rebuilding demonstration project and published a manual on how such rebuilding could be done safely. Together with the National Environmental Health Association, it launched the nation's first "Healthy Homes Specialist Credential," and its healthy homes training network spanned dozens of universities and other training providers. It trained tens of thousands of workers in lead-safe work practices after the EPA's RRP regulation took effect.

The Center led the efforts to establish the Find It Fix It Fund It campaign in 2015 with many others to bring lead hazard control activities into mainstream housing transactions. The Center also published the first State of the Nation's Healthy Housing report, which ranked the major metropolitan areas in how healthy their housing stock was. It produced the first national healthy housing model code, with the American Public Health Association. Together with Pew Charitable Trusts and Robert Wood Johnson Foundation, it released a major cost-benefit analysis of lead policies. It also broadened its work in later years to include healthy housing for the elderly, childcare centers, asthma, allergen sampling methods, lead dust contamination of porches, healthcare financing for healthy homes, ventilation, and weatherization and conducted studies on the effectiveness of window replacement.

A timeline covering the Center's history from its beginning in 1992 is available.[137] In 2014, the Center was recognized as the Collaborating Center for Healthy Housing Research and Training in the United States by the World Health Organization and the Pan American Health Organization. It played a key role in the publication of the first international housing and health guidelines,[138] as well as a major WHO report titled "Environmental Burden of Disease Related to Inadequate Housing" (Chapter 8).[139] Like the Alliance, the Center had extremely dedicated talented multidisciplinary staffs that together with many others led the nation's implementation of Title X. Both the Alliance and the Center were unique in bringing together the two worlds of housing and health in this era.

3.13 Confidence emerges

The passage of Title X, the various reports to Congress, the improved collaboration among CDC, HUD, and EPA, the creation of both advocacy and scientific nonprofit organizations with well-respected leaders in housing, health, environment, and policy, together with the development of a lead hazard inspection and abatement industry association, a well-trained workforce and increased public awareness and political will all resulted in greater confidence that the nation was finally coming to grips with the lead paint problem.

But implementing the new law resulted in new political and technical barriers.

References

1. Jacobs D.E. Address before the American Industrial Hygiene Conference and Exposition Roundtable on Lead Paint. June 3, 1991 (the University of Illinois Chicago Special Collections & University Archives, School of Public Health, "David E. Jacobs papers" the University of Illinois Chicago School of Public Health library).
2. Rosner DK, Markowitz G. A 'gift of God'? The public health controversy over leaded gasoline during the 1920s. *American journal of public health*. 1985;75(4):344−352.
3. Letter for William Duvall, President of the International Brotherhood of Painters and Allied Trades to William Raub, Acting director, National Institutes of Health, April 9, 1991 (the University of Illinois Chicago Special Collections & University Archives, School of Public Health, "David E. Jacobs papers" the University of Illinois Chicago School of Public Health library).
4. Jacobs D.E. A Call to Form the American Association for Lead Paint Abatement, 1989. The University of Illinois Chicago Special Collections & University Archives, School of Public Health, "David E. Jacobs papers" the University of Illinois Chicago School of Public Health library.
5. Personal communication, Lorne Garrettson with David Jacobs. 1986.
6. Elemental Green Website. How to find reclaimed home building materials. 2020 https://elemental.green/how-to-find-reclaimed-home-building-materials/.
7. EPA. Integrated Exposure Uptake Biokinetic Model. US Environmental Protection Agency. Lead at Superfund Sites: Software and Users' Manuals. https://www.epa.gov/superfund/lead-superfund-sites-software-and-users-manuals 2020.

8. HUD Guidelines for the Evaluation and Control of Lead-Based Paint Hazards in Housing, Department of Housing and Urban Development, Washington DC, HUD-1539. 1995

9. Opening address to Georgia Tech lead inspection courses, David Jacobs. 1991. Georgia Institute of Technology. The University of Illinois Chicago Special Collections & University Archives, School of Public Health, "David E. Jacobs papers" the University of Illinois Chicago School of Public Health library.

10. Estimates of the Nature and Extent of Lead Paint Poisoning in the United States. US Department of Commerce, National Bureau of Standards, NBS Technical Note 746, December 1972, the University of Illinois Chicago Special Collections & University Archives, School of Public Health, "David E. Jacobs papers" the University of Illinois Chicago School of Public Health library.

11. Section 118(f) of the Superfund Amendments and Reauthorization Act (SARA). Public Law. 99–499. US Congress, Washington DC. 1986.

12. The Nature and Extent of Childhood Lead Poisoning: A Report to Congress. Agency for Toxic Substances and Disease Registry. US Public Health Service, July 1988. https://stacks.cdc.gov/view/cdc/13238.

13. Authors protest report on lead poisoning: Researchers resign, call changes misleading. By Michael Weisskopf. Washington Post, June 13, 1987, p. A4.

14. Housing and Community Development Act of 1987. Public Law 100-242, U.S Congress, Washington DC. Feb 5, 1988. https://www.congress.gov/bill/100th-congress/house-bill/1698?s = 1&r = 51.

15. Stewart B. McKinney Homeless Assistance Amendments Act of 1988. Public Law 100-628. US Congress. Nov. 7, 1988. https://www.govtrack.us/congress/bills/100/hr4352/text.

16. Testimony of Miles Mahoney, Executive Director, Brookline Housing Authority on behalf of the Council of Large Public Housing Authorities before the Committee on Banking, Finance and Urban Affairs, Subcommittee of Housing and Community Development, US House of Representatives, US Congress. Sept 22, 1988.

17. Testimony of James Keck, Department of Housing and Community Development, Baltimore MD before the US House of Representatives, Committee on Banking Finance and Urban Affairs, Subcommittee on Housing and Community Development, US Congress. Sept 20, 1988. Also see Testimony from James McCabe at the same hearing.

18. Jacobs DE, Clickner RP, Zhou JY, Viet SM, Marker DA, Rogers JW, Zeldin DC, Broene P, Friedman W. The prevalence of lead-based paint hazards in U.S. housing. *Environ Health Perspect*. 2002;110(10):A599–A606. Available from: https://doi.org/10.1289/ehp.021100599. Available from: 12361941.

19. Statement of David L Echols, Executive Director Housing Authority of New Haven on behalf of the National Association of Housing and Redevelopment Officials, testimony before the Subcommmtitee on Housing and Community Development, Committee on Banking, Finance and Urban Affairs, US House of Representatives, Sept 22, 1988.

20. Section 1088(b)(3). Stewart B. McKinney Homeless Assistance Amendments Act of 1988. US Congress. Public Law 100-628—November. 7, 1988. https://www.govtrack.us/congress/bills/100/hr4352/text

21. Housing Authority Risk Retention Group Inc. https://www.bloomberg.com/profile/company/1269283D:US

22. Letter from John Salisbury, (Housing Authority Insurance) to Arthur Newberg, Director, HUD Office of Lead Abatement, Jan 23, 1992, the University of Illinois Chicago Special Collections & University Archives, School of Public Health, "David E. Jacobs papers" the University of Illinois Chicago School of Public Health library.

23. HUD Notice PIH 92-44 (PHA), Lead based paint risk assessment protocol, Sept 30, 1992. The University of Illinois Chicago Special Collections & University Archives, School of Public Health, "David E. Jacobs papers" the University of Illinois Chicago School of Public Health library.

24. Farfel M, Chisolm JJ. Health and Environmental Outcomes of Traditional and Modified Practices for Abatement of Residential Lead-Based Paint. *Am J Public Health*. 1990;80:1240–1245.

25. L. Vostal, E. Taves, J.W. Sayre, and E. Charney Lead Analysis of House Dust: A Method for the Detection of Another Source of Lead Exposure in Inner City Children. Environ Health Perspect. May 1974 p. 91–97.

26. ASTM E1792-96a. Standard Specification for Wipe Sampling Materials for Lead in Surface Dust. Available from: https://www.astm.org/DATABASE.CART/HISTORICAL/E1792-96A.htm.

27. GAO, Lead Based Paint Hazard Abatement Standards are Needed to Ensure Availability of Insurance. Government Accountability Office, Washington DC. July 1994. GAO/RCED-94-231

28. Farfel M. Evaluation of health and environmental effects of two methods for residential lead paint removal. Thesis. Johns Hopkins University, Baltimore Maryland 1987. https://catalyst.library.jhu.edu/catalog/bib_91837.

29. Commonwealth of Maryland Assembly. Law: Annotated Code of Maryland, Department of the Environment, COMAR · 26.02.07: Procedures for Abating Lead Containing Substances from Buildings, Effective Date, August 8, 1988. Also see: The Commonwealth of Massachusett. Law: 454 CMR 22.00: Deleading Regulations. Effective Date, November 25, 1988; and 105 CMR 460.000: Lead Poisoning Prevention and Control, Effective Date, September 26, 1988.

30. Ahrens KA, Haley BA, Rossen LM, Lloyd PC, Aoki Y. Housing Assistance and Blood Lead Levels: Children in the United States, 2005-2012. *Am J Public Health.* 2016;106(11):2049–2056.

31. Chiofalo JM, Golub M, Crump C, Calman N. Pediatric Blood Lead Levels Within New York City Public Versus Private Housing, 2003-2017. *Am J Public Health.* 2019;109(6):906–911. Available from: https://doi.org/10.2105/AJPH.2019.305021.

32. US Department of Housing and Urban Development. American Healthy Housing Survey II. Office of Lead Hazard Control and Healthy Homes. Washington DC. 2021. https://www.hud.gov/program_offices/healthy_homes/ahhs_ii.

33. Agreement by and between the U.S. Department of Housing and Urban Development, the New York City Housing Authority, and New York City. Section I(5). January 31, 2019.

34. United States District Court Southern District Of New York United States of America, Plaintiff, Against New York City Housing Authority, Defendant. Case 18cv5213 Opinion & Order. William H. Pauley III, Senior United States District Judge. November 14, 2018.

35. Neal Freuden, personal communication with David Jacobs, May 6, 2020.

36. Rabito FA, Shorter C, White LE. Lead levels among children who live in public housing. *Epidemiology.* 2003;14 (3):263–268, PMID. Available from: 12859025.

37. J. Willis. New Orleans Litigation History, Carter and Willis, handout for Georgia Tech training course, July 2, 1991 (the University of Illinois Chicago Special Collections & University Archives, School of Public Health, "David E. Jacobs papers" the University of Illinois Chicago School of Public Health library).

38. Statement of Judy A. England-Joseph, Director, Housing and Community Development Issues, Resources, Community, and Economic Development Division. G A O, Testimony Before the Subcommittee on Housing Opportunity and Community Development, Committee on Banking, Housing, and Urban Affairs, House of Representatives, July 8, 1996. HUD's Takeover of the Housing Authority of New Orleans. GAO/T-RCED-96-212.

39. Jacobs D.E. Status of HUD's Lead-Based Paint Program: A Critical Evaluation. Prepared for the Housing Authority Risk Retention Group. 1990. The University of Illinois Chicago Special Collections & University Archives, School of Public Health, "David E. Jacobs papers" the University of Illinois Chicago School of Public Health library.

40. H.R.5158 - Departments of Veterans Affairs and Housing and Urban Development, and Independent Agencies Appropriations Act, 1991. Senate Report 101-474. US Congress. November 5, 1990 https://www.congress.gov/bill/101st-congress/house-bill/5158.

41. The HUD-Independent Agencies Appropriations Act for 1989 (P.L. 100-404, H.R. 4800, House Report 100-701, Senate Report 100-401, Conference H. Rept. 100-817). US Congress. August 19, 1988.

42. Department of Housing and Urban Development Appropriations Bill, 1988. House of Representatives Report 100-701. June 14, 1988. Also Senate Appropriations report 100-401, June 24, 1988.

43. Jacobs D.E. Comments on Lead-Based Paint Matters in the Draft Report 102-000, to Accompany H.R. 2519, undated. Comments send to Congressional Staff. The University of Illinois Chicago Special Collections & University Archives, School of Public Health, "David E. Jacobs papers" the University of Illinois Chicago School of Public Health library.

44. VA-HUD-Independent Agencies Appropriations Act for FY1992 (Public Law 102-139, H.R. 2915, House Report 102-94, Senate Report 102-107, Conference H. Rept. 102-226). Signed By President George H. W. Bush, October 28, 1991.

45. Departments of Veterans Affairs and Housing and Urban Development and independent agencies appropriation bill 1992, Ms. Mikulski from the Committee on Appropriations, submitted the following report to accompany 2519, Report 102-000, draft, p. 50. This draft report also provided the HUD Lead Paint Office with 45 staff, 20 for headquarters and 25 for HUD field offices.

46. Memorandum for the Secretary from Jim Tarro, Assistant Secretary for Administration, Oct 17, 1991 (the University of Illinois Chicago Special Collections & University Archives, School of Public Health, "David E. Jacobs papers" the University of Illinois Chicago School of Public Health library).

47. VA-HUD-Independent Agencies Appropriations Act for FY1992 (P.L. 102-139, H.R. 2915, House Report 102-94, Senate Report 102-107, Conference H. Rept. 102-226).

48. Procedures for lead paint removal and detoxification: Guidelines and Attributes. National Bureau of Standards Report 10-658, Dec 10, 1971.

49. L. Chadzynski. Manual for the identification and abatement of environmental lead hazards. University of Illinois, grant from the US Public Health Service. June 1986. (the University of Illinois Chicago Special Collections & University Archives, School of Public Health, "David E. Jacobs papers" the University of Illinois Chicago School of Public Health library).

50. Review of lead paint abatement and current HUD requirements, prepared by Arthur D Little, Cambridge MA, Aug, 1988. the University of Illinois Chicago Special Collections & University Archives, School of Public Health, "David E. Jacobs papers" the University of Illinois Chicago School of Public Health library.

51. Update on HUD Guidelines, Baltimore Lead Letter Vol 2, No. 2 July 1989.

52. House and Senate Conference Report HUD Appropriations. Report 100-817, Aug 3, 1988.

53. Lead Based Paint: Hazard Identification and Abatement in Public and Indian Housing. Office of Public and Indian Housing, US Department of Housing and Urban Development, Sept 15, 1989.

54. Appendix B. Minority Opinions. Lead Based Paint Testing, Abatement, Cleanup and Disposal Guidelines. National Institute of Building Sciences. Document number 5047, March 14, 1989.

55. Letter from David Jacobs to Stan Lewis, Department of Housing and Urban Development. July 7, 1989, the University of Illinois Chicago Special Collections & University Archives, School of Public Health, "David E. Jacobs papers" the University of Illinois Chicago School of Public Health library.

56. Flash! Congress Forces HUD's Hand. Baltimore Leadletter Vol 2, No. 3, Sept 1989. the University of Illinois Chicago Special Collections & University Archives, School of Public Health, "David E. Jacobs papers" the University of Illinois Chicago School of Public Health library.

57. Lead paint poisons, despite 1971 removal law, by William E Schmidt, New York Times. Aug 26, 1990, p. 1

58. Lead-Based Paint: Interim Guidelines for Hazard Identification and Abatement in Public and Indian Housing, September 1990, Office of Public and Indian Housing, Department of Housing and Urban Development, Washington DC.

59. EPA 40 CFR 745, 2019. Dust-Lead Hazard Standards; Definition of Lead-Based Paint.

60. HUD Policy Guidance Number: 2017-01 Rev 1 Date: February 16, 2017. Revised Dust-Lead Action Levels for Risk Assessment and Clearance; Clearance of Porch Floors.

61. HUD Notice PIH 95-8, Quality control procedures for on-site lead-based paint testing activities. Feb 9, 1995. The University of Illinois Chicago Special Collections & University Archives, School of Public Health, "David E. Jacobs papers" the University of Illinois Chicago School of Public Health library.

62. Personal communication. Senator Jack Reed and David Jacobs. US Senate. 1996.

63. Aschengrau A, Beiser A, Bellinger D, Copenhafer D, Weitzman M. The impact of soil lead abatement on urban children's blood lead levels: phase II results from the Boston Lead-In-Soil Demonstration Project. Environ Res. 1994;67(2):125−148. Available from: https://doi.org/10.1006/enrs.1994.1069.

64. Comprehensive and Workable Plan for the Abatement of Lead-Based Paint in Privately Owned Housing Report to Congress. Department of Housing and Urban Development Washington, DC December 7, 1990.

65. Shabecoff P. US closes unit that cited health effects of lead. NY Times, July 26, 1982. P A-8.

66. L. Thompson. A Short History of HUD. 2006. P. 40. https://monarchhousing.org/wp-content/uploads/2007/03/hud-history.pdf.

67. Childhood Lead Poisoning Prevention: Blueprint for Prevention. Alliance to End Childhood Lead Poisoning, Washington DC. Undated ca. 1990. the University of Illinois Chicago Special Collections & University Archives, School of Public Health, "David E. Jacobs papers" the University of Illinois Chicago School of Public Health library.

68. Rogers D. The Lives of Two Insiders Turned Outsiders Reflect a Hunger to Find New Answers to Old Problems. Wall Street Journal. 1991;A14. Aug 21.

69. Alliance to End Childhood Lead Poisoning. A framework for action to make private housing safe: A proposal to focus national attention. Washington, DC: Alliance to End Childhood Lead Poisoning; 1993.

70. A. Guthrie and M. McNulty, Making the Most of Medicaid, Alliance to End Childhood Lead Poisoning, April 1993. the University of Illinois Chicago Special Collections & University Archives, School of Public Health, "David E. Jacobs papers" the University of Illinois Chicago School of Public Health library.

71. National Center for Healthy Housing. History of the Alliance for Healthy Homes. 2022. https://nchh.org/who-we-are/afhh/.

72. Alliance to End Childhood Lead Poisoning. International Action Plan for Preventing Lead Poisoning. Revised 3rd Edition, undated ca 1990, Washington DC. the University of Illinois Chicago Special Collections & University Archives, School of Public Health, "David E. Jacobs papers" the University of Illinois Chicago School of Public Health library.

73. Alliance to End Childhood Lead Poisoning, Global Dimensions of Lead Poisoning: The First International Prevention Conference, Final Report 64 (1994). the University of Illinois Chicago Special Collections & University Archives, School of Public Health, "David E. Jacobs papers" the University of Illinois Chicago School of Public Health library.

74. Alliance to End Childhood Lead Poisoning, Preventing Lead Poisoning in the Americas: Health, Environment and Sustainable Development, April 17-18, 1998, Santiago, Chile. The University of Illinois Chicago Special Collections & University Archives, School of Public Health, "David E. Jacobs papers" the University of Illinois Chicago School of Public Health library.

75. The Massachusetts Lead Law. 2020. https://www.mass.gov/the-massachusetts-lead-law Also see: Robert Klein, Lead poisoning prevention in Massachusetts. Letter to the editor, New England Journal of Medicine, Aug 23, 1973, p. 428

76. First national lead conference. Alliance to End Childhood Lead Poisoning. (1992). Preventing childhood lead poisoning – The first comprehensive national conference (October 6, 7 & 8, 1991 Washington, D.C. the University of Illinois Chicago Special Collections & University Archives, School of Public Health, "David E. Jacobs papers" the University of Illinois Chicago School of Public Health library.

77. Secretary Louis Sullivan is quoted in: Los Angeles Times. "New Lead Danger Level Set for U.S. Children: Health: Revised standard reduces the tolerable amount in the blood to less than half of previous limit." By Marlene Cimons, Oct. 8, 1991.

78. Personal communication Don Ryan and Ellis Goldman, 1991.

79. CDC Preventing Lead Poisoning in Children October. Centers for Disease Control and Prevention, US Department of Health and Human Services. Atlanta, Georgia. 1991. https://wonder.cdc.gov/wonder/prevguid/p0000029/p0000029.asp#head001000000000000.

80. CDC Strategic Plan to Eliminate Childhood Lead Poisoning, Centers for Disease Control and Prevention, US Department of Health and Human Services. Atlanta, Georgia. Feb 1991. https://wonder.cdc.gov/wonder/prevguid/p0000029/p0000029.asp.

81. LA Times Ibid Oct 8, 1991.

82. CDC: Lead levels still poisoning kids, by Charles Seabrook, Atlanta Journal Constitution, July 17, 1990.

83. Press Release, Alliance to End Childhood Lead Poisoning, HHS Strategic Plan for Eliminating Childhood lead poisoning, Feb 22, 1991.

84. Richard Jackson written testimony before the Senate Banking Subcommittee on Housing and Transportation, June 5, 2002.

85. Screening Young Children for Lead Poisoning: Guidance for State and Local Public Health Officials Centers for Disease Control and Prevention (CDC). Nov 1997. https://www.cdc.gov/nceh/lead/publications/screening.htm.

86. Directive from Sally Richardson, Director Center for Medicaid and State Operations, Department of Health and Human Services to all Health Care Financing Administrators, April 13, 1998. the University of Illinois Chicago Special Collections & University Archives, School of Public Health, "David E. Jacobs papers" the University of Illinois Chicago School of Public Health library.

87. US General Accounting Office. Medicaid: elevated blood lead levels in children. GAO publication no. GAO/HEHS-98-78. Washington, DC: US General Accounting Office; 1998.

88. Wengrovitz A, Kulkarni M. Strategies to Improve Medicaid Screening and Treatment for Lead Poisoning. *Clearinghouse Review Journal of Poverty Law and Policy*. 2005;26–36. May-June.

89. National Research Council. *Lead in the Human Environment. National Academy of Sciences*. Washington DC: National Academy Press; 1980.

90. Brown MJ, Mason G, Cosgrove E. Lead Poisoning: A New Approach to an Old Problem, Massachusetts. *Journal of Community Health*. 1986-1987;3:28–32.

91. Dignam TA, Evens A, Eduardo E, et al. High-intensity targeted screening for elevated blood lead levels among children in 2 inner-city Chicago communities. *Am J Public Health*. 2004;94:1945–1951.

92. CDC. *Screening Young Children for Lead poisoning: Guidance for State and Local Public Health Officials.* Atlanta: CDC; 1997.

93. Lead Screening in Children, https://www.ncqa.org/hedis/measures/lead-screening-in-children/.

94. National Committee for Quality Assurance. The state of health care quality, 2010. Reform, the quality agenda and resource use. Washington, DC: National Committee for Quality Assurance; 2010. Available at: http://www.ncqa.org/portals/0/state%20of%20health%20care/2010/sohc%202010%20-%20full2.pdf.

95. Lead Screening in Children, National Committee for Quality Assurance (NCQA). 2021. https://www.ncqa.org/hedis/measures/lead-screening-in-children/.

96. Boronow KE, Perovich LJ, Sweeney L, Yoo JS, Rudel RA, Brown P, Brody JG. Privacy Risks of Sharing Data from Environmental Health Studies. *Environ Health Perspect.* 2020;128(1):17008. Available from: https://doi.org/10.1289/EHP4817.

97. US EPA, Strategy for Reducing Lead Exposure, Feb 21, 1991. US Environmental Protection Agency, Feb; 21; 1991. https://nepis.epa.gov/Exe/tiff2png.cgi/9101KTPY.PNG?-r + 75 + -g + 7 + D%3A%5CZYFILES%5CINDEX%20DATA%5C91THRU94%5CTIFF%5C00002794%5C9101KTPY.TIF.

98. Roundtable, American Industrial Hygiene Conference, 1991. Getting the Lead Out: Industrial Hygiene and Lead-Based Paint Abatement. American Conference of Governmental Industrial Hygienists and the American Industrial Hygiene Association. The University of Illinois Chicago Special Collections & University Archives, School of Public Health, "David E. Jacobs papers" the University of Illinois Chicago School of Public Health library.

99. Housing and Community Development Act of 1992. Title X. Public Law 102-550.

100. Markowitz and Rosner. Lead Wars: The Politics of Science and the Fate of America's Children. California/Milbank Books on Health and the Public. 2014.

101. Warren C. *Brush with Death: A Social History of Lead Poisoning.* Johns Hopkins University Press; 2000.

102. Lead Testing Urged After Boy's Death, Washington Post, March 29, 1991. https://www.washingtonpost.com/archive/politics/1991/03/29/lead-testing-urged-after-boys-death/8cc66e93-c953-4474-b7e4-ce7bbd4ff61e/.

103. National Affordable Housing Act Amendments of 1992, report of the Senate Committee on Banking, Housing and Urban Affairs, to accompany S. 3031, Report 102-332, July 22, 1992, p. 108.

104. Personal communication, Don Ryan with David Jacobs. 2021.

105. The Pervasive Problem of Lead Paint, April 7, 1991. By Thomas J. Lueck. NY Times Section 10, p. 1. April 7, 1991. https://www.nytimes.com/1991/04/07/realestate/the-pervasive-problem-of-lead-paint.html.

106. Urgent Lead Paint Hazard Prevention Act, Senator Alan Cranston. Congressional Record Vol 137, No. 143, Oct 8, 1991.

107. Testimony of David Jacobs before the Subcommittee on Housing and Urban Affairs, US Senate Committee on Banking, Housing and Urban Affairs, Oct 17, 1991.

108. Testimony of David Jacobs before the Subcommittee on Housing and Community Development, US House of Representatives, April 29, 1992.

109. Testimony of David Jacobs before the Subcommittee on Housing and Urban Affairs, US Senate Committee on Banking, Housing and Urban Affairs, March 19, 1992.

110. Letter from John Rosen to Don Ryan, Oct 25, 1992. The University of Illinois Chicago Special Collections & University Archives, School of Public Health, "David E. Jacobs papers" the University of Illinois Chicago School of Public Health library. Also see Ryan's response stating that Title X required both abatement and interim controls. The University of Illinois Chicago Special Collections & University Archives, School of Public Health, "David E. Jacobs papers" the University of Illinois Chicago School of Public Health library.

111. National Affordable Housing Act Amendments of 1992. Report of the Senate Committee on Banking, Housing and Urban Affairs, Report 102-332, July 23, 1992 P. 116.

112. Title X, section 1004 (15). Public Law 102-550. US Congress, Washington DC. October 1992.

113. Public Law 91-695-Jan. 13, 1971 The Lead-Based Paint Poisoning Prevention Act. Federal Register p 2080. US Congress. January 13; 1971.

114. Public law 93-151. 1973 Amendments to the Lead Based Paint Poisoning Prevention Act. US Congress. November 9, 1973.

115. U.S. Consumer Product Safety Commission. 1977. Ban of Lead-Containing Paint and Certain Consumer Products Bearing Lead-Containing Paint. 16 CFR 1303. Fed Reg 42:44199.

116. Recommendations for the Prevention of Lead Poisoning in Children, National Research Council, National Academy of Sciences, NAS/ACT/P-831, July 1976.

117. Section 566 of the Housing and Community Development Act of 1987 (Pub. L. 1 0 0 - 242 approved February 5, 1988).

118. Letter from Michael Varner (Southern Earth Sciences, Inc.) to Jacob Savage (Biotech Research Laboratories), June 19, 1991.

119. National Lead Detection Services Inspection reports. 1988. Georgia Institute of Technology. Atlanta, Georgia. The University of Illinois Chicago Special Collections & University Archives, School of Public Health, "David E. Jacobs papers" the University of Illinois Chicago School of Public Health library.

120. Letter from David Jacobs, Director, HUD Office of Lead Hazard Control, Nov. 4, 1998. The University of Illinois Chicago Special Collections & University Archives, School of Public Health, "David E. Jacobs papers" the University of Illinois Chicago School of Public Health library.

121. H.R.2840 - Lead Contamination Control Act Amendments of 1991. Rep. Waxman, Henry A., Introduced 07/10/1991. https://www.congress.gov/bill/102nd-congress/house-bill/2840?r = 86&s = 1 Committees: House - Energy and Commerce.

122. Personal communication. Henry Waxman to David Jacobs. 1991.

123. Statement of Steve Driesler, National Association of Realtors on the Impacts of Lead Paint on our Nation's housing industry, Subcommittee on Housing and Urban Affairs, Senate Committee on Banking, Housing and Urban Affair, Oct 17, 1991.

124. Disclosure Rules Proposed on Lead Hazards. New York Times, Oct 28, 1994, p A-25. https://www.nytimes.com/1994/10/28/us/disclosure-rules-proposed-on-lead-hazards.html.

125. Jacobs D.E. Lead-Based Paint Presentation to the National Association of Realtors, Las Vegas, NV, November 7-8, 1991. The University of Illinois Chicago Special Collections & University Archives, School of Public Health, "David E. Jacobs papers" the University of Illinois Chicago School of Public Health library.

126. EPA. Lead Renovation, Repair and Painting Rule. April 22, 2008. Docket number: EPA-HQ-OPPT-2005-0049; FRL-8355-7; RIN: 2070-AC83. https://www.epa.gov/lead/lead-renovation-repair-and-painting-program-rules#rrp.

127. Stephan Johnson Hearing as EPA Administrator. The University of Illinois Chicago Special Collections & University Archives, School of Public Health, "David E. Jacobs papers" the University of Illinois Chicago School of Public Health library.

128. National Conference of the Lead and Environmental Hazards Association, 2011. The University of Illinois Chicago Special Collections & University Archives, School of Public Health, "David E. Jacobs papers" the University of Illinois Chicago School of Public Health library.

129. National Association of Home Builders; Hearth Patio and Barbecue Association; National Lumber and Building Material Dealers Association; and Window and Door Manufacturers Association vs. US Environmental Protection Agency. July 8, 2010. http://op.bna.com.s3.amazonaws.com/env.nsf/r%3FOpen%3djstn-877p5h.

130. National Center for Healthy Housing. History of NCHH. National Center for Healthy Housing. 2009 https://nchh.org/who-we-are/history-and-accomplishments/2000-2009/.

131. Home Depot to Pay $20,750,000 Penalty for Nationwide Failure to Follow Rules for Conducting Renovations Involving Lead Paint, US Department of Justice Press Release, Dec 17, 2020. https://www.justice.gov/opa/pr/home-depot-pay-20750000-penalty-nationwide-failure-follow-rules-conducting-renovations.

132. B. McLean, Fannie Mae's Last Stand. The Economist. Feb 2009. https://www.vanityfair.com/news/2009/02/fannie-and-freddie200902-2.

133. A Joint Proposal of the Enterprise Foundation and the Alliance to End Childhood Lead Poisoning to the Fannie Mae Foundation. National Center for Lead Safe Housing, Sept 1, 1992. P. 1. The University of Illinois Chicago Special Collections & University Archives, School of Public Health, "David E. Jacobs papers" the University of Illinois Chicago School of Public Health library.

134. Memorandum from Don Ryan Oct 18, 1992. The University of Illinois Chicago Special Collections & University Archives, School of Public Health, "David E. Jacobs papers" the University of Illinois Chicago School of Public Health library.

135. Jacobs D.E. Lead Abatement In New York City: A Plan to Prioritize Response Based on Risk. Feb 14, 1992. The University of Illinois Chicago Special Collections & University Archives, School of Public Health, "David E. Jacobs papers" the University of Illinois Chicago School of Public Health library.

136. Letter from Nick Farr to Don Ryan, November 18, 1992. The University of Illinois Chicago Special Collections & University Archives, School of Public Health, "David E. Jacobs papers" the University of Illinois Chicago School of Public Health library.

137. National Center for Healthy Housing timeline. https://nchh.org/sample-shortcodes/sample-timeline/.

138. World Health Organization. *WHO Housing and Health Guidelines 2018*. Geneva: World Health Organization; 2018, Licence: CC BY-NC-SA 3.0 IGO. Available from: http://www.who.int/sustainable-development/publications/housing-health-guidelines/en/.

139. Braubach M, Jacobs DE, Ormandy D. (eds). Environmental burden of disease associated with inadequate housing: A method guide to the quantification of health impacts of selected housing risks in the WHO European Region (book). World Health Organization (Europe). June 2011.

Growing pains—new regulations, enforcement, capacity, and proof emerge

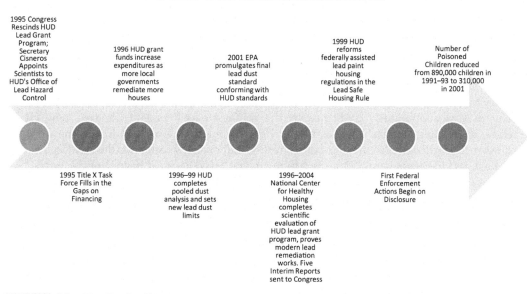

1995 Congress Rescinds HUD Lead Grant Program; Secretary Cisneros Appoints Scientists to HUD's Office of Lead Hazard Control

1996 HUD grant funds increase expenditures as more local governments remediate more houses

2001 EPA promulgates final lead dust standard conforming with HUD standards

1999 HUD reforms federally assisted lead paint housing regulations in the Lead Safe Housing Rule

Number of Poisoned Children reduced from 890,000 children in 1991–93 to 310,000 in 2001

1995 Title X Task Force Fills in the Gaps on Financing

1996–99 HUD completes pooled dust analysis and sets new lead dust limits

1996–2004 National Center for Healthy Housing completes scientific evaluation of HUD lead grant program, proves modern lead remediation works. Five Interim Reports sent to Congress

First Federal Enforcement Actions Begin on Disclosure

FIGURE 4.0 Timeline for Chapter 4.

If simply passing legislation was all that was needed, then 1992 might have been the end of the story. But getting large agencies such as the Department of Housing and Urban Development (HUD), the Centers for Disease Control and Prevention (CDC), and the Environmental Protection Agency (EPA), as well as local governments and the private sector to implement the new Title X law posed new scientific and policy obstacles.

The solution that emerged in the mid to late 1990s was threefold:

- practical, scientifically validated regulations and policies that were enforced;
- knowledgeable, committed, and qualified people in key positions of power; and
- engaged citizens groups to create new political will.

This chapter focuses on the first two; the story of how citizen and parent groups flexed their muscle is told in Chapter 5.

Fifty Years of Peeling Away the Lead Paint Problem
DOI: https://doi.org/10.1016/B978-0-443-18736-0.00013-3

113

4.1 The 1995 rescission and bringing science to the Department of Housing and Urban Development

By 1995 HUD's new lead hazard control grant program had awarded $280 million to 56 state and local governments to remove lead paint hazards in about 30,000 privately owned housing units with the most serious lead paint hazards. Reflecting the much larger need, between 42% and 82% of grant applications from local governments could *not* be funded due to insufficient Congressional appropriations.[1] The awards went to those jurisdictions with the best applications through a competitive process, although many of those not funded had applications that would have been funded if there been more funding.

But most of those initial awards remained unspent even a few years after they had been awarded. As a result, lead paint remediation was proceeding at a snail's pace under the new HUD grant program. This created even more consternation in Congress, which rescinded the lead paint appropriation in 1995 (as part of a larger budget rescission bill)[2] until HUD and its local government grantees could make more progress in fixing houses and spending the funding that had already been provided.

Bruce Katz, who became HUD Chief of Staff in 1993, had been one of the key architects of Title X while he was staff director for the powerful Senate Subcommittee on Banking, Housing, and Urban Affairs. He was now charged with implementing that law by HUD Secretary Henry Cisneros.

Cisneros was unanimously confirmed by the US Senate as President Clinton's new HUD Secretary, succeeding Secretary Jack Kemp in 1993. Kemp had done next to nothing on lead paint, other than grudgingly carrying out Congressional orders to establish the new lead paint office. Katz and Cisneros would first need to overcome HUD's sordid history of the 1980s.

In the 1980s, under the Reagan and first Bush Administrations, HUD was widely regarded as one of the most corrupt and ineffective federal agencies, prompting Congress to pass the HUD Reform Act of 1989.[3] Under President Reagan's HUD Secretary, Samuel Pierce Jr., several HUD officials were sent to prison as a result of a special prosecutor's investigation of wrongdoing and rampant mismanagement. It was Samuel Pierce who had been named in the paralyzing *Ashton v Pierce* class action lead paint lawsuit.

During Pierce's tenure, HUD illegally distributed millions of dollars in housing subsidies to prominent Republican consultants at a time when President Reagan's administration sharply reduced the agency's budget. Under Reagan, annual spending on subsidized housing programs dropped to only $8 billion from $26 billion,[4] explaining in part why HUD did so little on lead paint during the 1980s.

Some consider Cisneros to be among the most successful of the seventeen HUD Secretaries of Housing and Urban Development since the cabinet agency was formed in 1965,[5] and he is widely credited with reforming HUD in the early 1990s. He and Katz also reformed HUD's lead paint office. Two other future HUD Secretaries, one a Republican (Mel Martinez) and another Democrat (Shaun Donovan) would also play major lead paint poisoning prevention roles in future years, demonstrating the bipartisan support for lead paint poisoning prevention.

The existing staff in the new HUD lead paint office in the early 1990s came mostly from HUD's Policy Development and Research (PD&R) office, with training in housing or economics, but not physical, biological, or chemical science.

For example, Art Newberg, before becoming the new lead paint office's director under Kemp, designed and built custom homes, but he had no health or scientific background, with a degree in business. Before being appointed to run the HUD lead paint office, he had been working in HUD's manufactured housing program (such houses had no lead paint).

One of the key individuals within HUD who had worked in the Office of Public and Indian Housing on lead paint issues, Carolyn Newton, decided *not* to join the new office, because she thought it would fail and more could be done within the public housing office.[6] She had also helped to teach the Georgia Tech courses and ensured that virtually all housing authority executive directors completed lead paint training, despite their resistance. Five years after it was formed, she finally joined the new lead paint office, but only after being convinced it would take real action and had the full support of Secretary Cisneros.

Some questioned Senator Mikulski's wisdom in setting up a separate lead paint office, even if it was located within the immediate powerful office of the Secretary because that might take pressure off the various HUD program offices to fulfill their duties on lead paint. The challenge before the new lead paint office was to assume command of HUD's varied lead paint efforts across its many housing programs without replacing the day-to-day duties of the various program offices.

HUD was a vast organization, essentially structured in five main areas:

1. Public and Indian Housing, which funded local housing authorities (now through either the office of public housing or the office of Native American programs);
2. Community Planning and Development, which funded block grants to local governments and certain housing repair programs;
3. The Office of Housing, which funded large privately owned low-income housing developments (this also included the powerful Federal Housing Administration that operated very large mortgage and other housing finance systems);
4. The Office of Fair Housing and Equal Opportunity, which was responsible for combating housing discrimination and enforcing certain civil rights laws and housing requirements for those with disabilities; and
5. The Office of Policy Development and Research, which was responsible for conducting certain housing market economic research.

None of these had health and lead poisoning prevention as its mission and none employed scientists. For example, the American Housing Survey, the nation's largest ongoing survey of housing quality and an important scientific undertaking, was run by the Department of Commerce (Census Bureau) under contract to HUD.

In early 1995 the HUD Inspector General (IG) recommended that the entire HUD lead paint program be transferred to EPA "to spearhead lead-based paint activities nation-wide."[7] Strangely, the IG did not even mention Title X. The IG recommendations directly contradicted Congress and the law. It was an indication of the HUD career staff's continued unwillingness to address lead paint. This would be the first of several IG attempts to end HUD's lead paint program altogether (Chapter 5), all of which ultimately failed but not without further political interference.

With the notable exception of HUD IG Susan Gaffney during the latter part of Secretary Henry Cisneros' tenure, the HUD IG has historically *not* been truly independent of the HUD Secretary. Gaffney later battled another HUD Secretary, Andrew Cuomo, who had also attempted to interfere with the new lead paint office. Other HUD Inspectors General ignored or impeded the Congressional determination that HUD should assume leadership in lead paint poisoning prevention, perhaps explaining why consistent Congressional direct oversight was essential. In theory, Inspectors General were supposed to be independent, according to the IG Act of 1978,[8] but HUD was an exception, as we shall see.

Only a few years after the new HUD lead private housing grants had started in 1990, with millions remaining unspent, Congress was upset at the lack of progress. In a hearing on April 26, 1994, Congressman Louis Stokes, a Democrat from Cleveland and Congressman Jerry Lewis, a Republican from California both took HUD to task, grilling the lead paint office Acting Director (Ron Morony) who had come from HUD's Office of Policy Development and Research, where lead paint had languished for decades:

> Cong Stokes: We note that your request for the lead-based paint abatement program is $100 million … a decrease of $50 million. Can you tell us why the Department has reduced its support for an important program?

> Acting Lead Paint Office Director Ron Morony: We have made awards to some 29 localities. Only one of those is starting work…. The reason we slowed down the program was to let the capacity build.

> Congressman Lewis: And here we are four years later. During that period of time, we expended none of the money and you still have children at risk. Doesn't that leave you a little frustrated?

> Morony: Well, we are trying.

> Cong Stokes: Don't you think though that the Department should have done more to have pressed in this area?

> Morony: The decision has been made that the States should be the main people that regulate lead-based paint abatement…. And therefore it takes longer to do it that way.

> Cong Lewis: Well, I am just sharing your frustration, but it is very disconcerting. And those kids who in the meantime weren't served, God knows what the impact has been on their health.[9]

Katz realized that the new lead paint office was failing and that changes were needed. He wanted someone who knew the lead paint field and could work with the local grantees to get the money spent to fix houses with lead paint hazards to protect children, as Congress had ordered in Title X. He encouraged many to apply for the job of directing the office, including Jacobs, then at the National Center for Lead-Safe Housing. Katz told him, "Look, you helped get all this started, now you have to help us fix it."[10]

Cavanaugh and Mahoney also thought that someone with more scientific expertise and connections to the lead inspection and remediation field was needed, based on their experience with him in the public housing insurance program. But he had no experience in running a government organization. Others told him that the federal bureaucracy will "eat you alive" and recommended he could make a greater difference working outside government and staying focused on the science.

In his work at the Center and at Georgia Tech, Jacobs had built strong ties with many lead programs and professionals across the country and had testified before Congress during the Title X hearings. He was the principal author of the 1995 HUD guidelines and had been deeply involved in designing the Congressionally mandated scientific evaluation of the HUD lead hazard control grant program, both of which meant working with many of the initial HUD lead paint grantees. He also benefited from the government experience of Mahoney, Farr and others who had run public housing authorities and other national housing programs.

Cisneros thought the additional scientific expertise was an important development and announced it at the Washington Press Club in 1995, where he said, "Dave Jacobs brings both the scientific expertise and the practical field experience needed by the Department to make the nation's housing safe."[11] A physician in the audience, John Rosen, reflected the medical profession's annoyance with HUD, angrily asking why HUD was not doing more, to which Cisneros replied, "I do want to do more, that's why I'm announcing this change today. We need your help to protect as many children as fast as we can."

But bureaucratic inertia was substantial. Katz, and his Deputy Chief of Staff Amy Liu, worked behind the scenes to ensure that the office received the needed resources. Dwight Robinson, the HUD Deputy Secretary, who controlled HUD administrative operations under Cisneros, took a keen interest in the new opportunity to finally get HUD on board with lead poisoning prevention and ordered the HUD bureaucracy to provide the resources.

Michael Hill, who had directed the powerful Office of Administration at HUD in the late 1980s under the former HUD Secretary Jack Kemp (and who had been reassigned to the HUD lead paint office in the Clinton Administration as a "punishment") wrote:

> The Office of Lead Based Paint suffers from neglect, both from the Department and the Congress. During the Kemp years, it was treated as something of a nuisance, forced on the Department by Senator Mikulski—one of Kemp's least favorite persons. It was never given the appropriate number of staff to do anything but the minimum effort.... Except for Bruce Katz's efforts I don't see any substantive change in the Department's attitude towards the Office or the [lead poisoning] issue.[12]

Scientists were hired to staff the office, such as Warren Friedman, Eugene Pinzer, David Leavitt, and Ellen Roznowski-Taylor, all from the occupational industrial hygiene field, and others with an environmental research background, such as Peter Ashley, a scientist from Johns Hopkins who was familiar with the valuable research that had been done there by Mark Farfel, J Julian Chisolm and others. People with a background in science and health policy were hired, many of whom would go on to take leading roles, such as Matt Ammon. He conducted some of the nation's first enforcement actions of the new lead disclosure rule (described later in this chapter) and became the HUD lead paint office director in 2015, rising to become Acting HUD Secretary briefly in 2021 before President Biden's nominee was confirmed.

Rebecca Morley also joined the office as a Presidential Management Intern with a Georgia Tech Science and Technology Policy degree. Georgia Tech not only trained abatement contractors and inspectors but also offered a graduate degree that taught how science and policy can work together. Few institutions of higher learning train scientists in policy and even fewer train policy analysts in science, especially in the housing field.

As an intern, Morley was briefly stationed within the office of Senator Jack Reed from Rhode Island as part of a series of rotations that all such interns complete. She convinced him of the need to do more on lead paint poisoning. Years later, another HUD Secretary (Mel Martinez) would appear before Senator Reed, and after being asked tough questions on lead paint, remarked, "I get it: Senator Reed is Senator Lead." Morley later left HUD to succeed Farr as the second director of the National Center for Lead-Safe Housing, later transforming it to address other housing-related diseases and injuries as the National Center for Healthy Housing.

Public education expertise through community groups was important, not only the science. Dolline Hatchett was hired to direct the office's public outreach efforts, and she went on to play an important role in HUD's support for parent groups and the ill-fated Campaign for a Lead-Safe America (Chapter 5).

But it was not only about bringing in newly qualified and committed people to HUD. Existing personnel who knew how HUD did (and did not) work played a key role, including Ellis Goldman and Steve Weitz, who would run the new lead grant program and research efforts, respectively. New ties were created between the lead paint office and HUD's lawyers, which would be important in both reforming all of HUD's regulations as Title X required, as well as conducting some of the nation's first federal enforcement of lead paint regulations. Two HUD lawyers, John Shumway and John Kennedy, played important roles in both regulatory reform and enforcement, as did other key HUD legal staff named later in this book.

CDC also helped by detailing one of its staffers to the new HUD office, Stan Galik, an example of the growing ties between HUD and CDC. Goldman's and Galik's method of grants management proved to be particularly effective. Their view was that federal government personnel should help local programs succeed, as well as to conduct audits and oversight. Goldman was deeply admired by HUD's lead grantees across the country. Steve Weitz also was particularly effective in managing HUD's growing research agenda and led HUD's efforts to reform all its lead paint regulations during the 1990s to ensure the new regulations were supported by science.

But other existing HUD staff posed obstacles, some sleeping on the job or engaging in petty accusations to foment needless divisions that detracted from the office's effectiveness. Although some may dismiss it as mere "office politics," some offenses were extreme. For example, one was accused of biting another; one was accused of threatening another with a gun; another of urinating on chairs; still others resented the incoming scientists. This would ultimately lead to several being dismissed from the Federal service and early retirement for others.[13]

It was even necessary to issue an official policy on office courtesy,[14] including an explicit order from the Office's leadership that "rudeness, disrespect, snide and demeaning remarks about others...is not conducive to efficient office operations ... and I order it stopped."[15] This new no-nonsense personnel approach would eventually earn respect. One remarked after a problem staffer had been successfully fired, "I guess we all knew it had to happen."[16]

Obtaining the new staff was not simple, even with the support of Katz, Liu, Robinson, and Cisneros. The existing deficiencies in 1995–96 included inadequate technical assistance to local governments and local HUD staff, limited ability to conduct research and public education, inadequate oversight on fraudulent lead paint testing, and poor messaging to tell the public about what HUD was doing.[17] On top of that, a hiring freeze was in place due to Congressional antipathy to HUD in particular and government in general, but Katz and Robinson provided an exemption, authorizing additional staff for the lead paint office.[18,19]

The office was reorganized to emphasize strategic planning, customer service, team-building, dealing with discipline problems, new pathways for career development, and most importantly increasing commitment to the mission.[20]

A few years later, when Katz and Liu departed HUD to run the Brookings Institution Metropolitan Policy Program, Katz said, only half in jest:

> I now understand what my real job was as Chief of Staff. It was to keep the Congress and the HUD mafia (aka the bureaucracy) off your back so you can get something important done.[21]

In the mid-1990s, the lead paint office was forced to run the program from Washington with no staff in local HUD offices around the country, unlike other housing programs. For the first few years, the office only had 20–25 people to administer grants totaling hundreds of millions of dollars; reform HUD's (and other federal agency's) housing regulations related to lead-based paint; conduct research; and carry out enforcement, public education, and training. Toward the latter part of the 1990s, staff in local offices were added to increase the Department's ability to act on lead paint at the local level. By 2022 the office staff numbered about 50, which helped to administer record-high appropriations for the lead and healthy homes grant programs.

Another major change in the office's composition involved the recognition that effective public communications on the importance of lead paint were essential. Sharon Maeda, a special assistant to Secretary Cisneros was assigned to the new lead paint office, recognizing that public education would need to be a key component.

The newly reorganized office, with its improved expertise and authority, still faced the thorny problem of how to get the lead paint grants moving at the local level. In most cases this was simply due to a lack of capacity in local jurisdictions. Where would the new lead inspectors, risk assessors, and abatement contractors come from to carry out the work? HUD launched a series of efforts to conduct training and technical assistance around the country to stimulate the more active engagement of the private sector, a theme that would be repeated again and again as HUD sought to stimulate the market to improve X-ray fluorescence (XRF) instruments, laboratory performance and the number of private contractors available to do inspections and abatement. Gordon McKay, Dick Kennedy, and other HUD veterans were essential in helping local jurisdictions coordinate lead paint with many other HUD programs, such as Community Development Block Grants.

But in addition to building capacity, the lead paint office began a new management effort aimed at helping local grantees who were too slow in producing lead-safe homes and spending their money. In some cases, this would mean calling the local grantee into Washington to threaten that HUD was prepared to take the grant away if progress was not forthcoming.

It also included sometimes meeting with Congressional representatives to demand more progress, such as Representative Louise Slaughter from Rochester NY, one of the rare members of Congress with an advanced degree in public health. Sometimes, this meant making local personnel changes. Sometimes it meant moving the grant from a health or housing department into another local department. In more than one case (e.g., Detroit and Washington, DC), the grant was moved out of a local housing or health agency into a Mayor's office so that the Mayor could administer it directly.

Certainly, no Mayor, Governor or local Congressional representative wanted to be accused of failing to spend HUD money to protect children.

For example, Buffalo New York had been issued a HUD lead paint grant in 1995, but 2 years later, the city had abated only one housing unit. After HUD officials traveled to Buffalo and met with city officials and the Mayor, the city's lead grant was designated as "high-risk." It worked. Within a year, the program had abated over a hundred housing units, with hundreds more to follow in future years, including a new lead paint ordinance.[22] The approach worked not only in Buffalo but in many other local jurisdictions that had failed to produce safe housing. Community advocates and parents played an increasingly important role in getting local governments on board (Chapter 5).

Historically, HUD had often been focused on simply criticizing local grantees instead of helping them to find solutions. Secretary Cisneros told a Congressional committee in 1996 that at the time was attempting to abolish HUD entirely, "Look, I was the Mayor of San Antonio and I share the committee's concerns; every time I went to HUD for help in eradicating blight or overcoming some housing problem, the only answer I ever got was 'no' for some bureaucratic reason or another." The effort to abolish HUD went nowhere and it was Cisneros who led the effort to carry out reforms.

Cisneros' decision to increase scientific capacity within HUD led to a new appreciation of the importance of the HUD office in the lead poisoning prevention community. One news account stated:

> Because [the HUD lead paint office director] came to government with a background shaped by private industry training and expertise, he was a breath of fresh air in a government bureaucracy that was running out of ideas on how to manage the lead hazards it was charged to regulate.[23]

The new office also enlisted the help of Doug Farquhar at the National Conference of State Legislatures to engage with local elected representatives, governors, and others. He played a leading role in getting the HUD grants to operate more effectively as well as ironing out differences between HUD's regulations and local laws. He helped HUD to effect change at the state level and bridge the geographic divide among HUD (city-oriented), CDC (state-oriented), and EPA (region-oriented).

An example of the divide between HUD and a few states involved the Governor of Michigan, John Engler, mentioned briefly in the previous chapter. He sought to privatize many state-government functions, including the licensing of lead inspectors and abatement contractors. HUD had awarded a lead paint grant to the state, with the stipulation that a state law be passed and signed into law to authorize licensing of lead contractors, which the Governor had agreed to do in the state's grant application, as required by Title X. But in August 1995 the Governor abruptly reversed course and refused to sign legislation to set up a licensing program for lead inspectors and abatement contractors to ensure HUD-funded work was done properly.

The State health commissioner, James Haveman Jr. threatened to return the lead paint money to HUD,[24] although it was later made known that the Governor did not really want to return $4.9 million that would protect children, fearing the bad press.[25] A state senator wrote, "We encourage HUD to also look past some of the licensure issues to find alternatives."[26] Haveman wrote a letter to the Detroit Free Press, attacking HUD for "insisting on an elaborate bureaucracy...and an inflexible demand for more regulation."[27]

HUD shot back:

> HUD does not require a new bureaucracy ... existing mechanisms can and should be used. More than 20 states have already enacted legislation, including some of Michigan's neighbors. In order to win the grant, the Governor in 1992 agreed to license lead paint inspectors and abatement contractors, writing: "I will work with the Michigan Department of Public Health to establish ... training and certification [licensing] of personnel who conduct lead abatement." There are many other unfunded qualified applicants who have adopted certification procedures and where the work can proceed safely.[28]

The threat to send the money elsewhere worked. The Governor backed down, agreeing to create an enforceable certification regulation.[29] This paved the way for other states to create ways to ensure that the work was done properly. By 2022 the number of states with certification systems for lead paint inspectors, risk assessors and abatement contractors rose to 39 (including Michigan), as well as the District of Columbia, Puerto Rico and 4 Tribes.[31]

Michigan later emerged as one of the nation's leaders in the lead paint and healthy homes field, under the able leadership of Wes Priem, Courtney Wisinski, Carin Speidel, and many others. Priem and his colleagues established a "healthy homes university" that enabled the state to not only address lead hazards but other housing health hazards like asthma (Chapter 8).[32]

The lead paint local grantees also learned from each other. This occurred primarily at conferences of the emerging lead inspection and remediation industry, the affordable housing industry, and various public health and environmental conferences. They established the Lead Paint Grantees Association at a Lead and Environmental Hazards Association (LEHA) conference in 1997, which was led by Brenda Reyes, Paul Kowalksi, Betsy Mokrzycki, Chris Corcoran and others.

The association usually met during lead conferences, which published proceedings and newspapers. For example, LEHA had regular regional and national conferences, sometimes co-located with HUD conferences. Steve Weil, LEHA's executive director, continued to do so through 2022. Weil published a lead inspection and abatement industry newspaper "Deleading" from 1986 to 1997 that chronicled the growth of the lead inspection and abatement industry.[33]

There were also regular "Lead Tech" conferences, organized by another newspaper, "Lead Detection and Abatement Contractor," published by IAQ (Indoor Air Quality) Publications. It published proceedings for 5 years before ceasing its operations. The paper's advisory board included Kenn White and Steve Hays, both of whom had helped teach the Georgia Tech courses, plus Doris Adler who organized major training operations, Nick Farr from the National Center for Lead-Safe Housing, and many other leaders in the lead poisoning prevention field.[34]

These concerted efforts worked. Local governments became more productive as local capacity improved and local HUD staff became more involved in finding ways to help make the new program effective. By 1997 the lead paint office was recognized by the

White House Office of Management and Budget (OMB) as one of the best-run offices within HUD with one of the lowest unexpended balances of grant funding, second only to the HUD HOME program.[35] By 1997, the lead paint grantees were getting the work done and the money was finally being spent, a sharp turnaround from the 1995 rescission.

As a result, funding for the HUD lead hazard control grant program was restored. In an important development, the lead grant program became a separate HUD budget line item in 1996, engineered by Bruce Katz, who as a former Congressional staffer and now HUD Chief of Staff, knew how government funding worked. This proved to make future budget requests for lead paint more robust. Before the creation of the separate budget line item, the lead paint grant program had variously been included in the section 8 program, or the community planning and development budget or in the public and Indian housing program budget, a reflection of HUD's failure to make lead paint a priority. Making it a separate budget line item also elevated the importance of the program both in Congress and within HUD.

But by 1996 Congress had changed hands, adopting an anti-HUD stance under the House of Representatives Speaker Newt Gingrich's leadership. The focus now became cutting domestic programs like the lead paint grant program under the guise of "balancing the budget." At one hearing, then-Congressman John Kasich (a Republican who would later run for President and serve as Governor of Ohio) asked the HUD director of the lead paint office, "you don't really need a separate lead paint office anymore do you, because you have succeeded in reforming all the other HUD programs?" But veiled threats to abolish the lead paint office ultimately went nowhere and Kasich and many other Republicans and Democrats alike saw the lead paint program as a needed contribution to children's health and housing stability and improvement. Even in this more contentious environment, lead paint poisoning prevention retained its bipartisan support.

However, the Clinton Administration after 1996 and HUD under Secretary Andrew Cuomo was focused on balancing the budget and eliminating the deficit. The budget for the lead grant program was reduced, along with many other HUD programs. Despite protests, the staffing for the Lead Paint office declined from 31 in 1996 to only 22 staffers in 1997, about where it had been in the early 1990s.[36] Funding for the lead paint grant program from Congress also stagnated, falling from $150 million in FY 1994 to only $60 million in FY 1997 and 1998. Clinton did balance the budget, but in the case of lead paint, it came at a terrible price: decreased protection for children. For his part, HUD Secretary Cuomo also attacked the lead paint office and supported reduced budgets (Chapter 5).

Yet the new scientific and program expertise at HUD proved to be resilient. Staffing and funding both improved in later years (Chapter 5). The science continued to advance and was reflected in new HUD guidance beyond the interim public housing guidelines HUD had published in 1990.

4.2 The 1995 Department of Housing and Urban Development lead paint guidelines

As directed by Title X, HUD finalized its lead paint guidelines in 1995, which were written by the newly founded National Center for Lead-Safe Housing under a HUD grant. The research and practical experience had finally been distilled to produce an evidence-based set of procedures.[37] Although still without the force of law or regulations on its own, the 1995 guidelines were referenced in most other federal, state, and local regulations as a "documented methodology" that could be enforced.[38]

The authorization for its publication was technically restricted to "federally assisted housing" (not only public housing), but in effect they covered all housing. The document was informed by a formal call for scientific evidence. Surprisingly, this was the first time HUD had issued such a request.

Three years in the making and overcoming the failures of the earlier NIBS and Public Housing Interim Guidelines, these new ones established standard procedures that interrupted the exposure pathways, which had been discovered in the late 1980s (Chapter 3). They also avoided needless testing. For example, although the source of lead in dust or soil could sometimes be from nonlead paint sources (a nearby polluting industry for example), HUD said lead-contaminated dust "...means dust with lead above certain levels *regardless of the source*"[30] (emphasis added). This was done to avoid the prospect of expensive and usually inconclusive tests to determine the source of lead in dust, even though in most cases it was highly correlated with lead paint.

Although this grouping of lead-contaminated dust and soil with the term "lead paint hazard" struck some as counter-intuitive (because dust and soil are not paint), it was consistent with Title X, which had defined a lead-based paint hazard to be:

> ...any condition that causes exposure to lead from lead-contaminated dust, lead-contaminated soil, lead-contaminated paint that is deteriorated or present in accessible surfaces, friction surfaces, or impact surfaces...[39]

The 1995 HUD guidelines were prepared principally by experts in occupational and environmental health, inspection, abatement, training, housing (including public housing), engineering, exposure assessment, housing law, worker protection, public health, medicine, epidemiology, contracting, and other related fields. This multidisciplinary approach marked a departure from the previous medical focus (Chapter 2).

The transition from health to housing was now complete, with HUD finally taking on the leadership role that Congress had first envisioned for it in the 1973 and 1988 amendments to the 1971 Lead Paint Poisoning Prevention Act and then ultimately with the passage of Title X in 1992.

These guidelines included detailed procedures on interim controls, the very methods that had been completely ignored in the 1990 HUD Public Housing Guidelines. Because they were "interim," specific reevaluation schedules were included to help ensure they remained effective and that hazards did not reappear. Formal standards on maintenance were issued in a companion document.[40]

Although some initially claimed that interim controls were merely repainting, they were far more complex. In the 1995 guidelines, interim controls included:

- Paint stabilization (use of a compatible two-coat paint system) for deteriorated paint;
- Soil covering with vegetation, mulch, or other materials with drainage controls;
- Correction of friction surfaces (such as painted window sashes) known to release lead dust;
- Dust lead removal; and
- Correction of impact surfaces (such as doors) known to release lead paint chips.

The guidelines specified instances when such interim controls should *not* be used, principally in the case of poor property management.

There were new procedures for renovation, maintenance, encapsulation (use of a specialized long-lasting warranted coating), the impact of historic preservation regulations, and a much more detailed discussion of what to do to evaluate a home with a lead-poisoned child. This last chapter was drafted primarily by CDC personnel. Historic preservation in some areas later emerged as a significant problem, because some thought that even a poisonous substance like lead should be "preserved." Many local jurisdictions passed more informed procedures, most recently in Madison, WI.[41]

The 1995 guidelines offered a detailed description of the pros and cons of various approaches for specific types of building components, part of which is reproduced below (Table 4.1).

The description included health protection, capital expenses, maintenance, worker protection, amount of finish work to be expected, relative durability, product availability, labor intensiveness, and cost.

Recognizing the importance of cleaning and dust removal, the 1995 guidelines provided an evidence-based recommendation for a three-step process: intial HEPA vacuuming to remove debris, wet washing to dislodge dust and smaller debris, and a final HEPA vacuuming to remove what was left. Although the guidelines initially recommended the use of tri-sodium phosphate detergent in the wet cleaning step, it also recognized that phosphate detergents were being phased out due to environmental concerns and recommended other detergents.[42] Today, trisodium phosphate detergents are no longer used.

The 1990 Interim Public Housing Guidelines had very little on clearance dust testing to determine if a home that had undergone either abatement or interim controls was safe for children to occupy, a reflection of the failure to grasp the importance of lead-contaminated dust. For the first time, the 1995 guidelines did this by establishing detailed "clearance" procedures that involved a visual examination and a determination if the work had been completed adequately. If the home passed the visual assessment, then dust wipe samples were collected, analyzed, and compared to clearance dust guidance values (actual government standards were still 4–6 years away). If clearance was not achieved, then the contractor would need to repeat the cleaning until it did. Both interim controls and abatement could only be complete if both the visual assessment and dust testing were found to be acceptable; in addition, the contractor would not be paid until both the visual and dust testing was complete. This was similar in concept to a "punch-list," a common construction practice that is used to determine adequate completion of a project, a sign of integrating lead hazard control into normal work practices.

This modification in dust lead guidance was supported by a 1994 EPA guidance memo from Lynn Goldman,[43] a high-ranking EPA official who had previously served on the CDC panel for their 1991 blood lead statement. She was criticized internally by EPA lawyers and some of her own staff, who thought she was authorized to only issue regulations, not "guidance" and that the guidance memo might conflict with Superfund litigation. As a result, the guidance memo went into great detail about how it was only "preliminary," and was not based on children's health considerations, pending further research.

She issued the memo because she was convinced that a formal regulation would be years away, which proved true. Title X had ordered these standards to be done by 1994, but they

TABLE 4.1 Advantages and disadvantages of lead hazard control methods.

Attributes	Method										
	Removal							Enclosure			
	HEPA needle gun	Heat gun	HEPA vacuum blasting	HEPA sanding	Remove/replace	Caustic paste	Offsite stripping	Plywood paneling	Gypsum	Prefab metal	Wood, metal, vinyl, siding
Skill level	High	Moderate	High	Moderate	High	Moderate	Moderate	Moderate	Moderate	High	Moderate
Esthetics	Erodes surface	Gouges	Erodes surfaces	Gouges/roughens	Good	Gouges	Good	Good	Good	Good	Good
Applicability	Very low, limited to metal and masonry	Wide, can damage some components	Very low: limited to metal and masonry	Low: limited by surface contour	Wide: dependent on skill level	Wide, can damage some building components	Low: limited to removable building components	Wide, walls	Wide, walls and ceilings	Varied, limited by components	Wide, walls
Lead presence	Removed	Largely removed	Largely removed	Largely removed	Removed	Largely removed	Largely removed	Remains	Remains	Remains	Remains
Hazardous waste generation	Moderate	Moderate	Moderate	Moderate	Potentially high, pending TCLP test	High	High, but maintained offsite	Low	Low	Low	Low
Weather limitations	Moderate	High	Moderate	Moderate	Minimal	High	None	Minimal	Minimal	Minimal	Minimal
Applicable to friction surface	Some	Yes	Some	Some	Yes	Yes	Yes	No	No	Yes	No
Speed	Moderate	Slow	Slow	Slow	Moderate	Very slow	Can be slow, requires coordination	Moderate	Moderate	Moderate	Moderate
Training required	High	Moderate	High	Moderate	High	Moderate	Moderate	High	High	High	High

From HUD Guidelines for the Evaluation and Control of Lead Based Hazards in Housing, 1995.

were not finalized until 2001. She courageously issued the guidance anyway, overruling EPA lawyers, which put EPA and HUD on the same page in the mid-1990s. Indeed, the Government Accountability Office issued a scathing report in 1994, stating:

> EPA did not issue the standards defining hazardous levels of lead in paint, dust and soil by the legislatively imposed April 1994 deadline...While EPA expects to issue the standards by late 1995, the date appears optimistic...The delays that EPA has experienced...may have serious environmental and health consequences...To address this problem in the interim, we believe that EPA should promote the general public use of the existing guidelines established by HUD.[44]

Some of the strongest advocates for a more detailed dust clearance process were surprisingly from construction contractors themselves, like John Zilka. They wanted certainty that they had completed the job adequately so any future liability would be minimized. He stated:

> Clearance is a good thing for contractors, because it proves we did the job right. It's not a headache and not a barrier to getting paid.

The Alliance to End Childhood Lead Poisoning also recommended that the EPA guidance be used until final regulations could be issued, criticizing the agency for delaying its Title X mandate.[45]

As these important technical procedures were being finalized, so too were important remaining policy questions.

4.3 The Title X task force fills in the gaps

Despite Title X's comprehensive directives to federal agencies, Congress decided not to address in legislation several controversial areas. Instead, Title X directed the HUD Secretary to appoint a national task force to help resolve them, including:

- establishing clear standards of care for rental property owners;
- the concept of "safe harbor" for owners who could document compliance;
- increasing the availability of liability insurance;
- alternative systems for victim compensation;
- reflecting lead safety in property appraisals;
- incorporating insurance and mortgage underwriting standards; and
- expanding private financing.

On November 8, 1993, 1 year after Title X became law, HUD Secretary Cisneros appointed 39 members to the Task Force on Lead-Based Paint Hazard Reduction and Financing based on the categories of membership Title X had established. Importantly, Congress excluded representation of the lead, pigment and paint industries.

Secretary Cisneros appointed Cushing Dolbeare as Chair, which proved an inspired choice. Dolbeare, who had founded the National Low-Income Housing Coalition, was widely respected throughout the housing world for her knowledge of affordable housing, her integrity, and her skills in finding common ground and listening, perhaps strengthened by a Quaker upbringing.

Dolbeare audaciously appointed advocates for lead safety to head each of the Task Force's three committees. The liability standards and insurance committee was headed by Stephanie Pollack (who had worked on the Massachusetts lead abatement law, one of the strictest in the country). The financing committee was headed by Nick Farr from the Enterprise Foundation and the National Center for Lead-Safe Housing. The Implementation Committee was chaired by Don Ryan from the Alliance to End Childhood Lead Poisoning.

The Task Force also included two parents of lead-poisoned children, including Maurci Jackson who would go on to help found United Parents Against Lead and was one of the most compelling advocates. Other members included leaders from the environmental science and public health worlds, such as Ellen Silbergeld from the University of Maryland, Alina Fernandez, a clinician and Professor from the University of Illinois who served as the vice chair of the finance committee, and David Jacobs. It also included lead paint inspectors, abatement contractors, and unions. Miles Mahoney and a director of a public housing authority were included, as were housing professionals from the private and nonprofit sectors, such as construction, mortgage financiers (including representatives of Fannie Mae and Freddie Mac), low-income housing providers, developers, insurers, and real estate. Local legislators were included. The Task Force also included Karen Florini, an attorney with the Environmental Defense Fund as well as representatives of HUD and EPA and two legal services attorneys.

At Dolbeare's suggestion, the Title X Task Force established the ground rule that its recommendations required at least two-thirds support from this disparate group. The Task Force worked for 15 months to forge an agreement on 59 recommendations, many of which were adopted unanimously.

In brief, the Task Force adopted a tiered approach to lead safety, recognizing that the risks posed by lead-based paint in millions of US homes and apartments varied widely. It opened by stating what had by this time become obvious to all except the lead, paint and pigment industries:

> ... the current system is not working for parents, children, private property owners, lenders, insurers, or the nation at large. ... The Task Force envisions a system that will protect children from developing elevated blood lead levels, preserve our stock of affordable housing, and wisely invest scarce resources, both public and private.[46]

Depending on the degree of risk posed by different classes of housing, the Task Force adopted national standards for maintenance and lead hazard control in private housing to make clear property owners' responsibilities, some of which were later codified in the HUD lead-safe housing rule described later in this chapter. For example, the Task Force recommended a set of "essential maintenance practices" for all properties that contain or might contain lead paint along with more aggressive "standard treatments" for properties deemed higher risk.

It also called for expanded private financing for lead inspection and hazard control and targeted public financing for hazard controls in economically distressed housing occupied by low-income families.

The Title X Task Force recommended changes to the liability and insurance systems to incentivize hazard control and provide compensation to poisoned children. Specifically, it asked State legislatures to consider a system of insurance and liability limitations for complying owners and an optional, no-fault alternative to the tort system. This was similar in concept to a "worker's compensation" system for childhood lead exposure. Although Maryland did enact such a system, it was later ruled unconstitutional, because the court believed children should have the right to appear in court if they are injured from lead poisoning. Worker's compensation is grounded in the idea that in exchange for liability protection, employers will provide regulated compensation for workplace injuries.[47] This idea ultimately failed in the lead paint context, continuing most courts' failure to provide meaningful progress.

Finally, the Task Force recommended new public awareness initiatives (which would later be manifested in the Campaign for a Lead-Safe America) and a new research agenda.

The Task Force report provided detailed cost estimates for each of these systems for both high- and low-risk housing with differing options. In this way, the Task Force attempted to integrate lead paint into the housing finance systems already in place, showing that the incremental costs of protecting children were feasible.

Although the Task Force achieved remarkable consensus, several members expressed minority views, which were published in the final report.[48] For example, one nonprofit housing provider of older low-income housing worried that public subsidies would not be adequate, and without them "standard treatments" could not be implemented; another thought the standard treatments were too expensive for all older housing, much of which had been well-maintained in his view. One of the parents (Maurci Jackson) maintained that not enough parents were represented on the Task Force and that the recommendations were too complicated. She also thought that only training maintenance supervisors (instead of the maintenance workers too) would not be adequate in light of her own child's poisoning from sloppy maintenance workers in her Chicago home.

Others said the Task Force's view that the tort system was not working was wrong. Instead, they suggested that children did not have access to enough lawyers. They also thought that there had been only one scientist on the Task Force and that industry had dominated, although in fact there were several scientists and the lead industry had been excluded. Perhaps their overall complaint was that it thought the Task Force had decided that abatement was too costly to implement. They stated: "on cost-benefit grounds, complete abatement is demonstrably feasible," citing the 1991 CDC cost-benefit analysis as proof. But the CDC analysis they pointed to stated that its cost-benefit analysis used data from three so-called "abatement" programs that consisted of "scraping, spackling, and repainting areas with deteriorated lead-containing paint."[49] These methods were not actually abatement and were not interim controls either and were more like the paint removal methods that had previously been so widely discredited in the 1980s. Nevertheless, they agreed the Task Force recommendations were a substantial improvement.

In a separate rebuttal,[50] Dolbeare and four other task force members, including Alina Fernandez, Don Ryan, Stephanie Pollack, and Karen Florini, stated that since the three committees had been chaired by advocates, the industry had not in fact dominated. They thought the evidence showed the recommendations would in fact protect children, because proof of the effectiveness of both abatement and interim controls had been published and indeed had been included in Title X. They said the recommended standards in the Task Force report were actually more protective than almost all of the prevailing local lead paint laws currently on the books, which mostly called for the dangerous simplistic lead paint "removal" discussed in earlier chapters, such as power sanding, open flame burning, and other methods that are now banned.

4.4 The fight over lead dust standards[a]

The untold story of the lead dust standards involved controversy about degree of protection and which dust collection, lab method, and units of measurement were best. Recognition that lead dust is the primary pathway of most children's exposure begged the question of how to measure such exposures. This raised a multitude of thorny scientific questions, including where and how to collect and measure samples, feasibility, and how dust sampling would be used to assess risks and confirm acceptable cleanups. A years-long standoff between EPA and HUD delayed settling these questions, but ultimately science and common sense prevailed to produce methodologies that were reliable and practical to use across the nation and protected millions of children.

HUD wanted a simple reliable health-protective dust measurement method that predicted risk, correlated with children's blood lead, and could be used in millions of homes. EPA wanted a method it could use in its Superfund program. There had been extensive litigation by companies with a history of lead pollution, which (like the paint, pigment and lead industries) held that wherever the lead dust or contaminated soil had come from, it was not "their" lead, and they should not have to pay to clean it up. In short, the effectiveness of lead paint remediation could not be established without agreement on a standard lead dust sampling method that produced reliable results and that could be consistently compared to protective, measurable and feasible standards.

In the beginning, it boiled down to a seemingly esoteric debate on sampling methods. In the end, milllions of children's futures and billions of dollars were at stake.

Emission standards and the Superfund program related to lead were driven in large part by one of the world's largest lead mines in Kellogg, Idaho known as Bunker Hill. Ian von Lindern, who studied this site stated, "There simply was no other situation comparable to Kellogg in those days." One paper described it as "the dirtiest town in the country" and in 1974, 98.9% of the children had blood lead levels above 40 μg/dL.[51]

---

[a] An author's note: Some reviewers of this book recommended deleting this section because they thought dust sampling methods were esoteric. I chose to include it to help readers grasp the scientific complexity and its resolution. If nothing more, it is a window into the scientific enterprise and the resulting policy choices. In the end, consensus was achieved, but not without debate.

Yet EPA's IEUBK model which had been used extensively in the Superfund program regarded dust ingestion to occur only as a subset of soil ingestion. Lead paint concentration or condition was not in the model at all.[52] Of course, housing is not the same as a hazardous waste site. Senator Edward Kennedy had attempted in the late 1980s to use EPA's authority to address lead paint in housing as a "hazardous waste." This was rejected, because paint was not thought to be a "waste". But his efforts resulted in the EPA's "Three City" soil lead study in Boston, Baltimore, and Cincinnati, which Congress had mandated in the Superfund Amendments and Reauthorization Act of 1986.[53] Because the hazardous waste laws required a determination of who had produced the contamination and because it was not possible to prove which company had sold lead paint to a specific home and whether that paint was applied to a specific surface, Sen. Kennedy's effort proved fruitless.

On the other hand, it *was* possible to determine whether an industry emitting lead pollution had contaminated nearby residential yards. As a result, EPA's Superfund program was focused on soil contamination. After conducting a site-specific "Record of Decision" for that industry, EPA typically would excavate lead-contaminated soil.

But the Superfund program did not address lead paint, because for paint, there was no "Principal Responsible Party (PRP)." EPA defined such a party to be:

> Current owners and operators of a facility; past owners and operators of a facility at the time hazardous wastes were disposed; generators and parties that arranged for the disposal or transport of the hazardous substances; and transporters of hazardous waste that selected the site where the hazardous substances were brought.[54]

In the case of lead paint in housing, paint would need to be regarded as a "waste" and a "facility" would need to include a home. Both were problematic because some argued that lead paint was a "product" and not a "waste" and a "facility" did not mean a home. In the end, the litigious nature of the Superfund program made it largely inconsequential to lead paint in housing, despite attempts to "marry" the Superfund and HUD lead paint programs.

A few local Superfund programs *did* attempt to remove lead-contaminated dust from home interiors because they held that the contamination had come from the exterior soil (and the nearby industry had been proven to contaminate that soil) and the dust could therefore be considered to be a "waste." Some suggested that Superfund could also abate exterior lead paint to ensure that the soil excavation did not fail by re-contaminating the newly excavated soil. But most Superfund programs did not address lead paint and interior house dust at all.

HUD initiated an attempt to bridge the divide between EPA's Superfund program and HUD's lead paint program in the mid-1990s by focusing on the new Title X definition of a lead paint hazard, which included not only paint but also interior settled contaminated house dust and bare contaminated soil. With Congressional approval, HUD created a special category within its lead paint grant program to initiate a new joint effort with EPA that attempted to address residential paint, dust, and soil at residential hazardous waste sites. But the local polluting industry would sometimes attack the EPA's hazardous waste site investigations by saying that paint (not contaminated soil) was the real problem for children, and therefore they should not be held accountable for paying any portion of the bill for the pollution they had caused.

Of course, this led to confusion among local residents. If EPA treated the soil under the hazardous waste Superfund program, then it would be left to HUD to address the paint. But residents were upset that EPA had only completed the soil part, with HUD later attempting to complete the paint remediation. Why would there need to be a whole separate effort to address lead in paint they wondered? Were they still at risk and had the hazardous waste remediation only been half done?

HUD thought this misunderstanding could be overcome if there was a coordinated response. The HUD program would pay for paint remediation and EPA could order the polluting industry to pay for the soil and dust remediation. For example, HUD awarded a lead paint grant to Palmerton, PA, a small town that had been plagued by a lead/zinc smelter for a century.

Palmerton had the largest lead smelter in the world in 1898 and remained in operation until 1980. In 1982 EPA determined after extensive testing that the area qualified as a Superfund site, but this would launch seemingly endless lawsuits by the owner, Horsehead Industries (also Gulf and Western and Viacom Industries). After 10 years of litigation, the first 24 Palmerton homes had lead dust removal in 1992, but in 1995 EPA decided that it would need to do yet another risk assessment of the site, which confirmed the importance of lead-contaminted house dust. The EPA cleaning was essentially limited to HEPA vacuuming with no soil or paint mitigation. HUD's lead paint efforts in Palmerton began in July 1997, 16 years after the area's first designation as a Superfund site,[55] but only 30 housing units were addressed for lead paint hazards, despite a community task force that concluded all sources of lead exposure should be addressed.[56]

Despite efforts to bring them together, the EPA Superfund and HUD lead paint programs remained largely separated. The attempt to achieve a more holistic solution ultimately failed, and HUD was forced to end its special program to unify Superfund and lead paint hazard control.

This divide persisted. For example, much of the city of Omaha was identified as a lead hazardous waste site as late as 2022 due to historic pollution from a smelter, and although much of the soil was remediated, lead paint in housing remained a work in progress, with thousands of homes left with lead paint hazards.[57] The Superfund program remained unable to address lead paint or contaminated house dust.

This Superfund history is important in understanding the battle over the lead dust standard. The final promulgation of lead dust standards in housing would require overcoming the limitations of the IEUBK model, which required lead dust *concentration* inputs, not lead dust *loading* from the accepted wipe sampling method. Some Superfund sites also relied on a determination of how much of the lead was "bioavailable," even though no bioavailable lead dust exposure limits existed. And the lead dust standards needed to apply to all homes, regardless of whether they were in Superfund sites.

The method used to collect and measure the lead dust needed to be standardized for a validated dust lead limit. All exposure limits have an associated validated specific environmental sampling method, a basic principle that a later court decision failed to grasp (Chapter 6). The specific technical and policy issues to be elucidated by scientific research included the following:

- Vacuum versus wipe sampling;
- Lead dust results expressed in loading ($\mu g/ft^2$) versus concentration (ppm);
- Total versus "bioavailable" lead extraction in the laboratory;

- Dust sample collection efficiency for different sampling methods;
- Ability to accurately predict risk (blood lead); and
- Ease of use and practicality.

To start to answer these questions, a pilot study was conducted in 1993 that examined five different dust sampling methods, carried out with both public and private funding by the new National Center for Lead-Safe Housing.[58] The five pilot methods included an experimental vacuum using a cassette, a second vacuum using a cyclone, an experimental wipe method, and a routine wipe method that had been used in earlier lead abatement studies (this last method included two different laboratory extraction methods, one for "total" lead and the other for "bioavailable" lead). The pilot study did not attempt to correlate the dust results with blood lead levels because that occurred a few years later in a larger study in Rochester NY, described later in this chapter, also funded by the National Center for Lead-Safe Housing.

Two ways of expressing the amount of lead in dust demonstrate the intersection between scientific research and practical solutions. Similar to paint, lead dust could also be expressed as "loading" (measured as the weight of lead per surface area in micrograms of lead per square foot of surface area, $\mu g/ft^2$). Or lead dust could be measured as "concentration," (the portion of lead in the dust, often expressed as parts per million by weight, ppm).

Title X had settled this issue for paint as we have seen. But for dust, concentration had been a standard practice for many environmental contaminants for years and lead was no exception. EPA's IEUBK model used at Superfund sites required lead dust concentration (not loading) as one of its inputs and made assumptions about how much dust a child typically ingested to predict the resulting blood lead level. HUD thought measurements of dust and blood lead could be made directly with fewer assumptions and greater reliability than that offered by the IEUBK model.

There was also a very practical policy aspect to the debate on loading versus concentration. Because cleaning methods could not necessarily reduce lead concentration unless they preferentially removed lead from the dust, cleaning after lead mitigation might never be able to comply with clearance dust standards if they were expressed as a concentration. Or worse, concentration could be confounded by the total amount of dust present.

For example, in the case of a slumlord who wanted to minimize the chances that a lead dust hazard would be detected, simply sprinkling around some nonleaded dust (like sand) would dilute the concentration expressed as parts per million (ppm). But adding nonleaded dust could not change the amount of lead dust present if it was expressed in $\mu g/ft^2$, because the dilution effect did not exist for loading. This was the same as the case for paint. Loading, not concentration, was a more stable and informative metric. But it was not the norm for environmental measurements. A collision between the environmental and housing disciplines ensued.

EPA wanted lead dust concentration to run its IEUBK model. The only way to do that was to collect dust by vacuum methods over a measured surface area. The total dust collected then could be weighed and the amount of lead in the dust determined in a laboratory. Such results could be expressed in *both* loading and concentration. At first, this seemed like an ideal compromise as results could be expressed both ways, but later work showed the vacuum methods to be impractical for use in millions of homes.

The pilot study examined two vacuum methods that could in theory do that, plus an experimental wipe method that used a preweighed wipe that could also result in both loading and concentration metrics. The typical wipe sampling method was only capable of measuring loading, not concentration because it was wiped over a measured surface area but could not be weighed.

In addition to the loading and concentration dilemma, there were also other scientific issues to resolve. Collection efficiencies over measured surface areas might not be consistent or reliable for the different methods. For wipe samples, the amount of pressure applied could in theory vary and maybe interfere with collection efficiency. Loss of dust on the wipe might occur on the way to the lab. Inadvertent contamination of one sample by another through wiping using bare hands or movement of dust caused by tracking on the shoes of sampling personnel were other issues requiring resolution.

On top of all that, there were two methods being used to analyze lead dust in laboratories in this time period. The first involved the use of dilute acid to dissolve the lead in the dust to mimic stomach acid (in theory), the idea being to determine how much lead in the dust was "bioavailable" and absorbed into a child's body. The second was to simply use strong acids to recover virtually all the lead dust present in the sample.

Three important questions needed to be resolved: which methods best-predicted blood lead levels in children; which were most reproducible and reliable; and which could be used in millions of homes? Failure to answer these questions would cast doubt on the most important pathway of lead in paint to house dust to children's hands to subsequent ingestion and ultimately blood lead. HUD had previously employed a "blue nozzle" vacuum method in its national housing survey in 1990 reported in its Comprehensive and Workable Plan, but that method suffered from poor collection efficiencies on most household surfaces and was not correlated with children's blood lead level; it was ultimately rejected. One study showed that wipe sampling produced dust lead results that were 3.4–5.6 times higher than the Blue Nozzle samples.[59]

Another dust collection method was rejected because of uneven performance. This was the use of "sticky tape" that was applied to surfaces, then peeled away and sent to a laboratory to measure the lead on the tape. This method could only capture lead that was on the very top of the surface and missed lead particles that could be stacked one on top of the other, yet still be available to children.[60]

The pilot study findings led to the rejection of two of the five remaining methods: The experimental wipe procedure could not reliably collect dust from the typical uneven surfaces in most housing. The bioavailable method was also rejected, primarily because there was no way to determine if laboratories had sufficient quality control to collect reproducible amounts of lead dust.

This left three candidate methods: A wipe method Fig. 4.1(C) and two vacuum methods. One of the vacuum methods used a modified cyclone from another more complicated dust sampler (the High-Volume Surface Sampler). The modification was developed by Johns Hopkins University, with the cyclone attached to a small household vacuum (Fig. 4.1A). Another dust vacuum method (DVM sampler) used a typical air sampling pump with a 37-mm filter cassette fitted with a special collection nozzle at the intake that had been developed at the University of Cincinnati (Fig. 4.1B). The Hopkins method, known as the Baltimore Repair and Maintenance study method (BRM), would be challenged later in a lawsuit alleging

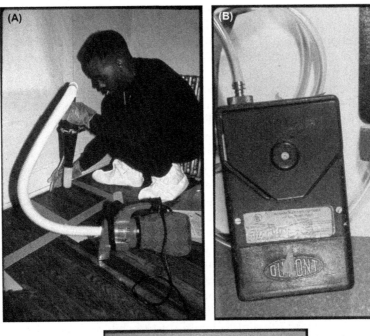

FIGURE 4.1 Vacuum and wipe sampling for lead dust. (A) The cyclone Baltimore Repair and Maintenance (BRM) method, (B) the Dust Vacuum Method (DVM), (C) The wipe method. Source: *Mark Farfel (vacuum method), David Jacobs (wipe method).*

that it was a "nonstandard" method (Chapter 6), even though there were no standard methods at the time, the very reason research was needed.

To conclusively establish which of the three methods was best, the National Center for Lead-Safe Housing funded the University of Rochester in New York to conduct a study that first collected dust using each of the three remaining methods and then compared each to blood lead levels in 205 children who lived in the homes, controlling for other confounding factors, such as season, age, and lead in paint, water, and soil.

The results showed that dust lead loading was a better predictor of blood lead compared to lead dust concentration (as measured by the two vacuum methods). The cyclone BRM vacuum method and the dust wipe method both correlated with blood lead equally well, but the DVM method showed a poor correlation. Because the cyclone method was difficult to use and required extensive cleaning after each sample, the wipe sample became recognized as the best method, combining the ability to reliably predict blood lead (health protection) with ease of use (feasibility) and laboratory measurability.[61]

However, using the wipe method precluded the use of the IEUBK model, which remained a point of contention between HUD and EPA for years.

4.4.1 "Model" wars

Scientists debated what all this meant, and policymakers struggled with how to set the enforceable lead dust standards the nation so desperately needed. Much of the scientific debate swirled around two types of statistical models.

The IEUBK is an example of a "mechanistic" model, which holds that a complex system can be understood by examining the workings of its individual parts and the way they are related to each other. The alternative is an "empirical" model, which examines how lead in children's blood is predicted from lead in paint, dust, soil, and other factors based on direct measurements in numerous replicated studies, instead of the more indirect measurements necessitated by a mechanistic model.

The IEUBK model contained many variables that attempted to predict how external lead exposure would interact with internal stores of lead (such as bone) to estimate the "exposure, uptake, biokinetic, and probability distribution" of lead in the body, with an output reflected in a predicted blood lead level. The environmental factors in that model included soil, house dust, drinking water, air, and food, but not paint. The IEUBK enabled each Superfund site to have its own target level of lead in soil and house dust to meet. But it was obvious that it would be impossible to set individual dust and soil standards for each individual house in the country, although the Office and Management and Budget would propose such an approach in HUD's reform of federal housing regulations (described later in this chapter).

Noticeably absent from the IEUBK model were lead paint and dust lead loading from wipe samples. EPA thought that the contribution from paint would be reflected entirely in the lead dust input term in the model. But the dust input was expressed as a dust lead concentration, not loading. Because lead dust was measured by wipe sampling, it would be necessary to somehow convert loading into concentration. Although attempts to do that were made, many believed it introduced huge sources of error and was at best not well established by the science.

Inevitably, the two units of measure, loading ($\mu g/ft^2$) and concentration (parts per million by weight), are fundamentally different and cannot be converted one into another reliably. The same is also true for wipe and vacuum sampling, a fact ignored by several courts, with disastrous legal and policy consequences.

Although EPA favored the mechanistic IEUBK model, HUD favored the empirical approach. HUD commissioned a new study that pooled many previous empirical studies to help inform what the new dust lead standard should be.

But EPA opposed HUD's empirical approach. The director of EPA's lead program at the time, John Melone, fearful that the federal government might end up with differing lead dust standards, pleaded with HUD not to undertake the new pooled study. But the new HUD scientific staff and many other leading research scientists, including some from CDC and even one from EPA, were convinced that the empirical model was the superior scientific approach and would be more practical in the field.

HUD proceeded with the empirical study in 1996, reasoning that EPA's efforts to date had failed to produce a proposed rule or even a theoretical framework for how to proceed. Title X required EPA to issue the lead dust in 18 months, which would have been April 1994. HUD thought the new study could end the delay. Although HUD and CDC invited EPA to participate in the new pooled analysis, EPA refused.

But one EPA scientist did choose to participate, despite disapproval from her superiors—Kathryn Mahaffey, who had previously worked at the Food and Drug Administration and was responsible for the first National Health and Nutrition Examination Survey (NHANES) blood lead measurements back in the 1970s. Initially, some thought that Mahaffey should be excluded from the HUD study, given her personal relationship with the HUD lead paint office director. But she was outraged by the suggestion, arguing (correctly) that she knew a great deal about the connection between lead in paint and children's blood lead. She had been a principal author on a National Academy of Sciences report on this subject and many others. Independent government ethics officials determined that there was no valid reason to exclude her, and she contributed greatly to the final HUD pooled study.

But so antagonistic was the rest of EPA that the Agency even refused to allow some of the data it had paid Johns Hopkins to collect from being used in the pooled analysis.

HUD commissioned the foremost lead researchers in the country to carry out this complex pooled empirical modeling analysis. The study was led by Bruce Lanphear, who had previously carried out the lead-in-dust study in Rochester that among other things had proven the superior wipe sampling method in predicting children's blood lead levels.[61] He was joined by Tom Matte from CDC, researchers from the National Center for Healthy Housing; the University of Cincinnati, which had been helping the Center carry out the Evaluation of the HUD lead hazard control grant program; Johns Hopkins University, which had also conducted path-breaking work on lead dust sampling and hazard control methods; Woods Hole Institute; Harvard University; Westat (a statistical consulting firm); and HUD staff.

This pooled analysis (completed in 1997 and published in 1998)[62] formed the foundation for both HUD and EPA lead dust standards. It was by far the most comprehensive epidemiologic analysis of childhood lead exposure at the time and confirmed the earlier 1980s studies that lead-contaminated house dust was indeed the major pathway of lead exposure for children. The results also demonstrated that a strong relationship between interior lead dust and children's blood lead levels persisted at levels considerably below the 1995 guidelines from both EPA and HUD (namely, 100 μg/ft^2 on floors).

The scientific process of pooling the raw data from 12 different studies was daunting because some of them used different sampling and laboratory methods, did not always adjust in the same way for other sources of lead, and included children of different ages. These barriers were overcome by adopting specific criteria for which studies could be included and which had to be excluded. For example, each study had to have well-defined protocols for blood, dust, paint, soil, timing of data collection, young children, and not restricted only to children with high blood lead levels to minimize bias. Some studies, such as an Australian study, could not be included because they did not have paint measurements. The final criterion was the availability of raw data that could be pooled into a new dataset for subsequent statistical analysis.

Among the variables ultimately included were interior floor dust lead loading, exterior lead exposure from perimeter soil, play area soil or exterior dust, paint lead content, household drinking water lead, degree of lead paint deterioration, child's race, child's age, socioeconomic status, child's mouthing behavior, and median national values for dust, paint, soil, and water.

What did this mean for establishing a scientifically valid, protective, and practical lead dust standard? The first step was selecting what a target blood lead level would be. The second involved a perhaps more difficult but necessary policy choice—what percentage of children should be protected from developing that target blood lead level? As shown in Chapter 1, once lead had been released into the environment, it was impossible to create a zero-risk world, meaning 100% of children could not be protected no matter how protective the standard was. But it *was* possible to create a low-risk world. Perhaps the best articulation of this key principle is from the National Academy of Sciences.[63] Two key steps were needed to establish the lead dust standard.

First, the target blood lead level: Between 1991 and 2012, CDC defined a blood lead level at or above 15 μg/dL as the trigger warranting an environmental investigation and remediation.

Second, the percentage: To protect most children (95%) from developing a blood lead level above the target level of 15 μg/dL.

Together, these two targets allowed a determination of a resulting floor dust lead level from the new pooled empirical model.

Using these two targets, 40 μg/ft^2 was determined to be the lead dust standard for floors, after controlling for other variables. Because this new standard included both a target blood lead level ($<$15 μg/dL) and how many children were protected (95%), it was health-based for its time.

The struggle to reform federally assisted housing regulations is described later in this chapter. Those 1999 regulations were the first time an enforceable lead dust standard had been promulgated by the federal government. In addition to the health criteria, the ability to measure and attain the standard were important in determining feasibility.

For example, some who commented on HUD's proposed regulation thought it should be limited to "bioavailable" lead. But none of the research had ever shown that so-called "bioavailable" lead was more highly correlated with blood lead than "total" lead. In theory, there are different salts of lead in paint, with different solubility and maybe "bioavailability." Lead chromate may be least soluble, lead oxide and lead carbonate more soluble, and lead acetate most soluble of all. But in practice, solubility was not the same as bioavailability and it simply meant using a weak instead of strong acid to perform laboratory analysis. It had never been shown that using such weak acid extraction was somehow better able to predict a child's risk of elevated blood lead. For years, the industry had attempted to use a similar argument at mining Superfund sites, arguing that lead ore (which is often lead sulfide (galena)) was insoluble and not "bioavailable" (and thus would not be absorbed into the body), absolving them of having to pay for cleanup.

On the other hand, numerous studies showed a good correlation between total lead and children's blood lead, including the pooled analysis, the Rochester lead-in-dust study, and others. HUD and EPA both responded in the preamble to the final regulation on lead dust standards that the agencies have "concluded that the language of Title X supports an interpretation that lead-contaminated dust and soil are covered regardless of the source of the lead." Several housing industry groups, including the National Multi-Housing Council, the National Apartment Association, and the National Leased Housing Associated sued EPA, arguing that this rule would make landowners liable for lead problems from industrial pollution, but the courts rejected it.[64]

In Australia and elsewhere, specialized studies using stable isotope ratios[65,66] did in fact show a significant contribution from paint to children's blood lead. But the method proved to be complex and expensive and sometimes the results were inconclusive because lead with different isotopic ratios had been mixed in the production of lead paint.

In the final regulation, HUD also adopted clearance lead dust standards for almost all federal housing programs, because there had been numerous studies showing that

> without clearance testing and without adequate dust-lead standards, children's blood lead levels may worsen as a result of lead-based paint hazard control work in housing. Therefore, HUD has provided for clearance testing when lead hazard control work is done in housing covered by this rule.[67]

In establishing what the dust lead standard should be, HUD referred to the CDC Statement of 1991, which required an intervention in a home when a child's blood lead exceeded $15 \mu g/dL$. The pooled analysis[62] showed that using a floor dust lead standard of $40 \mu g/ft^2$, more than 95% of children would have a blood lead less than $15 \mu g/dL$. Table 4.2 shows 4.7% of children would have a blood lead level above $15 \mu g/dL$. Although some later suggested the dust lead standard was not "health-based," in fact it was, because the standard was tied to the blood lead intervention level in effect at the time.

In addition to this health criterion, HUD also examined the feasibility of such a standard for both clearance and risk assessment, looking at both the ability to comply as well as the ability to measure accurately. It cited data from HUD's large-scale evaluation of its lead hazard control grantees (described in the next section of this chapter):

> The average dwelling unit undergoing lead hazard control had a median floor dust-lead level of $17 \mu g/ft^2$ immediately following hazard control work. That level declined to $14 \mu g/ft^2$ six months later and remained at the same level one year following the work. Therefore, it is feasible to reach and maintain a floor dust-lead standard of $40 \mu g/ft^2$.

TABLE 4.2 Health basis of HUD dust lead standard, 1999 (15 μg/dL was the CDC environmental blood lead environmental intervention trigger at the time).

Floor dust-lead loading (μg/ft^2)	Percentage of children with blood lead levels greater than or equal to 15 μg/dL (95% confidence intervals)
1	0.1 (0.0–0.6)
5	0.7 (0.4–2.6)
10	1.4 (0.4–4.6)
20	2.7 (0.9–7.8)
25	3.2 (1–9)
40	4.7 (2–13)
70	7.2 (3–18)

From Federal Register, 24 CFR Part 35, September 15, 1999. Preamble to HUD Lead Safe Housing Rule.

Laboratory capabilities were also examined in the preamble to the HUD rule:

> A standard cannot be set at a level that cannot be measured reliably. Many analytical laboratories currently report method detection limits of 25 μg/wipe. For floors, this means a method detection limit of 25 μg/ft^2.... A method detection limit at least 4 times lower than the regulatory standard is desirable to ensure reliable results. For all laboratories in the HUD Evaluation Study, the average method detection limit is currently 11 μg/wipe. Therefore, HUD believes that laboratories will be able to report detection limits of 10 μg/wipe without having to resort to more sensitive and more expensive types of analytical procedures. In short, no increase in analytical cost is expected in order to achieve a detection limit of 10 μg/wipe, which is one-fourth the new floor dust lead standard of 40 μg/ft^2. This will ensure that reliable measures of dust lead loading can be made. A floor dust-lead standard of 5 μg/ft^2 is well below method detection limits reported by most laboratories and is therefore not feasible to implement at this time.

EPA would later adopt the same dust lead standard in 2001. For its part, EPA produced its own analysis,[68] which showed that 95%–99% of children would be protected from developing a blood lead level of 10 μg/dL (not 15) at a floor dust lead level of 40 μg/ft^2 using wipe sampling.[69] The Agency primarily used its mechanistic IEUBK model, arriving at much the same answer as the HUD empirical analysis.

An EPA lead dust panel for the agency's Science Advisory Board a decade later in 2009 analyzed newer data from dust sampling in homes of NHANES children and essentially confirmed the findings from the HUD empirical analysis and the EPA mechanistic analysis.[70]

In short, the policy and science converged at both agencies to produce a new lead dust standard, a historic achievement.

Because the target blood lead level was reduced by CDC to 5 μg/dL in 2012, HUD later issued a new floor dust lead guidance of 10 μg/ft^2 in 2017[71] and EPA, under court order, issued its revised lead dust hazard standard in 2019 to match HUD's.[72] Other studies had also demonstrated the new standard was achievable and that laboratories could measure it reliably.

TABLE 4.3 Floor dust lead standards 1990–2019.

Year	Source	Floor dust lead ($\mu g/ft^2$ by wipe sampling)
1990	HUD Interim Public Housing Guidelines	200
1995	HUD Lead Paint Guidelines and EPA Guidance	100
1999	HUD Regulations	40
2001	EPA Regulations	40
2017 and 2019	HUD Guidance and EPA Regulations	10

Much like the continuing reduction in the CDC blood lead trigger values, the floor dust lead allowable levels were also reduced over time (Table 4.3).

Lead paint inspections, which sometimes included dust samples, were yet another point of contention between HUD and EPA. Paint lead measurements were required for the empirical model. EPA lawyers insisted that HUD's definition of a lead paint inspection required a statutory change in Title X from Congress. This was later shown to be unnecessary, and HUD's position was ultimately accepted by EPA. HUD conducted research to show that the inspection process could be made shorter and more efficient (e.g., by not necessarily taking readings in every single room or not taking multiple readings on the exact same spot), which was reflected in a 1997 revision of the inspection chapter in the 1995 HUD lead paint guidelines. According to one account:

> EPA caved into HUD, but maybe EPA woke up and realized the changes might actually benefit the lead [inspector] and that a protracted fight could have seriously damaged the whole [inspection] profession.[73,74]

The final promulgation of enforceable lead dust and soil standards by the end of the 1990s at both HUD and EPA was a major achievement. The appearance of uniform, scientifically validated lead paint risk assessment and inspection procedures marked another.[75] It overcame the bureaucratic inertia of competing agencies and fulfilled one of the key mandates of Title X.

4.5 Do the new remediation methods work?

The end of the fight to establish lead dust standards was an important milestone, as was the development of inspection and risk assessment standard protocols. But perhaps even more important was answering the question for which Congress had mandated research in Title X and appropriations acts: did the hazard control methods being used by the new HUD lead paint grantees really work? Or were they doing more harm than good, like the early paint removal efforts in the 1970s and early 1980s? Congress mandated a major scientific evaluation with regular interim reports.[76] It wanted increased certainty that the taxpayer dollars appropriated were doing what they were intended to do—protect children.

Most of the earlier studies on lead hazard control had been small and were confined to a research setting, usually including only a hundred or so homes. Carried out by the National Center for Lead-Safe Housing, the University of Cincinnati and the first two waves of local HUD lead paint grantees, this new Congressionally mandated study was the largest ever, covering over 3000 housing units in 14 jurisdictions across the United States.[77] It also was done over the longest time period: 12 years following hazard control for a subset of four jurisdictions.[78,79]

The Evaluation of the HUD Lead Hazard Control Grant Program measured not only effectiveness but also had new cost estimates, the first since HUD had published its Comprehensive and Workable Plan in 1990. The Evaluation also had new data on durability of the interventions, particularly important because interim controls were included in the mix of treatments used by the grantees.

There were five interim reports, which HUD sent to Congress:[80] 1995 and 1996,[81] another in 1996,[82] 1997,[83] and 1998.[84] The final report was released in 2004,[77] and there were at least 14 peer-reviewed publications.[78,79,95,192–202]

The main results showed that for young children who lived in the remediated homes,

- blood lead levels declined by 37%; and
- lead dust levels improved even more, by 66%–95% over the initial 2-year follow-up period, controlling for age and many other confounders.[85]

Dust lead levels improved more than blood lead because the latter included other sources of lead exposure, like playgrounds or other houses that the children might visit where there had been no remediation. Plus, body stores of lead in bone contributed to blood lead, even if all housing exposures had been eliminated.

The initial Evaluation was to last only 1 year, but it was extended to 2 and 3 years to determine how long-lasting the methods would be. Years later, the HUD IG attempted to politically interfere again and falsely accused the lead paint office of improperly extending the study (Chapter 5).

Extending the study proved to be important: Although the initial improvement in blood lead levels was 18%–30% lower after 1 year, the improvements were even higher at 2 years (37%). This was the first study to show that even for children with lower beginning blood lead levels at the CDC level of concern at the time (10 µg/dL), blood lead levels still improved significantly following remediation. Some earlier studies had seen blood lead improvements, but only when they were above 15 or 25 µg/dL at the beginning (these other studies are examined in Chapter 6).

Although the methods used by most of the grantees were a mixture of both interim controls and permanent abatement, they were also found to be durable. Twelve years after the remediation in a subset of 4 of the original 14 jurisdictions, the homes continued to have much lower dust lead levels. (The homes in the other 10 jurisdictions could not be followed). For houses that had all the lead-contaminated windows replaced, the floor lead dust improved by 41%. For those with only some windows replaced the dust improved by 28% compared to the homes with no windows replaced. Even in homes where none of the windows had been replaced, the floor dust lead improved, declining from 10.8 to 1.7 µg/ft^2 from clearance to 12 years later.[79] Windows had the highest lead paint and lead dust levels compared to any other building component before hazard control, and the importance of window replacement was confirmed in other later studies.[86,87]

How did occupants of these homes respond? When a researcher knocked on their door and said, "We did some work to fix lead paint hazards in your home 12 years ago, and we want to see if it is still working," the participants were surprised, but most readily agreed to enroll in the extended evaluation.

This collaboration between occupants and scientists marked one of the major successes during this time period. Houses are a place of refuge and privacy, but the prospect of contributing to the effort to protect children was compelling, enough to convince most to welcome the researchers into their homes.

Initially, the Evaluation and its long-term follow-up study posed important scientific design issues that seemed daunting. Although none of the grantees were using the now-banned paint removal methods like power sanding, abrasive blasting, or torching, they were all doing slightly different things, often due to variations in local law and type of housing. For example, some treated lead paint that was slightly lower than in other places, for example, Maryland used a lead paint trigger level of $0.7 \, mg/cm^2$ but the rest used $1 \, mg/cm^2$. Some favored window repair instead of window replacement. Some favored cleaning with limited painting and a few others chose the more extensive abatement as was done in public housing.

Ideally, there would be a comparison group and the different hazard control methods would be randomized to reduce inadvertent bias in the results. But a true control group would have meant that lead hazards would be measured, and no hazard control implemented, a clearly unethical approach that no one wanted or followed. And randomizing houses to receive different kinds of hazard control proved to be problematic because owners needed to agree to whatever was to be done in the houses they owned. In the end, a pre—post-design was used, which meant that each of the houses would serve as its own control, a well-accepted practice.[88] Other smaller studies did have comparison groups, as we shall see later. These other studies had the same results, that is, the treatments were durable and significantly reduced both blood lead and dust lead.

But the grantees tended to be more focused on meeting the grant goals with HUD than they were on collecting scientific data. They wanted to avoid another Congressional budget rescission. Many had not previously participated in a research project and were unfamiliar with adherence to data collection protocols, quality control procedures, approval from Institutional Review Boards and other research requirements. Although HUD required all the first 11 grantees to participate under the Title X authority, many were reluctant to do so, fearing even more delays.

But they ultimately recognized the importance of proving that what they were doing was effective. Another three grantees agreed to join the Evaluation a year later. The final jurisdictions included a good national representation, with a focus on the highest risk areas in the East, Midwest, and West: Alameda County, CA; Baltimore, MD; Boston, MA; California; Chicago, IL; Cleveland, OH; Massachusetts; Milwaukee, WI; Minnesota; New Jersey, New York, NY; Rhode Island; Vermont; and Wisconsin. The South was not represented, largely because it had chosen not to apply for HUD grants. The data collection began in January 1994.

Designing the study began shortly after the passage of Title X. It was led by the new National Center for Lead-Safe Housing and its research director Warren Galke. The University of Cincinnati was the data coordinating center. Scott Clark, Sandy Roda, Bill Menrath, Paul Succop, and others at the university played important roles in ensuring the validity and analysis of the Evaluation data.

The University of Cincinnati was one of 10 occupational health centers in the nation that had experience in dust, soil, and paint measurement and had been involved in the earlier lead-in-dust study in Rochester. The Cincinnati evaluation team was led by Scott Clark, an occupational hygiene Professor. He would later go on to be among the first to show that new lead paint was still being made in other countries. Together with Jacobs, he worked to pass a new policy to ban all nonessential uses of lead (including new lead paint), which was passed by the American Public Health Association.[89]

To complete the Evaluation, a truly multidisciplinary team was needed. The Center already had experts in epidemiology, housing, environmental sampling, lead hazard control, local lead program administration, statistics, nursing, insurance, and construction. CDC strongly aided the work, offering the expertise of a medical doctor and epidemiologist, Tom Matte, who was experienced in validated study design and ways to control for confounding variables. Cincinnati brought data collection, quality control, analytical modeling, and other expertise to the table. HUD's staff was also engaged, including both its new scientists and existing program managers, including Steve Weitz, Warren Friedman, and Peter Ashley and others.

None of it could happen without the active support of the grantees, which were not being paid extra money to participate. The Center sought to engage them right from the very beginning in the design and execution of the study. But mistakes would be made, some downright comical. For example, one of the preliminary meetings featured a lunch menu designed by the Center's director, Nick Farr. After a morning of deliberation, the grantees filed into the lunchroom. They chose either the soup or salad at the table settings, assuming it would simply be the first course. They expected to be wined and dined to encourage their participation in the study.

But that was all the food there was to be, with Farr later explaining that he did not want them to fall asleep after lunch. Despite his history as a mortgage banker (where he later said he was "paid way too much"), he developed a reputation for minimizing expenses. The Center's initial grant from the Fannie Mae executive would last for nearly a decade as a result. The grantees took the limited food all in stride, and would play a leading role in carrying out the Evaluation. But they vowed to never let Farr do the menu again.

The "glue" that held the study together over the decade between its start and the publication of the final report was Jonathan Wilson. Hired initially as an intern at the Center, he was a quick learner and grasped the many nuances such a study would entail, ultimately writing much of the final report. Galke, who had led much of the study in its earlier years, had polio and passed away before the study could be completed. His contributions to this evaluation were manifest. Sherry Dixon from the Center and Paul Succop at the University of Cincinnati would be key to the complex statistical analyses. Pat McLean and others also played large roles.

The enrolled participating homes in the Evaluation were older, had higher lead paint levels, housed more people per home, were more deteriorated, and had a lower market value than average homes in the United States. This was not surprising since the grant program was targeting high-risk homes. The people served by the program were lower income and were more likely to be of color than the national average, showing the program was well targeted to those needing it the most.

The different jurisdictions also had differing lead paint levels and deterioration (Fig. 4.2). Those homes that had more hazards would also tend to have more intensive interventions. Because of the absence of a control group, some researchers believed the study was inherently flawed, and some so-called "systematic" reviews of the evidence did not include it.[90,91] But its findings were supported by other studies that were able to include randomization and other comparison groups that were possible in smaller studies.[92,93]

One of the challenges in the Evaluation was to find a good way to group the many different hazard control methods used by the grantees. With the help of construction experts like Bob Santucci and Armand Magnelli, work specifications were developed that aided the grantees in obtaining proper bids. These specifications enabled the hazard control methods to be grouped into various "strategies." The most common involved replacement of windows and doors. Interim controls (paint stabilization) were often used on walls, ceilings, and floors, with some enclosure of other surfaces. Table 4.4 shows which methods were most frequent. Large-scale paint removal had been largely abandoned by this time, replaced by building component replacement, enclosure, and paint stabilization.

The cost of each type of treatment ranged from only $430 for cleaning and minor repainting to full lead abatement at $9570 (1995 $US dollars). The Evaluation also tracked exterior costs and strategies, as well as soil interventions.

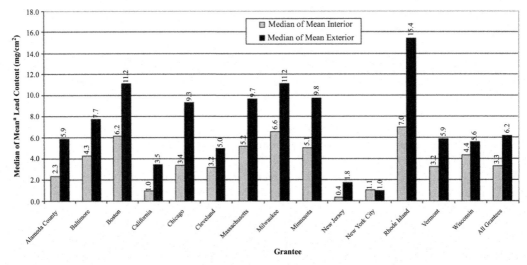

FIGURE 4.2 Paint lead levels by jurisdiction. Source: *From Evaluation of the HUD Lead Hazard Control Grant Program; 1995–2004.*

TABLE 4.4 Lead paint hazard control methods used by HUD grantees.

Component	Paint stabilization	Paint removal	Enclosure	Replacement
Wall/ceiling	✓		✓	
Floor/stair	✓		✓	
Doors	✓			✓
Trim	✓	✓		✓
Windows	✓		✓ (jambs)	✓
Exterior	✓			

From NCHH, Evaluation of the HUD Lead Hazard Control Grant Program.

Which of these many strategies worked the best? Although the focus of the study was to determine if the different methods together protected children, this seemed like a natural question to ask.

Attempting simplicity, the final evaluation report stated, "No differential interior strategy effect was noted for declines in blood lead."

In subsequent court cases, the lead and paint industries would seize on this statement and argue it meant that simple cleaning was all that was needed in their attempt to settle major lawsuits against them. Such an interpretation would reduce their costs if they were ordered to help pay for eliminating lead paint hazards. In later years, a jury in Rhode Island found that they should pay, but that was later overturned by a state appeals court. A judge in California also ruled that they should pay, which was upheld in all higher courts. The Evaluation would figure prominently in these and other court cases described later in this book.

Despite the paint companies' misrepresentations, the report *really* said:

> For the four interior strategies that were examined in the one-year post-intervention blood lead models, window sill and window trough dust lead loadings were significantly lower in dwellings where windows were abated.

Window replacement was also found to be the most durable intervention.

One-time cleaning alone, favored by the lead, paint and pigment industries, which was done in a small subset of the Evaluation homes was found to be *not* effective for very long. Although it did initially reduce dust lead levels, the reductions did not persist at 6 months and 1 year later. After initially large reductions at clearance right after the limited cleaning, dust lead increased markedly across floors, sills, and window wells. The average increases following the clean-only strategy were very high, increasing from 101 μg/ft^2 to 5500 μg/ft^2 at 6 months and 5790 μg/ft^2 at 1 year.[94]

Cleaning alone was a clear failure over the long term, although professional cleaning did have its place as a way to minimize exposures for a month or so until more extensive methods could be put in place. This was not much of a surprise, because failing to correct deteriorating lead paint resulted in higher lead dust.

At the time, most grantees used a floor dust lead clearance value of $200\,\mu g/ft^2$ based on the 1990 HUD Interim Public Housing Guidelines, much higher than the $100\,\mu g/ft^2$ used in the 1995 HUD guidelines and the concurrent EPA guidance cited previously.

But the clearance dust lead levels were much lower than $200\,\mu g/ft^2$. The average (geometric mean) interior floor dust lead loading after cleaning (i.e., at clearance) was only $12\,\mu g/ft^2$, and after 3 years, it declined even further, to only $9\,\mu g/ft^2$.

About three-fourths (76%) of all homes treated in the Evaluation were able to achieve clearance on the first try and of those that failed only one additional cleaning was required in most cases. This important finding helped to show that the methods were feasible, measurable, and protective. And later, HUD was able to use these data to support its reform of housing regulations in 1999.

Perhaps more importantly, there was an overall *reduction* in blood lead levels of 37%, with the majority seeing major improvements. Yet 9.3% of the children in the Evaluation still had a rise of $5\,\mu g/dL$ or more.[95] This was expected because the HUD lead paint grant program could not eliminate all lead exposure outside the child's home. None of the individual hazard control strategies appeared to be associated with the increase, which meant that the exposures had come from elsewhere. The age of the child (children between 1 and 3 years old typically see increases), season of blood lead collection (higher in summer), and parental education level were all variables most likely to be associated with blood lead increases. It was also possible that these children were exposed to lead in other settings, such as playgrounds or in other homes they visited.

Finally, the Evaluation showed a much more complicated picture of various pathways of exposure, with many more variables involved than just the "lead paint to dust to hand to ingestion to blood lead" that the earlier studies in the 1980s had documented (Chapter 1), although dust lead remained central (Fig. 4.3). Importantly, it showed that the "house to blood lead pathway" can be interrupted (depicted by the dotted line) by using the modern HUD hazard control methods used by the grantees. HUD research was also able to show that other pathways were not important, such as lead dust coming from forced air ducts.[96]

The evidence was clear: the program was working.

The interim Evaluation reports to Congress increased confidence among policymakers that the new HUD lead hazard control grant program was indeed protecting children, another example of how good policy (and good government spending) had been informed by science.

In 2022 the Center completed another pathway study in about 350 homes undergoing remediation with the help of Michigan lead paint grantees, including Courtney Wisinski, Carin Speidel, Michael Jacobson, Samantha Crisci, and others, to determine if the pathways had changed over the previous decades. Although lead levels were in fact much lower, the pathways remained essentially the same[97] (Fig. 4.4). Endogenous (dependent) variables are in ovals. Significant exogenous (independent) variables are in rectangles. Solid red lines and arrows indicate statistically significant pathways. Blue dotted lines and arrows indicate that the pathway was not statistically significant.

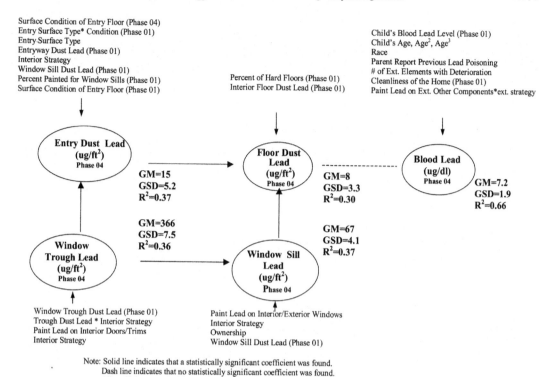

FIGURE 4.3 Pathways of housing and blood lead, 2004 (*dotted line* shows no relationship to blood lead following intervention). GM and GSD are Geometric Means and Geometric Standard Deviation, respectively. *From: NCHH. Evaluation of the HUD Lead Hazard Control Grant Program; 2004.*

4.6 The struggle to reform all federal housing lead paint regulations

The results of the Evaluation and other similar studies such as the Baltimore Lead Paint Abatement and Repair & Maintenance study (Chapter 6) informed how housing regulations should be changed to be consistent with the new science. Title X mandated that HUD and other federal agencies with housing programs reform their lead safety regulations and by 1999 that had been (mostly) completed.

But it almost never happened due to last-minute political threats from Congress and HUD Secretary Andrew Cuomo.

HUD had many different housing programs, as did other agencies such as the Departments of Agriculture (rural housing), Defense (military housing), Interior (historic and park housing) and State (diplomatic housing). Even the US Marshals Service in the Department of Justice had housing, such as those seized in drug enforcement actions, which would sometimes remain occupied by families for years until court action was completed.

Before Title X, all the federal housing programs treated lead paint in widely different ways, none based on a scientific understanding of how exposures actually occurred, and how best to

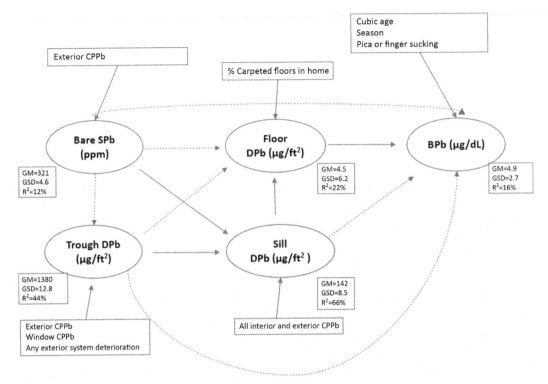

FIGURE 4.4 Pathways of lead exposure, 2022. Endogenous (dependent) variables are in ovals. Significant exogenous (independent) variables are in rectangles. Solid red lines and arrows indicate statistically significant pathways. Blue dotted lines and arrows indicate that the pathway was not statistically significant. *BPb*, blood lead; *DPb*, dust lead; *CPPb*, condition-weighted paint lead; *SPb*, Soil lead; GM and GSD refer to Geometric Mean and Geometric Standard Deviation, respectively. Source: *From Wilson J, Dixon SL, Wisinski C, Speidel C, Breysse J, Jacobson M, Jacobs DE. Current pathways and sources of lead exposure: The Michigan Children's Lead Determination Study (the MI CHILD study). Environ Res (accepted, August 21, 2022) 2022.*

prevent those exposures. Some programs treated only interiors, others treated only exteriors; some addressed only "accessible" surfaces, with differing definitions of what that meant (4 or 5 ft up from the ground). Still others only addressed so-called "chewable" (protruding) surfaces like windowsills. Many programs had a simple ill-defined "broom sweep" cleanup that typically left high lead dust levels behind. None required a dust clearance test to be sure the cleanup had been done correctly, although the 1995 HUD guidelines had recommended it.

The government staff in each program believed they knew what worked best, even without evidence to back it up. Changing that mindset, emphasizing the importance of lead dust, and most important, overcoming fears that the new methods would impose huge costs on each program were all barriers. The Title X task force had helped to reduce the fear of huge costs with its detailed economic analysis, augmented by a new one for the reformed regulations. Initially suspicious of the new lead paint office, most HUD programs came to accept that it had the evidence of what really worked. The reformed regulation came to be known as the "HUD Lead-Safe Housing Rule."[101]

An initial approach in the reform effort could have been to simply adopt a "one-size-fits-all" approach. Indeed, the Title X Task Force in 1995 had presented "standard treatments." Standard treatments were defined as measures aimed "at controlling lead hazards that can be performed by a trained maintenance crew with sufficient knowledge of lead-based paint hazards." The standard treatments consisted of six steps:

1. deteriorated paint repair;
2. making horizontal surfaces smooth and cleanable;
3. eliminating friction on surfaces like windows that produce lead dust;
4. eliminating impact surfaces like poorly hung doors that produce paint chips;
5. covering bare soil; and
6. specialized cleaning and dust tests.[98]

Its chief advantage was that the cost of doing a risk assessment to find out where lead hazards were located could (in theory) be avoided. But this in effect made all paint into lead-based paint since any deteriorated paint was assumed to be lead-based paint. Because the main expense was on the hazard control side, not the inspection side, and because most paint, even in older housing was not actually lead paint, housing programs did not adopt the standard treatment option.

At the outset, there were more than 20 different strategies in existing housing programs before reform. This would ultimately be reduced to seven that were based on the extent and longevity of each program's subsidy (the larger and longer the subsidy, the more intensive the strategy).

The seven strategies promulgated in the Lead-Safe Housing Rule, in order of increasing intensity, were as follows:

1. Safe Work Practices during Rehabilitation;
2. Ongoing Maintenance;
3. Visual Assessment and Paint Stabilization;
4. Risk Assessment and Interim Controls;
5. Inspection, Risk Assessment, and Interim Controls;
6. Risk Assessment and Hazard Abatement; and
7. Lead Paint Inspection and Abatement of all lead paint.

The reformed regulation's goals were stated in the rule's preamble:

> to ensure that housing receiving Federal assistance and federally owned housing that is to be sold does not pose lead-based paint hazards to young children ... HUD does not believe Congress intended that Federal funds be used to subsidize housing that can poison children.

The reform used the experiences of local governments charged with implementing the lead hazard control grants,[99] as well as "the results of new scientific and technological research and innovation on the sources, effects, costs, and methods of evaluating and controlling lead hazards."[100]

All of HUD's lead paint regulations were consolidated, instead of scattered throughout the Code of Federal Regulations. Some program officials thought the lead requirements should be in specific program requirements. But that meant the lead requirements were buried in the many other program requirements, making compliance and enforcement difficult. On top of that, some homes were funded through more than one program, such as home repair and rental subsidy, which made it hard to know which requirements applied.

Perhaps the reform effort's most significant achievement was the increased recognition of the importance of lead dust. Whenever paint was significantly disturbed, the new rule required that clearance dust lead testing was done before children reoccupied the home to be sure it was safe for them. Ongoing lead-based paint maintenance was also required in rental housing, although HUD's authority could only cover those instances in which there was a continuing federal relationship (subsidy) with the property. Scheduled reevaluation was required whenever interim controls were done, because it was still unclear how long interim controls would remain effective, despite the early evidence from the Evaluation study that they were quite durable. And new special procedures were required whenever a child was identified with an elevated blood lead level to end the typical practice of just moving the child into another home instead of fixing the hazards in the current home.

The reform effort was headed by Steve Weitz, a division director within the new lead paint office who had previously worked at HUD's PD&R, Warren Friedman, a scientist and occupational health scientist, and Jacobs (the office's director). Together with others they led the effort to convince the many other offices within HUD and other federal agencies to accept the scientifically validated changes in their programs.

It was up to each program office, HUD's legal department, the Secretary himself, and ultimately the White House through the Office of Management and Budget (traditionally no friend of regulation) that would all need to agree, a process that took years. Title X was passed in 1992, and the new reformed regulation was proposed in 1996 and finalized in 1999. Another 2 years were needed to enable all local programs to comply.

Why did it take so long?

The process included massive public involvement, including public participation in forming the HUD guidelines, the Title X Task Force report, public meetings, and nearly a hundred detailed comments on the proposed rule.

After HUD issued the proposed regulation in 1996, the public comments were quite diverse. In addition to the inevitable concerns over cost, some argued that it set the wrong priorities, spent limited resources unwisely, should stress permanent abatement more (or less), was inadequately focused on units currently occupied by children, paid either insufficient (or too much) attention to lead-contaminated soil, did not defer enough to EPA and/or to private-sector standards, and stressed liability risk-management over health-based hazard control standards. Some also argued that because funding had been historically inadequate for all housing programs, it was an "unfunded mandate," an argument that would figure prominently in last-second attempts to prevent its completion.

Most importantly, the new regulations operationalized Title X's new definition of "lead-based paint hazard." The regulations included two methods: risk assessments that included dust wipe and other environmental sampling to identify lead-based paint *hazards* and inspections to determine the *presence* of lead-based paint (or a combination of inspections and risk assessments). For the first time, HUD also regulated not only permanent abatement (which was mostly under EPA jurisdiction) but also interim controls, which EPA did not regulate at all.

The HUD regulation also benefited from EPA's Title X mandate to require certification of trainers, risk assessors, inspectors, and abatement contractors that had been finalized in 1996, but would not take full effect until 1999, the same year the new HUD regulation became final.[100]

Rather than attempt to regulate program-by-program, the new HUD rule was structured around the *type* of federal assistance that was received, such as housing rehabilitation, mortgage insurance, rental subsidy, public housing, and property sale (disposition). In some cases, it was also structured around the *amount* of assistance, such as for rehabilitation programs. If there was a greater or longer-term investment of taxpayer dollars, the stringency of the requirements also increased. Public housing modernization had the largest funding for rehab and thus required abatement at that time, as we have seen.

For other federally assisted housing rehabilitation receiving more than $25,000 in assistance, hazard abatement would be required; for $5000–$25,000 interim controls would be required (at a minimum); and for rehab getting less than $5000 in assistance safe work practices and clearance (at a minimum) would be required. Attempting to avoid the more expensive lead abatement requirements, some local programs later attempted to reduce the amount of rehab assistance provided to less than $25,000, but most did not. These dollar amounts were specified in Title X itself, based on Congress' determination of what a "high" or "low" level of public investment meant.

A summary of which specific housing programs had which specific requirements is available elsewhere.[101]

HUD's largest housing program, the tenant-based section 8 rental assistance voucher program (in later years called Housing Choice Vouchers) for privately owned subsidized housing, was a particular point of contention. This program relied on low-income tenants to find their own housing from private owners, which had to pass a Housing Quality Standard inspection before the landlord could receive the subsidy. But accepting the voucher was voluntary for landlords and HQS inspections did not include lead paint.

Some argued that HUD did not have legal authority to reform the voucher program, saying Title X was "silent" on it because such housing passed in and out of HUD subsidies, depending on which families rent the home. Others argued that Title X covered all forms of "housing assistance," including rental voucher assistance. And still others argued that the 1937 Act provided HUD with the legal authority to provide "decent safe and sanitary housing," which obviously could be construed to include "lead-safe housing."

To the disappointment of the HUD lead paint office, the argument that in Title X Congress had not authorized reform of the Housing Quality Standards (HQS) that govern the section 8 voucher program won. Those HQS standards only provided for a visual examination for deteriorated paint (but not testing that paint to see if it was really lead paint) and no upfront dust or soil testing and no up-front risk assessment. But the lead paint office did win an important compromise, clearance dust testing.

Such testing would now be required if deteriorated paint was found during the initial HQS inspection. HUD agreed to pay for the cost of those clearance tests through administrative fees charged by its public housing authorities, which was how they paid for the HQS inspections.[102] Therefore "lead-safe housing" units could be obtained in the section 8 program, because many of them had deteriorated paint that would trigger the dust clearance testing requirement following work to make the paint intact.

Conducting upfront lead paint risk assessments in the voucher program would have been more consistent with most other subsidized housing programs, and Congress called for HUD to do so years later in 2014,[103] when Congress ordered HUD to end the

Housing Quality Standards in favor of other more rigorous inspection standards used in other programs (known as Uniform Physical Condition Standards), but HUD did not comply as of 2022.[115]

The Housing Choice Voucher program was one of the few programs for which the lead requirements were restricted only to units with young children. All the other programs covered virtually all federally assisted units built before 1978 when new lead paint was banned, because which homes would have young children is often unknown. Housing reserved exclusively for the elderly or housing in which children were not allowed, such as zero-bedroom units were excluded under Title X. But covering only units with children could trigger discrimination against families, which, although illegal, was often hard to prove.

Furthermore, the regulation exempted properties that were shown not to have lead paint, either because they were built after new lead paint had been banned or they were proven by an inspection to be free of lead-based paint.

However, in recent years it has been suggested that a few houses without lead paint could still have high dust or soil lead levels, either from lead paint that previously existed in the home or from some other source. HUD's national survey in 2000 showed that about 5% of homes with no lead paint still had lead dust hazards, climbing to 61% for homes that had significantly deteriorated lead paint.[104] But including homes with no lead paint was at odds with targeting homes with the highest risks. Identifying homes without lead paint proved to be important in rebuffing last-minute attempts to sabotage the new reformed HUD regulation.

HUD also exempted unoccupied housing that was slated for demolition, because the housing would never have any occupants and thus no lead poisoning. In recent years, demolition of older housing likely to have lead paint has emerged as a significant concern and recent protocols have been developed to minimize the dispersal of lead dust during demolition, typically using fire hoses to wet down the building.[105,106] Demolition remains a largely unregulated source of community lead exposure today,[107,108] although some cities have begun the process of implementing safer demolition of homes with lead paint.[109]

HUD was also faced with the issue of what to do when the amount of deteriorated lead paint was very small: Should a pinhole or other minor deterioration constitute a "lead-based paint hazard?" HUD chose to adopt "de minimis" (minor) areas of less than 2 square feet on interiors or less than 10 square feet on exteriors. In such cases, the rule said the small areas of deterioration must be corrected but would not trigger the need for clearance testing and other requirements, because the amount of lead dust produced would be small. Other practical issues included the following: adverse weather, or a property held only for a short time, or emergency responses during disasters, or exempting housing rehab that did not disturb any paint, such as a furnace replacement.

One creative aspect of the new regulation was a requirement that local housing and health agencies share information on children with elevated blood lead levels and whether their residences were receiving federal assistance. This was intended to end the claim made by some assisted housing providers that they had not known a poisoned child had been identified in one of their units.

In some cases, they had a good point. Because blood lead level information is technically confidential protected health information that is typically provided only to the child's parent or legal guardian, who may or may not pass on the information to the owner, there was some validity to their claim. For their part, housing agencies also closely guarded which units were subsidized to avoid objections in some communities that federally assisted housing should not be allowed in their neighborhoods.

To overcome such obstacles, the reformed regulation required data sharing of the two lists between local housing and health agencies to determine when a "match" occurred between a poisoned child and a subsidized housing unit. CDC and HUD jointly issued a rather unique letter detailing how privacy matters could be both protected but also advance the cause of childhood lead poisoning prevention by data matching.[110]

Similarly, the rule clearly stated that it was illegal for housing providers to discriminate against families with children to avoid complying with the rule. Citing the Fair Housing Act, HUD stated:

> it is illegal for owners of housing to discriminate against families with children, or [elevated blood lead level] EBL children, even if the unit is known to have lead-based paint hazards. Therefore, no renter or buyer may be asked to sign a statement that a child, or EBL child, is not expected to reside in the dwelling.[111]

Some landlords even went so far as to (illegally) require a blood lead test before renting to a family with young children.

There were many other challenges to the final regulation, both within HUD and elsewhere. One major sticking point was whether HUD or federally insured mortgages should be covered by the rule, including properties that HUD would own for a limited time. Title X language was ambiguous because it had included requirements for these programs, but only when appropriations were "sufficient," without defining exactly what that meant.

The multifamily housing mortgage insurance staff within HUD agreed to adopt fairly rigorous requirements for that program: either an inspection or risk assessment followed by abatement or interim controls would be required, depending on the type of insurance.

But for single-family housing mortgage insurance, the HUD program staff refused to update their requirements at all. They argued that if the two other main entities providing mortgage insurance, Fannie Mae and Freddie Mac, which were quasi-independent entities, were not required to do lead inspections and hazard control, then why should the HUD Federal Housing Administration (FHA) be required to do so?

Title X was unclear on Fannie Mae and Freddie Mac. Some argued that because their mortgage insurance was backed by the federal government, it constituted "housing assistance" and was therefore covered. Others argued that Title X had not specifically mentioned Fannie and Freddie, so it was *not* covered. Although at the time HUD in theory regulated Fannie, Freddie, and FHA, in fact they were essentially beyond HUD's ability to do so in any significant way (Since 2008 Fannie and Freddie have been regulated by the Federal Housing Finance Agency).

The single-family mortgage insurance office within HUD did not allow the rule to be released "until the Office of Lead Hazard Control eliminates its requirement pertaining to exclusively FHA single family insured mortgages as opposed to all mortgages."[112] A year earlier, HUD staff in the single-family program office had blocked the rule, stating: "the final rule is unacceptable."[113] Fannie Mae and Freddie Mac still have little in the way of lead requirements as part of their underwriting standards, despite the recommendations in the Title X task force, although there were attempts within Fannie Mae to require lead inspections in 1995.[114] There were repeated calls for them to do so by 2016.[115] But as late as 2022, the only underwriting requirements related to lead paint in Fannie Mae single family housing was a request to determine if there was "visible or documented evidence of peeling paint." There was no requirement to conduct a lead paint inspection or risk assessment.[116]

The FHA Commissioner, who some suggest is the second most powerful figure within HUD next to the Secretary, at the time of the initial development of the rule was Nic Retsinas during the Cisneros regime. He was sympathetic to doing more on lead paint. But he left shortly after the proposed rule was issued in 1996, and the next FHA Commissioner, William C. Apgar, refused to overrule his single-family insurance office staff.

The HUD lead paint office wrote to Apgar:

> It is very likely the Department will be attacked for putting out a final rule that *weakens* existing lead-based paint requirements. The rationale that "Freddie and Fannie don't do it, so why should we?" is not defensible. This would leave the Single Family Mortgage Insurance Program as the *only program HUD administers without any lead-based paint requirement*...The [lead paint] office has received a number of complaints from purchasers who asked how HUD could provide mortgage insurance on a house that had paint chips all over the floor and was unsafe for children...[Fannie Mae and Freddie Mac] indicated that while they do compete with FHA on a number of issues, they have no desire to compete over a children's environmental health issue. They indicated they wanted a standard practice on lead paint in the mortgage industry. HUD can give them that in this rule.[117] (emphasis in original)

The HUD staff in the single-family office replied:

> The [final] rule would place FHA mortgage insurance products at an extreme disadvantage in the conventional marketplace.... We object to the draft final rule because we are concerned that these higher standards will lead to higher costs to the FHA mortgage insurance funds.[118]

Apgar went along with the HUD staff in the single family housing office, and the lead paint office was not powerful enough to overcome the FHA commissioner. In an attempt to get the final rule done after 3 years of negotiations, the lead paint office sent the final rule to the White House Office of Management and Budget for final approval, even though the single-family office had refused to support it. One HUD lawyer wrote:

> I've been informed by OMB that the Office of Lead Hazard Control has addressed all issues with OMB and is very close to clearing this rule. The problem for us internally [at HUD] is the OLHC and Housing have not yet resolved the issues ... for single family housing.... The urgency now is that OMB is close to clearing the rule, and we hope not to have the rule delayed because of internal matters.[119]

The HUD lead paint office decided it could wait no longer. In an email, it stated:

> If [single family] Housing ever does come aboard, fine. But it would be irresponsible to wait any longer. Since a "drop dead" meeting with the Deputy Secretary doesn't appear to be in the cards, we'll let the Secretary make a final decision.... We are going forward with the [final] rule with subpart E [on single family mortgage insurance] reserved.[120]

The HUD Secretary at the time, Andrew Cuomo, yielded to Apgar and the mortgage banker influence within HUD, overruling the lead paint office. The result was that the single-family mortgage insurance program remained the one HUD program that was never reformed on lead paint (the final regulation "reserves" Subpart E for this purpose). The existing requirement for this program (predating Title X) is simply to "cover or remove" deteriorated paint. There are no cleanup, dust clearance testing, or occupant or worker protection requirements for that program.

This would prove to have tragic consequences, resulting in the poisoning of two young children whose parents had moved into a single-family home with an FHA-insured mortgage in Maine. The episode was recounted in a HUD investigation[121] and also was the subject of a Senate hearing in 2000, a year after the final rule was published.[122] Senator Susan Collins from Maine asked the parent (Cindy Pratt):

Sen. Collins: Your home was actually purchased through an FHA loan; is that correct?

Mrs. Pratt: Yes

Sen Collins: As a requirement of that loan, FHA required inspections of the home?

Mrs. Pratt: Yes

Sen Collins: But never did those inspections identify the lead hazards in your home?

Mrs. Pratt: No, ma'am

Sen Collins: That suggests to me that there is a real problem with some of our Federal housing programs as well. It seems to me there needs to be a better job done when a federally secured or provided loan is involved.... And as you said, had you known, you never would have purchased the house.[122]

HUD eventually provided the family with funding to remediate their home; its internal investigation concluded, "The resulting public ill will and Congressional oversight and response will also be costly to the program and to the Department."[121]

Indeed, the same problem had been highlighted in Congress' deliberations leading to Title X. For example, a Senate committee report stated:

Antionette and Timothy Burke, homeowners from Pennsylvania, testified ... that HUD approved their house for FHA mortgage insurance despite the presence of defective lead-based paint. HUD indicated ... the Department made no special effort to inform lenders and prospective borrowers about their ability to use insured funds for lead paint activities.[122]

Although multifamily mortgage insurance had modern extensive lead paint requirements, the much larger single-family mortgage insurance program had antiquated weak requirements that were never updated.

In hindsight, it might have been better to appeal to Retsinas to overrule his staff in the single-family mortgage insurance office before he departed, another example of the importance of having committed people in key positions of power at the right time.

One newspaper reported:

> The decision to exclude single-family homes [with federal mortgage insurance] from the regulation has angered public health advocates and members of the remediation community alike. "HUD should be embarrassed that the rule doesn't cover single-family homes," said Kenn White. "There is a glaring omission in the single-family category," said Don Ryan. "Are people living in single-family homes somehow unaffected by lead?" said White. ... The real reason, some suspect is HUD's competition with other companies in the secondary mortgage market (Fannie Mae and Freddie Mac). ... Chip Coffay of Fannie Mae's single family business operations ... said, "Our policy is to make the appraisal as cheap and quick as possible. ... We would not want the market to see a lead paint regulation as punitive."[123]

Reform of this program, as well as Fannie Mae and Freddie Mac underwriting standards, was never completed and is discussed further in the conclusion of this book.

Another objection to the final rule came from the White House Office of Management and Budget (OMB). Although it thought the cost-benefit analysis[124] had been very well done, it suggested fundamental changes to the regulation that were both impractical and not consistent with the science. OMB guidance[125] held that the standard criterion for deciding whether a government program was justified economically is the discounted monetized value of expected return on investment.

This involves assigning monetary values to benefits and costs, discounting future benefits and costs using certain discount rates (3% or 7%), and subtracting costs from discounted benefits. (A discount rate is employed in an attempt to value the increased benefit of a dollar today versus the benefit of a dollar at some point in the future; discount rates are supposed to be tied to Federal Treasury interest rates, which have been much lower than 3% or 7% in recent years.)

The OMB guidance stated that programs with a positive net present value increase social resources and are generally preferred, but programs with a negative net present value should generally be avoided. In the case of lead poisoning, the costs were immediate, but the benefits accrued to children, most of which are not realized until they grow up decades later. Using a purely economic rule and discounted benefits, no parent would "rationally" choose to invest in their children, because the parents do not get the discounted monetized benefits.

Most economists agree that for intergenerational benefit transfer scenarios like childhood lead poisoning, the discounting process breaks down due to the large number of years involved. Nevertheless, the OMB criterion on benefits outweighing costs needed to be proven, even though Congress had already required the rule.

The cost-benefit analysis (regulatory impact analysis) that the economist Rick Nevin completed for HUD said:

> Employing a three percent discount rate for the lifetime earnings estimates, the Regulatory Impact Analysis concluded that benefits of first-year activities [in the final rule] were $1.143 billion, while costs were only $253 million. Thus, the estimated net positive benefit was shown to be $890 million. If a seven percent discount rate was used for lifetime earnings benefits, the present value of the benefits associated with first year activities is estimated to be $324 million, and estimated costs remain at $253 million. The final rule would therefore entail net benefits of $71 million using the seven percent discount rate.[124]

The benefits included the avoided cost of medical care for lead-poisoned children, avoided special education for children suffering from reduced IQ, and increased lifetime earnings (because people with higher IQ's make more money over time), as well as a number of factors that were impossible to quantify, such as the benefit of avoided litigation and stress on parents and children and others. This made the quantified net benefit likely to be an underestimate.

Some of these additional benefits were updated in 2009,[126] and again in 2017[127] showing that if anything, the benefits were even higher than previously estimated.

Although the net benefits of the reformed regulation were lower at the 7% discount rate, they were still positive for almost all programs. But for some programs, such as HUD-owned single-family and multifamily housing, housing with project-based assistance, single-family housing receiving rehabilitation assistance of more than $5000 per unit, and housing receiving assistance for acquisition, leasing, support services, or operation, there was a net *cost* using the 7% discount rate, but a net *benefit* at 3%. Some feared that this would cause OMB to reject the new regulations for those programs, but it chose not to because of the explicit language that Congress had included in Title X.

But OMB said it wanted to "optimize" the requirements to increase the benefits even further, but it did so by suggesting that only pre-1940 housing units be covered, which would have prevented many homes with lead paint from being addressed at all. This also conflicted with Title X's explicit language covering pre-1978 housing. Randall Lutter at OMB suggested that because both dust lead and soil lead were correlated with blood lead, one way to optimize remediation would be to trade one off for the other. For example, in homes with higher dust lead but lower soil lead, only the dust would be remediated; for homes with higher soil lead but lower dust lead, only soil would need to be remediated.

Aside from being impractical, such a strategy would have placed children at needless risk. Instead of the integrated approach that had been shown to work so well in the Evaluation of the HUD lead grant program where paint, dust, and soil were all addressed, Lutter's proposed approach would use an unproven strategy that in many cases would have left behind dangerous dust or soil lead.

It also created a seemingly endless proposition: should paint not be abated if the lead in water was high? Should paint not be addressed if air or water levels were high? It was similar to the old lead paint industry "shell game" about other sources of lead being more important that had led only to paralysis in the 1970s and 1980s. This approach was eventually rebuffed as well, and OMB eventually approved the final regulation with no significant changes. But the debate dragged on for many months.

One of the more extreme comments on the HUD regulation came from the lead and paint industries, which also attacked the cost-benefit analysis, suggesting that HUD had overestimated the benefits. They said that because children living in HUD-assisted housing were more likely to earn less than the average income, the calculated loss in lifetime earnings was too high. Ethically, this was dubious at best and callous at worst. Reflecting the consensus, HUD stated it was "not appropriate to declare that the value of damage to children in one socioeconomic group is less than the value of damage to children in another socioeconomic group."[128]

One of the limitations of all cost-benefit analyses is that it does not determine *who* pays and *who* benefits, it only compares costs and benefits.

Some housing industry groups thought the new regulations would "break the bank." But the economic analysis showed that as a percentage of revenue, the rule would only have compliance costs between 1%–7% of assisted privately owned housing revenue (Table 4.5). This was similar to what the Title X Task Force had shown previously.

TABLE 4.5 Cost of lead hazard control in federally assisted housing.

Program group	Average incremental compliance cost per unit, all properties	Average incremental compliance cost per unit, high-cost properties	Average annual total revenue per unit, all properties	Average incremental compliance cost as a percent of revenue, all properties	Average incremental compliance cost as a percent of revenue, high-cost properties
Pre-1960 Housing w/Multifamily (MF) Mortgage Insurance	$414	$1120	$8000	5.2%	14%
Post-1959 Housing w/MF Mortgage Insurance	0	0	$8000	0	0
MF Housing w/Project-Based Assistance. >$5K/Unit	$255	$1120	$10,000	2.6%	11%
MF Housing w/Project-Based Assistance. <$5K/Unit	$60	$570	$6000	1.0%	9.5%
Singlefamily (SF) Housing w/Project-Based Assistance	$82	$870	$6500	1.3%	13%
MF Housing w/Tenant-Based Rental Assistance	$59	$560	$6200	1.0%	9.0%
SF Housing w/Tenant-Based Rental Assistance	$103	$870	$6200	1.7%	14%
MF Public Housing	$311	$1120	$7400	4.2%	15%
SF Public Housing	$511	$2095	$7400	6.9%	28%

From Federal Register 24 CFR Part 35, 1999. HUD Lead Safe Housing Rule Preamble.

4.6.1 Cold feet

But perhaps the biggest challenge to the rule came from some in Congress and a HUD Secretary, Andrew Cuomo. By the late 1990s, a wave of antiregulatory fever had swept in, led by House Republicans. Rep. Jim Walsh, a Republican from upstate New York, had been lobbied by large housing industry groups such as the National Association of Home Builders, which had been trying to get Title X repealed ever since it was passed in 1992.

Other powerful interests, including the US Conference of Mayors, the Association for Local Housing Finance Agencies and the National Association of Counties sent a letter to Walsh:[129]

> Although HUD has tried to help us implement the rule and address concerns with training and other efforts ... the new [lead paint] requirements ... will substantially increase the cost of rehabilitation of units, and without sufficient funding to pay for these increased costs, states and localities will be forced to reduce the number of units they will be able to rehabilitate.

In addition to costs, the letter said there were problems with a lack of certified workers and laboratories, unanticipated impacts on the elderly, expenditure rate delays, and other issues.

Although these groups opposed the regulatory reform, others supported it. Jefferson County in Kentucky, which had one of the nation's oldest lead poisoning prevention programs, wrote to Secretary Andrew Cuomo to thank him, and invited him to speak at a celebration where children could "thank our public officials for making their housing lead-safe."[130]

Other housing groups wrote to Walsh and other Members of Congress urging that the regulation not be delayed.[131] The Alliance to End Childhood Lead Poisoning played a key role in getting the groups who supported the rule better organized. Over 51 groups signed onto their statement arguing "HUD's Lead Safety Regulation Must Not Be Delayed,"[132] including the National Low Income Housing Coalition (founded by Cushing Dolbeare who had chaired the Title X Task Force), the National Community Development Association, the National Association of Housing and Redevelopment Officials, the American Public Health Association, the Council of State Community Development Agencies, the Children's Defense Fund, and others.[133]

But HUD Secretary Andrew Cuomo, who was about to leave HUD to launch a successful election campaign for governor of New York, began to get cold feet about issuing the regulation at the last minute in 1999. (Cuomo would be forced to resign from being the New York Governor in disgrace in 2022).

Earlier, he had supported the rule. He had conducted several lead poisoning prevention events, including a visit to an elementary school in DC to encourage parents to have their children tested for lead exposure. He even had his own blood tested in front of the children to show how important it was.[134] But before that happened, what if Cuomo's blood lead level turned out to be high? That could become the headline, not the importance of getting children tested. So, a CDC doctor, Gary Noonan, tested Cuomo's blood before the event (it turned out to be low). Noonan later joked that the ability to "stick" (draw blood from) a cabinet-level official not once but twice was one of the more unusual parts of his career.

Despite Cuomo's initial support, he could not afford to incur the wrath of an upstate New York Congressman that might cost him votes in his race for Governor. Cuomo said he might have to retract the final regulation at the last minute.

But Shaun Donovan, then an assistant FHA commissioner who later became HUD Secretary himself under President Obama, believed the reformed rule was a good one. He said, "Let's see if we can make a deal." He and the HUD lead paint office director went to meet with Valerie Baldwin, a House Committee staffer working for Walsh, who also thought the rule was a good one. She agreed to permit HUD to reallocate funds to pay for the inspections of project-based section 8 housing, an unusual move.

This changed an "unfunded mandate" that the section 8 project-based owners were complaining about into something that was needed to protect children. Later to become known as the "Big Buy," HUD paid for combination lead inspections and risk assessments. George Caruso, the Executive Director at the National Affordable Housing Management Association, played a key role in gaining the support of the project-based section 8 housing owners, other housing associations and working with both HUD and Congress to complete the testing.[135] The testing results showed that there were a few homes that had lead paint in the project-based section 8 housing stock, but most were in fact free of lead paint and the hazards that were identified were less severe than in older housing (much of the project-based housing stock was built after 1960).

Whether this was really a good use of funds (testing homes unlikely to have lead paint) was arguable. But the deal did assuage both Cuomo and Walsh, and the reformed HUD Lead-Safe Housing regulation became final. Had it not been for the intervention of Donovan and Baldwin, it might have never seen the light of day. Again, good people in the right position at the right time mattered.

Although the rule became final and housing agencies were given a year to get ready, some complained that even after a year they were still not able to comply. HUD adopted a tailored approach, so that if local agencies could prove they did not have the needed capacity, they could get additional time to comply, as well as technical assistance to help them. The methods they had to use to prove this were quite detailed.[136]

More than 700 jurisdictions initially filed Statements of Inadequate Capacity with HUD's Office of Healthy Homes and Lead Hazard Control, which qualified them for a 6-month transition period in implementing HUD's lead safety regulation and made them eligible for training subsidized by HUD. This included 42 states (all except AL, FL, MS, NH, OK, RI, TN, and WV). Some large cities with extensive older housing also filed inadequate capacity statements, the notable exceptions including Baltimore, Denver, Los Angeles, Miami, New Orleans, San Francisco, St. Louis, and Washington DC.[137]

This would be extended one more time for an additional 30 days, but only for a few jurisdictions. By January 10, 2002, no further extensions were granted.[138]

In addition, HUD funded nationwide training for workers, created a Lead Paint Compliance Assistance Center, and provided $84 million for clearance testing in Section 8 tenant-based voucher rental assistance program, as well as other funding.[139]

This tailored and phased approach worked. For example, St Paul Minnesota wrote:

As a result of our planning, training, and communication between agencies, we now feel more comfortable and confident that we can successfully implement this regulation.[140]

Many housing organizations launched their own training programs, as did HUD, to enable local programs to comply, including the National Apartment Association[141] and the National Multi-Housing Council.[142] The National Center for Lead-Safe Housing, Quantech, the Alliance and others helped to coordinate the training and technical assistance across the nation.

Part of the problem with low capacity was due to the way training was structured. EPA approached lead paint training as a small, highly specialized activity, with the result that only a few thousand abatement contractors, risk assessors, and inspectors had been trained in 1999. Millions of workers were involved in housing repair and painting work in federally assisted housing, but they had virtually no lead paint training at all. There was considerable argument over what constituted "abatement" (which would require extensive training) and what constituted ordinary repairs. Much of this revolved around the "intent" of the work. For example, a job replacing windows would ordinarily not be considered to be "abatement," yet if the same window had been inspected and lead paint was found, then some argued that this made it an abatement project. The effort to train the millions of workers engaged in rehab gathered steam after HUD reformed its regulations.

The Alliance and others argued that it would be best to streamline the training and end the confusion over who was supposed to do what.[143] A later HUD Secretary, Julian Castro, noted that this system seemed needlessly complicated. He asked, "Is it really necessary to have 7 different lead paint disciplines? Do we really need separate training for lead paint risk assessors, inspectors, sampling technicians, abatement contractors, project designers, workers trained in 'lead-safe work practices,' and workers trained in lead paint renovation, repair and painting?"[144]

A training dispute erupted between the Alliance and the National Paint and Coatings Association (NPCA, the lead, paint and pigment industry's main trade group) as the final rule was released. The NPCA had just released a new brochure, "Preventing Childhood Lead Exposure." The Alliance said that the NPCA's materials should not be used because they did not discuss how surface preparation should be done (this is the step that often releases the most lead dust), did not say that painters should be trained in lead-safe work practices (as the new HUD rule required), and that dust testing need not be done after cleanup was completed (the HUD rule required such clearance dust lead testing).[145]

In response, Steve Sides (an NPCA occupational health specialist), wrote that the industry was "taken aback" by the Alliance attack, saying that the Alliance was a "mouthpiece for trial lawyers" attempting to sue the lead paint companies.[146] Sides also noted that the industry had supported ClearCorps, which was an ill-fated attempt to use AmeriCorps youngsters to perform lead hazard control work (described in more detail later).

A year earlier, the Alliance requested that the Lead Industries Association stop distributing a video featuring a cast character on a children's program, Sesame Street, mainly because it said that just better handwashing and diet would stop lead poisoning. The cast character was "Susan" (Loretta Long) who in the early seasons played a housewife and nurse. In a joint press release between the Alliance and the Environmental Defense Fund, Karen Florini, said:

> The Lead Industries Association should be spending their money to make high-risk housing safe for children, not on industry PR [public relations].[147]

Opposition to the reformed regulation also threatened the very existence of the HUD lead paint office. In the same wave of antiregulatory fervor that gripped the Congress in the late 1990s and 2000s, some suggested that the office was no longer needed, because the various program offices could perform their duties under the new regulation. For example, Representative John Sununu, a Republican Congressman from New Hampshire asked Secretary Martinez at a HUD budget hearing on May 24, 2001 whether the office could be eliminated, because "there are too many HUD programs."

Martinez, also a Republican, pushed back, pointing out that the office was the only source of federal funds targeted at privately owned low-income housing, demand for the grants to local governments was quite large, the office was a center of scientific expertise that did not exist in other HUD program offices, the office was tasked with enforcement, and finally that the Appropriations Act of 1991 had permanently established the office within the Office of the Secretary.[148]

There were two important needs not addressed in the final HUD Lead-Safe Housing Rule: updating the single-family mortgage insurance antiquated requirements and requiring upfront risk assessments in the tenant-based housing choice voucher section 8 program. These were recommended in 2016 but not included in the 1999 final HUD regulation (see the conclusion to this book). Some health entities wanted the final rule to have more abatement requirements. And many housing organizations did not believe additional resources needed for compliance would be forthcoming.

Neither side was entirely happy, but many ultimately felt their needs had been resolved. In 2001 after the George W. Bush Administration assumed office, the National Association of Realtors and the Institute of Real Estate Management wrote:

> HUD's assistance has helped us educate our members…we have reached a constructive resolution to all of our outstanding questions and concerns.[149]

The completion of the reform of all federally assisted housing regulations was an important milestone. The regulation withstood the test of time, with only one relatively minor technical update between 1999 and 2022.

4.7 Improved lead paint testing—how government stimulated private innovation

Part of the regulatory reform effort relied on better measurement. Throughout the lead paint experience before the 1990s, measurement technology for blood, dust and soil suffered from accuracy and precision problems. Lead in paint was no exception.

There had been longstanding problems with the two brands of instruments that were in use to measure lead paint using XRF technology, Princeton Gamma Tech, and Warrington Micro Lead, which had been developed with HUD funding during the 1970s.[150] Although simple to use, research at Georgia Tech and elsewhere had found these instruments produced highly variable results even when multiple readings were taken on the same exact spot. Some of this was due to electronic instability within the instrument, some was due to differing amounts of moisture in the paint film (which interfered with the results) and some was also due to differences in the material underneath the paint. For example, paint on metal would not necessarily produce the same result if the same lead paint was on plaster, wood or drywall. Even the instrument manufacturers admitted in the early 1990s that much of the physics and electronics technology dated back to the 1950s.

Some interim solutions were put in place to try to minimize these sources of error. The 1990 HUD Interim Guidelines had recommended taking three readings on the same spot and then averaging them. If the instrument reading was close to the federal standard of $1 \, mg/cm^2$, then the guidelines recommended scraping off the paint to take additional readings on the bare surface. This "substrate correction factor" (known as a "substrate-equivalent lead measurement") was then subtracted from the reading with the paint present to estimate the actual amount of lead in the paint itself. Clearly, this complicated inspections.

The guidelines also recommended checking the instrument calibration using an instrument manufacturer-supplied paint film with a known amount of lead paint. But the instrument could only be calibrated in the factory, not in the field, unlike most other environmental testing equipment. HUD had opted not to allow such field calibrations in the design of the instruments in a belief that this would make them "idiot proof" and less subject to tampering from lead inspectors, who before 1996 were still mostly unregulated, unlicensed, and poorly trained.

But these corrective measures proved to be only partially effective, and it increased the time, complexity, and cost of doing lead inspections. And no one really knew how well the instruments were performing in the field in a quantitative way.

One of EPA's (and OSHA's) strengths was in standardizing measurement methods for environmental contaminants. For example, validated air sampling methods were developed for use in both ambient air and the workplace to enable compliance with the Clean Air Act and OSHA's lead occupational health standards, both of which were established during the 1970s. Had HUD worked more closely with these two agencies during that decade, perhaps XRF lead paint analyzers might also have been better than they were.

Starting in the mid-1990s HUD and EPA worked together to create a level playing field among the XRF instrument manufacturers by developing more robust quality control mechanisms.

First, HUD paid the National Institute of Standards and Technology (NIST) to develop independent, validated paint films with differing known amounts of lead paint concentrations so that XRF instrument calibrations could be reliably checked across a range of concentrations by inspectors in the field.

Second, HUD also paid to create a national archive of real-world paint film samples that had been collected carefully, usually from public housing authorities. The archive samples were specially tested so that the concentrations of real-world lead paint on differing building materials like metal and wood were known with a high degree of confidence.

Third, in the mid-1990s EPA and HUD conducted research showing that K-shell XRF instruments performed better than did L-shell XRFs because the L-shell X-rays did not penetrate all layers of paint. It also showed that chemical spot test kits were unreliable.[151]

Finally, HUD and EPA obtained some of each commercially available brand of XRF instruments and used them on the archive samples. The results on how well they performed were published for all to see in XRF "Performance Characteristics Sheets" (PCS). The PCS was an independent assessment of how well each instrument had performed on which specific substrates. More importantly, the PCS contained specific procedures to be used to minimize the errors of each specific manufacturer's XRF brand. HUD and EPA required that PCSs be used by lead paint inspectors working in the HUD lead paint grant program and certified under EPA authority, as did the states that licensed lead paint inspectors.

Initially, manufacturers resisted having the government publish estimates of how well each of their instruments performed, especially the two older companies. But new XRF companies now entered the field, sensing a more reliable market in which they could fairly compete. Eventually, all the XRF instrument manufacturers recognized the value of the standardization embodied in the PCSs and the neutral role government played.

Instrument manufacturers in later years regularly showcased their instruments at lead paint conferences, often organized by the Lead and Environmental Hazards Association (LEHA), which brought private and public sector lead poisoning prevention professionals together.[33] Leaders of LEHA included many of the nation's most skilled inspectors, abatement contractors, community groups, training providers, and many others. Key leaders of LEHA included Doris Adler, Richard Baker, Larry Brooks, Vincent Collucio, Neal Freuden, Kate Kirkwood, Tom Laubenthal, John MacIsaac, Jack Paster, Ron Pike, Susan Rosmarin, Michael Sharp, Howard Varner, Lee Wasserman, Steve Weil, Kenn White, John Zilka, and many others.

The XRF companies showed how well their respective instruments performed on the latest PCSs at the conferences. For example, Jack Paster and Sia Afshar, Hal and Lee Grodzins, Bill Radosevitch, John Pesce, Scott Clark, and others were particularly effective in articulating the need for reliable testing technologies and creating successful business and testing models to enable the lead testing field to expand and mature.

Not only did all this serve to improve the accuracy, precision, and reliability of lead paint measurements, but it also stimulated innovation by instrument manufacturers. Now that there was a clear standardized measurement protocol in place, manufacturers could claim that their newer models were better than the old ones (and they could prove it). Using newer technologies and the government testing program, private-sector competition improved, and more importantly, instruments became faster and more reliable, driving down the cost of inspections and improving their reliability.

Further government action also streamlined lead paint inspection protocols. By 1997 it had become clear that most homebuyers were not taking the opportunity to have lead paint inspections under the disclosure rule. Some of this was due to the perceived high cost of lead inspection, which required changes to the HUD guidelines and EPA regulations. One expert stated:

> HUD has correctly realized that a major reason no one is having their home tested for lead-based paint is that the inspection is very time-consuming and very expensive.[152]

HUD also did a study to see how all this was working by determining how much variability there was from one inspector to another. It found that the rates of false negatives and false positives were both below 10%, which created even more certainty in the private market.[153]

The lead paint archive still existed in 2022 and was still used to produce PCSs, although it was almost abandoned when HUD initially reduced funding to maintain it, but then reversed course, recognizing it as an invaluable national resource. In fact, the archive was almost lost when leaks caused major damage in 1997.[154]

The effort to determine how well XRF instruments worked was a demonstration of how government standards can benefit the private sector. Today, calls on the government to reduce regulatory burdens ignore positive examples like this of how the two can work together to serve public needs and create the conditions for successful private companies to form, compete and prosper. One such example of a public/private partnership was the lead paint testing experience from the mid-1990s.

4.8 National lead laboratory accreditation program

The same problem existed for laboratory-based testing. Virtually all laboratories rely on standard operating procedures to ensure their results are reliable. But for those analyzing lead in paint chips, dust, and soil, during the early 1990s there was no independent and validated way for them to assess whether their reported results reflected the real amount of lead present, that is, whether their results were reliable.

When the Housing Authority Risk Retention Group began its program to reduce liability exposures for public housing authorities in the late 1980s, it had to create its own independent effort to manufacture wipes with known amounts of lead dust using Georgia Tech's labs and a powdered lead paint Standard Reference Material made by NIST at the Department of Commerce and a special sample preparation and weighing protocol. This was subsequently augmented by reference labs, such as the University of Cincinnati, Johns Hopkins University, and the Wisconsin Occupational Health Laboratory at the University of Wisconsin-Madison.

The use of nonanalytical grade baby wipes also hindered laboratories' ability to dissolve all lead dust. Most forms of lead dust are highly insoluble, which is why strong acids and heat are needed to get it into the solution. If the lead dust is not fully dissolved, then the amounts reported by instruments such as atomic absorption spectroscopy or inductively coupled plasma emission spectroscopy, which analyze the amount of lead in aqueous solutions, will be underestimated.

This problem was first recognized for lead dust wipe samples by the Wisconsin Occupational Health Laboratory. The Georgia Tech connection, which operated the OSHA compliance assistance program for the state, proved to be essential in developing the necessary lab methods.

Many of the existing laboratories were doing a very poor job of analyzing dust wipe samples. Levels of lead dust reported by the labs in 1991 ranged from 47% to 150% of the true value.[155]

Working with the National Institute for Occupational Safety and Health and the American Industrial Hygiene Association, a protocol was developed to assess laboratory performance in 1994, with both EPA and HUD support.[156] Title X mandated that both EPA and HUD establish procedures to increase the nation's capacity, not only by increasing the number of lead inspectors and abatement contractors but also by ensuring the nation's laboratories were capable.

The result was the creation of the National Lead Laboratory Accreditation Program (NLLAP) in 1997.[157] Administered by EPA, it defined the minimum requirements and abilities that a laboratory must meet to attain EPA recognition as an accredited environmental lead testing laboratory. EPA and HUD and other federal and state agencies established NLLAP to recognize laboratories that demonstrated the ability to accurately analyze paint chips, dust, or soil samples for lead.

Each lab went through on-site assessments. It also had to successfully perform within established tolerance limits on a continual basis in the Environmental Lead Proficiency Analytical Testing (ELPAT) Program, also administered by the American Industrial Hygiene Association to ensure that laboratories accurately analyzed samples for lead. The ELPAT program supplied samples with known amounts of lead in paint, dust, and soil (much as the public housing insurance program had done years earlier) to determine if the reported amounts had sufficient recoveries and reliability.

In the 2000s, this became a new international standard, the International Organization for Standardization and International Electrochemical Commission (ISO/IEC) Standard 17025:2005 (E) "General requirements for the competence of testing and calibration laboratories." The experience from the occupational hygiene field proved to be an essential ingredient.

Two organizations initially accredited laboratories, the American Industrial Hygiene Association and the American Association for Laboratory Accreditation, under EPA's authority; in practice, there were few differences between the two, although it did spark competition.[158] By 2022 the organizations that accredited these labs for EPA are the American Industrial Hygiene Association, American Association for Laboratory Accreditation, Perry Johnson Laboratory Accreditation, and the ANSI-ASQ National Accreditation Board. EPA published lists of approved laboratories, so consumers and inspectors and housing owners could get accurate and precise results.[159] The program ensured that variability between and within laboratories in analytical results was negligible.[160]

Laboratories analyzing blood lead specimens participated in a quality control program that predated Title X. Established in 1988 under the Clinical Laboratory Improvement Amendments, laboratories participating in the CDC program today accurately measure blood lead with a success rate of 96%.[161]

All these improvements were essential in overcoming the measurement problems that Congress had identified in the late 1980s (Chapter 3) and further increasing confidence.

4.9 First enforcement actions

For years, some worried that rigorous enforcement of national and local lead laws would cause property abandonment and/or reduce housing values. As early as 1988, this theory had been debunked, with one housing price study showing that abandonment and housing increases from lead abatement did not occur.[162]

At the federal level, new enforcement actions began with the disclosure rule, one of the first to appear under the Title X authority. The legal structure of the disclosure law was daunting when the first attempts were made to enforce it. Some government lawyers initially thought that the legal settlements with those who had violated it had to be restricted to simply getting an agreement for better disclosure going forward. But others, including

John Shumway, John Kennedy, and Jacobs at HUD, together with other senior Department of Justice lawyers (such as Lois Schiffer and Bruce Nilles), and EPA enforcement officials (such as Eric Schaeffer), argued successfully that disclosure was not an end to itself. Instead, its intent was to stimulate the production of housing without lead hazards. They argued that settlements could do exactly that. EPA and HUD issued a joint letter stating that penalties for owners' noncompliance could be reduced "by taking reasonable steps to identify and address such [lead paint] hazards."[163]

Over a hundred thousand housing units would be abated with private funding from owners as a result of enforcement and settlements with landlords during the 1990s, described later.

There were also important revisions between the proposed disclosure rule and the final one that in many ways made it more enforceable. For example, the proposed rule would have treated lead paint as a real estate "contingency," meaning that if lead paint was found, a sales or lease contract could be invalidated. Instead, the final rule treated a lead inspection just like any other home inspection. The proposed rule also required that the pamphlet that was to be distributed at sales and leasing transactions be read "and understood" by the buyer. But of course, there was really no way to prove if the buyer or renter had really understood the pamphlet and the final rule only required that it be distributed.[164]

The requirements in the disclosure regulation were straightforward. In addition to distributing an educational pamphlet (Fig. 4.5), the lease or sales contract had to have a specific Congressionally written lead hazard warning short statement. It also required that a document would need to be signed by both the owner (or a designated agent) and with the buyer/renter stating that disclosure had in fact occurred.

The fines for noncompliance could be substantial. There were 11 possible violations, and each carried a fine of up to $11,000 (1996 US$) for each housing unit for each sale or lease. For a property owner with hundreds of units rented or sold multiple times, the potential fines could quickly amount to millions of dollars.

The final rule was released in 1996. The press conference announcing it featured unlikely partners: Margaret Sauser, the mother of a lead-poisoned child and one of the founders of United Parents Against Lead joined with the President of the National Association of Realtors. After recounting her story of how her two children had been poisoned, Sauser said that "this law is going to stop [lead poisoning] from happening to another family."[165] Art Godi, President of the National Association of Realtors said the rule was consistent with their policy to disclose all material property defects. "Realtors want homeowners to live in safe homes and are ready to do their part."[166] He later would work to train all real estate agents in the nation to comply. HUD Secretary Henry Cisneros and EPA Administrator Carol Browner also spoke in support from a housing and environmental perspective. Browner said the law was a "right to know."

A press release issued by the Alliance to End Childhood Lead Poisoning in 1996 said,

> The American dream of owning a home turns into a nightmare for too many American families when lead poisons their children ... everybody wins with the disclosure of lead hazards.[167]

Because it was a joint regulation, it was unclear which agency would enforce which parts of disclosure. Initially, EPA suggested that HUD should cover federally assisted

FIGURE 4.5 Lead paint disclosure pamphlet and warning language. Source: *From EPA, CPSC and HUD, Protect Your Family; 1996.*

housing, most of which already had lead paint notice and disclosure requirements predating the new law, and EPA would enforce it for all other homes. HUD rejected that idea because federally assisted housing was at most only about 4% of the nation's housing stock, EPA was much more susceptible to endless litigation, and HUD had more experience in real estate transactions.

Both agencies worried that enforcement of what some thought to be a mere paperwork requirement would appear as a heavy-handed government requirement with no benefit to children.

HUD already enforced the Real Estate Settlement Procedures Act, which required full disclosure of all costs during real estate transactions. With its new lead grantees, HUD was also better connected to local health departments than EPA, which tended to operate through its 10 regional offices across the country. HUD was concerned that EPA, with its history of narrow legal interpretations and litigation, might not be able to enforce the law fully.

On the other hand, HUD had little experience in enforcing a health law and only EPA had subpoena authority. It would take both agencies to achieve effective enforcement. HUD and EPA enlisted the Department of Justice (DoJ), the nation's main enforcement authority when it came to national laws like the disclosure rule. DoJ had recently formed a new environmental crimes section under the Clinton Administration, and the lawyers there proved eager to help. One of them, Bruce Nilles, who would later play an important role in the President's Task Force report on lead in 2000, remarked: "This is why I went to law school—to help people, not write arcane legal opinions on things that don't matter."[168]

Nilles and his colleagues, together with HUD and EPA decided their first enforcement actions should focus on local landlords in DC, because DC had many older homes and lead-poisoned children. They identified some likely noncompliant landlords, who in some cases had hundreds of housing code violations and had poisoned large numbers of children over decades. One of the first cases involved a landlord who lied to federal officials that he had complied when in fact he had not (Fig. 4.6). He knew about the lead disclosure rule and had 2283 housing code violations. He pled guilty to obstructing justice and making false statements to conceal his failure to disclose, marking the first criminal case in the United States involving lead paint disclosure. He served 2 years in jail, was fined $250,000 for each of six felony counts, and agreed to do lead abatement in his properties.[169,170] Other cases would soon follow.

Strong penalties were imposed in the hopes that it would send a message to other landlords that disclosure mattered. In the DC case, the owner had backdated disclosure forms. In a New England case where a young child had died from lead paint exposure, forgeries of tenant's signatures, backdated forms, and attempts to throw records away were discovered.[171]

These enforcement efforts received important support from the US Attorney General, Janet Reno, as well as Eric Holder, who would later become Attorney General himself. Both recognized that lead poisoning was not only a health problem but also an important civil rights and environmental justice issue.

HUD's lawyers, including John Shumway, John Kennedy, Lee Ann Richardson, Kevin Sheehan, and Teresa Payne, together with HUD staffers Matt Ammon, Carolyn Newton, Tara Jordan-Radosevitch, and others worked with Department of Justice lawyers, EPA lawyers, local grand juries and local prosecutors to win numerous cases.

These and other cases would result in a Special Commendation for Outstanding Service from the US Department of Justice (Fig. 4.7).[172]

Landlord jailed, fined in lead-based paint warning case

ANY doubts over the federal government's seriousness about enforcing lead-based paint regulations could have dissolved with the recent criminal conviction of a Maryland landlord for brushing over the truth with the Department of Housing and Urban Development.

The 65-year-old landlord was sentenced to two years in prison and fined $50,000 for obstructing a HUD investigation and making false statements to investigators. He lied to hide his failure to tell tenants about the lead-based paint in his low-income apartment buildings in suburban Washington.

It was the first criminal prosecution related to failure to give lead hazard warnings, which are required by the federal Lead Hazard Reduction Act of 1992, according to the Department of Justice, which brought the case in federal court in Greenbelt, Md.

As part of a plea agreement, the landlord has provided tenants with notices about lead paint assessments performed by an independent contractor, the Justice Department said in a news release posted on its Web site, www.usdoj.gov/.

The owner "admitted that he had notice of actual lead-paint hazards in one of his apartment buildings from District of Columbia lead inspectors, who informed him that they found lead in the building. However, (he) failed to disclose actual and potential lead hazards before leasing to tenants," the Justice Department said.

The landlord admitted trying to obstruct a HUD investigation "by backdating his signature, backdating tenant signatures and directing tenants to backdate forms by entering the date they moved into their apartments, rather than the date they were actually warned about health risks, which was after they had moved in," the Justice Department said. "In some cases, the tenant signatures were signed by ... resident property managers."

The Apartment Association of Central Okla-

RICHARD MIZE

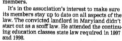

REAL ESTATE

homa takes steps to make sure its member landlords follow the law, a spokeswoman said. The association even sells the required forms to its members.

It's in the association's interest to make sure its members stay up to date on all aspects of the law. The convicted landlord in Maryland didn't start out as a scoff law. He attended the continuing education classes state law required in 1997 and 1998.

Owners of older duplexes and smaller apartment houses, with no association to help look out for them, could be breaking the law — out of ignorance, or out of carelessness.

And, of course, that's bad news for people living in their buildings, who are the ones potentially exposed to the dangers of lead-based paint.

The Justice Department took the opportunity of the conviction to remind landlords and interested tenants, of the basics:

"The Lead Hazard Reduction Act requires landlords to give tenants warnings, which can be done by using a standard disclosure form, about actual and potential lead paint hazards present in the property, and an Environmental Protection

Agency pamphlet about how to minimize the dangers to children. The law also directs landlords to document their compliance with the law by keeping lead disclosure forms and tenant signatures on file."

All of which is regulation. Let's not forget the real reasons for it:

"Lead poisoning is a significant health risk for young children," the Justice Department said. "Although ingesting lead is hazardous to all humans, children under 6 are at the greatest risk of lead poisoning because their bodies are still developing and because ordinary hand-to-mouth activity brings them into frequent contact with lead in paint chips, dust and soil.

"Lead adversely affects virtually every system of the body, and it can impair a child's central nervous system, kidneys and bone marrow. At high levels, lead poisoning can cause coma, convulsions and death. Lead poisoning is especially acute among low-income and minority children living in older housing."

Real Estate Editor Richard Mize can be reached by e-mail at richardmize@oklahoman.com or by phone at (405) 475-3518.

FIGURE 4.6   One of the first lead paint disclosure enforcement cases. *Reproduced with permission. The Oklahoman and Richard Mize, April 6, 2002. p 2C.*

FIGURE 4.7   Recipients of Award from Department of Justice for Lead Paint Disclosure Rule Enforcement. From left to right: Teresa Baker Payne (HUD), David Jacobs (HUD), Rachel Jacobson (DoJ), John Kennedy (HUD), Lois Schiffer (DoJ), not identified (DoJ), Carolyn Newton (HUD), Janet Reno (US Attorney General), Bruce Nilles (DoJ), Mark Sabath (DoJ), Eric Holder (DoJ), John Shumway (HUD), David Topol (DoJ). Not pictured: Matt Ammon (HUD). Source: *Environment and Natural Resources Division, US Department of Justice, April 28, 1999.*

Predictably, landlords were unhappy about the prospect of enforcement of the disclosure rule, particularly in DC where some of the first actions had been taken. The delegate to Congress for DC, Eleanor Holmes Norton, organized an evening session in the basement of the Capitol to respond to some of her constituents, which was attended by over a hundred very upset landlords.

After HUD explained what the law required, they screamed, complaining that they had not been made aware of its new requirements and that DC landlords were being unfairly targeted. They also attacked the disclosure law as just more bureaucracy and paperwork that would do no good.

But Delegate Norton rose to the challenge. She said,

> Look, I used to run the US Equal Employment Opportunity Commission,[173] and here is how it works. HUD told you more than a year ago what you had to do to get ready. Then they told you again when the regulation was first issued, and the clock started ticking. Then they told you yet again when it was time to comply a year later. Now they are telling you that time is up, and they are coming after you if you do not comply with the law. Look, it's simple: hand out the pamphlet, get the forms signed and provide the information. It protects all our children.

After that, the angry words stopped; they knew she was right.

Even the main lawyer at the National Association of Realtors knew better. She wrote to her membership:

> Compliance is in your control...Compliance is not difficult...comply with the law and protect yourself, clients, and customers without slowing down transactions.[174]

Later, Delegate Norton confided: "You know they were desperate when they came to me for help, because they know I'm a tenant's rights advocate." Unknown to both Norton and Jacobs at the time, they were alumni of the same undergraduate college, Antioch, although they had attended at different time periods.

It was one thing to levy and collect fines, but what about fixing the lead paint problems in the homes where the violations had occurred? HUD and DoJ proposed that although some small fines would need to be issued because there had been a violation, much of the money could be devoted to actually fixing hazards in homes where disclosure had not occurred if the owners agreed.

But a few EPA lawyers initially disagreed, saying that the disclosure law did not require correcting any hazards, only disclosure. The best that could be hoped for they said was compliance going forward.

But HUD and DoJ lawyers thought that in a settlement with a noncompliant owner or agent there was much more flexibility and an opportunity to create homes safe for children. HUD and DoJ lawyers argued that landlords could be given a choice: either pay the fine or fix your housing to protect children. Most (but not all) voluntarily chose the latter. As a result, in the early years of enforcement, there would be agreements to test and abate 166,000 privately owned homes, worth $22 million in fines, inspections and abatement in 47 states and DC. Eric Schaeffer, then the director of EPA's Office of Enforcement and Compliance Assistance, who worked closely with HUD and DoJ to bring about the settlements, paved the way for many other EPA attorneys such as Mary McAuliffe, Stephanie Brown and others, to vigorously pursue enforcement of the disclosure rule.[175]

But a few landlords went to court. In one of the first such cases, HUD's lawyers looked like they were hopelessly outnumbered. HUD had only two lawyers dedicated to the case and HUD's lead paint office director was their only expert witness. On the other side, representing owners were a dozen highly priced private lawyers, armed with their own experts. The case was against American Rental Management. The Administrative Law Judge in that case found that the company's "violations can best be understood as a manifestation of slip-shod management practices resulting from a failure to take the [disclosure] law seriously," ruling in HUD's favor.[176]

At the first joint lead paint press conference between HUD, the Department of Justice, and EPA, the efforts to target DC landlords were publicized because they had clearly worked. The settlements resulted in the inspection and abatement of lead paint in 12,000 homes in DC alone, worth about $2.6 million, plus $181,000 in community projects to reduce lead poisoning and over $200,000 in fines.[177]

Many other enforcement actions quickly ensued. A case in Minnesota resulted in nearly 4500 apartments to be remediated, costing the owners nearly $1 million and paying a fine of $10,500. A community advocate there, Sue Gunderson, said,

> I want to thank HUD, EPA, and the US Attorney's office for their strong leadership. The big winners in this legal action are the children of Minnesota.[178]

Another DC case involved two of the district's biggest landlords, resulting in a settlement to abate lead paint hazards in 4500 homes. There had been many poisoned children in these properties and the action also triggered reform of the local lead poisoning prevention law.[179]

After the initial DC cases, some of the other first enforcement cases were brought against the US Navy, which owned housing occupied by young children, a real estate company in Oklahoma, two landlords in Philadelphia where a home with lead paint had been condemned as unfit for habitation, and another landlord who had agreed to address lead paint.[180]

A new division within the HUD lead paint office was created to enforce the disclosure rule. Although it had only limited staff, it successfully brought new cases in the years to come. This was temporarily delayed as Secretary Cuomo attempted to move enforcement out of the lead paint office and the HUD legal office into a new Departmental "Enforcement Center." Both Matt Ammon from the Lead Paint Office and John Shumway from the Office of General Counsel (HUD's legal office) were transferred briefly to that new Center, which was headed by Jon Gant, who later directed the lead paint office.

The Enforcement Center turned out to be ineffective, primarily because it was an attempt to separate enforcement from program offices, where much of the expertise was present. In a memo from HUD's General Counsel to Secretary Cuomo, the results of the unique collaboration between the HUD Lead Paint Office, the Office of General Counsel and the Department of Justice were highlighted:[181]

1. An enforcement protocol ... to target the most egregious nondisclosure cases;
2. An agreement with the HUD Real Estate Assessment Center to routinely check for disclosure in public housing and project-based section 8 housing;
3. An agreement from the Assistant Attorney General for the Environment and Natural Resources Division of DoJ to provide additional assistance from US Attorneys nationwide;
4. Meeting with the National Association of Attorneys General to explore state attorney general participation;
5. An interagency agreement for funding for investigations and establish a hotline for tips and complaints of violations; and
6. The presentation of the Assistant Attorney General's Award for Environmental Justice to the Joint HUD/DoJ team.

This failed model of a separate enforcement center was similar to that of EPA, which had developed its Office of Enforcement and Compliance Assistance (OECA), separated from the EPA lead paint program office. In 2001 the HUD Enforcement Center proposed to hire 28 additional staffers to enforce the lead paint regulation, but the Office of General Counsel succeeded in convincing Secretary Martinez that enforcement would be better handled by the HUD lead paint office, with support from HUD and DoJ lawyers.[182,183]

Perhaps there are merits to separating enforcement from program offices. But in the case of lead paint, the separation of program expertise from enforcement proved to be a mistake within both EPA and HUD. The number of lead enforcement cases fell precipitously after the move to the HUD Enforcement Center. EPA's OECA staff lawyers in Washington headquarters sometimes worked to *prevent* settlements with those who had violated the disclosure rule, in the belief that the rule could not allow abatement as part of a settlement. But after appeals to Eric Schaeffer, the OECA Director, more homes were abated as settlements of disclosure violations. This was memorialized in guidelines of civil penalties issued jointly by HUD and EPA.[184] Some lawyers in EPA regional offices, such as Mary McAuliffe in Chicago and her colleagues, agreed with HUD and DoJ and achieved important settlements with some of the nation's largest landlords and their agents that produced thousands of abated housing units.

Defining how the rule was to be enforced was complex. It was documented in a enforcement handbook[185] and a "Federal Lead-Based Paint Enforcement Bench Book"[186] aimed at judges and attorneys. The latter was written largely by Stephanie Brown at EPA. Both documents included specific procedures on how to respond to tips and complaints, tenant affidavit forms (including a medical release form if elevated blood lead levels were involved), sample warning letters to owners and landlords, targeting procedures, standard operating procedures for inspections to determine compliance, frequently asked questions to guide compliance inspections, how to refer inspection findings to local and federal law enforcement personnel, how to review leases and sales contracts, subpoena forms, how to undertake administrative law proceedings, draft settlement procedures, draft Administrative Law judge findings, consent decrees, criminal referral letters, abatement work plans, monitoring the abatement work to ensure it complied with settlement agreements, and other matters.

The confidential nature of elevated blood lead levels also posed a barrier to targeting enforcement of the disclosure rule to those properties where the risks were greatest. Some local health departments thought that HUD should not be allowed to receive such information, because it was not a public health agency.

This was overcome by a unique joint letter from CDC and HUD, which interpreted the Privacy Rule to include HUD. It stated:

> For those agencies and institutions that are "covered entities" under the Privacy Rule, the [HUD Office of Healthy Homes and Lead Hazard Control] is functioning as a public health authority as defined by the Rule (45 CFRR 164.501)...Therefore you may disclose to OHHLHC, without authorization, the information that [includes]...reporting the address where there is a history of lead-based paint hazards and/or children with elevated blood-lead levels...[110,186]

By 1999 HUD had developed a system to obtain lists of lead-poisoned children to target its enforcement work, including 40 cases in 20 cities.[187]

HUD also developed an enforcement system that integrated disclosure compliance as part of its normal inspections of federally assisted housing, mostly conducted by HUD's Real Estate Assessment Center.[188]

Another benefit was the extensive education on lead that resulted from the disclosure rule. One study suggested that 45% of renters had not received their lead paint disclosures in 1997, but a year later about 85% of renters had received their disclosures.[189]

HUD also launched its Campaign for a Lead-Safe America, with the wife of the vice president, Tipper Gore, as its spokesperson (the Campaign is described in detail in Chapter 5). The campaign was initially tied to disclosure rule enforcement efforts, as well as partnerships with four major hardware store chains, and HUD distributed over 9000 books on lead paint safety to libraries across the country. HUD and State Health Departments, such as Rhode Island's, created memoranda of understanding to ensure that children living in federally subsidized homes were matched with lead-safe housing units.[190]

The Campaign would later be sabotaged by Secretary Andrew Cuomo and others, but initially there were many partners who committed to supporting it, including:

> Amerispec (one of the largest home inspection services); Amerigroup (a Health Maintenance Organization serving Medicaid populations); American Academy of Pediatric Dentists; American College of Nurse Midwives; Bank of America; Connect for Kids; Duron Paint and Wallcoverings; Enterprise Foundation; Home Improvement Lenders Association; Kiwanis International; Mets Laboratories; National Association of Housing Cooperatives; National Association of Realtors; National Association of the Remodeling Industry; National Association of Women and Infant Care Directors; National Center for Lead Safe Housing; National Hispanic Housing Council; National Housing and Rehabilitation Association; National Paint and Coatings Association; National Renovation Lenders Association; National Rural Health Association; Painting and Decorating Contractors of America; and Univision (a Spanish language Television network).

The Enforcement Center at HUD disappeared after Secretary Cuomo left HUD in 2001. Not a single lead paint disclosure case was ever brought by the Enforcement Center.

4.10 Mustering the proof

By the end of the 1990s much of what Title X mandated had been accomplished:

- The HUD lead paint office acquired more highly skilled staff, including policy specialists, better managers, and scientists;
- Congress ended the grant program rescission of 1995 and appropriated additional funds with the increased assurance and evidence that monies would be spent (and most importantly spent well);
- The scientific foundation for quantified lead dust standards was completed and final rules that used those standards were promulgated;
- The evidence that the new lead hazard control methods really worked was published, principally through the Evaluation of the HUD Lead Hazard Control Grant program and other smaller studies in Baltimore and elsewhere;
- HUD completed the reform of its regulations on lead paint for virtually all federally assisted housing;
- New and better lead testing instruments and better lab testing quality controls were put in place;
- Large public education efforts had been conducted; and
- Enforcement began in earnest, resulting in many lead-safe homes through settlements with landlords who had violated the disclosure law.

But where was the evidence that all this was actually working? HUD turned to the two main sources of evidence at the time, the number of poisoned children (from CDC's National Health and Nutrition Examination Survey) and the number of homes with lead paint (from HUD's new National Survey of Lead and Allergens in Housing, which later became the American Healthy Housing Survey).

NHANES showed that the number of children with elevated blood lead levels (above 10 µg/dL, the CDC level of concern at the time) had declined from 890,000 in 1991−94 to 310,000 in 1999−2003.[191] This was still far too many, but it meant that about 500,000 children each year had been kept safe. NSLAH showed that the number of housing units with lead paint, initially estimated at 64 million in 1990, declined to 38 million by the end of the decade.[104] These were remarkable successes. It was backed by the evidence and that made all the difference in creating confidence and the foundations for still further progress.

But with hundreds of thousands of poisoned children and millions more at risk in 38 million homes with lead paint more was needed. Parents and community groups increasingly found their voice and would play a key role.

References

1. Moving Toward a Lead-Safe America A Report to the Congress of the United States. U.S. Department of Housing and Urban Development Office of Lead Hazard Control. Washington DC. February 1997.
2. Public Law 19.104. US Congress. Emergency Supplemental Appropriations for Additional Disaster Assistance, For Anti-Terrorism Initiatives, For Assistance in The Recovery From The Tragedy That Occurred At Oklahoma City, And Rescissions Act, 1995. https://www.govinfo.gov/content/pkg/PLAW-104publ19/html/PLAW-104publ19.htm.
3. HUD Reform Act of 1989. US Congress. HR 3101. August 15, 1989. https://www.congress.gov/bill/101st-congress/house-bill/3101?s = 1&r = 100. Also see HUD timeline: https://www.huduser.gov/hud_timeline/.
4. Shenon P. Samuel R. Pierce Jr., Ex-Housing Secretary, Dies at 78. New York Times. Nov. 3, 2000. < https://www.nytimes.com/2000/11/03/nyregion/samuel-r-pierce-jr-ex-housing-secretary-dies-at-78.html/ >.
5. Rusk D. *Inside Game/Outside Game: Winning Strategies for Saving Urban America.* Washington DC: Brookings Institution Press; 1999.
6. Personal communication. Carolyn Newton (HUD Office of Public and Indian Housing and David Jacobs (Director of EPA Southern Lead Paint Training Consortium, Georgia Institute of Technology). 1991.
7. HUD Office of Inspector General, Audit 95-AO-179-0802. Memorandum to Dwight Robinson, HUD Deputy Secretary from Janice LeRoy, HUD IG, June 14, 1995. The University of Illinois Chicago Special Collections & University Archives, School of Public Health, "David E. Jacobs papers" the University of Illinois Chicago School of Public Health library.
8. US Congress. Inspector General Act of 1978. Public Law. 95–452 October 12, 1978.
9. Hearings before a subcommittee of the Committee on Appropriations, House of Representatives, 103rd Congress second session, Part 6, p. 128-129. September 22, 1994.
10. Personal communication, HUD Chief of Staff Bruce Katz and David Jacobs (Director HUD Office of Lead Hazard Control), 1995.
11. HUD Press Release Number 95-162, Aug 22, 1995. US Department of Housing and Urban Development. Washington DC. The University of Illinois Chicago Special Collections & University Archives, School of Public Health, "David E. Jacobs papers" the University of Illinois Chicago School of Public Health library.
12. Memorandum for David Jacobs (Director HUD Lead Paint Office) from Michael F Hill (Senior Advisor to the Director of the HUD Office of Lead Hazard Control). Lead Abatement Tax Credit, Aug 21, 1995. The University of Illinois Chicago Special Collections & University Archives, School of Public Health, "David E. Jacobs papers" the University of Illinois Chicago School of Public Health library.
13. Memorandum for Bruce Katz (HUD Chief of Staff) and Amy Liu (HUD Deputy Chief of Staff) from David Jacobs (Director HUD Office of Lead Hazard Control). Lead Office Staffing, Feb 2, 1996. The University of Illinois Chicago Special Collections & University Archives, School of Public Health, "David E. Jacobs papers" the University of Illinois Chicago School of Public Health library.
14. Memorandum from David Jacobs (Director HUD Office of Lead Hazard Control) to the Staff of the HUD Office of Lead Hazard Control. June 3, 1996. The University of Illinois Chicago Special Collections & University Archives, School of Public Health, "David E. Jacobs papers" the University of Illinois Chicago School of Public Health library.
15. Memorandum from Ron Morony (Deputy Director, HUD Office of Lead Hazard Control) to the Staff of the Office of Lead Hazard Control, Nov 14, 1996. The University of Illinois Chicago Special Collections &

University Archives, School of Public Health, "David E. Jacobs papers" the University of Illinois Chicago School of Public Health library.

16. Personal communication, Dorothy Allen to David Jacobs (HUD Office of Lead Hazard Control Administrative Officer) and David Jacobs (Director of the HUD Office of Lead Hazard Control) 1997.

17. Memorandum from David Jacobs (Director HUD Office of Lead Hazard Control) to Bruce Katz (HUD Chief of Staff). Probable effects of not staffing the lead office, 1995. The University of Illinois Chicago Special Collections & University Archives, School of Public Health, "David E. Jacobs papers" the University of Illinois Chicago School of Public Health library.

18. Memorandum for Marilynn Davis (HUD Assistant Secretary for Administration) from Dwight Robinson (HUD Deputy Secretary), Approval of Hiring Freeze Exemption for the Office of Lead-Based Paint. 1996. The University of Illinois Chicago Special Collections & University Archives, School of Public Health, "David E. Jacobs papers" the University of Illinois Chicago School of Public Health library.

19. Memorandum for Dwight Robinson (HUD Deputy Secretary) from David Jacobs (Director, HUD Lead Paint Office). Aug 4, 1995. The University of Illinois Chicago Special Collections & University Archives, School of Public Health, "David E. Jacobs papers" the University of Illinois Chicago School of Public Health library.

20. Update from consultants Sandy Panchek and Lauretta Grier, to Office of Lead Hazard Control, Jan 7, 1997 and Notes from Office Retreat, Nov 9-10, 1996. The University of Illinois Chicago Special Collections & University Archives, School of Public Health, "David E. Jacobs papers" the University of Illinois Chicago School of Public Health library.

21. Personal communication, 1996. Bruce Katz (HUD Chief of Staff) to David Jacobs (HUD Director of the Lead Hazard Control Office).

22. Letter from David Jacobs (Director HUD Office of Lead Hazard Control) to David Pawlik (Buffalo Department of Community Development) May 25, 1998. The University of Illinois Chicago Special Collections & University Archives, School of Public Health, "David E. Jacobs papers" the University of Illinois Chicago School of Public Health library.

23. The Top Ten. Lead Detection and Abatement Contractor, July 1997, p. 8, The University of Illinois Chicago Special Collections & University Archives, School of Public Health, "David E. Jacobs papers" the University of Illinois Chicago School of Public Health library.

24. Letter from Michigan Dept of Public Health Acting Director James Haveman Jr to David Jacobs (Director HUD Office of Lead Hazard Control), Aug 31, 1995. The University of Illinois Chicago Special Collections & University Archives, School of Public Health, "David E. Jacobs papers" the University of Illinois Chicago School of Public Health library.

25. Memorandum from Doug Farquhar, National Conference of State Legislatures to HUD lead paint office, Aug 24, 1995. The University of Illinois Chicago Special Collections & University Archives, School of Public Health, "David E. Jacobs papers" the University of Illinois Chicago School of Public Health library.

26. Letter from Jon Cisky, Michigan State Senator to David Jacobs (Director HUD Office of Lead Hazard Control, March 13, 1996 "We encourage HUD to also look past some of the licensure issues to find alternatives." The University of Illinois Chicago Special Collections & University Archives, School of Public Health, "David E. Jacobs papers" the University of Illinois Chicago School of Public Health library.

27. HUD puts rules ahead of public health, James Haveman, Acting Director Michigan Dept of Public Health, Detroit Free Press Sept 14, 1995. The University of Illinois Chicago Special Collections & University Archives, School of Public Health, "David E. Jacobs papers" the University of Illinois Chicago School of Public Health library.

28. Letter from David Jacobs to James Haveman, Oct 5, 1995. The University of Illinois Chicago Special Collections & University Archives, School of Public Health, "David E. Jacobs papers" the University of Illinois Chicago School of Public Health library.

29. Letter from James Haveman to David Jacobs, April 5, 1996 and April 12, 1996. The University of Illinois Chicago Special Collections & University Archives, School of Public Health, "David E. Jacobs papers" the University of Illinois Chicago School of Public Health library.

30. Guidance on the Lead-Based Paint Disclosure Rule: Part II. US Department of Housing and Urban Development, Office of Lead Hazard Control. December 5, 1996. https://www.hud.gov/sites/documents/DOC_12349.PDF

31. EPA. Lead-Based Paint Abatement and Evaluation Program: Overview. Environmental Protection Agency 2022. https://www.epa.gov/lead/lead-based-paint-abatement-and-evaluation-program-overview.

32. Largo TW, Borgialli M, Wisinski CL, Wahl RL, Priem WF. Healthy Homes University: a home-based environmental intervention and education program for families with pediatric asthma in Michigan. *Public Health Rep.* 2011;126(Suppl 1):14−26. Available from: https://doi.org/10.1177/00333549111260S104. PMID: 21563708. PMCID: PMC3072899.

33. Lead and Environmental Hazards Association. 2022. https://leha.us/.

34. Lead Detection and Abatement Contractor, Jan 1998, p 2. The University of Illinois Chicago Special Collections & University Archives, School of Public Health, "David E. Jacobs papers" the University of Illinois Chicago School of Public Health library.

35. Program Assessment and Rating Tool. US Office of Management and Budget. 1997 The University of Illinois Chicago Special Collections & University Archives, School of Public Health, "David E. Jacobs papers" the University of Illinois Chicago School of Public Health library.

36. Memorandum to Todd Howe, Deputy Chief of Staff from David Jacobs, Director HUD Office of Lead Hazard Control, Request to Fill Positions in the Lead Paint office, March 12, 1998. The University of Illinois Chicago Special Collections & University Archives, School of Public Health, "David E. Jacobs papers" the University of Illinois Chicago School of Public Health library.

37. Guidelines for the Evaluation and Control of Lead-Based Paint Hazards in Housing, US Department of Housing and Urban Development. Washington DC, HUD-1539. 1995. https://www.hud.gov/program_offices/healthy_homes/lbp/hudguidelines1995.

38. 40 CFR 745.227. Lead; Requirements for Lead-Based Paint Activities in Target Housing and Child-Occupied Facilities; Final Rule. US Environmental Protection Agency, August 29, 1996. 45778 Federal Register, Vol. 61, No. 169.

39. US Congress. Title X of the 1992 Housing and Community Development Act. Public Law 102-550 Section 1004. Definitions (15). October 1992.

40. National Institute of Building Sciences, Lead Based Paint Operations and Maintenance Work Practices Manual for Homes and Buildings, May 1995, Supported by HUD, NIBS Document number 5401-2. The University of Illinois Chicago Special Collections & University Archives, School of Public Health, "David E. Jacobs papers" the University of Illinois Chicago School of Public Health library.

41. Personal communication. Joe Schirmer (former epidemiologist with the Wisconsin State Department of Health) and David Jacobs, 2022.

42. Guidelines for the Evaluation and Control of Lead-Based Paint Hazards in Housing, US Department of Housing and Urban Development., Washington DC, HUD-1539. Chapter 14. 1995.

43. Goldman L. EPA Assistant Administrator, Guidance on Residential Lead-Based Paint, Lead Contaminated Dust and Lead Contaminated Soil, US Environmental Protection Agency, Washington DC. July 14, 1994. The University of Illinois Chicago Special Collections & University Archives, School of Public Health, "David E. Jacobs papers" the University of Illinois Chicago School of Public Health library.

44. US General Accounting Office. Status of EPA's Efforts to Develop Lead Hazard Standards. B-256299. GAO/RCED-94-114, May 16, 1994.

45. Alliance to End Childhood Lead Poisoning. EPA's National Guidelines for Lead Hazards in Dust, Soil and Paint: A Summary and Analysis. August 1994. The University of Illinois Chicago Special Collections & University Archives, School of Public Health, "David E. Jacobs papers" the University of Illinois Chicago School of Public Health library.

46. Putting the pieces together: controlling lead hazards in the nation's housing. Report of the Lead-based Paint Hazard Reduction and Financing Task Force, 1995. US Department of Housing and Urban Development. Washington, D.C. http://hdl.handle.net/2027/mdp.39015034868367.

47. Guyton GP. A brief history of workers' compensation. *Iowa Orthop J.* 1999;19:106−110.

48. Appendix C "Other Views" in: Putting the pieces together: controlling lead hazards in the nation's housing. Report of the Lead-based Paint Hazard Reduction and Financing Task Force, 1995. US Department of Housing and Urban Development. Washington, D.C. http://hdl.handle.net/2027/mdp.39015034868367.

49. Strategic Plan for the elimination of childhood lead poisoning. Centers for Disease Control and Prevention, US Department of Health and Human Services. Altanta, Georgia. 1991. Appendix 2, page 12. https://stacks.cdc.gov/view/cdc/44719.

50. Dolbeare C, Fernandez A, Florini K, Pollack S, Ryan D, Widess E. Title X Task Force Deserves Advocates' Strong Support. Unpublished response to Title X Task Force report. July 10, 1995. (The University of Illinois Chicago Special Collections & University Archives, School of Public Health, "David E. Jacobs papers" the University of Illinois Chicago School of Public Health library.)

51. Living in Bunker's Shadow. The Spokesman-Review, Jan 9, 1983, p. 10. The University of Illinois Chicago Special Collections & University Archives, School of Public Health, "David E. Jacobs papers" the University of Illinois Chicago School of Public Health library.

52. United States Environmental Protection Agency, Office of Superfund Remediation and Technology Innovation EPA 9285.7-42 540-K-01-005. User's Guide for the Integrated Exposure Uptake Biokinetic Model for Lead in Children (IEUBK) Windows®, Prepared for The Technical Review Workgroup for Metals and Asbestos (TRW) Syracuse Research Corporation. May 2007.

53. US Environmental Protection Agency. Urban Soil Lead Abatement Demonstration Project. Volume I: Integrated Report. EPA 600/AP-93/001a. July 1993.

54. US Environmental Protection Agency. Types of Superfund Liable Parties. Webpage. 2022. https://www.epa.gov/enforcement/superfund-liability#:~:text = There%20are%20four%20classes%20of,of%20the%20hazardous%20substances%2C%20and.

55. Superfund Town Begins Final Residential Cleanup, Lead Detection and Abatement Contractor newspaper, Oct 1997, p. 6 and 16.

56. Borough of Palmerton — Category B HUD Grant final report, Sept 8, 2001. The University of Illinois Chicago Special Collections & University Archives, School of Public Health, "David E. Jacobs papers" the University of Illinois Chicago School of Public Health library.

57. Omaha Receives Federal Grant for Lead Hazards. Jan 27, 2019. https://www.cityofomaha.org/latest-news/520-omaha-receives-federal-grant-for-lead-hazards.

58. Jacobs DE, Menrath WB, Succop PA, Cohen R, Clark CS, Bomschein RL. A Comparison of Five Sampling Methods for Settled Lead Dust: A Pilot Study. The National Center for Lead-Safe Housing. June 15, 1993. The University of Illinois Chicago Special Collections & University Archives, School of Public Health, "David E. Jacobs papers" the University of Illinois Chicago School of Public Health library.

59. US Environmental Protection Agency. Repair and Maintenance Pilot study. Nov 29, 1994. The University of Illinois Chicago Special Collections & University Archives, School of Public Health, "David E. Jacobs papers" the University of Illinois Chicago School of Public Health library.

60. Side-by-side analysis of wipe and sticky tape dust samplers. Letter from DE Jacobs to Terry Burke (Wisconsin Occupational Health laboratory). May 28, 1992. The University of Illinois Chicago Special Collections & University Archives, School of Public Health, "David E. Jacobs papers" the University of Illinois Chicago School of Public Health library.

61. Lanphear BP, Emond E, Weitzman M, Jacobs DE, Tanner M, Winter N, Yakir B, Eberly S. A Side-By-Side Comparison of Dust Collection Methods for Sampling Lead-Contaminated House Dust. *Environ Res.* 1995;68:114—123.

62. Lanphear BP, Matte TD, Rogers J, Clickner RP, Dietz B, Bornschein RL, Succop P, Mahaffey KR, Dixon S, Galke W, Rabinowitz M, Farfel M, Rohde C, Schwartz J, Ashley PJ, Jacobs DE. The contribution of lead-contaminated house dust and residential soil to children's blood lead levels: A pooled analysis of 12 epidemiologic studies. *Environmental Research.* 1998;79(1):51—68.

63. National Research Council. *Science and Decisions: Advancing Risk Assessment.* Washington, DC: The National Academies Press; 2009. Available from: https://doi.org/10.17226/12209.

64. National associations ask appeals court to allow challenges on part of recent rule. Email between David Jacobs, Director of HUD Office of Lead Hazard Control and Joe Ventrone, Special Advisor to HUD Secretary Mel Martinez. April 11, 2001. The University of Illinois Chicago Special Collections & University Archives, School of Public Health, "David E. Jacobs papers" the University of Illinois Chicago School of Public Health library.

65. Jacobs DE. Lead-Based Paint as a Major Source of Childhood Lead Poisoning: A Review of the Evidence in Lead In Paint, Soil and Dust: Health Risks, Exposure Studies, Control Measures and Quality AssuranceIn: Beard ME, Allen Iske SD, eds. Philadelphia: American Society for Testing and Materials; 1995:175—187.

66. Gulson B. Stable lead isotopes in environmental health with emphasis on human investigations. *Science of the Total Environment.* 2008;75—92.

67. US Department of Housing and Urban Development. Preamble. Requirements for Notification, Evaluation and Reduction of Lead- Based Paint Hazards in Federally Owned Residential Property and Housing Receiving Federal Assistance. 24 CFR Part 25 Federal Register, Vol. 64, No. 178. September 15, 1999. P. 50140.
68. US Environmental Protection Agency Risk Analysis to Support Standards for Lead in Paint, Dust, and Soil. Volumes I and II. EPA 747-R-97-006. December 1997.
69. US Environmental Protection Agency Lead; Identification of Dangerous Levels of Lead. Federal Register, Vol 66. pages 1205–1240. 40 CFR Part 745.
70. US Environmental Protection Agency, Science Advisory Board. SAB Review of EPA's Approach for Developing Lead Dust Hazard Standards for Residences. U.S. Environmental Protection Agency. EPA-SAB-11-008, July 7, 2001.
71. US Department of Housing and Urban Development. Revised Dust-Lead Action Levels for Risk Assessment and Clearance; Clearance of Porch Floors. Policy Guidance Number: 2017–01. January 31, 2017.
72. US Environmental Protection Agency. Hazard Standards and Clearance Levels for Lead in Paint, Dust and Soil. 40 CFR 745. June 21, 2019.
73. Inside Scoop, Lead Detection and Abatement Contractor Newspaper, Nov 1997, p. 11. The University of Illinois Chicago Special Collections & University Archives, School of Public Health, "David E. Jacobs papers" the University of Illinois Chicago School of Public Health library.
74. HUD Forges Ahead on Chapter 7 Revisions, Lead Detection and Abatement Contractor newspaper, Oct 1997, p. 5. The University of Illinois Chicago Special Collections & University Archives, School of Public Health, "David E. Jacobs papers" the University of Illinois Chicago School of Public Health library.
75. Collucio V, ed. *Lead-based paint hazards*. NY: Van Nostrand Reinhold; 1994.
76. Title X, Section 1052. Also see: U.S. Congress (1991). Senate Appropriations Committee Report No. 102–107, pp. 40–54. U.S. Congress, Washington, DC.
77. Evaluation of the HUD Lead Hazard Control Grant Program. National Center for Healthy Housing and University of Cincinnati. 2004. https://nchh.org/research/eval-of-the-hud-lead-hazard-control-grant-program/.
78. Wilson J, Pivetz T, Ashley PJ, Strauss W, Jacobs DE, Menkedick J, Dixon S, Tsai HC, Brown V. Evaluation of HUD-Funded Lead Hazard Control Treatments at Six Years Post-Intervention. *Environ Res.* 2006;102 (2):237–248.
79. Dixon S, Jacobs DE, Wilson J, Akoto J, Clark CS. Window Replacement and Residential Lead Paint Hazard Control 12 Years Later. *Environ Res.* 2012;113:14–20. Available from: https://doi.org/10.1016/j.envres.2012.01.005.
80. Galke W, Clark CS, Wilson J, Jacobs DE, Succop P, Dixon S, Bornschein RL, McLaine P, Chen M. Evaluation of the HUD Lead Hazard Control Grant Program: Early Overall Findings. *Env Res Section A.* 2001;86:149–156. Available from: https://doi.org/10.1006/enrs.2001.4259.
81. National Center for Healthy Housing, Evaluation of the HUD Lead-Based Paint Hazard Control Grant Program, First and Second Interim Reports, 1995 and 1996. The University of Illinois Chicago Special Collections & University Archives, School of Public Health, "David E. Jacobs papers" the University of Illinois Chicago School of Public Health library.
82. National Center for Healthy Housing, Evaluation of the HUD Lead-Based Paint Hazard Control Grant Program, Third Interim Report, March 1996. The University of Illinois Chicago Special Collections & University Archives, School of Public Health, "David E. Jacobs papers" the University of Illinois Chicago School of Public Health library.
83. National Center for Healthy Housing, Evaluation of the HUD Lead-Based Paint Hazard Control Grant Program, Fourth Interim Report, February 1997. The University of Illinois Chicago Special Collections & University Archives, School of Public Health, "David E. Jacobs papers" the University of Illinois Chicago School of Public Health library.
84. National Center for Healthy Housing, Evaluation of the HUD Lead-Based Paint Hazard Control Grant Program, Fifth Interim Report. March 1998. The University of Illinois Chicago Special Collections & University Archives, School of Public Health, "David E. Jacobs papers" the University of Illinois Chicago School of Public Health library.
85. National Center for Healthy Housing and University of Cincinnati, Evaluation of the HUD Lead-Based Paint Hazard Control Grant Program, Final Report 2004. https://nchh.org/research/eval-of-the-hud-lead-hazard-control-grant-program/.

86. Jacobs DE, Tobin M, Targos L, Clarkson D, Dixon SL, Breysse J, Pratap P, Cali S. Replacing Windows Reduces Childhood Lead Exposure: Results from a State-Funded Program. *J Public Health ManagPract*. 2016;22(5):482−491. Available from: https://doi.org/10.1097/PHH.0000000000000389. PubMed PMID 26910871.

87. Nevin R, Jacobs DE. Windows of opportunity: Lead poisoning prevention, housing affordability and energy conservation. *Housing Policy Debate*. 2006;17(1):185−207.

88. Matthew S. These. Observational and interventional study design types; an overview. *Biochem Med (Zagreb)*. 2014;24(2):199−210. Available from: https://doi.org/10.11613/BM.2014.022. PMCID: PMC4083571 PMID. 24969913.

89. Calling for a Global Ban on Lead Use in Residential Indoor and Outdoor Paints, Children's Products, and All Nonessential Uses in Consumer Products, Oct 28 2008. Policy Statement Number:20084.

90. Cantor AG, Hendrickson R, Blazina I, Griffin J, Grusing S, McDonagh MS. Screening for Elevated Blood Lead Levels in Childhood and Pregnancy: Updated Evidence Report and Systematic Review for the US Preventive Services Task Force. *JAMA*. 2019;321:1510−1526.

91. Jacobs DE. Invited Commentary on lead screening update from the US Preventive Services Task Force. *Journal of Pediatrics*. 2019;243. Sept.

92. Farfel M. Lead-Based Paint Abatement and Repair and Maintenance Study in Baltimore: Findings Based on Two Years Of Follow-Up. US Environmental Protection Agency. Report No. 747-R-97-005 December 1997.

93. Braun JM, Hornung R, Chen A, Dietrich KN, Jacobs DE, Jones R, Khoury JC, Liddy-Hicks S, Morgan S, Vanderbeek SB, Xu Y, Yolton K, Lanphear BP. Effect of Residential Lead-Hazard Interventions on Childhood Blood Lead Concentrations and Neurobehavioral Outcomes: A Randomized Clinical Trial. *JAMA Pediatr*. 2018;172(10):934−942. Available from: https://doi.org/10.1001/jamapediatrics.2018.2382. 30178064. PMCID: PMC6233767.

94. Tohn ER, Dixon SL, Wilson JW, Galke WA, Clark CS. An Evaluation of One-Time Professional Cleaning in Homes with Lead-Based Paint Hazards. *Applied Occupational and Environmental Hygiene*. 2003;18(2):138−143. Available from: https://doi.org/10.1080/10473220301437.

95. Clark S, Galke W, Succop P, Grote J, McLaine P, Wilson J, Dixon S, Menrath W, Roda S, Chen M, Bornschein R, Jacobs D. Effects of HUD-supported lead hazard control interventions in housing on children's blood lead. *Environ Res*. 2011;111(2):301−311. Available from: https://doi.org/10.1016/j.envres.2010.11.003, Epub 2010 Dec 22. PMID 21183164. www.sciencedirect.com/science/journal/00139351/111/2.

96. VanOsdell D. Lead contaminated dust emissions from forced air ducts. US Department of Housing and Urban Development Cooperative Agreement No., NCLHR0027-97, RTI Project No. 91U-706-003, July 22, 1999. The University of Illinois Chicago Special Collections & University Archives, School of Public Health, "David E. Jacobs papers" the University of Illinois Chicago School of Public Health library.

97. Wilson J, Dixon SL, Wisinski C, Speidel C, Breysse J, Jacobson M, Jacobs DE. Current Pathways and Sources of Lead Exposure: The Michigan Children's Lead Determination Study (the MI CHILD study). *Env Res*. 2022; (Accepted September 2022).

98. US Department of Housing and Urban Development Secretary Henry Cisneros. Controlling Lead Hazards in the Nation's Housing. Report of the Lead-Based Paint Hazard Reduction and Financing Task Force. 1995. https://babel.hathitrust.org/cgi/pt?id = mdp.39015034868367&view = 1up&seq = 5.

99. Lessons Learned in Designing Local Lead Hazard Control Programs. National Center for Lead Safe Housing. 1997. The University of Illinois Chicago Special Collections & University Archives, School of Public Health, "David E. Jacobs papers" the University of Illinois Chicago School of Public Health library.

100. US Environmental Protection Agency. Lead-Based Paint Activities. Final rule. 40 CFR part 745, subparts L and Q, 61. Federal Register 45777−45830). August 29, 1996.

101. US Department of Housing and Urban Development. Lead-Safe Housing Rule Requirements Summary https://www.hud.gov/program_offices/healthy_homes/enforcement/lshr_summary.

102. US Department of Housing and Urban Development. Housing choice voucher program and rental certificate program: PHA Administrative fees for lead-based paint hazard clearance tests and risk assessments. Notice PIH 2000-49 (HA), Oct 27, 2000.

103. Senate bill for HUD's 2014 appropriations directed HUD to "... move to a consistent inspection standard across housing assistance programs, as well as [for] oversight of Section 8 units". See page 100 of The Senate Subcommittee Report 113-45, June 27, 2013. https://www.gpo.gov/fdsys/pkg/CRPT-113srpt45/pdf/CRPT-113srpt45.pdf.

104. Jacobs DE, Clickner RL, Zhou JL, Viet SM, Marker DA, Rogers JW, Zeldin DC, Broene P, Friedman W. The Prevalence of Lead-Based Paint Hazards in U.S. Housing. *Environ Health Perspect*. 2002;110:A599–A606. Sept 13.

105. Jacobs DE, Cali S, Welch A, Catalin B, Dixon SL, Evens A, Mucha AP, Vahl N, Erdal S, Bartlett J. Lead and other heavy metals in dust fall from single-family housing demolition. *Public Health Rep*. 2013;128(6):454–462. Available from: https://doi.org/10.1177/003335491312800605. 24179257. PMCID: PMC3804089.

106. Farfel MR, Orlova AO, Lees PS, Rohde C, Ashley PJ, Julian Chisolm Jr. J. A study of urban housing demolition as a source of lead in ambient dust on sidewalks, streets, and alleys. *Environ Res*. 2005;99(2):204–213. Available from: https://doi.org/10.1016/j.envres.2004.10.005. Epub 2004 Dec 15. PMID. 16194670.

107. Farfel MR, Orlova AO, Lees PS, Rohde C, Ashley PJ, Chisolm Jr. JJ. A study of urban housing demolitions as sources of lead in ambient dust: demolition practices and exterior dust fall. *Environ Health Perspect*. 2003;111 (9):1228–1234. Available from: https://doi.org/10.1289/ehp.5861. 12842778. PMCID: PMC1241579.

108. Bowie J, Farfel M, Moran H. Community experiences and perceptions related to demolition and gut rehabilitation of houses for urban redevelopment. *J Urban Health*. 2005;82(4):532–542. Available from: https://doi.org/10.1093/jurban/jti075. 15958787. PMCID: PMC3456674.

109. East Baltimore Development, Inc. (EBDI). Operations Protocol for Salvage, Deconstruction, Demolition and Site Preparation Activities. April 2006. The University of Illinois Chicago Special Collections & University Archives, School of Public Health, "David E. Jacobs papers" the University of Illinois Chicago School of Public Health library.

110. US Department of Housing and Urban Development and US Centers for Disease Control and Prevention. Joint Letter. Confidentiality of Childhood Lead Poisoning Data. Mary Jean Brown and David Jacobs, Directors. 2002. https://www.cdc.gov/nceh/lead/docs/policy/HUD_letter.pdf.

111. US Department of Housing and Urban Development, Office of Fair Housing and Equal Opportunity. Memorandum from Susan M. Forward, Deputy Assistant Secretary for Enforcement and Investigations to Directors, Fair Housing Enforcement Centers Directors, Program Operations and Compliance Centers. Requirements Concerning Lead-Based Paint and the Fair Housing Act. August 1, 1997. https://www.equal-housing.org/wp-content/uploads/2014/09/1997-Lead-Based-Paint-and-Fair-Housing.pdf.

112. US Department of Housing and Urban Development. Memo from Frederick Douglas Jr, Deputy Assistant Secretary for Single Family Housing to William Apgar, Federal Housing Commissioner, Subject: Lead Paint Proposed Rule, March 9, 1999. The University of Illinois Chicago Special Collections & University Archives, School of Public Health, "David E. Jacobs papers" the University of Illinois Chicago School of Public Health library.

113. US Department of Housing and Urban Development. Memorandum for Art Agnos, Acting General Deputy Assistant Secretary for Housing-Federal Housing commissioner, June 8, 1998. The University of Illinois Chicago Special Collections & University Archives, School of Public Health, "David E. Jacobs papers" the University of Illinois Chicago School of Public Health library.

114. Draft lead paint underwriting standard, fax from Win Hayward (Fannie Mae) to David Jacobs, Aug 1, 1995 and letter from Don Ryan to Win Hayward Aug 24, 1995. The University of Illinois Chicago Special Collections & University Archives, School of Public Health, "David E. Jacobs papers" the University of Illinois Chicago School of Public Health library.

115. National Center for Healthy Housing Find It Fix It Fund It. A Lead Elimination Action Drive. 2016. https://nchh.org/build-the-movement/find-fix-fund/.

116. Federal National Mortgage Association (Fannie Mae) Selling Guide, Single Family Housing April 7, 2021. Chapter E-2. P. 1118. https://singlefamily.fanniemae.com/media/25551/display.

117. US Department of Housing and Urban Development. Memorandum from David Jacobs (HUD Lead Paint Office Director) to William Apgar (Federal Housing Authority Commissioner). Housing's Non-concurrence on the lead paint final rule. July 20, 1998. The University of Illinois Chicago Special Collections & University Archives, School of Public Health, "David E. Jacobs papers" the University of Illinois Chicago School of Public Health library.

118. US Department of Housing and Urban Development. Memo from Office of Housing to Lead Paint Office, Aug 21, 1998, the University of Illinois Chicago Special Collections & University Archives, School of Public Health, "David E. Jacobs papers" the University of Illinois Chicago School of Public Health library.

119. Email from Camille Acevedo (HUD Office of General Counsel) to Vance Morris (HUD Office of General Counsel), April 30, 1999, the University of Illinois Chicago Special Collections & University Archives, School of Public Health, "David E. Jacobs papers" the University of Illinois Chicago School of Public Health library.

120. Email from David Jacobs to Camille Acevedo and others, May 10, 1999, the University of Illinois Chicago Special Collections & University Archives, School of Public Health, "David E. Jacobs papers" the University of Illinois Chicago School of Public Health library.

121. US Department of Housing and Urban Development. Memo to File from Bill Burney (Maine Field Office Director), January 11, 2000. The University of Illinois Chicago Special Collections & University Archives, School of Public Health, "David E. Jacobs papers" the University of Illinois Chicago School of Public Health library.

122. US Congress. Subcommittee on Public Health Senate Health, Education, Labor and Pensions Committee. Lewiston, Maine Field Hearing. November 15, 1999.

123. Omission of single-family homes from HUD lead paint rules leaves industry frustrated. Indoor Environment Business, Nov. 1999, p. 7-8. The University of Illinois Chicago Special Collections & University Archives, School of Public Health, "David E. Jacobs papers" the University of Illinois Chicago School of Public Health library.

124. U.S. Department of Housing and Urban Development. Economic Analysis of The Final Rule on Lead-Based Paint: Requirements for Notification, Evaluation And Reduction Of Lead-Based Paint Hazards In Federally-Owned Residential Property And Housing Receiving Federal Assistance. September 7, 1999. https://www.hud.gov/sites/documents/DOC_25478.PDF.

125. White House Office of Management and Budget. Circular A-94 Guidelines and Discount Rates For Benefit-Cost Analysis Of Federal Programs. 1995.

126. Gould E. Childhood lead poisoning: conservative estimates of the social and economic benefits of lead hazard control. Environ Health Perspect. 2009;117(7):1162–1167. Available from: https://doi.org/10.1289/ehp.0800408. 19654928. PMCID: PMC2717145.

127. Pew Charitable Trust and Robert Wood Johnson Foundation. 10 Policies to Prevent and Respond to Childhood Lead Exposure. Aug 31, 2017. https://nchh.org/resource/report_10-policies-to-prevent-and-respond-to-childhood-lead-exposure_english/.

128. US Department of Housing and Urban Development. Lead-Safe Housing Rule preamble. Federal Register, Vol. 64, No. 178 Wednesday, September 15, 1999. P 50190.

129. US Conference of Mayors, the Association for Local Housing Finance Agencies and the National Association of Counties. Letter to Honorable James T Walsh (Congressman, US House of Representative), May 12, 2000. The University of Illinois Chicago Special Collections & University Archives, School of Public Health, "David E. Jacobs papers" the University of Illinois Chicago School of Public Health library.

130. Letter to US Department of Housing and Urban Development Secretary Andrew Cuomo from Paula Staley, Jefferson County, KY, Aug 2, 2000. The University of Illinois Chicago Special Collections & University Archives, School of Public Health, "David E. Jacobs papers" the University of Illinois Chicago School of Public Health library.

131. Letter from Northern Manhattan Improvement Corporation to Representative Joe Walsh, May 19, 2000. The University of Illinois Chicago Special Collections & University Archives, School of Public Health, "David E. Jacobs papers" the University of Illinois Chicago School of Public Health library.

132. Alliance to End Childhood Lead Poisoning Press Release. Public Health, Children's Welfare, Environmental Protection and Affordable Housing Groups Come Together to Urge Congress: HUD's Lead Safety Regulations must not be delayed. May 19, 2000. The University of Illinois Chicago Special Collections & University Archives, School of Public Health, "David E. Jacobs papers" the University of Illinois Chicago School of Public Health library.

133. US Department of Housing and Urban Development. HUD announces $105 million to help communities protect children from the dangers of lead. Press Release and HUD News, No. 00-227, Aug 24, 2000.

134. National Affordable Housing Management Association. NAHMA News Vol 12, No. 1, Jan/Feb 2001, p. 3. The University of Illinois Chicago Special Collections & University Archives, School of Public Health, "David E. Jacobs papers" the University of Illinois Chicago School of Public Health library.

135. Getting the Lead Out by George Caruso, NAHMA News, National Affordable Housing Management Association, Vol XII, No. 1, Jan/Feb 2001. The University of Illinois Chicago Special Collections & University Archives, School of Public Health, "David E. Jacobs papers" the University of Illinois Chicago School of Public Health library.

136. US Department of Housing and Urban Development, Office of Healthy Homes and Lead Hazard Contro. Extensions to the Transition Assistance Period for HUD's Lead Safe Housing Rule, Program Notice OHHLHC-01-02, March 7, 2001. The University of Illinois Chicago Special Collections & University

Archives, School of Public Health, "David E. Jacobs papers" the University of Illinois Chicago School of Public Health library.

137. Alliance to End Childhood Lead Poisoning. Alliance Alert. November/December 2000. P. 1. The University of Illinois Chicago Special Collections & University Archives, School of Public Health, "David E. Jacobs papers" the University of Illinois Chicago School of Public Health library.

138. Final Extension to the Transition Assistance Period for HUD's Lead Safe Housing Regulation, Program Notice OHHLHC-01-05, August 5, 2001.

139. Cities to get clarification of lead paint regs for housing. HUD will assist local governments, by Cameron Williams, Nation's Cities Weekly, Sept 11, 2000.

140. Letter from Jim Yannarelly (St Paul) MN to Nick Farr (National Center for Lead Safe Housing), Sept 26, 2000. The University of Illinois Chicago Special Collections & University Archives, School of Public Health, "David E. Jacobs papers" the University of Illinois Chicago School of Public Health library.

141. National Apartment Association. NAA Prepares You for Lead Compliance, Units Magazine, Vol 24, No. 8 Oct 2000. The University of Illinois Chicago Special Collections & University Archives, School of Public Health, "David E. Jacobs papers" the University of Illinois Chicago School of Public Health library.

142. Lead-Based Paint Regulations. National Multi Housing Council Newsletter, March 9, 2001. The University of Illinois Chicago Special Collections & University Archives, School of Public Health, "David E. Jacobs papers" the University of Illinois Chicago School of Public Health library.

143. Alliance to End Childhood Lead Poisoning. Critical Choices Memorandum. Jan 5, 2001. The University of Illinois Chicago Special Collections & University Archives, School of Public Health, "David E. Jacobs papers" the University of Illinois Chicago School of Public Health library.

144. Personal communication, US Department of Housing and Urban Development Secretary Julian Castro to David Jacobs, 2016.

145. Alliance to End Childhood Lead Poisoning. A Critique of the National Paint and Coatings Association's Lead Paint Educational Materials. The University of Illinois Chicago Special Collections & University Archives, School of Public Health, "David E. Jacobs papers" the University of Illinois Chicago School of Public Health library.

146. Sides S. National Paint and Coatings Association Statement on the Alliance to End Childhood lead poisoning Critique of its lead-based paint education materials. Aug 28, 2000. The University of Illinois Chicago Special Collections & University Archives, School of Public Health, "David E. Jacobs papers" the University of Illinois Chicago School of Public Health library.

147. Environmental Defense Fund and Alliance to End Childhood Lead Poisoning. Press Release. Groups Call for Halt of Misleading Lead Poisoning Video, June 3, 1999. The University of Illinois Chicago Special Collections & University Archives, School of Public Health, "David E. Jacobs papers" the University of Illinois Chicago School of Public Health library.

148. US Department of Housing and Urban Development. Memo from Oscar Anderson (HUD Congressional Affairs office) to Marshall Bass (Staffer in the Office of John E Sununu, US House of Representatives (undated)). The University of Illinois Chicago Special Collections & University Archives, School of Public Health, "David E. Jacobs papers" the University of Illinois Chicago School of Public Health library.

149. National Association of Realtors. Letter from Lee Vertandig, Government Affairs senior vice president, National Association of Realtors to US Department of Housing and Urban Development Secretary Mel Martinez, Sept 4, 2001. The University of Illinois Chicago Special Collections & University Archives, School of Public Health, "David E. Jacobs papers" the University of Illinois Chicago School of Public Health library.

150. Dekovsky R. Developing Chemical Instrumentation for Environmental Use in the Late Twentieth Century: Detecting Lead in Paint Using Portable X-Ray Fluorescence Spectrometry. *Ambix.* 2009;56(2):138–162.

151. Midwest Research Institute and Quantech Inc. US EPA and HUD. A field test of lead-based paint testing technologies: Summary report. EPA-747-R-95-002a. May 1995.

152. Inside Scoop, Lead Detection and Abatement Contractor newspaper, July 1997 The University of Illinois Chicago Special Collections & University Archives, School of Public Health, "David E. Jacobs papers" the University of Illinois Chicago School of Public Health library.

153. Dewalt G, Cox D, Friedman W, Jacobs D. Field Evaluation of Lead-Based Paint Inspections, Volume 1: Results, US Department of Housing and Urban Development Office of Lead Hazard Control Sept 30, 1998.

The University of Illinois Chicago Special Collections & University Archives, School of Public Health, "David E. Jacobs papers" the University of Illinois Chicago School of Public Health library.

154. Lead paint archive inspection May 7, 1997. The University of Illinois Chicago Special Collections & University Archives, School of Public Health, "David E. Jacobs papers" the University of Illinois Chicago School of Public Health library.

155. Letter to Joe Breen (Environmental Protection Agency) from David Jacobs (Georgia Tech Research Scientist). October 22, 1991. The University of Illinois Chicago Special Collections & University Archives, School of Public Health, "David E. Jacobs papers" the University of Illinois Chicago School of Public Health library.

156. Esche CA, Groff JH, Schlecht PC, Shulman SA. Laboratory evaluations and performance reports for the proficiency analytical testing (PAT) and environmental lead proficiency analytical testing (ELPAT) programs. National Institute for Occupational Safety and Health. Division of Physical Sciences and Engineering. Department of Health and Human Services publication number NIOSH 95-104 November 1994. https://stacks.cdc.gov/view/cdc/6240.

157. US Environmental Protection Agency. The National Lead Laboratory Accreditation Program (NLLAP). 2020. https://www.epa.gov/lead/national-lead-laboratory-accreditation-program-nllap.

158. AIHA or A2LA? Labs Must Now Choose, Few Differences between NLLAP Accrediting Groups, by David Rasmussen, Lead Detection and Abatement Contractor newspaper, Oct 1997. The University of Illinois Chicago Special Collections & University Archives, School of Public Health, "David E. Jacobs papers" the University of Illinois Chicago School of Public Health library.

159. National Lead Laboratory Accreditation Program List. https://www.epa.gov/lead/national-lead-laboratory-accreditation-program-list.

160. Schlecht PC, Song R, Groff JH, Feng HA, Esche CA. Interlaboratory and intralaboratory variabilities in the Environmental Lead Proficiency Analytical Testing (ELPAT) Program. *Am Ind Hyg Assoc J.* 1997;58 (11):779–786. Available from: https://doi.org/10.1080/15428119791012270. PMID. 9373923.

161. Caldwell KL, Cheng PY, Vance KA, Makhmudov A, Jarrett JM, Caudill SP, Ho DP, Jones RL. LAMP: A CDC Program to Ensure the Quality of Blood-Lead Laboratory Measurements. *J Public Health ManagPract.* 2019;25 (Suppl 1):S23–S30. Available from: https://doi.org/10.1097/PHH.0000000000000886. PMID. 30507766.

162. Ford DA, Gilligam M. The Effect of lead paint abatement laws on rental property values. *The American Real Estate and Urban Economics Association Journal.* 1988;16(1):84–94.

163. Joint EPA and HUD letter to property owners, Eric Schaeffer and David Jacobs, May 2, 2001. The University of Illinois Chicago Special Collections & University Archives, School of Public Health, "David E. Jacobs papers" the University of Illinois Chicago School of Public Health library.

164. Memorandum from Bruce Katz (HUD Chief of Staff) and David Jacobs to Sally Katzen (White House Office of Management and Budget), Oct 24, 1995. The University of Illinois Chicago Special Collections & University Archives, School of Public Health, "David E. Jacobs papers" the University of Illinois Chicago School of Public Health library.

165. Meckler L. Lead Paint Disclosure Law, Associated Press. March 7, 1996. The University of Illinois Chicago Special Collections & University Archives, School of Public Health, "David E. Jacobs papers" the University of Illinois Chicago School of Public Health library.

166. National Association of Realtors Embraces New Lead Paint Regulations. News Release. March 6, 1996. The University of Illinois Chicago Special Collections & University Archives, School of Public Health, "David E. Jacobs papers" the University of Illinois Chicago School of Public Health library.

167. Alliance to End Childhood Lead Poisoning News Release, March 6, 1996. The University of Illinois Chicago Special Collections & University Archives, School of Public Health, "David E. Jacobs papers" the University of Illinois Chicago School of Public Health library.

168. Personal communication, Bruce Nilles (US Department of Justice Attorney) with David Jacobs (Director of US Department of Housing and Urban Development Office of Lead Hazard Control). 1999.

169. Landlord gets jail for lying over lead, by Jerry Seper, Washington Times, July 12, 2001, p. C1.

170. Landlord admits not warning DC tenants about lead paint, by Ruben Castaneda, Washington Post July 12, 2001, p. B4.

171. Manchester Man Admits Guilty in Lead Poisoning Case; Plea to Forging Lead Hazard Disclosure Documents is Precedent Setting. US Environmental Protection Agency Press Release. Release Date: 12/19/2001, https://archive.epa.gov/epapages/newsroom_archive/newsreleases/a6d4a21fa2ccb9a185256b270079ed89.html.

172. Special Commendation for Outstanding Service in the Office of the Assistant Attorney General, Environment and Natural Resources Division, US Department of Justice Awards Ceremony. Washington DC April 28, 1999.

173. The US Equal Employment and Opportunity Commission (EEOC) is responsible for enforcing federal laws that make it illegal to discriminate against a job applicant or an employee because of the person's race, color, religion, sex (including pregnancy, gender identity, and sexual orientation), national origin, age (40 or older), disability or genetic information. 2022. https://www.eeoc.gov/overview.

174. Janik L. Lead paint crackdown targets real estate industry. Realtor magazine. Oct 1999, p. 54-55.

175. Memo from Eric Schaeffer (Director of the US Environmental Protection Agency Office of Enforcement and Compliance Assistance to David Jacobs (Director of the US Department of Housing and Urban Development Office of Lead Hazard Control (undated). "The collaborative efforts by EPA and HUD in settling this (AIMCO) case resulted in a remarkable outcome, of which both agencies can be proud." The University of Illinois Chicago Special Collections & University Archives, School of Public Health, "David E. Jacobs papers" the University of Illinois Chicago School of Public Health library.

176. US Department of Housing and Urban Development Office of Administrative Law Judges, in the matter of American Rental Management Company, HUD ALJ 99-01-CMP, decided May 26, 2000.

177. US Department of Justice and US Department of Housing and Urban Development. Press conference. July 15, 1999. Attorney General Janet Reno and HUD Secretary Andrew Cuomo. The University of Illinois Chicago Special Collections & University Archives, School of Public Health, "David E. Jacobs papers" the University of Illinois Chicago School of Public Health library.

178. U.S. Settles Cases Against Minnesota Landlords - Nearly 4,500 Apartments In Four States To Become Lead Free. Agreements signal stepped up enforcement of Lead Disclosure Rule. HUD Press Release No. 04-063, July 1, 2004. https://archives.hud.gov/news/2004/pr04-063.cfm.

179. Lead paint in apartments costs landlords $540,000, by Sewell Chan, Washington Post, Oct 5, 2000. https://www.washingtonpost.com/archive/local/2000/10/05/lead-paint-in-apartments-costs-landlords-540000/f287c1bb-2086-46d0-9300-4c89d15795e8/.

180. EPA Press Release. EPA imposes first civil penalties for failure to disclose information on lead-based paint, July 29, 1998. The University of Illinois Chicago Special Collections & University Archives, School of Public Health, "David E. Jacobs papers" the University of Illinois Chicago School of Public Health library.

181. US Department of Housing and Urban Development. Memorandum from Gail Laster, HUD General Counsel to Andrew Cuomo, HUD Secretary, June 17, 1999. The University of Illinois Chicago Special Collections & University Archives, School of Public Health, "David E. Jacobs papers" the University of Illinois Chicago School of Public Health library.

182. US Department of Housing and Urban Development. Email from George Weidenfeller, HUD Office of General Counsel to Joe Ventrone, HUD Deputy Chief of Staff, Feb 28, 2001. The University of Illinois Chicago Special Collections & University Archives, School of Public Health, "David E. Jacobs papers" the University of Illinois Chicago School of Public Health library.

183. US Department of Housing and Urban Development. Memorandum to HUD Deputy Secretary Alphonso Jackson from David Jacobs through Jon Gant (HUD Department Enforcement Center Director). Responsibility for Enforcement of Lead Paint Disclosure Regulation. Oct 3, 2001. The University of Illinois Chicago Special Collections & University Archives, School of Public Health, "David E. Jacobs papers" the University of Illinois Chicago School of Public Health library.

184. Schaeffer E, Jacobs D. Guidelines for Assessment of Civil Penalties for Violations of Section 1018 of the Residential Lead-Based Paint Hazard Reduction Act of 1992, undated. The University of Illinois Chicago Special Collections & University Archives, School of Public Health, "David E. Jacobs papers" the University of Illinois Chicago School of Public Health library.

185. Section 1018 Enforcement Handbook. US Department of Housing and Urban Development. The University of Illinois Chicago Special Collections & University Archives, School of Public Health, "David E. Jacobs papers" the University of Illinois Chicago School of Public Health library.

186. Brown, S. Federal Lead-Based Paint Enforcement Bench Book. National Center for Healthy Housing. 2009. https://nchh.org/resource-library/federal-lbp-enforcement-bench-book_2009.01.23.pdf.

187. Memorandum from Gail Laster, HUD General Counsel to Andrew Cuomo, HUD Secretary, June 17, 1999. The University of Illinois Chicago Special Collections & University Archives, School of Public Health, "David E. Jacobs papers" the University of Illinois Chicago School of Public Health library.

188. Memorandum from Gail Laster, HUD General Counsel and David Jacobs to HUD Secretary Andrew Cuomo, Lead Based Paint Disclosure Enforcement Strategy. Jan 7, 2000. The University of Illinois Chicago

Special Collections & University Archives, School of Public Health, "David E. Jacobs papers" the University of Illinois Chicago School of Public Health library.

189. S. Ciochetto, Bureau of the Census; Barbara A. Haley, U.S. Department of Housing and Urban Development. How Do You Measure "Awareness"? Experiences with the Lead-Based Paint Survey. Census Working Paper. 1997. https://www.census.gov/content/dam/Census/library/working-papers/1995/adrm/sm9501.pdf. S. Ciochetto, Bureau of the Census, Center for Survey Methods Research, Washington, D.C. 20233-9150.

190. Jacobs D. Congressional Testimony before the Senate Committee on Banking, Housing and Urban Affairs. Providence RI, May 11, 1998. The University of Illinois Chicago Special Collections & University Archives, School of Public Health, "David E. Jacobs papers" the University of Illinois Chicago School of Public Health library.

191. Brody D, Brown MJ, Jones RL, Jacobs DE, Homa D, Ashley PJ, Mosby JE, Schwemberger JG, Doa MJ. Blood Lead Levels- United States, 1999-2002, U.S. Centers for Disease Control and Prevention. *Morbidity and Mortality Weekly Report.* 2005;54(20):513–516. May 27.

192. Clark S, Menrath W, Chen M, Succop P, Bornschein R, Galke W, Wilson W. The Influence of Exterior Dust and Soil Lead on Interior Dust Lead Levels in Housing That Had Undergone Lead-Based Paint Hazard Control. *Journal of Occupational and Environmental Hygiene.* 2004;1(5):273–282. Available from: https://doi.org/10.1080/15459620490439036.

193. Dixon S, Wilson J, Kawecki C, Green R, Phoenix J, Galke W, Clark S, Breysse J. Selecting a Lead Hazard Control Strategy Based on Dust Lead Loading and Housing Condition: I. Methods and Results. *Journal of Occupational and Environmental Hygiene.* 2008;5(8):530–539. https://doi.org/10.1080/15459620802219799, http://nchharchive.org/Portals/0/Contents/Article0765.pdf.

194. Breysse J, Dixon S, Wilson J, Kawecki C, Green R, Phoenix J, Galke W, Clark S. Selecting a Lead Hazard Control Strategy Based on Dust Lead Loading and Housing Condition: II. Application of Housing Assessment Tool (HAT) Modeling Results. *Journal of Occupational and Environmental Hygiene.* 2008;5(8):540–545. https://doi.org/10.1080/15459620802222587, http://nchharchive.org/LinkClick.aspx?fileticket = HmG4aNBOokM%3D.

195. Dixon SL, Wilson JW, Clark CS, Galke WA, Succop PA, Chen M. The Influence of Common Area Lead Hazards and Lead-Hazard Control on Dust Lead Loadings in Multi-Unit Buildings. *Journal of Occupational and Environmental Hygiene.* 2005;2(12):659–666. Available from: https://doi.org/10.1080/15459620500403737.

196. Dixon SL, Wilson JW, Clark CS, Galke WA, Succop PA, Chen M. Effectiveness of Lead-Hazard Control Interventions on Dust Lead Loadings: Findings from the Evaluation of the HUD Lead-Based Paint Hazard Control Grant Program. *Environmental Research.* 2005;98(3):303–314. Available from: https://doi.org/10.1016/j.envres.2005.02.002.

197. Galke W, Clark S, McLaine P, Bornschein R, Wilson J, Succop P, Roda S, Breysse J, Jacobs D, Grote J, Menrath W, Dixon S, Chen M, Buncher R. National Evaluation of the HUD Lead-Based Paint Hazard Control Grant Program: Study Methods. *Environmental Research.* 2005;98(3):315–328. Available from: https://doi.org/10.1016/j.envres.2004.12.011.

198. Dixon SL, Wilson JW, Succop PA, Chen M, Galke WA, Menrath W, Clark CS. Residential Dust Lead Loading Immediately After Intervention in the HUD Lead Hazard Control Grant Program. *Journal of Occupational and Environmental Hygiene.* 2004;1(11):716–724. Available from: https://doi.org/10.1080/15459620490520792.

199. Clark S, Grote J, Wilson J, Succop P, Chen M, Galke W, McLaine P. Occurrence and Determinants of Increases in Blood Lead Levels in Children Shortly After Lead Hazard Control Activities. *Environmental Research.* 2004;96(2):196–205. https://doi.org/10.1016/j.envres.2003.11.006, https://nchh.org/resource-library/Article0048.pdf.

200. Clark S, Chen M, McLaine P, Galke W, Menrath W, Buncher R, Succop PA, Dixon S. Prevalence and Location of Teeth Marks Observed on Painted Surfaces in an Evaluation of the HUD Lead Hazard Control Grant Program. *Journal of Applied Occupational and Environmental Hygiene.* 2002;17(9):628–633. Available from: https://doi.org/10.1080/10473220290095952.

201. Clark S, Chen M, McLaine P, et al. Prevalence and location of teeth marks observed on painted surfaces in an evaluation of the HUD lead hazard control grant program. *J Appl Occup Environ Hyg.* 2002;17(9):628–633. Available from: https://doi.org/10.1080/10473220290095952.

202. Clark S, Galke W, Succop P, et al. Effects of HUD-supported lead hazard control interventions in housing on children's blood lead. *Environ Res.* 2011;111(2):301–311. Available from: https://doi.org/10.1016/j.envres.2010.11.003. Available from: http://www.sciencedirect.com/science/journal/00139351/111/2/.

The Nation Acts: community organizing, a 10-year solution from the President's Cabinet, and political sabotage

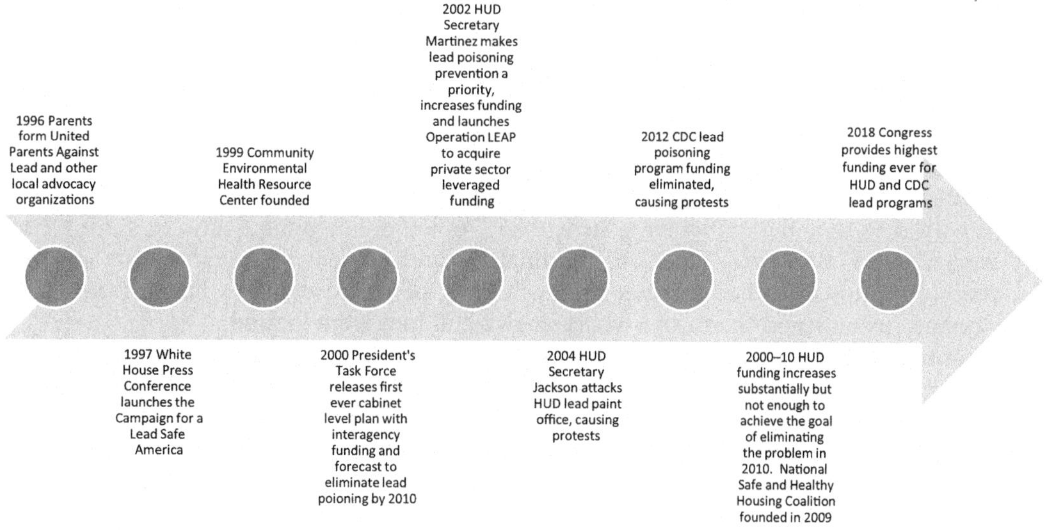

1996 Parents form United Parents Against Lead and other local advocacy organizations

1999 Community Environmental Health Resource Center founded

2002 HUD Secretary Martinez makes lead poisoning prevention a priority, increases funding and launches Operation LEAP to acquire private sector leveraged funding

2012 CDC lead poisoning program funding eliminated, causing protests

2018 Congress provides highest funding ever for HUD and CDC lead programs

1997 White House Press Conference launches the Campaign for a Lead Safe America

2000 President's Task Force releases first ever cabinet level plan with interagency funding and forecast to eliminate lead poioning by 2010

2004 HUD Secretary Jackson attacks HUD lead paint office, causing protests

2000–10 HUD funding increases substantially but not enough to achieve the goal of eliminating the problem in 2010. National Safe and Healthy Housing Coalition founded in 2009

FIGURE 5.0 Timeline.

Much of the architecture envisioned by Title X was in place by the end of the 1990s. The Department of Housing and Urban Development (HUD) lead paint grants for low-income privately owned homes were in operation across the country, and the money Congress had appropriated was being spent more expeditiously and wisely. Federal regulations had been reformed to reflect the science, most states had created a sizeable well-trained workforce with help from EPA, CDC and other agencies, procedures to measure and control lead paint

Fifty Years of Peeling Away the Lead Paint Problem
DOI: https://doi.org/10.1016/B978-0-443-18736-0.00012-1

hazards had been tested and validated, enforcement began in earnest, CDC and HUD national blood lead and housing surveys demonstrated progress, and qualified, committed personnel occupied key positions within both national and local governments.

5.1 Parents and communities

None of that would have been possible without growing and increasingly better-organized parent and community groups that began to spring up around the country. In the 1960s there had been some attempts to mobilize citizens, but they were not sustained. For example, allied with medical and legal professionals in groups such as Physicians for Social Reform, the Welfare Rights Organization, and Community Legal Services, some community groups like the Young Lords in New York City conducted lead poisoning awareness efforts in the 1960s that helped create the political will to pass the 1971 Lead-Based Paint Poisoning Prevention Act.[1]

Larger and more robust community and parents' groups emerged in the 1980s and 1990s. These new groups were loosely coordinated in a network anchored by the Alliance to End Childhood Lead Poisoning, and many became local chapters of United Parents Against Lead (UPAL). HUD, CDC, and EPA all viewed these activities as part of the national effort to build capacity and awareness. All three agencies funded such activities mostly as part of their grants to city, county, or state governments. Philanthropy and sometimes the private sector provided minimal funding to the groups. However, most of the groups operated through dedicated volunteers, typically without pay. This hampered their effectiveness.

One of the reasons HUD had initial difficulties in getting local governments to spend the money Congress had appropriated in the early and mid-1990s was that many communities did not have organized groups to build local political commitment for the new efforts. In some cases, professional grant writers hired by local governments had written the rather complicated applications to HUD, but the actual local staff that took on the work had not been involved in how the grants were structured, sometimes creating delays in getting hazard control work underway and in coordinating with parent and community groups. Community groups often had a better understanding of where the most hazardous houses were located, who owned them, and where at-risk children were located.

Without such local advocacy efforts, local governments that won lead hazard control grants sometimes were unable to enroll enough housing units, despite the tremendous need and large numbers of poisoned children. These local governments sometimes lacked political will. In one egregious example, in 1996 the city government of Cincinnati returned about half a million dollars unspent with no prior warning to HUD, stating that the city no longer wanted to address homes with lead hazards, even though there was plenty of time and homes left to do so. Although a local university protested, arguing that they could get the city turned around and on board, it was too late, and HUD was forced to return the money to the Treasury unspent. HUD staff and parents were outraged by such a waste after so much hard work to get the program up and running.[2]

EPA and CDC were also hampered by the lack of community group involvement in the early 1990s. For example, EPA relied mostly on delegating certification (licensing) of abatement contractors to the states. Title X had given EPA the authority to run such programs if states chose not to. Initially, some states resisted taking on the certification effort (even though it would mean revenue from licensing fees), mostly due to bureaucratic inertia and lack of priority from political leaders. By 2022 the program had expanded to 39 states, DC, Puerto Rico, and 5 tribes, which all took on the certification function; EPA ran the programs for the remaining states (the largest to not have its own certification program was New York State).[3] If citizen groups were not organized to advocate for states to run such programs, there would likely be fewer certified contractors to do the work properly.

Similarly, CDC relied on states to report blood lead data and funded many of them to do so. But for lead poisoning, such reporting was voluntary,[4] although recommendations to make it a requirement appeared in 2016.[5] Citizen groups were important in advocating for physicians and laboratories to report their data to CDC. Because much of these data were for children served by Medicaid or other subsidies, it only seemed fair that taxpayer-supported tests get reported. In 2020 blood lead data were reported to CDC for 32 of the 50 states and the District of Columbia, leaving a significant gap.[6]

For HUD's program, starting in the 1990s, the absence of community groups in some areas prompted new efforts to ensure that citizens were aware their local governments had won a lead paint remediation grant, so they could ensure their local leaders followed through in ensuring it was actually implemented. It was clearly one thing to win the funding, and it was quite another to get it spent to fix hazards in homes. Title X had put the policies in place but putting those policies into actual practice proved to be just as important.

Many groups sprang up around the country spontaneously, consisting of parents whose children had been poisoned, local health and housing professionals, concerned citizens, churches, advocates for children's health, tenant rights, affordable housing and environmental justice, and others. A directory of such groups issued by the Alliance to End Childhood Lead Poisoning in 1995 listed nearly 100 groups,[7] with a second edition 4 years later listing over 300 groups (both directories were funded by CDC). The Alliance helped parents change the conversation from their child's story to what policymakers needed to do, led by Rachel Herzog, the Alliance's community organizer.

Title X mandated that only cities, counties, and states were eligible to receive the HUD lead hazard control grants. But recognizing the importance of such local groups, HUD awarded extra points in scoring those local government applicants that included community groups as a subgrantee. HUD also funded a few such groups directly. These community groups were involved in helping to recruit landlords and others to enroll in the program and educating local leaders and the public on the need for stronger local laws. For the first time, many of the groups were able to not only rely on unpaid volunteers, but instead could be paid through the HUD grant to the local government.

The advocacy for stronger local laws sometimes created friction between local governments and "rabble-rousing" citizen groups. But the better-performing grantees, such as Milwaukee under Amy Murphy's leadership, understood their value and helped to ensure that they were active and effective. Local efforts are detailed later in this chapter.

HUD's lead hazard control grant program historically had many more applicants than it could fund. As a result, the application process was very competitive, so every point counted in scoring the applications and making final award decisions. The Notice of

Funding Availability[9] issued by HUD was organized around five main factors, some of which highlighted the importance of community groups:

1. Capacity—Were the right personnel in place or ready to be hired and were management systems in place? Did the local government have experience in housing rehab and/or lead poisoning prevention? This factor specifically mentioned community groups, asking applicants to:

   "Describe how grassroots community-based nonprofit organizations, including faith-based organizations, will be involved in your grant program's activities. These activities may include outreach, community education, marketing, program sustainability activities and lead-based paint inspections/risk assessments, and lead hazard control work. If you [the local government grant applicant] do not describe *strong engagement with external nonprofit organizations . . . you will not receive full points*" (emphasis added).

2. Need—Were there significant numbers of lead poisoned children and high-risk low-income housing documented? Community groups could sometimes help to determine this if reporting to local government was inadequate.

3. Approach—Were the right partners committed to the program including health and housing providers and local community groups and was there a mechanism to provide funding to such partners? HUD specifically stated, "You must describe how the intended education program(s) will be culturally sensitive, targeted, and linguistically appropriate and identify the means available to supply the educational materials in other languages common to the community," all of which were clearly things that a community group could do well. HUD also required a description of "how your program will work to adopt lead and healthy homes housing policy at the agency, neighborhood, city, county or state level," clearly a nod to community groups.

4. Budget—This was required to have a 10%–25% match from local funds. Community advocates often played a key role in ensuring that their local governments provided the match funding to be eligible for the larger HUD lead grant program. Such local funding was often scarce and limited, and without advocates the funds would sometimes not be made available for lead paint remediation.

5. Evaluation—Were there systems in place to ensure that the program goals were being met? This section specifically asked that grantees monitor "requirement of and course completion of *community stakeholders* and employees/participants for lead," another nod to the importance of community groups.

For its part, CDC required the development of local strategic plans, which included community groups in their formulation. These were part of CDC's blood lead surveillance grants,[10] which mostly went to state agencies and a few large cities.

EPA also funded community groups, as part of its efforts to encourage states to run their own lead certification (licensing) programs. In some cases, community groups would provide training for lead professionals, landlords, and others (examples included United Parents Against Lead in Virginia, the Childhood Lead Action Project in Rhode Island, and the Healthy Homes Collaborative in California). Both HUD and EPA also funded local groups to work with local real estate professionals in ensuring that owners and agents were aware of the new lead disclosure regulation.

In some other cases, local governments used community groups to provide direct inspection and hazard control services and others provided referrals. Examples included Cleveland (Environmental Health Watch), Washington DC (Lead-Safe Washington), Baltimore (Coalition to End Lead Poisoning), Rhode Island (Childhood Lead Action Project), Minneapolis (Sustainable Resource Center), and others.

But this system had flaws. On more than a few occasions, community groups complained to HUD that although they had helped draft a local jurisdiction's HUD application and were initially shown as a subgrantee in the proposed budget, when the actual award was successful the local government took a different direction and decided *not* to fund them. There was little that a federal agency could do to force a grantee to fund a local group, even if they had initially proposed to do so. Sometimes there were local procurement requirements that were a barrier. In other situations, local advocates had become strident opponents of the local government, resulting in antipathy that made funding one's adversary unlikely. In still other cases, local community groups attempted to eliminate local government's role completely, attempting to take over the entire grant, also creating antipathy.

In later years, some questioned whether Title X's requirement that only local governments were eligible to receive the HUD lead hazard control grants should be revised to enable local community groups to also be eligible to run federally funded lead hazard identification and remediation efforts. Although this could result in two or more organizations within the same jurisdiction doing the same thing, perhaps causing duplication of effort, by 2016 the idea had started to gain more traction.

Churches also became involved, including the US Conference of Catholic Bishops,[11] and the United Church of Christ (UCC) (Fig. 5.1). In fact, the UCC had long played an important role, issuing a major report on hazardous waste sites near Black and other minority communities in 1987, "Toxic Wastes and Race."[12]

The director of the UCC's Environmental Justice efforts for 15 years in the 1980s and 1990s was Charles Lee. He went on to direct the environmental justice office at EPA and helped support federal lead poisoning programs.[13] In his earlier years, he had worked with the New Jersey Committee on Occupational Safety and Health, yet another example of the intersection of occupational safety and health with lead poisoning prevention. At

FIGURE 5.1 Cleveland New Year's resolution from the United Church of Christ on lead paint. From *Brooks Berndt Resolution for the New Year Protect the Children. United Church of Christ, December 27; 2018. https://www.ucc.org/resolution_-for_the_new_year/.*

EPA, he later commissioned a paper on disparities in housing and lead poisoning for an Environmental Justice Conference in 2011.[14]

EPA also awarded a grant to the National Safety Council. Among other things, this work involved the National Council of LaRaza to produce a Spanish language campaign on lead poisoning prevention. Together with HUD and CDC, EPA funded the National Lead Information Center, which established a nationwide hotline and resources to help ensure outreach, inspections, remediation, and enforcement were all optimized.[15]

Throughout the late 1990s and 2000s, there were attempts to consolidate the various parent and community groups that were focused on lead poisoning prevention. One such attempt was UPAL. Yet none of them succeeded in establishing a single truly nationwide organization, although UPAL still existed in a few cities in 2022, such as Richmond Virginia under the leadership of Queen Zakia Shabazz.

Parents of lead-poisoned children were vocal and often had compelling stories to tell. However, many were initially ill-prepared for public speaking, and some had to overcome an initial sense of guilt that somehow, they had done something wrong; that if only they had done something differently, their child's lead poisoning would have been prevented. Some parents recognized that through peer-to-peer training, other parents could overcome this misplaced sense of blame, learn to speak publicly, and tell their stories in compelling ways to advance policy change. This became one of the main functions of community and parent organizations.

One example of an effective parent speaker was Susan Thornton from Maine, whose 2-year-old had been poisoned. She told a Congressional committee that instead of just educational pamphlets about cleaning and diet:

> We needed help getting my house fixed. ... Cleaning was the only preventive advice we received. I felt I was a terrible parent. It's like being caught speeding without knowing the speed limit, or what speeding is.[16]

Zakia Shabazz from Richmond Virginia told how her son's lead poisoning could have been prevented if there were more funding. Her son in later years ran a successful lead paint training operation. Shabazz's work focused on high-risk populations, especially African American children and HUD and other officials spoke before their conferences.[17]

Whitlynn Battle ran the Children's Lead Education and Poisoning Prevention program in Birmingham Alabama, a jurisdiction that did not even have a housing code until 2015. Her daughter, Destiny, suffered lead poisoning. The group offered the "Destiny Award" to those who had worked to end lead poisoning. She became an advocate for other families facing lead poisoning, including one that was particularly tragic. A 19-month-old child, Kewan Smalls died in a fire after the family was forced to leave their HUD-subsidized section 8 home in Birmingham because it had serious lead paint hazards and many other housing problems, even though it had "passed" the local housing authority's inspection. The family had been forced to move into a homeless shelter where the fire broke out. At Battle's urging, HUD investigated and found the local housing authority had failed to conduct proper inspections, not only in that house but many others as well. HUD revoked the housing authority's inspection authority and ordered an independent inspection firm to carry them out instead.[18] Hundreds of subsidized housing units were later found to be deficient, and landlords were ordered to repair them. Battle and the Alliance to End Childhood Lead Poisoning issued a press release, which stated:

> The underlying causes of this tragedy are rooted in the grim realities that low-income Birmingham families face in finding housing, the ever-worsening national crisis in affordable housing, and the challenges posed by lead paint.[19]

After the HUD lead paint office and the local HUD office had completed their investigations and taken action, Battle wrote:

> ...large government agencies like [HUD] seem so enormous and unapproachable that I sometimes forget that within the makeup of those agencies are kind human beings that care greatly about communities like my own and are working to solve this countries [sic] lead problem.[20]

Liz Colon from Rhode Island and Janet Phoenix from Washington DC both showed how their experiences with their own lead-poisoned children had led them to work with Sesame Street, a public TV children's show, to create "Lead Away."[21] Colon said:

> Lead Poisoning is an issue that has affected me personally, and to be able to work on the issue through Sesame Street, a program that I enjoyed as a child, as well as with my children has been a wonderful opportunity.[22]

Another community-based nonprofit organization in Rhode Island, the Greater Elmwood Neighborhood Services, conducted lead abatement, and hosted a press conference celebrating the 100th abated house in the neighborhood in 1998, featuring parents, the Mayor, the State health department director, and HUD officials.[23] The Rhode Island group also worked to secure the support of Rhode Island's Senator Jack Reed, who emerged as a long-time supporter of the HUD lead program, due in part to the strong community advocacy efforts in that state. The other Rhode Island Senator, Sheldon Whitehouse, previously served as the State's Attorney General and won an important jury verdict against the lead and paint companies (Chapter 7).

In Chicago, which had perhaps the largest number of lead-poisoned children compared to any other city in the 1990s, the Lead Elimination Action Drive helped get the city to take more action,[24] and actively helped recruit homes into the program.[25] It also produced a 1998 report on the "silent epidemic" that stimulated a great deal of press.[26]

In Alameda County California, which became one of the more innovative and successful lead and healthy housing programs in the country by the late 2010s, citizen movements played a major role as well. In 1987 a study showed that one in five children tested for lead in the flatlands of Oakland had lead poisoning and that housing had high levels of lead in the paint and soil. In response to the study, People United for a Better Oakland launched a series of "Community Lead Action and Information Meetings" and later filed a class action suit to enforce the state's blood lead testing requirement. Together with public health, environmental health, housing and political leaders, they fought to form a County Lead Abatement Program and the County Health Officer formed a task force that proposed the development of a unique new multidisciplinary agency. By 1991 the Alameda County

Board of Supervisors passed a resolution officially establishing the Alameda County Lead Abatement Program, which was funded in part by assessing an annual $10 fee on all residential dwellings constructed before 1978. The cities of Oakland, Berkeley, and Alameda were the first to participate in the program and the city of Emeryville joined in 1992.

Led by Steve Schwartzberg and later Larry Brooks, the Alameda County Lead Abatement Program became the agency responsible for the case management of all lead-poisoned children in Alameda County. Its program was one of the first in the nation to become an accredited lead-in-construction training provider, and to receive one of the new lead hazard control grants from HUD in 1992. It was one of the first lead programs to expand its mission to other housing-related health hazards in the late 1990s (Chapter 8). The program began integrating healthy homes messages into education and outreach, and training public health home visitors, housing, program staff, and others in the principles of a healthy home. In 2002 it received one of the first Healthy Homes grants from HUD to focus on repairs in homes of children with asthma.[27]

Yet another example of an effective parent's voice included Nancy Pavur from New Orleans. She lived in a large home where painting contractors had decided to power sand much of the building exterior, which as we have seen became a prohibited method of paint removal. The result was that her two children were hospitalized, and the family dog died from lead poisoning (pets sometimes exhibit earlier signs of lead poisoning as a result of their licking behavior).

She decided to tell her story and publish it. Working with a researcher from Tulane University and HUD, a peer-reviewed journal article documented this tragedy and how she had spent nearly $200,000 (1999 dollars) to clean up the mess in her own home. Had the project been done properly, it might have cost about $10,000, not to mention avoiding the tragedy.[28]

5.2 The Campaign for a Lead-Safe America

Margaret Sauser, a national founder of UPAL and who led that organization in Michigan in the 1990s and 2000s, addressed the hard-nosed Washington press corps at a White House Press conference in 1997. That press conference launched the Campaign for a Lead-Safe America. Sauser shared the stage with the wife of the vice-president Tipper Gore, HUD Secretary Andrew Cuomo, EPA Administrator Carol Browner and the Mayor of Baltimore, Kurt Schmoke.[29]

Sauser spoke about how preparing the nursery for her second child had involved removing old lead paint by scraping. She said:

> I should have seen a flashing red light from the lead dust and paint chips, my pediatrician should have seen it, and the hardware store where I bought the paint should have seen it, but nobody did. We should have known, but none of us did.

The press release that launched the Campaign provided more tragic details on the Sauser family's experience with lead paint:

> Jon and Margaret Sauser's sons ... suffered lead poisoning after the couple renovated a 67-year-old home they purchased in 1990 in Kalamazoo, MI. The Sausers didn't know that the old layers of paint they

removed from the home and garage contained lead and spread invisible lead dust through the home. They were also unaware that old lead pipes in their home contaminated their water with lead. The lead poisoning caused Jonathan to experience behavior problems, learning difficulties, insomnia, stomach problems, and other ailments. Cameron, who was born into the lead-contaminated home, experienced slowed growth, difficulties with speech and motor skills, and other problems. After learning of the lead contamination of their home three years after they moved in, the Sausers moved and declared bankruptcy when they were told they would be unable to sell the contaminated home. "If only we had known in 1990 what we know now about lead, our sons would never have been poisoned," Mrs. Sauser said. "No parent and no child should have to go through this. By making more parents aware of the dangers of lead, this new federal initiative will benefit children around the nation. Our precious youth must no longer remain our lead detectors. We must find the lead before it finds our children."[29]

When she was done, there was not a dry eye in the room.

Sauser went on to help write an important booklet (Fig. 5.2), published by the State of Michigan with HUD and EPA funding, "Coping with your Child's Diagnosis of Lead Poisoning," in which she spoke to other parents:

I wish [this booklet] would have been around six years ago when all the emotions ... were drowning me when no one—not even family—could understand my grief! I am so pleased that [the government] felt the need to create this booklet. Many parents and children will benefit from its insight.[30]

FIGURE 5.2 Coping with your child's diagnosis of lead poisoning. From: *United Parents Against Lead of Michigan and Michigan. Department of Community Health (undated). The University of Illinois Chicago Special Collections & University Archives, School of Public Health, "David E. Jacobs papers" the University of Illinois Chicago School of Public Health Library.*

It discussed the various stages parents go through when first being told their child has been poisoned: denial, anger, guilt, grief, acceptance, all followed by "learning to cope" by becoming empowered and advocating for change.

The press release for the Campaign for a Lead-Safe America quoted the vice-president's wife, Tipper Gore, who had agreed to chair the campaign:

> If we can keep 1 million children from suffering from lead poisoning by the simple act of removing lead-based paint hazards and educating Americans everywhere, this Administration can continue our progress on behalf of our children's health.[31]

EPA's Administrator Carol Browner said:

> Today's announcement is based on a central philosophy ... the American people have a right to know about any health hazards in their home and in their communities. Given the tools to make their own decisions, we believe parents will take the right course of action in protecting their children from lead poisoning. The new enforcement strategy and public education campaign will give parents those tools.

HUD Secretary Cuomo took the opportunity to announce another round of lead-based paint grants to local governments that would make thousands of homes safe for children. Yet he later ended the Campaign before it could get fully organized.

At the White House press conference, the Campaign promised public service advertising in publications throughout the nation; videos with Sesame Street characters; educational materials distributed by major hardware retailers (Lowe's, Home Depot, Sears, and ACE hardware) in over 6000 stores; distribution of a new interactive video training course for maintenance workers to teach them to do their work safely; and other activities involving the National Association of Realtors, the Consumer Federation of America, the Women's Basketball Association and other groups. Spokespersons included two Pittsburgh celebrities: Mr. Rogers from the Public TV show "Mr. Roger's Neighborhood" and a Pittsburgh Steelers football quarterback.[32] The Campaign developed radio scripts, brochures, print ads, transportation ads, and a partnership program. The celebrities who had agreed to be part of the Campaign included Magic Johnson, Tim Allen, Candice Bergen, Jamie Lee Curtis, Gloria Estefan, Michael J Fox, Ron Howard, Tom Selleck, Bob Vila, Dick Van Dyke, and Sigourney Weaver.

Parent participants in the Campaign included Whitlynn Battle from Birmingham Alabama, Maurci Jackson of Chicago, and many others, who each had tragic yet hopeful stories of what they had endured and how they had worked to prevent the problem for thousands of other families. Maurci Jackson was one of the very first parents to play a key national role, serving on the Title X Task Force in the mid-1990s and helping to organize United Parents Against Lead. She passed away at age 40 in 2002. A newspaper reported:

> Until the day in 1990 when Maurci Jackson's daughter toddled up to her with a half-eaten chip of lead paint clasped in her hand, the South Side woman always thought she would follow her parents into the restaurant business. "She didn't go out to be an activist, she went out to find information," said her mother, Maura Jackson. "And then when she realized this information wasn't getting out, she said 'If I don't know, how many other parents don't know?'" ... Ms. Jackson [became] a crusader who opened a wide public discussion on lead poisoning prevention. In 1992, she helped form a support group called

Parents Against Lead. "I think she was essential as a lead poisoning advocate in Chicago," said Anne Evens, [then] director of the Childhood Lead Poisoning Prevention Program at the Chicago Department of Public Health. "She was really key in empowering parents to help their children, but also in giving parents a voice in setting the agenda."[33]

The Campaign also included the distribution of an illustrated book called Maintaining a Lead-Safe Home[34] to 3500 libraries, which was illustrated by Dennis Livingston, a gifted artist from Baltimore, who had also conducted many lead hazard control jobs in inner city homes and across the nation.

United Parents of Lead in Michigan also released a compelling graphic, depicting children as "canaries in a coal mine," arguing that children should not be used as detectors of lead paint hazards through elevated blood lead levels (Fig. 5.3).

Following the initial launch of the Campaign, the HUD lead paint office awarded a $1 million grant through a competitive process to Vanguard Communications, a small woman-owned business to carry it out, with expert advisors from the Alliance. The Campaign targeted the public and three key audiences: parents, grandparents, and parents-to-be; maintenance, rehab, and home improvement contractors; and owners and managers of multifamily properties. Its two key messages were that lead-contaminated dust from paint and unsafe repainting could cause problems and children could be protected by using lead-safe work practices to address lead paint and contaminated dust. The $1 million budget included only $228,000 in direct labor, demonstrating its promise and cost-

FIGURE 5.3 Children as lead detectors and "Canaries in the Coal Mines." From: *United Parents Against Lead of Michigan (undated). The University of Illinois Chicago Special Collections & University Archives, School of Public Health, "David E. Jacobs papers" the University of Illinois Chicago School of Public Health Library.*

effectiveness. Vanguard Communications had also obtained support from Bank of America, Kiwanis International, the National Rural Health Association, and 800 others.[35]

HUD also awarded smaller additional public education grants, again through a competitive process, to Baltimore, Philadelphia, Kansas City, Consumer Action, Esperanza Housing Corporation, La Clinica del Pueblo, West Harlem Action, Consumer Research Council, National Safety Council, Children's Television Workshop, and others.

The HUD IG questioned whether the lead paint office had the legal authority to provide grants to local community and parents groups, yet another IG attempt to end the HUD lead paint efforts. But the HUD Office of General Counsel had already ruled on this question in 1994 and it did so again in 1998 when it stated in a memo to the HUD IG:

> Section 1011 (g) of Title X provides that the Secretary shall develop the capacity of eligible applicants. ...
> Clearly under this provision, the Lead [Paint] Office is authorized to make grants to recipients other than
> State and local governments to develop the capacity.[36]

This reappeared in 2004 under yet another IG audit, described later in this chapter. Between 1970 and 2020, the HUD IG with one notable exception repeatedly emerged as a formidable barrier, attempting to sabotage the HUD program, directly contradicting, or ignoring Congress, Title X, and HUD's legal authority. In the case of lead paint, the HUD IG repeatedly failed to perform its independent oversight duty during these years.

Jon Cowan replaced Bruce Katz as Secretary Cuomo's new Chief of Staff. He chose to prevent Vanguard Communications from carrying out its Campaign duties in 1998. A year earlier, Cuomo and Cowan were irked by an article on lead paint that appeared in Parade,[37] a national Sunday magazine with a large circulation. Cuomo asked why he had not been quoted instead of HUD's lead paint experts. The answer was that the article had been written before Cuomo had become Secretary, but his appetite for publicity became legendary.

Under orders from Cuomo, Cowan moved the Campaign and other public education out of the lead paint office and into the HUD Office of Public Affairs, with disastrous consequences.

Cowan issued an order to stop the Campaign on October 12, 1999, overruling the HUD Office of Lead Hazard Control.[38] The Federal Housing Administration (FHA) Commissioner at the time, Bill Apgar, had separately issued a contract to do public education on other housing issues, such as home mortgage insurance, rehab, a home buying guide, a program to provide homes to police officers, and home injuries. None of this was related to lead poisoning. The FHA campaign produced a few TV ads but a little more.

The results of this political interference were predictable. For example, in Allentown PA, the local newspaper complained that HUD had not delivered promised brochures and other public education materials.[39] Ralph Scott of the Alliance (which was an advisor to Vanguard) said in that article:

> We've obviously lost an opportunity to follow through on some real momentum. .... Everyone is clamor-
> ing for materials from us. We're trying to get them free from HUD's clutches as fast as we can.

A staffer for Senator Susan Collins said, "We're committed to this [lead paint] issue, and it's unfortunate that HUD wasn't able to coordinate its public relations campaign."[39]

The Alliance to End Childhood Lead Poisoning and Cushing Dolbeare, who had chaired the Title X task force chair, urged "HUD officials not to disrupt or terminate the Campaign for a Lead Safe America or place it under new management."[40]

But Cowan wanted the Campaign to be limited to television ads. Earlier, he had also opposed Secretary Cuomo's appearance at the White House Press Conference, but other advisers such as Chris Lehane and Cheryl Fox (who had also helped craft Title X with Katz and Ryan), prevailed.

But the damage wrought by Cuomo had been done. Following the White House Press conference, the Campaign for a Lead Safe America, despite considerable promise at its launch, ultimately failed due to his political interference.

The end of the Campaign for a Lead Safe America also highlighted other problems faced by the HUD lead paint office at the time. Appropriations for lead paint were at their lowest level between 1997 and 2000 (except for the 1995 rescission), between $60 and $70 million per year.

Working behind the scenes, Saul Ramirez, who was the Deputy Secretary during Cuomo's tenure, attempted to restore staffing in the lead paint office, increase funding, and limit the political interference. He was a strong supporter of HUD's lead paint program and would go on to direct the National Association of Housing and Redevelopment Officials. He would also serve on the board of directors of the National Center for Healthy Housing in future years. But he was unable to stop the Cuomo juggernaut.

Although Katz and HUD Secretary Cisneros (Cuomo's predecessor) had increased staffing for the HUD lead paint office to 31 in the mid-1990s, during the Cuomo years, it declined to only 17 in 2000. The reductions could not have come at a worse time. Despite repeated requests for staffing increases[41] to manage the increasing workload from overseeing grants, enforcement of the disclosure rule, the Campaign and other public education, expanding research, and the issuance of the reformed HUD lead paint regulation for federally assisted housing, the Cuomo years at HUD marked a low point for lead poisoning prevention.

Increases in staff for the office would not be approved until 2001, when Michael Hill (who had been "exiled" to the lead paint office in the early 1990s) was appointed to HUD's Office of Administration and the incoming new HUD Secretary Mel Martinez again made lead and healthy homes a Departmental priority.[42]

Despite the political interference from the Cuomo Administration, many parents continued to tell their stories and did so courageously and creatively, creating a renewed political will. For example, Leann Howell from Ohio was one of the more flamboyant. She created a fictional character, "Leadie Gaga" (a take-off on the star Lady Gaga) and she dressed the part; she developed a hand puppet modeled after the Dutch Boy lead painter that was used in countless educational campaigns (Fig. 5.4). She was able to make a convincing case of the need to do more. Tragically, her son was poisoned with a blood lead level of 44 µg/dL at only 10 months of age, a story recounted elsewhere.[43] Her website stated, "I've already been through hell. Not only will I survive but I will win."[44] Another vocal parent was Tamara Rubin, who ran "lead-safe Momma."[45]

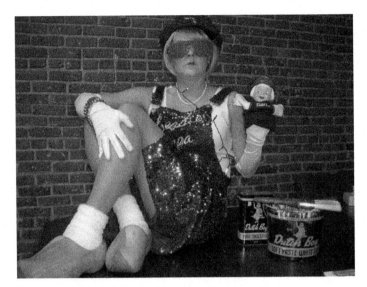

FIGURE 5.4 Picture of Leadie Gaga, posing with old Dutch Boy lead-based paint cans and the Dutch Boy hand puppet. *Reproduced with permission from: Leann Howell.*

Although many parents continued to tell their stories, some were rebuffed. For example, Secretary Andrew Cuomo was presented a handwritten letter from the mother of a lead-poisoned child in Syracuse in 1999 (Fig. 5.5), but he refused to read it to the assembled dignitaries. Yet it was compelling in a way that only a mother's voice can articulate.[46]

5.3 The Community Environmental Health Resource Center

The Community Environmental Health Resource Center (CEHRC) was a project of the Alliance for Healthy Homes (which had changed its name in 2001 from the Alliance to End Childhood Lead Poisoning to reflect a mission to address other housing hazards beyond lead paint). CEHRC created a collection of community-based organizations that trained local community members on how to document housing health hazards and housing code violations in high-risk housing and how to use the results to spur action by property owners and policy makers. Congress also supported the effort, stating: "The [Appropriations] Committee encourages HUD to evaluate a proposal to create the CEHRC."[47]

CEHRC's purpose was a "citizen science" effort that was formed "to empower communities at risk by providing access to new tools and strategies for holding landlords and government agencies accountable and winning needed policy changes."[48] The concept was hammered out at a 2-day meeting at the historic 16th Street Baptist Church in Birmingham Alabama where a Local Leadership Council was formed. (That church was a center for the civil rights movement; tragically the church was bombed and 4 young girls attending Sunday school died in 1962). Instead of relying solely on experts to assess housing-related health hazards, the group sought to reduce cost and complexity and enable community members to do some of their own inspections. CEHRC wanted to

I am writing this letter to thank you, and your staff, but most of all the lead embatement program. I want to let you know that without this program I don't know where me and my family would be wright now. My grandaughter Destiny, and her mother share a two family house with me. The house was in bad shape, and my grandaughter got lead posion from the paint and plaster that dated back almost 150 years ago. I Anastasia, own the house, but an accident left me disabled on a fixed income. My daughter Stacey, is struggling with a little girl with a lead level of 21. My daughter also only works part-time, so her income is very limited. The lead embatement program put all new thermo windows in my daughter's apartment. They sided the whole house, replaced old doors and carpeted my front porch floor. My house looks better then it ever did because of the lead embatement program. Also by getting

rid of the lead paint my grandaughter's lead is down to 15 and is getting better everyday. I hope the program continues to grow, and help other families like mine. Thank you all for everything.

Truly Yours

bridge the gap between science and communities at risk by providing tools, training, technical assistance, strategy advice, and grants to help trained local project staff identify and address health hazards in high-risk housing.

There were dilemmas posed by such an approach. Landlords might evict a family in retaliation for a notice from a community group about health hazards. Or they might simply abandon a property instead of making repairs. It could backfire if unsafe work practices were used to correct hazards or if the work was not completed. A resident might misunderstand the test results and abandon the property without good cause or adequate compensation. The family might be stigmatized after neighbors overheard information about the home. Or undocumented residents might be deported. Respecting and protecting tenants' rights was complex.[49]

To overcome some of these concerns, CEHRC adopted a guiding principle that the rights of individual residents whose homes were investigated were always paramount to the organizing and advocacy strategies and tactics designed to benefit these families, their neighbors, and the larger community.

CEHRC trained 200 community members in nine cities and regions across the country as housing hazard investigators; funded technical and strategy assistance; and supported evaluation and peer-to-peer communication. They inspected more than 2500 homes for hazards and then used the results to press for community-wide solutions for lead poisoning and other healthy housing problems.[50] The results also enabled local jurisdictions to identify needs as part of a larger grant application to HUD, consistent with Title X's mandate to build local capacity.

Other creative ways of focusing community and citizen attention were contained in a CDC publication developed by the Alliance.[8] For example, in Rochester NY, a vacant house built in 1894 was used to show what lead hazards looked like, and to demonstrate simple but effective hazard control treatments. It was used to conduct tours and open houses for officials and the public. Various "stations" in the house featured basic lead poisoning information, photos documenting the creation of the lead lab house, lead-safe cleaning tools, lead-safe work practices, brochures and materials, dust wipe sampling kits, and an XRF (X-Ray Fluorescence) portable lead paint analyzer. A portable blood lead analyzer allowed people to have their own blood lead levels checked. Visitors learned how to take a lead dust wipe sample in the house.

Visiting the demonstration home motivated several state and local officials to engage with advocates in substantial ways and pass new local laws. Local residents, property owners, code enforcement officers, and city officials became more aware of the problems of lead hazards and options for addressing them. HUD later adopted this model and purchased a mobile home outfitted in much the same way, sending it around the country.

Another CEHRC group in Greensboro, NC decided to engage with landlords to build public awareness and foster less adversarial, more supportive relationships. This included free training in lead-safe work practices; how hazard control interventions could reduce legal liability; sources of grants and no- or low-interest loans for rehabilitation and lead abatement; and information about other services such as low-cost clearance lead dust testing.

The Alameda County, CA health department also worked with CEHRC to increase testing rates for blood lead among pregnant women through educational materials for at-risk pregnant women. CDC later released an important document on how the thorny issue of testing blood lead levels in pregnant and lactating women might best be handled, a document that Kathryn Mahaffey helped to develop (and which is dedicated to her).[51]

The Alameda County effort succeeded in getting more women tested, explaining how lead gets into the body, how it could affect a developing fetus, and how to create a lead-safe home environment. Over 25,000 packets of culturally appropriate outreach materials were distributed to high-risk pregnant women and their families through community programs, including Women and Infant Care, Head Start, MediCal (the Medicaid system in California), Black Infant Health, and other agencies. Over 15,000 postcards urging people to "get tested now" were also mailed in high-risk areas.

Another CEHRC group in California, the Healthy Home Collaborative headed by former fashion designer Linda Kite and others organized "toxic tours" for policymakers to enable them to get a first-hand look at unhealthy housing conditions like lead paint hazards. The tours enabled them to visit homes with hazards (and some that had been repaired) and talk with residents and advocates about the problems and policy solutions. Policy makers were moved by the personal experience of seeing hazardous conditions first-hand and having face-to-face interaction with families directly affected, who wanted not only to tell their story but pushed for policy changes. First-year medical students were also given such tours.

5.4 Community groups and the press

Many community groups also stimulated press coverage. In May 2001, the Providence (Rhode Island) Journal ran a six-part series on lead poisoning that helped set the stage for new state legislation and regulations. A photojournalist (John Freidah) initiated the idea for the series, but it could not have happened without the commitment of the newspaper, the hard work of the Childhood Lead Action Project, the time and effort of many government officials and scientists who educated and provided information to reporter Peter Lord, and the willingness of many families to open their lives to the journalists and the public.[52]

The Detroit Free Press conducted an in-depth investigation of lead poisoning in Detroit and Michigan that began appearing on January 21, 2003.[53] The paper followed up with a 5-day investigative series and continuing coverage. The reporting and persistent work of community advocates resulted in lead poisoning becoming a top priority of the Governor and the state legislature. In addition, the series prompted EPA to order the removal of lead-contaminated soil in a Detroit neighborhood near a former lead smelter.

The Baltimore Sun also ran an in-depth series in 2000. It highlighted the large disparities that still existed:

> Few other places contain such a wealth of historic and medical documentation on the disease, and few bear so many scars. For if any city was in a position to save its children, it was Baltimore. And Baltimore did not.[54]

The Cleveland Plain Dealer worked with a local advocacy organization there, Environmental Health Watch, which was formed in the 1980s and led by Stu Greenberg and later Kim Foreman. The front-page articles[55] played a key role in passing a new lead paint ordinance in that city in 2020.[56]

Best practices by community groups in 2000 and 2002 were recognized in a catalog produced by UPAL of Michigan. They included the Silver Valley People's Action Coalition (a superfund site in Idaho), Improving Kids Environment (Indiana), the Maine Lead Action Project, Healthy Homes Detroit Project, Get the Lead Out in Grand Rapids, MI, Durham Parents Against Lead, UPAL of Michigan, UPAL of North Carolina, UPAL of Syracuse, UPAL of Philadelphia, Organization of Parents Against Lead, Lead Safe America Foundation, Childhood Lead Action Project (Rhode Island), Baltimore Parents Against Lead, Chicago Metropolitan Tenants Organization, Lead-Safe Washington, Omaha Healthy Kids Alliance, Association of Parents to Prevent Lead Exposure in Cleveland, Philadelphia Citizens for Children and Youth, Wisconsin Citizen Action, Ohio Healthy Housing Coalition, Sustainable Resources Center, and many others.[57,58]

The conclusion to this book describes how during the 2010s, many local parent and community groups became members of the National Safe and Healthy Housing Coalition, which in 2022 had over 600 members representing 400 organizations in all 50 states.[59] Carrying on the network the Alliance had begun in the 1990s, the Coalition became a broad umbrella group that was organized through the efforts of the National Center for Healthy Housing (NCHH) and others. NCHH also provided limited funding to local groups around the country,[60] as did the Green and Healthy Housing Initiative.[61] More recently, organizations have made progress in Lewiston (Maine), Buffalo (NY), Patterson, NJ, Grand Rapids (MI), Utica (NY), New Orleans (LA), Cleveland (OH), and Richmond (VA).[62] More detailed reviews of these and other community groups are available elsewhere.[63]

5.4.1 Advocates and scientists

One of the key lessons to emerge from the lead paint experience in this time period was that the most effective way to present the problem and its solution was through a collaboration of parent and community groups, together with housing and health professionals, scientists, and government. The statistics and the evidence could be presented, but it was also important to put a human face on the tragedy to promote evidence-based political will.

One example of the newfound collaboration was manifested through awards. For example, in 2000 the Alliance to End Childhood Lead Poisoning succeeded in establishing the David P Rall Award for Public Health Advocacy at the American Public Health Association to recognize leaders in public health. Rall was a key scientist in this time period, serving on the boards of both the Alliance and NCHH.

He was the first director of the National Toxicology Program from its inception in 1978 through 1990. Following that, he directed the National Institute for Environmental Health Sciences (NIEHS) (part of the National Institutes of Health), which has Environmental Health Sciences Core Centers across the country. He ensured that those cores all had community engagement as one of their four areas. As a scientist, administrator, and advocate, he educated scientists, governments, and their local collaborators to address lead and other environmental hazardous agents. He was able to bring together some of the best minds and hearts of his time to the cause of world health through a safe and clean environment. A building at NIEHS, which he directed, is dedicated to him.[64]

The first person to be given the Rall Award in 2000 was Eula Bingham, who had issued OSHA's first comprehensive health standard on lead as OSHA Secretary in 1978, another link to the occupational health community.[65] Bingham overcame industry opposition to the lead standard for industrial workers. It might not have passed at all if she had not publicly complained of the "palace guard" surrounding President Jimmy Carter. According to the NY Times, in her campaign for workplace safety, Bingham clashed with business, Congress, and other members of the Carter administration, but the President supported her and overruled his economic advisors.[66]

How the President came to support Bingham was a case study in how science can guide successful policy. The future director of HUD's lead paint office interviewed former President Carter after completing a course on archival research at the Carter Library in Atlanta in 1987. He asked the former President why he had backed health advisors like

Eula Bingham over the objections of economic advisors and why he had decided in favor of engineering controls instead of wearing respirators to protect worker health. The President answered, "She's great ... I think she, more than any single person, formed a proper working relationship between OSHA, employers and employees."

Cotton dust in textile mills brought the issue of science and policy to the nation's attention. The quarrel with the President's health and economic advisors was about whether it would be better to use ventilation and other engineering controls or whether it would be better to require workers to wear respirators (respirators being cheaper but far less effective, as shown in Fig. 1.11). The former President continued:

> Well, I think that was what Eula wanted. And it's what I thought was the only feasible approach. I ran a peanut shelling plant [before I became President].... In dry weather, clouds of impenetrable dust cover the people who are running the plant, including me and my brother and my employees.... If you're growing seed peanuts, then you treat them with chemicals to prevent rot in the ground.... We used a mercury-based treatment at the time. And all the warning labels say wear a respirator. I never could get an employee to wear a respirator for more than an hour or two.... And I never wore a respirator.... I knew there was no way you were going to make all the employees in a cotton mill wear respirators. I ran a cotton gin and broke my finger—I ran it myself and the air was filled with lint, but I would never wear a respirator.[67]

This preference for engineering controls over personal protective equipment played out in the lead paint field as well. Controlling lead paint hazards through abatement and interim controls (which were forms of engineering controls) was far more effective than screening children and educating parents to wash their hands more frequently (which were forms of personal protection and administrative controls).

The importance of having scientists who understood these differences and who could lead major health-oriented government agencies was manifested at OSHA. Bingham was an authoritative toxicologist and had direct knowledge of the harmful effects caused by exposure to hazardous agents, including lead. She served as the editor of a major toxicology textbook and later asked Jacobs to write the chapter on lead toxicology and methods of controlling exposures.[68] Many of the authors of chapters in the toxicology textbook had served in key national and international government positions, including Bailus Walker (a board member of the Alliance), Laura Welch, Philippe Grandjean, Bruce Fowler (a key author of the 1993 Academy of Sciences report on lead), Peter Infante, Celeste Monforton, Michael Kosnett (a member of the CDC advisory committee on childhood lead poisoning prevention), Richard Lemen, Frank Mirer, David Carpenter (active in the New York State lead program), and many others.

The difficulty in promulgating scientifically based regulations was shown by what happened to the industrial worker standard for lead which was similar to the difficulties faced by the HUD regulations on lead paint in housing. Although Bingham succeeded in issuing the lead standard for industrial workers in 1978, it faced many subsequent court challenges. In 1980 the DC Circuit Court upheld most of the standard, but ruled that "increased flexibility" be provided for 38 industries. In 1982 OSHA delayed the deadline for primary and secondary lead smelting, battery manufacturers, and others, but the United Steel Workers union contested. In 1984 the DC Circuit Court ordered OSHA to lift the stay for some industries and by 1988 a few industries still did not comply, but by and large the standard was considered by many to be a success in addressing lead poisoning in industry. But it would take a decade to get it fully in place, much like the delays in promulgating and implementing the HUD lead safe housing rule and other Title X mandates described in the previous chapter.

Title X required OSHA to extend these same protections to workers in construction, which included those engaged in lead hazard control work in housing. How odd that it took a housing law to protect construction workers. But it did not happen without a fight.

The National Association of Home Builders had opposed Title X from the very beginning, and it decided to go after the occupational lead standard as well. It argued that workers in construction were not being over-exposed to lead and that the construction lead standard was not needed, despite Congress' clear directive to the contrary.

In a rebuttal, the National Center for Lead-Safe Housing showed that workers in residential lead hazard control work did in fact have exposures above OSHA's Permissible Exposure Limit (which was an amount of lead in the air in a worker's breathing zone, 50 μg of lead per cubic meter of air, μg/m^3):

> In fact, the range of exposures documented during the HUD Demonstration project show that the highest exposure reported for encapsulation was 72 μg/m^3. Component replacement had a high exposure of 121 μg/m^3, and cleaning had a high exposure of 590 μg/m^3. Even preparation work (removal of carpeting) indicated a peak exposure of 206 μg/m^3 ... between 1984 and 1988 [OSHA data show] that 54.1% of the 122 air samples collected were above 100 μg/m^3 (twice the OSHA PEL for General Industry). This indicates widespread overexposure to lead in these trades.[69]

With the authority from Title X, the OSHA construction standard for lead was promulgated in 1996 and was not challenged in court; the Homebuilders knew they would lose.[70] It was reviewed by OSHA in 2007 and was not changed.[71] Others have noted the blood lead standards and other medical management requirements first established in the 1978 regulation were extremely outdated by 2006.[72] The National Institute for Occupational Safety and Health stated:

> Some studies suggest that the current OSHA Permissible Exposure Limit and the NIOSH Recommended Exposure Limit may be too high to protect against certain health effects.[73]

Parent and community groups, together with their allies in public and occupational health, housing, environment, and other scientific fields achieved important advances in increased public awareness, despite political interference. Yet children were not simply "small adults," and a new effort to coordinate different agencies took hold.

5.5 The President's Cabinet approves a 10-year strategy, 2000–10

In April 1997 President Clinton issued Executive Order 1304, "Protection of Children from Environmental Health Risks and Safety Risks."[74] It created the highest-level task force on lead paint yet, comprised of powerful cabinet-level officials.

Despite Congressional attempts in Title X to assign clear responsibilities, this was the first time *all* relevant federal agencies were brought together to develop a specific timeline, a multiagency integrated budget request, and a forecast on what it would take to eliminate lead poisoning from lead paint hazards from 2000 to 2010. Importantly, this Executive

Order spanned not only the Clinton Administration, but also the succeeding George W Bush Administration, which extended it. Both Administrations convened cabinet-level meetings on how to address lead paint.

The Executive Order was issued primarily because most previous federal action had erroneously assumed children were merely "small adults." But in fact, as the Order stated:

> [Children] suffer disproportionately from environmental health risks and safety risks. These risks arise because children's neurological, immunological, digestive, and other bodily systems are still developing; children eat more food, drink more fluids, and breathe more air in proportion to their body weight than adults; children's size and weight may diminish their protection from standard safety features; and children's behavior patterns may make them more susceptible to accidents because they are less able to protect themselves.[74]

The initial impetus for the Task Force's creation was the realization that car seats for infants had often been poorly designed based only on their smaller size instead of their unique characteristics. Its membership was at the second-highest level of the executive branch of government—the President's Cabinet (the President is the highest level). Cochaired by the Secretary of the Department of Health and Human Services (Donna Shalala at the time, who had previously served as an Assistant Secretary at HUD in the 1970s), and the EPA Administrator (Carol Browner, who had participated in the White House Press Conference on lead), it also included the Secretaries of Education, Labor, Energy, HUD, Agriculture, Transportation, as well as the Attorney General and others. Most importantly, it also included the White House OMB, which gave the Task Force real power over funding.

The Executive Order specifically required that the Task Force undertake "an identification of high-priority initiatives that the Federal Government has undertaken or will undertake in advancing protection of children's environmental health and safety," which obviously included lead paint. A "senior staff" planning committee was established among key federal agencies, which created a new dynamic. Each time the respective heads of their agencies were to meet, planning committee members sought to determine what each Cabinet official could report on their respective agency's progress.

Initially, the senior staff defined 4 main areas: asthma, unintentional injuries, developmental disorders, and cancer.[75] It appeared as though lead paint might not be a Task Force priority, since it was grouped under a more ambiguous category of "developmental disorders."

Indeed, much of the initial focus was the creation and implementation of a new and very large "National Children's Study." As originally conceived, it would look at children's health in a 50,000-member cohort from preconception to at least 20 years into the future. It was billed as another "Framingham Study," which discovered much of what is known about heart disease today. Starting in 1948 with 5209 adult subjects, by 2013 the Framingham study was on its third generation of participants.[76] But because funding was never provided, the National Children's study collapsed after a 5-year planning effort, what some critics called "a $1.3 billion scientific humiliation."[77,78]

However, unlike the National Children's Study, the Task Force action on lead paint marked a major advance. In the wake of all the citizen action, increased capacity, new science-based and practical regulations and standards, and funding that had been put in place throughout the 1990s on lead paint, the next step was for the Cabinet to state what was needed to end lead paint poisoning as a major public health problem over a discreet time period. The result was *"Eliminating Childhood Lead Poisoning: A Federal Strategy Targeting Lead Paint Hazards,"* released in February 2000.[79] Issued by the Task Force Cabinet Secretaries, its main authors were Jacobs, Tom Matte from CDC (who had worked on the CDC 1991 strategy, the HUD Evaluation and many other research projects), Bruce Nilles (who had been a leader at the Department of Justice in bringing the first enforcement actions on the lead disclosure rule), Joanne Rodman (from the EPA Children's Office) and Lynn Moos (from the EPA lead poisoning prevention program).

The report's release was announced by Tipper Gore, the wife of the vice president. Mayors from Baltimore and Milwaukee and elsewhere issued their own supporting press releases. Cushing Dolbeare, founder of the Low-Income Housing Coalition and other community groups such as Wisconsin Citizen Action and the National Coalition for Lead-Safe Kids did the same.

Having learned the importance of parents stories, the report began with this:

> Like any other parent, the most important priority in my life is to provide my three children, Damien, Samuel, and Nathan, with a happy and healthy home—a place where they can grow, learn, and develop into productive adults. What I didn't know was that our home would threaten my children's health. In April of 1996, my family and I managed to save enough to buy our own home. Within four months of moving in, our pride and joy evaporated when Samuel, then 10 months old, was diagnosed with a blood lead level of 32 µg/dL. Worse yet, a month later, Samuel's lead level had risen to 50 µg/dL. He was hospitalized that same afternoon and for three long agonizing days he stayed in the hospital and began treatment.... *Today, our house has new windows, and lead abatement has been completed on the interior and exterior of our home. Samuel's blood lead level has dropped.... I share my story with the hope that other families and their children will learn about the dangers of lead, and that one day soon, lead poisoning will be a disease of the past* (emphasis added).

This hopeful tone reflected the report's ambitious agenda. It was no longer just about how to manage the risks in the near term. Instead, it presented what it would take to finally solve the problem. The opening sentence read, "For the first time, [the report] presents a coordinated federal program to eliminate childhood lead poisoning in the United States."

It included an interagency budget request for the fiscal year 2001, a novelty. Normally, each agency prepared its own separate budget without knowing what the others were requesting, with the White House making the final call on what went into the President's budget request to Congress. In fact, HUD and the other agency budget directors warned the staffers developing the Task Force report not to tell each other what each agency was requesting, because that was normally a closely guarded secret. But because this was a Cabinet-level document, the "secrets" were shared.

None of the earlier strategic plans throughout the 1980s and 1990s had ever developed a multiagency budget, including the 1988 Agency for Toxic Substances and Disease Registry (ATSDR) report to Congress, CDC's 1991 strategic plan, HUD's 1990 Comprehensive and Workable Plan, the EPA 1990 strategy, or the 1995 Title X task force.

But this new President's Task Force authors overcame the agency silos. Even the Office of Management and Budget at the White House came to support the idea of such an integrated budget request and signed onto the final lead paint report. It included a specific 10-year plan covering 2000–2010, with new cost-benefit numbers showing a *net* benefit of at least $8.9 billion. It reiterated that lead poisoning was a completely preventable disease and showed how eliminating lead hazards in homes could be done in 10 years (2000–10).

The 2000 Task Force report articulated an inclusive vision:

Every child deserves to grow up in a home free of lead paint hazards.

The authors of the 2000 report debated whether the term "lead poisoning" should be used instead of "lead exposure." Some argued that poisoning means an acute problem resulting in hospitalization, but others knew that much of the harm from lead was asymptomatic and that most children who were poisoned were never hospitalized.

Although it mentioned other sources of lead and the success in banning lead from gasoline, new paint, and other sources, it focused on the new definition of lead paint hazards and older housing. The Task Force report, demonstrating more clearly than ever the housing-related nature of lead poisoning, used the most recent data from the National Health and Nutrition Examination Survey (NHANES) at the time. For all children, about 4.4% had blood lead levels above the CDC level of concern at the time (10 µg/dL). In older housing, that percentage increased to 9% and for African American children living in older housing, it reached an astonishing 21%. The report warned that:

without further action, over the coming decades large numbers of young children may be exposed to lead in amounts that could impair their ability to learn and to reach their full potential.

The report built on the Title X foundation: certified lead paint risk assessors, inspectors, abatement contractors, regulations, enforcement, and public education, all of which were developed during the 1990s. By 2000 there were housing-based lead paint hazard control programs in over 200 cities.

The Task Force developed a forecast on what would happen over the course of the next decade using three options: status quo (doing nothing more), the effect of HUD's newly reformed federally assisted housing regulation issued just the year before in 1999, and how a new 10-year plan could achieve the goal (Fig. 5.6). Reflecting the increased focus on science, the report contained a validated 20-year projection covering 1990–2010 to quantify both costs and benefits. To do that, it used data from the Evaluation of the HUD Lead Hazard Control grant program, the third NHANES for 1991–94 (the most current national blood lead data available at the time), the 1997 American Housing Survey, the 1999 Economic Analysis accompanying the HUD regulation covering federally assisted housing, the Residential Energy Consumption Survey, US Geological Survey data on the historical use of lead in paint, and the 1990 HUD National Survey of Lead paint in Housing.

If the first "do-nothing" approach was used, there would still be some reduction in the number of children poisoned and the number of homes with lead paint, because houses were continually being demolished or extensively rehabilitated, both of which gradually eliminated existing lead paint. The report contained numerical estimates of what that reduction was likely to be by examining what had occurred in the previous 10 years. Some critics of the report suggested that no one could really know how many homes would be demolished or rehabbed going forward, but they ultimately agreed that this would be the best approach, and that fine tuning could occur in the future.

To produce the numbers, HUD asked Rick Nevin again to help. He was the economist who had prepared the cost-benefit analysis of HUD's reformed lead regulations that had

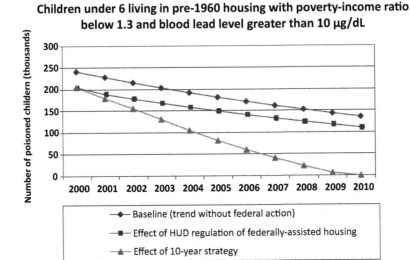

Children under 6 living in pre-1960 housing with poverty-income ratio below 1.3 and blood lead level greater than 10 µg/dL

— Baseline (trend without federal action)
— Effect of HUD regulation of federally-assisted housing
— Effect of 10-year strategy

FIGURE 5.6   Forecast showing number of lead-poisoned children from three scenarios between 2000 and 2010: baseline (ongoing demolition only); effect of Federally Assisted Low-Income Housing Regulation; and the effect of the new 10-year President's Task Force Strategy. From: *President's Task Force on Children's Environmental Risks and Safety Risks. Eliminating Childhood Lead Poisoning: A Federal Strategy Targeting Lead Paint Hazards; 2000.*

survived the rigorous White House OMB review. He worked to avoid attacks by using conservative economic assumptions. For example, he again used the 3% and 7% discount rates required by OMB and most other economists, even though he knew such discount rates were not valid for intergenerational benefit transfers like lead poisoning.

The economic analysis showed that between 2000 and 2010 total benefits were at least $11.2 billion at a 3% discount rate and $3.5 billion at a 7% discount rate. In both cases, net benefits were very large. The benefits were due to savings from avoided medical care, avoided special education, increased lifetime earnings due to increased cognition, and housing price benefits due to improvements in housing, such as window replacement, which increased home value and reduced energy costs. There were more intangible benefits that were not included, such as avoided hypertension in later life; improvements in children's height, hearing, and vitamin D metabolism; crime, and avoided emotional health costs involved in caring for poisoned children.

Because these intangible benefits could not be quantified reliably, the calculated benefits were an underestimate. The report stated:

> The quantifiable monetary benefit (which does not include all benefits) of eliminating lead paint hazards through interim controls in the nation's pre-1960 low-income housing stock over the next 10 years will be $11.2 billion at a 3% discount rate ($3.5 billion at a 7% discount rate). The net benefits of interim controls are $8.9 billion at a 3% discount rate and $1.2 billion at a 7% discount rate. The monetary benefit of abatement of low-income housing is estimated at $37.7 billion at a 3% discount rate [$20.8 billion at a 7% discount rate]. The benefit of permanently abating lead paint is considerably greater because more children would benefit over a considerably longer time span.

Cost estimates came from HUD's lead hazard control grant program, even though these were typically the worst-case houses and therefore the costs were higher than for more typical homes where lead paint hazards were less severe. This was yet another conservative assumption, but the cost data were the most reliable and documented. The report also assumed there would be no cost reductions over the 10 years, even though economies of scale, increased competition, and new technological advances would all likely lead to reduced costs. But again, there was no convincing way to quantify any of that, so the conservative approach ruled the day.

Nevin went on to publish other important work linking early childhood lead exposure to crime, murder, unwanted teen pregnancies, and other antisocial behavior.[80–82] He also wrote his own book, "The Lucifer Curves,"[83] in which he showed the word "Lucifer" came from a Hebrew word meaning the god "Venus." In Latin, "Lucifer" means "light-bringer" and was the name of the planet Venus. In perhaps the only book on interstellar lead, Nevin stated:

Lead sulfide is vaporized on the planet's surface, rises as a mist, condenses in the cooler Venus clouds, and settles as a "shiny, metallic frost on the tops of the mountains," making the Venus highlands more reflective than lower elevations. Lucifer is not the name of the devil … [but] is just the Latin name for a reflected image caused by a toxic lead-contaminated environment.

This put a new twist on the idea that lead poisoning and evil were connected.

The costs over the 2000–10 time period were presented in two ways: what they would be if interim controls (the more initially inexpensive method but with higher long-term costs) versus abatement (the more initially expensive but with lower long-term costs). Using data primarily from the HUD lead hazard control grantees, the annual costs were $230 million and $2.1 billion for interim controls and abatement, respectively, if only low-income homes were addressed. If all older homes were addressed, the annual costs were $1.84 billion and $16.6 billion for interim controls and abatement, respectively (Table 5.1).

All 4 estimates were considerably higher than HUD's annual lead paint appropriation at the time, which was only $60 million in 2000. Yet the net benefits were compelling. These new cost estimates were much lower than HUD's 1990 "Comprehensive and Workable" plan, which had been over $500 billion and had led to paralysis. As a result of the new data,

TABLE 5.1 Annual costs of lead paint remediation using four scenarios, 2000–10.

Pre-1960 Housing stock	Lead hazard screening and interim controls ($1000 per unit)	Inspection/risk assessment and abatement of lead paint ($9000 per unit)
All Pre-1960 Housing at Risk of Lead Paint Hazards (1.84 million units/year)	$1.84 billion	$16.6 billion
Pre-1960 Housing Occupied by Low-Income Families Not Covered by HUD Regulation (230,000 units/year)	$230 billion	$2.1 billion

From: President's Task Force on Children's Environmental Risks and Safety Risks. Eliminating Childhood Lead Poisoning: A Federal Strategy Targeting Lead Paint Hazards; 2000.

the President's budget request for the HUD lead program increased to $90 million. Although this was an improvement, it was still far short of the projections in the 10-year plan. Still, it marked a significant departure from the policy paralysis that had gripped the nation before 1990, when a solution to the lead paint problem appeared to not be feasible.

Beyond the HUD program, the 2000 President's Task Force report also examined the role of other lead poisoning prevention programs at CDC, EPA, Department of Justice, Department of Defense, Consumer Product Safety Commission, and the Occupational Safety and Health Administration. It also included those with little or no historical involvement with lead poisoning prevention, such as the Departments of Treasury and Energy. The Department of Energy operated weatherization programs, which sometimes disturbed lead paint and replaced windows, a key source of lead exposure. The Residential Energy Survey was used to estimate savings not only from lead poisoning prevention, but also reduced energy use due to better windows.

Although the Department of Treasury participated, its duties were ill-defined and in truth it never engaged in a meaningful way. The Task Force's Treasury recommendation simply stated that it should "explore the use of financial incentives (such as tax credits or deductions) or federal grants to control lead paint hazards in housing occupied by low- and moderate-income families with young children not served by HUD grants."

For example, there were discussions about whether the President or Congress could propose a tax credit or tax deduction for families with higher incomes than those served by the HUD program. It seemed unfair to permit landlords to deduct lead hazard control as a business expense, but homeowners could not. The Alliance also attempted to use the Community Reinvestment Act to stimulate more private sector investment in controlling lead paint hazards, which had some limited success in New Jersey and other areas.[84]

In later years, a lead-based paint tax credit bill was proposed by some members of Congress, but was never passed.[85,86] Similarly, proposals to impose a tax on lead production to establish a "lead abatement trust fund" were introduced in Congress, but never passed.[87] For example, then-Congressman Ben Cardin introduced the "Lead Based Paint Hazard Abatement Act of 1991," which would have generated $1 billion from a tax of 75 cents on newly mined lead and 37 cents per pound on recycled lead, but it never gained traction despite support from the Alliance to End Childhood Lead Poisoning and the Environmental Defense Fund.[88]

In 2000 the President's Task Force report concluded, "Further exploration on the specifics of the financial incentives would enable a careful weighing of the advantages and disadvantages of each proposal." The conclusion to this book outlines more specific financing recommendations.

The President's Task Force also provided a venue to assess different requirements among different agencies. For example, although the Department of Energy (DOE) sometimes replaced windows in the context of the weatherization program, there were disconnects between HUD and DOE. DOE issued a Notice in 2002, stating:

> Weatherization work may directly disturb lead-based paint, possibly creating hazardous conditions.... The Weatherization Assistance Program is not funded to do lead-based paint abatement work, nor to do lead-based paint hazard control or [paint] stabilization.... DOE funds may not be used for abatement, stabilization, or control of the lead-based paint hazard that is in the house.... Testing for lead-based paint is

not an allowable weatherization expense.... Clearance testing is not an allowable use of DOE funds.... States should develop a lead-based paint "deferral policy" ... to defer weatherization work in homes that either have tested positive or are assumed to have lead-based painted surfaces.[89]

Energy conservation and lead poisoning prevention remained in separate silos.

Although the DOE Notice called for "lead safe weatherization" (work practices that would be used when lead-based paint was disturbed during weatherization), it effectively prevented DOE funds from being used to replace lead-contaminated windows. For its part, HUD's lead programs sometimes installed windows that were not energy efficient, due to higher costs. Both agencies were carrying out their respective mandates to maximize the number of units that could be made energy-efficient or free from lead-based paint hazards. HUD made attempts to bridge the divide, writing to DOE:

> DOE's current weatherization policy and training materials are inconsistent with HUD's lead safe housing rule ... and does not implement DOE regulations.... Costs of eliminating health and safety hazards are allowable if elimination of the hazards is necessary before or because of weatherization materials...We request that DOE revise its Lead Safe Weatherization policy.[90]

The divide would not be reconciled for another decade when Congress ordered HUD and DOE to coordinate; most importantly funds were provided to make it a reality. By 2022 DOE permitted certain health and safety improvements,[91] although the opportunity was first articulated in 2006.[92]

For enforcement, the President's Task Force recommended concentrating on housing with a history of lead-poisoned children or had physical or management problems. Outreach and education recommendations were also included to ensure that all parents and property owners were involved. Increased blood lead screening was recommended because many children, especially high-risk children, were still not being tested. There were specific recommendations on compliance assistance, incentives, monitoring, and targeted actions. Perhaps stating the obvious, the report warned that "without this increased enforcement, the full benefits of lead paint regulations will not be realized."

Other recommendations included conducting a national survey of daycare centers, which was later completed in 2013,[93] as well as better coordination of weatherization and lead hazard control work, which started in earnest in 2015.[94]

The prospect of obtaining Medicaid waivers to fund lead hazard control was raised in this report. Rhode Island won such a Medicaid waiver for window replacement beginning in 1998[95] and again in 2018.[96] More recently, Michigan won $65 million in Medicaid/CHIP funding for lead paint abatement, lead pipe replacement, and other services, marking the first time that significant health care funding was used for lead remediation, starting in 2017.[97] Ohio, Maryland and other states also obtained such waivers a few years later.

The heart of the President's Task Force 10-year plan was funding. The projections showed there would still be hundreds of thousands of poisoned children annually by 2010 if the only options were ongoing demolition and rehab (Fig. 5.6). Enforcement of HUD's new rule in federally assisted housing would result in a large reduction in poisoned children as well. But the forecast showed that there would still be a large gap in 2010.

Increased HUD grants would be a big part of the solution if Congress and the President both were to request the additional funding recommended in the Task Force Report, but even that would still leave a large gap in 2010. The difference was to be made up by "leveraged" private sector and local government funding.

The Task Force highlighted the experiences in Milwaukee and Boston. At the time, these two cities had some of the strongest programs in the nation, under the capable leadership of Amy Murphy and Paul Hunter, respectively. They each had more local staffers than did HUD for the entire nation and an engaged citizenry. After receiving a $3 million lead hazard control grant from HUD, the City Council of Milwaukee passed a local lead poisoning ordinance to focus on all housing units in two high-risk neighborhoods. HUD funds and approximately $400,000 in leveraged private funds were used to partially defray landlords' costs of complying with the ordinance. Upon completion, the program made nearly 1000 homes safe for children. Boston leveraged $3.7 million in nonfederal funds with $7.7 million in HUD lead-hazard control grants.[79] These examples demonstrated that leveraged funding could play a significant role in closing the gap.

This idea of leveraged private funding came to partial fruition in the following Administration of President George W. Bush, a Republican who extended the Clinton Executive Order for the Children's Environmental Health Task Force. His first housing Secretary was Mel Martinez, a Republican from Florida. Upon being briefed about President Clinton's Task Force report and the need for more appropriations, at least doubling the 2001 appropriation from $100 million to the recommended minimum of at least $230 million, Martinez said, "Oh no program is going to get a *doubling* of their budget." But then he paused. Noting the HUD budget was more than $30 billion at the time, he then said that the increased funding request was really a "drop in the bucket."[98]

Martinez made lead paint one of his priorities, after his grandchild was almost poisoned. He called the HUD lead paint office director to his office on November 13, 2002, and asked him to speak with his adult daughter.[99] She said they had been looking for a new apartment because she was pregnant and had another young child and needed more room. In the process of looking for a different home, she had been provided with a lead paint disclosure pamphlet and with a lead paint disclosure form. But she said she had not received any lead paint disclosure when she had rented her existing apartment a year earlier. She wanted to know if her child and her pregnancy were at risk.

Secretary Martinez said he wanted this to be investigated using the normal process, which meant that Anne Evens, who ran the Chicago health department's lead program at the time, sprang into action, much as she did with other requests. They sent out a lead paint inspector to investigate, who found some minor lead hazards and issued a Mitigation Notice Letter.[100]

HUD also investigated the possible noncompliance with the federal disclosure rule.[101] The landlady said she was unaware of the disclosure rule and would comply in the future, stating she had many children of her own and wanted to do the right thing. All the hazards were abated through the action of the local health department and the landlady.

In later speeches, Secretary Martinez recounted this story, saying that "for a week or two we didn't know if my own grandchild had been poisoned. No family should have to go through that." Martinez also authorized the lead paint office to launch a massive training effort to promote compliance with the new reformed HUD lead paint regulations, with over 200 course offerings in 100 cities,[102] a marked contrast to the previous HUD Secretary's reluctance to implement the reforms.

No other previous HUD Secretary had formally made lead paint one of the Department's top priorities. Martinez proposed the Operation Lead Elimination Action Program (Operation LEAP), an effort to bring in other private dollars, as the President's Task Force had recommended a year earlier. HUD created a separate category within its lead grant program (with Congressional approval) to provide seed funding to those who could bring in leveraged additional funding. HUD testimony before the Congress stated, "Congress appropriated $6.5 million for (Operation LEAP in) FY 2002. We are hopeful the private sector will respond to this opportunity to help solve the problem."[103]

Although HUD provided grants to groups that attempted to bring in those additional private dollars, the effort failed to close the gap. Nevertheless, Martinez proposed (and Congress funded) a record level of HUD lead paint funding, reaching the highest amount since its beginning in 1991, $176 million in 2003. But even this amount was far less than what the President's Task Force had recommended as the bare minimum, $230 million per year for 10 years. It was not until 2018, a decade later, that $230 million would finally be appropriated by Congress.[104]

Also in 2018 President Trump's Administration released its own lead paint strategy. It began: "This is *not* a budget document" (emphasis added). The vision for ending the problem instead became "reducing" it. The Cabinet Secretaries were missing. Instead of "eliminating hazards," President Trump's report said: "identify lead exposed children and improve their health outcomes." Instead of a Campaign for a Lead-Safe America, there was an ambiguous goal to somehow "communicate more effectively."[105]

5.5.1 Lead paint industry interference

In 2001 the paint companies visited Secretary Martinez and offered their support, one possible avenue for the significant private funding the Task Force had called for a year earlier. Martinez asked them to work with his staff, including Phil Musser, who was then the Secretary's deputy chief of staff, and John Kennedy, a HUD lawyer who had earlier played a key role in the enforcement of the disclosure rule.

Tony Diaz, a lawyer representing Sherwin Williams from Jones, Day, Reavis, and Pogue (now called Jones Day), offered a 3-pronged effort:

> Provide significantly discounted paint to enable federal, state, and local funds to address more lead abatement needs more quickly; provide support for training programs; and participate in government-created public awareness.... We [the lead paint companies] also believe it is important for state and local governments to ensure that landlords and property owners ... responsibly maintain their property.[106]

The Benjamin Moore paint company offered a public awareness program it called, "Lead Free is Best for Me" in Newark NJ, Rochester NY, and New Haven CT. Its proposed plan said:

> Provided that Congress appropriates the necessary funds in FY 2002, Benjamin Moore & Company assumes HUD will issue an RFP [Request for Proposals].... Our understanding is that in the RFP, there

will be specific language that, in addition to the federal grant, there are *funds provided by Benjamin Moore* in the form of in-kind support of a public relations firm[107] (emphasis added).

But HUD thought the government should not be promoting specific paint companies:

> In meetings over the summer [with the paint companies], the industry candidly stated that the proposals were in response to lawsuits brought by cities, counties and one state [to force the companies to help pay for abatement] and a desire to be perceived as part of the solution, not part of the problem. Both entities [paint companies] indicated a desire for a public event with the Secretary.... HUD's General Counsel and the lead paint office urge the Department to avoid the appearance of taking sides in the lawsuits.[108]

Shortly after that, Phil Musser, a Deputy Chief of Staff for Secretary Martinez wrote in an email:

> I got some interesting feedback from some folks in the industry the other day that our lead office was too cozy with the Trail [sic] lawyers and their hunt to make this the next tobacco ... how would you respond to that?

The response was sent on the same day:

> There are those who feel we're too cozy with the industry, not the trial lawyers. It's impossible to please both sides on this issue, although we've been careful to work with both sides equitably. We [the HUD lead paint office] never made any statement regarding suits against the industry ... the invention of lead paint risk assessments for a large insurance company that covered public housing authorities who were sued by trial lawyers representing lead poisoned children was effective in preventing lead poisoning.... Even with increased funding, HUD dollars will never be enough alone to eliminate the problem.... It's probably not legal for the government to endorse a private company in a Notice of Funding Availability.... Is this [supposed coziness] somehow related to our report on CLEARCorps?[109]

The CLEARCorps story is described later in this section.

Sherwin Williams and Benjamin Moore paint company executives offered surplus paint, training, and some public relations services in exchange for an endorsement from HUD. When told of that offer, Secretary Martinez said:

> Is that all [they offered]? We won't take that deal; do you know how bad we would look?[98]

The industry's influence was also apparent elsewhere. For example, Kevin Kelly (a former aide to Senator Mikulski) suggested that Congress provide additional funding for high-risk cities. Many of these cities were becoming engaged in lawsuits to get the lead paint companies to contribute money to help fund abatement; some thought the new additional funding for high-risk cities may have been an attempt to shift the burden of future payment from the lead, paint and pigment companies to the taxpayers.

After leaving the Congressional staff, Kelly also worked as a lobbyist for the paint companies by promoting funding for CLEARCorps. The lead and paint industries had previously worked with the Community Lead Education and Reduction Corps (CLEARCorps), which began as the Coalition for Lead-Safe Communities, a group that formed in 1994 through the efforts of the Lead Industries Association, the National Multi Housing Council, and the

National Paint and Coatings Association. According to its newsletter, CLEARCorps also received financial support from paint and chemical companies, including BASF, Benjamin Moore & Co., DuPont, ICI Paints, PPG Industries, Sherwin-Williams, and Valspar.[110]

In 1995 the Shriver Center at the University of Maryland (Baltimore County) received a large grant from the Corporation for Public Service, a domestic "Peace Corps" known as AmeriCorps. Its basic idea was to pay young volunteers a living wage stipend to undertake some minimal residential lead hazard control activities in 10–15 cities.

The paint lobbyists succeeded in 1999 in getting a $2.5 million Congressional earmark, but the appropriators also asked for an evaluation of the program. Although the idea of using young idealistic volunteers to serve the community by reducing hazards seemed appealing, there was also concern that the youngsters might be putting both themselves and residents in the homes they attempted to improve at risk if the work was not done properly. Congress provided additional funding to CLEARCorps of $1 million in 2000 and 2001.[111]

The initial promise of CLEARCorps gave way to pessimism about its ability to succeed. The Lead Industry Association had already withdrawn its support in 1996 and the National Multi Housing Council attempted to reorganize its "fragments" into a new program.[112]

As a result of Congress' request for an evaluation, HUD completed two studies, one on management and the other on CLEARCorps effectiveness of the minimal lead hazard controls implemented by the volunteers. The evaluations showed that the program had completed 109 units in which "clean-only" had been done, and another 161 units involving other remediation, which seemed far too little for millions of dollars. On the management side, the report found that CLEARCorps had not completed policies and procedures manuals, did not have an adequate training program for its volunteers, did not do monitoring, oversight, or technical assistance to the various sites in the program, and did not track its costs.[111]

The evaluation of the environmental impacts of CLEARCorps also pointed to problems, as had the management report. Examining the program in 4 different cities, deteriorated lead paint had been left behind after work was completed in 20 out of 27 homes that were reviewed. Levels of lead dust were above clearance standards in 21 out of the 27 homes. Although lead dust was reduced by 89% from preintervention levels, 78% still had dust levels that failed the clearance standards at the time, suggesting that the cleaning had failed to remove enough lead dust to produce remediated homes that were safe for children. The report also documented large variability in each of the four cities examined.[113]

By 2003 the program ceased operations in many of its cities. CLEARCorps in 2020 may only be present in Detroit and Minnesota. In Detroit, the State of Michigan, which had been partially funding the group, issued a "stop work" order in 2019.[114] In Baltimore, CLEARCorps operations were transferred to the Baltimore Coalition to End Childhood Lead Poisoning to avoid duplication of effort, according to the local health department commissioner.[115] No additional Congressional earmarks for CLEARCorps occurred after 2001.

The lead paint industry's attempts to use well-meaning youth to do lead abatement failed. Later attempts by the lead paint industry to undertake other public relations efforts are examined in Chapter 7 following lawsuits that required them to help pay for remediation.

However, the lead paint companies were not the only ones failing.

5.5.2 Why the 2010 goal was not achieved

Was it wrong to suggest that the nation could eliminate childhood lead poisoning? In a speech before a thousand lead poisoning prevention and healthy homes professionals in 2004,[116] the history of other previous efforts to eliminate global epidemics was summarized. In 1909 the Rockefeller Foundation attempted to eliminate hookworm; the effort to eliminate yellow fever was abandoned in 1932. In 1955 the World Health Organization failed to eradicate yaws, a disease associated with over-crowding. During the 1950s and 1960s malaria eradication failed. But other diseases were either fully or almost eliminated, such as polio, smallpox, tetanus, hepatitis B, rubella, and others.[117]

Some of these successes were due to vaccines, and some re-emerged if vaccines and other public health measures were not consistently implemented, maintained, and enforced. The President's Task Force in 2000 believed that lead poisoning could join the ranks of the diseases that had been conquered, because its main source was finite (lead paint in older housing) and because the "vaccine" meant making those homes safe. The "pathogen" in the form of lead paint was still out there, but if hazards were eliminated, most of the disease could also end.

The 2004 speech honored some of those who had worked to eradicate the disease but passed away before the vision had been achieved: Ellis Goldman, who had overseen the HUD lead paint grant program from its beginning, adopted a focus on helping local governments overcome barriers instead of the more typical bureaucratic "gotcha" mindset that too often consumed oversight. Maurci Jackson, one of the first mothers of a lead-poisoned child to speak out nationally was described as one who forced decision-makers to pay attention to the tragedy of childhood lead poisoning (her mother Maura Jackson was a reviewer of this book). J Julian Chisolm at Johns Hopkins, who had conducted groundbreaking research on proper abatement methods was also honored. Ken Dillon (a mold expert who had worked to uncover mold-induced illnesses) was also remembered, part of the emerging healthy homes effort described later in this book.

The vision of the President's Task Force in 2000 to eliminate the problem by 2010 was clearly not achieved, although there had been progress. The geometric mean (average) blood lead level during this time period declined by 31%, from 1.9 µg/dL at the turn of the century in 2000 to 1.3 µg/dL during the 2007–10 time period.[118] There had also been progress on the housing side as well. HUD estimated that the number of homes with lead paint had declined from 64 million in 1990 to 38 million in 2006.[119,120]

Neither the Obama Administration nor the Trump Administration ever charged their cabinets to finish the job and neither renewed the vision of the eradication of lead poisoning. At this writing, the Biden Administration was presented with a draft Executive Order to re-establish the President's Cabinet Task Force and billions were proposed for both lead drinking water pipe and paint remediation.[121] Congress chose to fund lead drinking water pipe replacement yet ignored lead paint in its infrastructure law.

The 2000 President's Task Force quantified what it would take to solve the problem, what it would cost and how much the nation could save if the necessary resources were appropriated. A new cost-benefit analysis in 2017 showed that the nation would save at least $80 billion each year for each birth cohort if they were protected from lead poisoning.[122]

Put another way, deferring remediation meant that the costs of lead poisoning were still being paid, instead of reaping the benefits and savings of prevention. Although Congress has increased HUD lead paint funding to even higher levels, under mounting pressure from the lead problems in Flint and elsewhere,[123] the delay meant that the cost of remediation had grown to $55.8 billion (discussed further in the conclusion to this book).[124]

The President's 2000 Task Force developed a method to predict what would happen in the future. But was it true? Because it used housing data from HUD's 1990 housing survey and blood lead data from CDC's NHANES from 1991 to 1994, some questioned whether using those data to produce a forecast covering 1990–2010 was reliable. Later analysis proved that the estimates were incredibly on target, a tribute to Nevin's painstaking forecast.

A study found that empirical data released a couple of years after the 2000 President's Task Force report was released had in fact validated its forecast. For the year 2000, the Task Force had predicted 23.3 million pre-1960 housing units with lead paint hazards, close to the actual number from HUD's 1999 lead paint survey of 20.6 million units.[125] The small discrepancy was due mostly from newer homes built after 1960, which were under sampled in the 1999 HUD survey.

The 2000 President's Task Force also predicted 498,000 children with blood lead levels above $10 \, \mu g/dL$, compared to the CDC NHANES empirical estimate of 434,000 children for 1999–2004.[126] The Task Force predictions were well within 95% confidence intervals of empirical estimates for both residential lead paint hazards and blood lead outcomes.[127]

In short, the President's Task Force methodology worked. The goal was not achieved mainly because adequate funding was never appropriated and because the science was ignored. But it was not only about money. Political interference appeared again in 2004 and again in 2012, which also delayed progress.

5.6 Political sabotage

5.6.1 Attack and counterattack at HUD, 2004[1]

As important and effective as it was, not everyone in the conservative Bush Administration wanted HUD to fund parent and community groups. In 2004 Alphonso Jackson became the next HUD Secretary after his predecessor Mel Martinez departed to become a US Senator. Jackson had been Deputy Secretary under Martinez.

---

[1] A note from the editor/author. This section describes an attempt to eliminate HUD's lead paint program in the early 2000s and how those attempts were overcome. Because the author of this book was the director of that program at the time, and the attacks included an attempt to fire him, some readers may find the account biased. Yet not including it would render the history incomplete. Because the author's account is well documented, we chose to include it here. This section includes information on how funding was reduced for community groups, local governments and researchers in the latter part of the George W. Bush Administration. It also describes yet another attempt by the HUD Inspector General and a HUD Secretary (Alphonso Jackson) to eliminate the HUD lead paint program. The CDC director (Tom Friedan) and Congress also attempted to eliminate the CDC lead poisoning prevention program in 2012. This section describes efforts to defend both HUD and CDC lead poisoning prevention programs by citizens, parents, and scientists, Congress and local housing and health departments.

Four years later in the middle of the housing financial crisis of 2008, Jackson was forced to resign, because he had illegally awarded HUD contracts to his friends, denied others to his political opponents, and committed other violations of ethics regulations. He came to be regarded as one of the most corrupt HUD Secretaries, resulting in major federal investigations,[128,129] following in the footsteps of HUD Secretary Samuel Pierce during the Reagan years. After Members of Congress demanded the investigations, Jackson admitted, "I lied, and I regret having done that."[130]

Jackson wanted the lead paint program to return to the industry's idea that parents were the problem and mere education was the solution. In a 2006 national press release, he said:

> If we can educate young mothers before their child ends up in an emergency room, then we've done our job.[131]

Martinez had been a strong supporter of HUD's lead paint office. But Jackson under-took a years-long attempt to disable it, an effort that began when he was Deputy Secretary in 2002. Jackson inserted a political appointee, Emma Thomas, as the office's Deputy Director. She was a dentist with no background in lead poisoning prevention. She had her marching orders from Jackson, who granted her expanded control over the office's operations. Thomas stopped HUD funding for UPAL in Michigan, despite its well-documented progress.[132,133] She also stopped funding for other chapters of UPAL, many of which had already received funding from HUD or from HUD local grantees.

Upon becoming HUD Secretary in 2004, Jackson attempted to move the lead paint office out of the Office of the Secretary and into the Office of Public and Indian Housing, which would have effectively buried the office within the HUD bureaucracy, rendering it impotent, a return to the HUD of the 1970s and 1980s, when little was done on lead paint. Lawyers within HUD told Jackson that this was illegal because the original Congressional Act that created the office in 1991 required it to be located permanently in the Secretary's office[134] to ensure it had the power to create effective lead policies across the entire department. But Jackson pursued the relocation attempt anyway. He was forced to retreat after being scolded by Congress.

Reflecting the new attempts from Secretary Jackson to eliminate funding for parent and community groups, a HUD staffer conducted a review of UPAL's work. UPAL of Michigan's Director Margaret Sauser (the parent who had addressed the White House Press conference described earlier), wrote to HUD:

> I believe your treatment of us [UPAL] has been dishonest and unfair. Your "oversight" of our project has grown to the point of being abusive ... your findings in this report are ungrounded and unjustified ... cutting off our funding one year into a 3-year grant is not fair.[135]

The State of Michigan also weighed in, stating that UPAL had been appointed to serve on the Governor's lead poisoning prevention advisory committee, and that the State:

> ...is confident in the organization's ability to be effective [and use] funding appropriately, efficiently and productively. The [UPAL] organization displays devotion and progress towards its objectives.[136]

Thomas also stopped HUD funding for CEHRC, which had provided funding for nine community-based organizations working on lead-safe housing.

However, it was not only about community group funding. Thomas made antisemitic statements and berated staffers for their physical appearance. Overruling then-Deputy Secretary Jackson's objections, Secretary Martinez finally removed Thomas after her serious deficiencies were documented, including being late to work, stopping a prospective employee from being hired due to age, falsifying time and attendance records and travel vouchers, attempting to stop work with Catholic Charities by unsubstantiated allegations of corruption, and proposing grants to her former schools of dentistry in Georgia,[137,138] all ethical violations similar to those that would eventually force Jackson out of office.

A few months later in 2003, Jackson launched a "Quality Management Review (QMR)" of the lead paint office, even though the White House Office of Management and Budget had previously recognized the office as one of the most effective within HUD.[139] At his confirmation hearings in 2004 to become HUD Secretary, Senator Reed asked Jackson about this:

> Sen Reed: I have a strong interest in the issue of lead poisoning prevention. ... The employees [of the HUD lead paint office] have an excellent track record of performance. However, I have received reports about a recent, complex management crisis in this office. ... Specifically I am referring to a large number of employees who are leaving the office, 6 in the last month alone, a number of complaints to the Equal Employment Opportunity Commission and Inspector General's office and a complaint made directly to your office. Have you investigated this matter?
>
> Jackson: I agree that the employees and grantees of the office have an excellent track record. ... The office is regarded as the leading authority in the nation. ... I have expressed my support for the leadership the office has demonstrated over the years ... which makes the [lead paint and healthy homes grant program] one of the most effective in the Department.
>
> Sen Reed: The [White House] Office of Management and Budget has ranked the Lead Hazard Control grant program as one of the most effective HUD programs for the past two years. However, over the same period, HUD has consistently requested far less money than Congress has decided to appropriate. ... Why does HUD continue to undervalue the program?
>
> Jackson: Childhood lead poisoning prevention is one of my priorities and my budget request will reflect that.[140]

But in fact, Jackson proposed to cut $35 million from HUD's lead poisoning prevention program in FY 2005. This 20% reduction was opposed by key legislators, including both Republican Senator Christopher Bond and Democratic Senator Barbara Mikulski, who both said they would restore the funding. The Washington Post reported: "The [George W] Bush Administration's proposed budget would cut $35 million [from the lead paint program], a 20% reduction that is opposed by key legislators."[141]

In FY 05, the final enacted Congressional appropriation restored $28 million of the $35 million reduction Jackson wanted, from $174 million to $167 million. But this marked a long-term reduction starting in 2006 that would not end until increases started in 2017.

The QMR falsely alleged that the lead paint office had a high level of unexpended funds, which Jackson used to justify his reduced funding request. After all, why appropriate more funding if existing funding had not yet been used? But in fact, the lead paint program had the second-lowest unexpended balance of any other HUD program, shown in the bullets below.[142]

Yet the political attacks continued. When the QMR review failed to produce any significant findings, the HUD Inspector General launched yet another audit of the office.[143]

As we have seen in earlier chapters, the HUD IG had shown through the decades that it wanted the HUD lead paint office eliminated, and this proved to be the latest such effort.

The IG audit alleged that the office had made unsolicited grants without an evaluation, that grant amendments and extensions had no record of evaluation in the files, and that funds were not expended within an allowable timeframe, despite evidence that none of this had happened since the beginning of the office.[144] The audit specifically targeted grants to the Tides Foundation, which was a well-known and respected nonprofit organization that had been working with community groups for decades, as well as the Alliance to End Childhood Lead Poisoning, the National Center for Healthy Housing, and the University of Cincinnati. All of these were nonprofit organizations (except the University) and all had strong ties to community-based organizations. At the time, UPAL was the only national organization for parents of lead-poisoned children. With HUD's assistance, UPAL had grown to over 20 local chapters.

Before the IG audit began, files suddenly "disappeared," only to reappear later. For example, the office was accused of failing to keep a log of proposals, which a secretary later "found." The IG audit alleged the office administered two grant programs, when in fact there were seven Congressionally authorized grant programs for which the office was responsible. The audit alleged that the director (Jacobs) had "sole authority" over grant awards, when in fact final awards were always based on staff reviews. The IG was later forced to admit that they could not find a single instance in which the Director of the Office had overruled a staff decision on whether a grant application should be funded.[145]

The IG also alleged that Jacobs had written a UPAL grant application for them, citing a date when the electronic file containing the grant application had supposedly been created. Such documents are to be written by the applicant, not government staff. But that was impossible, because the HUD staffer reviewing the grant application had asked questions of the applicant about the proposal *before* the date the electronic file was supposedly created.[146]

Like the quality management review before it, the IG audit repeated the false allegation that the office had high unexpended balances of funds. Chapter 3 showed that indeed this was a problem a decade earlier in 1995 when local jurisdictions did not have the necessary capacity, but by 2000 it had been largely solved. The HUD Budget office provided evidence to the IG that in fact the lead paint office had the second-lowest unexpended balance ratio of all HUD programs. (A lower ratio means that there were less unspent funds.)

In order of increasing unexpended balance to annual appropriation ratios, the HUD budget office showed the following:

- Community Development Block Grants: 1.4;
- Lead Paint grants: 1.7
- HOME grants: 1.8;
- Homeless Assistance Grants: 2.6;
- Public Housing Capital Fund: 3.2;
- Brownfields Redevelopment: 3.5;
- Housing Opportunities for Special Populations with Aids: 5.9.[147]

This ranking was also confirmed by the White House Office of Management and Budget in 2004, which stated, "The Lead Hazard Control Grant program does not have an excessive amount of carryover [unexpended] funds."[148] Yet the IG and the QMR ignored the evidence.

After both the QMR and the IG audit failed, now-Secretary Jackson called Jacobs into his office in 2004 and demanded his resignation anyway. When asked why, Jackson threatened to launch yet another investigation. Jacobs refused to resign, believing that he had done nothing wrong, as the facts had demonstrated.

Jackson then suspended Jacobs' authority to run the office on Jan 8, 2004, placing a new Deputy director (Joe Smith, who had replaced Emma Thomas) in charge.[149] Jackson then proposed to fire Jacobs and force him out of the federal service. Because Jacobs was a career employee in the Senior Executive Service (not a political appointee), Secretary Jackson did not have the authority to immediately fire him. Instead, the Secretary had to state specific charges.

The first charge was "malfeasance" (mismanagement of grants), which alleged that Jacobs "approved 8 grant amendments between 1994 and 2002." But Jacobs did not begin working for HUD until 1995 and all the other amendments had been scored and recommended by office staff.

The second charge involved "Preferential Treatment and Failure to Act Impartially Concerning Grant Actions," alleging that Jacobs had "created and edited the underlying proposal document [for a grant to UPAL]," although the proposal had been reviewed by HUD staff *before* it was supposedly created by Jacobs.

The third and most serious charge involved the "Use of Public Office for the Private Gain," alleging that some grants were provided to "certain organizations with which [Jacobs] had nongovernmental affiliations," such as the Alliance, UPAL, the National Center for Healthy Housing and the University of Cincinnati.[150] Virtually the entire lead poisoning prevention movement was being attacked.

Jacobs' legal defense team wrote to the HUD Secretary:

> The assertion that grant funds were misspent because of Dr. Jacobs' supposed mismanagement of the grants program is not supported.... The IG auditors did not go out in the field or otherwise interview grantees or their beneficiaries to see whether the money was spent effectively.... [The charges] were not supported by a scintilla of evidence.[151]

The IG even refused to release documents pertaining to the investigation of Jacobs and the lead paint office.[152]

A wave of protest letters from local housing and health officials, scientists, advocates, and others expressed outrage over the attacks. Jackson finally backed down, stating, "I have decided to take no further action to remove you [Jacobs] from the Federal Service."[153]

One of the protest letters came from Dorr Dearborn, a renowned pediatrician who played a leading role in investigating infant fatalities in Cleveland associated with deficient ventilation and mold exposure (Chapter 8). He wrote to Alphonso Jackson:

> As a life-long Republican, I am outraged at your plans to remove David Jacobs as Director of the Office of Healthy Homes and Lead Hazard Control. After reviewing the Inspector General's Report, the apparent basis of your actions, it is readily apparent that the charges are minor, lack substance, may not be fully accurate, and represent an attempt to find excuses.... I have worked with Director Jacobs and his staff for over six years and can attest to his high moral character.... As a medical scientist working on housing as the major component of

the environmental health of our children, I would be very disappointed to lose a key government leader who has always used science as the basis of his decision making.... As a member of the Presidential Prayer Team, I feel I also need to bring this matter to the attention of President [George W.] Bush.[154]

Other protest letters came from housing officials, such as the General Manager of the Los Angeles Housing Department, who wrote about the "positive contributions to make Los Angeles safer for our children."[155]

A director of a community-based organization wrote:

I have known Dr. David Jacobs as an inspired leader, an honorable gentleman, and a scholar. He has been a great supporter of the concept that well-trained community members must be central to the development of scientific research [and that lead poisoning] cannot be addressed without the education and empowerment of community members. He has supported the concept that at-risk communities of color can and must be invested in.[156]

A homebuilder also wrote:

If there is one person I can point to in the last ten years who has done more to articulate and focus the effort to create more healthy housing, it is Dave Jacobs. It was Dave who invited several of us from the building community to share in a healthy housing protocol development session with medical professionals. From that day forward he has been instrumental in keeping housing and health professionals working together to produce healthier housing.[157]

Another city housing official from New York City stated:

[Jacobs] is one of the few people in the lead community to command respect from all sides of this contentious issue. His good judgment has enabled HUD to craft a program that balanced the protection of children with the interest of promoting affordable housing.[158]

Further descriptions of the letters are in this endnote.[159]

Debunking the HUD IG audit finding that HUD had awarded a grant without evaluating its proposal, Don Ryan wrote:

For the record, on 12/19/01 the Alliance received four pages of detailed comments and questions from the [HUD] Government Technical Representative assigned to evaluate the proposal submitted on 12/12/01. The Alliance met with the GTR and the HUD contract officer on 12/20/01 to answer questions.... The Alliance then amended and resubmitted our budget and proposal ... including audit information ... which was submitted on 2/13/02, followed by further submissions and agreements, resulting in the cooperative agreement that was signed on 3/8/02. Throughout this process, it was clear that ... HUD would not award this cooperative agreement unless and until the Alliance satisfactorily addressed HUD's concerns.[160]

After Emma Thomas had been forced out by Secretary Martinez, Jackson moved to minimize Jacobs' authority by appointing another Office Deputy Director, Joe Smith, who had previously worked in the office of Administration and had no experience in lead poisoning prevention. Jackson gave him responsibility for all "day-to-day" office operations.[161] Like Thomas, he carried out Secretary Jackson's efforts to sabotage the lead paint program, attacking many of the skilled scientists and staffers that had been assembled over the previous years.

Smith issued a sole-source contract (the very thing Jacobs had been falsely accused of) to firms with no experience to score lead paint grant applications, instead of relying on the analysis and recommendations of skilled government scientists, as had been done in previous years. He hired temporary workers based on key word searches on "monster.com" (a temporary job agency) none of whom had lead poisoning prevention experience. The results were predictable: applicants who *were* qualified were not awarded and those who *were not* received funding anyway.

This led to yet another wave of protests to Roy Bernardi, then the HUD Deputy Secretary under Jackson. In perhaps the most egregious example, one grant of $1.9 million was awarded to AIMCO, which a year earlier had been fined $129,580 for failure to comply with the lead paint disclosure law The company had pledged to abate its housing units *using its own funds* under a settlement agreement with the Department of Justice. It then attempted to use taxpayer dollars from HUD's lead paint program to fulfill its legal obligation, and Smith's flawed grant reviews had allowed it.

The lead poisoning prevention community was in an uproar. New charges were leveled by the Alliance to End Childhood Lead Poisoning to the HUD IG, not an "anonymous" source as in the previous audit. In its complaint, the Alliance charged that Joe Smith:

> improperly delegated authority to a contractor, whose reviewers may not have been properly trained or qualified.... As a result, the funding decisions that flowed from this flawed process do not reflect the merit of the applications submitted ... an abuse and mismanagement approaching dereliction of duty.[162]

The Washington Post reported:

> Two ineligible applicants received grants worth $5 million.... The IG found that one applicant was denied $365,736 in funds it was eligible to receive.... For the first time, HUD hired a contractor to read and rank grant applications.... But the review process fell apart almost from the start.[163]

The Alliance wrote:

> Recent management changes [the appointment of Smith] in the HUD lead paint office and its growing loss of technically qualified personnel threaten to stall national momentum in protecting children.... Excessive oversight and reporting systems are now stifling creativity, overburdening state and local project managers, and hurting dozens of programs across the country.[164]

Under the deluge of criticism, Smith became wildly abusive. He threatened to "shoot" staffers. In perhaps one of the most egregious episodes, Smith threatened a pregnant employee—the stress was so great that she thought she might miscarry her baby. A letter from the lead paint office staff to HUD Deputy Secretary Roy Bernardi complained:

> Since he was transferred to the Office in Oct 2002, Mr. Smith has created and perpetuated a hostile work environment that includes constant violent outbursts, intimidation, threats, harassment, bullying, and other disruptive behavior.... All of the directives and tasks Mr. Smith conceived and assigned were solely at his own direction and precluded input from office staff. He has completely alienated the entire staff to the extent that a number of stellar professionals no longer feel committed to our mission.... He stated, "I'll shoot you and you don't want to be shot." We understand that Mr. Smith's track record for creating hostile work environments is well documented.... He has set up the [lead paint] office for failure by setting goals and milestones based on neither past

performance or current resources.... It is critical that Mr. Smith be removed from his current job, duties and position within the Office as soon as possible...The tragic consequence is that, with Mr. Smith as Deputy Director of the Office of Healthy Homes and Lead Hazard Control, programs that have enjoyed success over the last 10 years through positive impacts on the health and residents, particularly children, will not continue.[165]

Secretary Jackson ignored the staff's letter, promoting Smith to Director after Jacobs was reassigned to another HUD office by Jackson in 2004. There were also other repercussions. For example, Smith decided to end the "Lead Listing," a website sponsored by HUD that enabled the public to find local lead inspectors, risk assessors, and abatement contractors. The website designer operating the listing wrote to HUD:

We are very concerned that a disruption of service on the Lead Listing (of unknown duration) is anticipated.... A disruption of service on such a long-standing website that is relied upon heavily by the regulated community will in our opinion impose hardships and generate significant complaints.[166]

Before being forced out of the lead paint office, Jacobs wrote to Jackson's aide Camille Pierce about Smith:

I need a deputy and administrative personnel who are committed to the mission and procedures of the office and the Department, not one whose stated goal is to drive out talented staff in one of the Department's most successful and highly regarded programs.[167]

As the former Mayor of Syracuse, Deputy Secretary Bernardi knew how lead poisoning prevention programs should work, because his city had one of the nation's best under the leadership of Betsy Mokrzycki. He knew things were wrong. He exonerated Jacobs after a hearing and consideration of the facts, recommending that Jacobs not be removed from the Federal Service. *Housing Affairs Newsletter* on Sept 10, 2004 on page 6 reported:

HUD Secretary Alphonso Jackson says Dave Jacobs...will not be fired after all. Instead, Jacobs has been assigned to a position in another [HUD Office]. Jackson tried to fire Jacobs in May after HUD's inspector general questioned five grants...But a point-by-point rebuttal of the IG's findings by Jacobs and an outpouring of support for him from HUD regional staff, state professionals, public health experts and community leaders forced Jackson to retreat. A mystifying HUD memo says the withdrawal of the firing is a vindication for Jacobs that the charges were unfounded...HUD officials gave no reason why Jacobs was removed as a supervisor following vindication...

Secretary Jackson reluctantly accepted the recommendation in light of the facts and the protests. In the wake of the Washington Post story about how grants had been scored incorrectly, Bernardi also removed Smith as the lead paint office director because of Smith's decision to use an unqualified contractor instead of government scientists. The newspaper quoted Jereon Brown, a HUD spokesman who said, "There is no doubt the credibility of the office is lacking." The Washington Post continued:

The [lead paint] office, meanwhile, appears to be in turmoil. The outgoing director, Joseph F. Smith, filed notices that could lead to the firing of two Healthy Home employees because of "inappropriate conduct," "lack of candor" and "undermining supervisory authority." Smith has been transferred elsewhere in HUD. Eddie Eitches, president of Local 476 of the American Federation of Government Employees, said the two [HUD lead paint office] employees have received outstanding job ratings and have no record of being disciplined. "It's my belief," Eitches said, "that the two employees are being scapegoated because they complained ... and those complaints have been validated by the IG report." He added: "These are completely bogus charges. Never in my history of federal service have I seen anything like this."[168]

The office's reputation had been compromised. On May 25, 2005, Bernardi appointed Michael Hill as Acting Director. Hill had worked in the lead paint office for more than a decade and knew the administrative duties.

Jackson's attempts at sabotage and the abusive Joe Smith ultimately were rebuffed. Jacobs was reassigned to another office within HUD and then required to move to Chicago in 2005. Although he remained at HUD for another year, in charge of supervising all environmental reviews for the Midwest, the commuting back to DC every weekend eventually took its personal toll and he left HUD in 2006. He subsequently became research director (and later Chief Scientist) at the National Center for Healthy Housing, associate Professor at the University of Illinois Chicago School of Public Health, faculty associate at the Johns Hopkins University Bloomberg School of Public Health and Director of the US Collaborating Center on Healthy Housing Research and Training for the World Health Organization and the Pan American Health Organization.

In the ensuing years, the HUD lead paint office was saddled with a series of directors, first Michael Hill and then Jon Gant, neither of whom had any lead poisoning prevention or related scientific experience, as had been the case before 1995. Hill and Gant served as capable caretakers after Smith was finally forced out, but no new initiatives emerged. Skilled personnel continued to leave and at first it appeared that Jackson and the IG had in the long run finally succeeded in their attempts to end HUD's lead paint program.

But Mathew Ammon became the lead paint office director in 2013 under President Obama's HUD Secretary, Julian Castro. With extensive training in environmental science, he enabled the office to regain its effectiveness and stature. Ammon had previously led the office's successful lead paint enforcement efforts and had successfully managed the HUD lead paint grant program. Ammon's background in environmental science was instrumental—yet another example of having qualified people in key positions of power at the right time. The office re-emerged as a center of excellence within HUD. Ammon briefly became Acting HUD Secretary in the early days of the Biden Administration until a new political appointee was confirmed as HUD Secretary. Ammon's appointment was a recognition of competence and scientific expertise.

5.6.2 Attempted elimination of the CDC lead program, 2012

The political interference during this time period also occurred at the Centers for Disease Control and Prevention, but in this case, it occurred under both Republican and Democratic Administrations.

A Congressional report in 2002 alleged that the George W. Bush Administration interfered with the appointment of scientists to the CDC Childhood Lead Poisoning Prevention Advisory Committee, and instead appointed biased members who had ties to the lead industry.[169] The report found that 3 renowned lead experts with decades of experience had been rejected. The Secretary of Health and Human Services, Tommy Thompson, overruled CDC staff recommendations for membership on this committee for the first time in its history.[170] The three experts rejected by Secretary Thompson over CDC staff objections were Michael Weitzman (from the University of Rochester), Bruce Lanphear (from the University of Cincinnati), and Susan Klitzman (from the City University of New York).

Instead, Thompson appointed William Banner, an expert witness who had served on the side of the industry in the Rhode Island lawsuit (Chapter 7). He stated that the blood lead level of concern should be "over 70 µg/dL and closer to 100 µg/dL,"[171] far above the level of concern at the time (10 µg/dL); it was as if he wanted to return to the 1970s. (A fatal dose is somewhere between 100 and 150 µg/dL.)

Secretary Thompson also appointed Joyce Tsuji, who had served as an expert for the American Smelting and Refining Company (ASARCO) and other companies that caused widespread lead pollution and were listed on superfund sites.[169] As a result of this political interference, the CDC lead poisoning advisory committee produced little of consequence in this time period.

Many involved in lead poisoning prevention were initially optimistic when President Obama was elected in 2008 because he had been a community organizer on the South side of Chicago, where lead poisoning was rampant. Some believed "he gets it" when it came to lead poisoning.[172] But this turned out to be mistaken. Funding for both HUD and CDC declined during his watch from 2008 to 2016.

In 2012 the Obama Administration's CDC director, Tom Friedan, decided to virtually eliminate the CDC lead program because he mistakenly believed lead poisoning was only a problem in a small number of "hot spots." His expertise was in communicable diseases, such as tuberculosis, and he had little interest in noncommunicable diseases like lead poisoning (limitations of the concept of communicable and noncommunicable diseases are discussed in Chapter 9).

Indeed, Friedan had little experience in environmental public health in general. He launched a new initiative, known as "Winnable Battles." He believed that CDC should not focus on a wide range of issues but should rather be limited to ten public health problems that could be solved quickly. Although the overall CDC budget increased, he repurposed 97% of the lead poisoning prevention funding to support this Winnable Battles initiative.[173]

Friedan initially proposed to combine the lead and asthma programs at CDC, cutting the funding for both in half. Many were perplexed about why CDC could not have both a lead and an asthma program, as had been the case for years.

Lobbyists for the CDC asthma program who had been hired with money from the huge tobacco settlement a few years earlier argued instead to Congress that the asthma program should be fully funded, and the money to do that should be taken from the lead poisoning prevention program. A young staffer on Sen Harkin's staff agreed to do that, later saying that she had never heard from the lead poisoning prevention people.

The tobacco-funded asthma lobbyists won. At the last minute, the 2012 budget for CDC's lead poisoning program was virtually eliminated. Although the President's proposed FY 2012 budget proposed merging CDC's $34 million Healthy Homes and Lead Poisoning Prevention Program with the $31 million Asthma Control Program, their combined budgets were cut in half, to only $30 million. The Senate then eliminated the CDC lead program entirely, and the House reduced it to only $2 million, barely enough to pay a few staff at CDC headquarters. The final 2012 appropriation enacted by the full Congress for the CDC lead poisoning program was slightly less than $2 million, far less than the $29 million the previous year and $35 million in previous years (Fig. 5.7).[174]

This last-minute maneuver produced enormous protests from around the country, which was covered by the press.[175] The protests came from state and local governments that depended on the CDC funding to carry out surveillance and case management activities, as well as advocates and scientists. A nationwide Declaration said:

> We are hundreds of delegates to the Lead and Environmental Hazards Association and the National Association of Lead and Healthy Homes Grantees representing thousands from all walks of life to assess the state of the nation's childhood lead poisoning prevention campaign.... Congress must restore the CDC program to its previous level of at least $29 million as a separate protected line item and restore its staff capacity ... we demand that the nation's budget not be balanced on the backs of our most vulnerable children. We demand that our children be protected and that our country's resources be properly allocated to restore the nation's capacity to prevent childhood lead poisoning.[176]

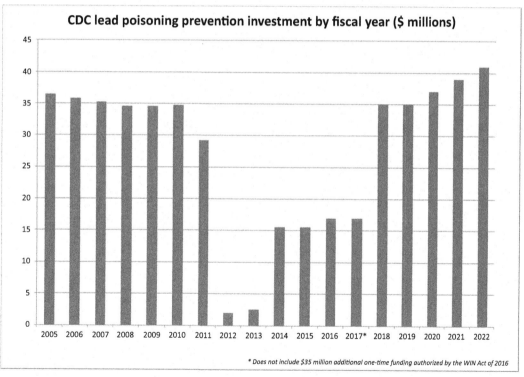

FIGURE 5.7 CDC lead poisoning prevention funding 2005–22. From: *National Center for Healthy Housing, Sarah Goodwin.*

The Children's Health Protection Advisory Committee for EPA wrote a blistering letter as well, outlining the retreats that occurred during the Obama Administration:

> The Children's Health Advisory Committee is concerned that both Congress and this Administration have recently abandoned the battle to protect children from lead poisoning.... The CDC program for 2012 has been largely eliminated and EPA and HUD programs have inadequate (and increasingly fewer) resources.... The [EPA] has taken only one or two enforcement actions to implement its Renovation, Repair and Painting Rule in the three years after it was promulgated. The Agency rejected a proposed rule to require dust lead testing following renovation to ensure cleanup is done properly and that children are protected as is already required in federally assisted housing and many local rules. EPA has not updated its dust lead standard, despite reports from its Science Advisory Board and well-documented evidence that the existing standards promulgated more than a decade ago do not protect children adequately.[177]

Some attempted to have local governments provide funds to make up for the CDC funding cut, but local resources were often not available.[178]

A few lead poisoning prevention organizations attempted to hire professional lobbyists, but it was too late. The National Safe and Healthy Housing Coalition also issued a position statement protesting the reductions.[179] It would take four more years, but eventually the CDC funding was restored in 2018, largely through the efforts of a revitalized and much larger national coalition of parents, community groups, citizens, and scientists (see Conclusion).

None of Friedan's 10 Winnable Battles were won. When the CDC funding to the state and local health departments lead poisoning prevention programs was withdrawn, all suffered severe losses in critical personnel and some local programs closed entirely. The consequences of political interference at CDC were later manifested in the failure of the Covid-19 pandemic response.

The Flint lead drinking water disaster is perhaps one of the best examples of the consequences of Frieden's political interference. Since the early 1990s states reported blood lead test results to the CDC,[180] whose epidemiologists reviewed them looking for areas with the greatest lead risk, emerging trends in risk for lead poisoning, and potential errors in data collection or transmission. For example, they were able to identify increased risk among newly arrived refugee children, areas where a small number of housing units repeatedly poisoned children, and invalid laboratory reporting of results.

The Michigan Childhood Lead Poisoning Prevention Program, which was funded by CDC had one of the best track records. However, when the funding was reduced in 2012, the Michigan lead poisoning prevention staff was cut to a single part-time nurse and one part-time data manager. As a result, in 2012 Michigan stopped reporting data to CDC and did not start reporting again until late 2016 after some funding was restored and personnel were again in place.

Without the CDC surveillance program, it would fall to Mona Hanna-Attisha, a Flint pediatrician, to notice an alarming increase in the number of children with blood lead levels above the CDC reference value starting around 2013 compared to previous years.[181] She and her colleagues found that as the lead in Flint drinking water increased, so did children's blood lead levels.[182] City residents and parents had already noticed problems in the water quality.

What ensued was a large-scale effort to test all Flint residents less than 18 years old, develop and deploy a plan to reduce children's exposure, institute academic, nutritional, and other supports to help mitigate the impact of the exposure, replacement of lead water pipes and a change in the water distribution system. All of this might have been prevented if the CDC program had been in operation.

CDC Director Frieden even encouraged President Obama to go to Flint in 2016 to drink the contaminated water to down-play the importance of the problem there. President Obama stated the children of Flint "would be just fine."[183] When asked why the CDC lead program had been virtually eliminated, one White House staffer in the office of the First Lady (Michelle Obama) hung his head in shame and said, "I don't know why we did that."

Friedan also formally disbanded CDC's childhood lead poisoning prevention advisory committee by not renewing its charter in 2013.[184] This advisory committee had been in existence since 1989. In 2012, it recommended that CDC lower the blood lead level that triggered action from 10 μg/dL to 5 μg/dL. It had previously recommended the level of concern in 1991 of 10 μg/dL. CDC grudgingly adopted the new reference blood lead value of 5 μg/dL in 2012, but many believed that the committee was disbanded for recommending the new reference blood lead level, which was contrary to Friedan's belief that lead poisoning had already been solved.[185] Friedan was also angered that the advisory committee vehemently opposed his funding cuts to the lead poisoning prevention funding. The well-respected director of the CDC's lead poisoning prevention program, Mary Jean Brown, who had many decades of lead poisoning prevention experience, was forced into retirement a few years later after being sidelined in 2012. Frieden's changes had made her position there untenable. The CDC lead program was relocated to ATSDR, which was focused primarily on hazardous waste sites, rather than responding to local and state lead poisoning prevention programs.

By 2020 CDC resurrected the CDC childhood lead poisoning prevention advisory committee, which held its first meeting in many years in early May 2020. The advisory committee chair was none other than Matt Ammon, who had become the director of HUD's office of lead poisoning prevention and healthy homes.

The political attacks at both HUD and CDC took their toll. Both were eventually rebuffed.

But an Appeals Court in Maryland also attacked, leveling false allegations that lead poisoning prevention programs were harming children, not helping them. That is the story of the next chapter.

References

1. Warren C. E-book *Brush with Death: A Social History of Lead Poisoning*. Baltimore: Johns Hopkins University Press; 2001.
2. HUD. Termination of Cincinnati Lead Grant; 1996. The University of Illinois Chicago Special Collections & University Archives, School of Public Health, "David E. Jacobs papers" the University of Illinois Chicago School of Public Health Library.
3. EPA Lead-Based Paint Abatement and Evaluation Program: Overview, 2022 <https://www.epa.gov/lead/lead-based-paint-abatement-and-evaluation-program-overview>.
4. Coates RJ, Jajosky RA, Stanbury M, Macdonald SC. Introduction to the summary of notifiable noninfectious conditions and disease outbreaks—United States. *Morb Mortal Wkly Rep.* 2015;62(54):1–4.
5. Find It Fund It Fix It Campaign; 2016. <https://nchh.org/build-the-movement/find-fix-fund/>.
6. CDC National Childhood Blood Lead Surveillance Data. <https://www.cdc.gov/nceh/lead/data/national.htm>, 2020
7. Alliance to End Childhood Lead Poisoning; 1995. *Directory of State and Local Childhood Lead Poisoning Prevention Advocates*. Washington, DC. <https://nchh.org/who-we-are/afhh/publications/>.

8. Building Blocks for Primary Prevention: Protecting Children from Lead-Based Paint Hazards. Centers for Disease Control and Prevention. Produced by the Alliance for Healthy Homes, Washington, DC; October 2005. The University of Illinois Chicago Special Collections & University Archives, School of Public Health, "David E. Jacobs papers" the University of Illinois Chicago School of Public Health Library.

9. HUD NOFA (Notice of Funding Availability); 2017. < https://www.hud.gov/sites/documents/2017 LBPHCNOFA.PDF > .

10. Childhood Lead Poisoning Prevention Projects, State and Local Childhood Lead Poisoning Prevention and Surveillance of Blood Lead Levels in Children CDC-RFA-EH18-1806, 2018.

11. Children at Risks From Lead Poisoning. United States Conference of Catholic Bishops, 2022. < http:// www.usccb.org/issues-and-action/human-life-and-dignity/environment/at-risk-from-lead-poisoning. cfm > .

12. Toxic Wastes and Race in The United States. A National Report on the Racial and Socio-Economic Characteristics of Communities with Hazardous Waste Sites. Commission For Racial Justice. United Church of Christ; 1987. <https://www.nrc.gov/docs/ML1310/ML13109A339.pdf>.

13. Morrison DS. Rallying point Charles Lee's long-standing career in environmental justice. *Am J Public Health.* 2009;99(suppl 3):S508–S510. Available from: https://doi.org/10.2105/AJPH.2009.178590.

14. Jacobs DE. Environmental health disparities in housing. *J Am Public Health Assoc.* 2011;101(suppl 1): S115–S122.

15. National Lead Information Center. 800-424-LEAD. Updated 2022. See: < https://www.epa.gov/lead/forms/ lead-hotline-national-lead-information-center > .

16. Collins calls lead paint major hazard for kids, by Chris Gosier. Bangor Daily News; November 16, 1999. P B1 regarding Congressional hearing on November 15, 1999.

17. United Parents Against Lead conference; May 29, 2004. The University of Illinois Chicago Special Collections & University Archives, School of Public Health, "David E. Jacobs papers" the University of Illinois Chicago School of Public Health Library.

18. Letter from Mack Heaton, HUD Office of Public Housing to Erskine Brown, Sr, Housing Authority of Birmingham; March 7, 2000. The University of Illinois Chicago Special Collections & University Archives, School of Public Health, "David E. Jacobs papers" the University of Illinois Chicago School of Public Health Library.

19. Alliance to End Childhood Lead Poisoning. The Story Behind a Birmingham Child's Death; November 22, 1999. The University of Illinois Chicago Special Collections & University Archives, School of Public Health, "David E. Jacobs papers" the University of Illinois Chicago School of Public Health Library.

20. Letter from Whitlynn T Battle to David Jacobs; January 14, 2000 (the University of Illinois Chicago Special Collections & University Archives, School of Public Health, "David E. Jacobs papers" the University of Illinois Chicago School of Public Health Library).

21. Sesame Street. Lead Awareness. Elmo's Tips for Going to the Doctor (undated) < https://www.sesamestreet. org/toolkits/leadaway > .

22. Rhode Island Childhood Lead Action Project. Liz Colon. 2022. https://www.facebook.com/childhoodleadac-tionproject/community/?ref = page_internal.

23. Elmwood Lightens Burden of Lead, by CJ Chivers, *Providence Journal Bulletin*; February 11, 1998.

24. Lead Poisoning Alarm Goes Out, by Polly Sheridan, Chicago Tribune; July 13, 1995:A3.

25. Getting the lead out, by Ed Finkel, Neighborhood Works; May/June 1998.

26. Goldman J. Silent Epidemic: Childhood Lead Poisoning in Cook County; March 1998.

27. History of Alameda Lead and Healthy Homes Program; 1992–2012 (undated). <https://www.achhd.org/ documents/History_of_ACHHD.pdf>.

28. Jacobs DE, Mielke H, Pavur N. The high cost of improper lead-based paint removal. *Environ Health Perspect.* 2003;111:185–186.

29. Mrs. Gore, Cuomo and Browner Announce Clinton Administration Campaign for A Lead-Safe America To Protect Nation's Children. HUD Press Release No. 97-274; November 19, 1997. < https://archives.hud.gov/ news/1997/pr97-274.cfm > .

30. Coping with Your Child's Diagnosis of Lead Poisoning. <https://www.hud.gov/sites/documents/DOC_ 12451.PDF>.

31. The Campaign for a Lead Safe America was later changed to a "Healthy Homes Campaign" under the slogan, "Get Tough, Stay Smart," but this campaign was short-lived.

32. New Anti-lead campaign has Pittsburgh connections, by Karen MacPerson Pittsburgh Post-Gazette; November 20, 1997:A12.

33. Maurci Jackson Obituary, Chicago Tribune, January 17; 2002. <https://www.chicagotribune.com/news/ct-xpm-2002-01-17-0201170100-story.html>.

34. Livingston Dennis. Maintaining a Lead Safe Home; 1997. <http://nchharchive.org/Portals/0/Contents/Maintaining_a_Lead_Safe%20Home_A_Do_it_Yourself_Manual.pdf>.

35. Campaign for a Lead Safe America grant application to HUD. The University of Illinois Chicago Special Collections & University Archives, School of Public Health, "David E. Jacobs papers" the University of Illinois Chicago School of Public Health Library.

36. Memorandum for Susan Gaffney. HUD Inspector General from Gail Laster, HUD General Counsel; April 16. In: *The University of Illinois Chicago Special Collections &*. University Archives, School of Public Health, "David E. Jacobs papers" the University of Illinois Chicago School of Public Health Library; 1998.

37. Parade Magazine; 1997. The University of Illinois Chicago Special Collections & University Archives, School of Public Health, "David E. Jacobs papers" the University of Illinois Chicago School of Public Health Library.

38. Letter to Karen Williams (HUD) from Joseph Kelly (Vanguard Communications); November 3, 1999. The University of Illinois Chicago Special Collections & University Archives, School of Public Health, "David E. Jacobs papers" the University of Illinois Chicago School of Public Health Library.

39. Missing Materials Poison Campaign Palmerton Group will promote lead awareness week without government brochures, David Slade, Allentown Morning Call; October 24, 1999.

40. Memorandum from Don Ryan. Alliance to End Childhood Lead Poisoning, to Cushing Dolbeare regarding Campaign for a Lead Safe America; October 13, 1999. The University of Illinois Chicago Special Collections & University Archives, School of Public Health, "David E. Jacobs papers" the University of Illinois Chicago School of Public Health Library.

41. Memorandum to V Steven Carberry, HUD Chief Procurement Officer and Victoria Bateman, Acting Chief Financial Officer from David Jacobs; September 26, 2000. The University of Illinois Chicago Special Collections & University Archives, School of Public Health, "David E. Jacobs papers" the University of Illinois Chicago School of Public Health Library.

42. Memorandum for David E Jacobs from Michael Hill, Acting Associate General Deputy Assistant Secretary for Administration; April 17, 2001. The University of Illinois Chicago Special Collections & University Archives, School of Public Health, "David E. Jacobs papers" the University of Illinois Chicago School of Public Health Library. The memo approves additional 7 staff to be hired.

43. Howell L. Becoming Unleaded. Book Forthcoming 2022.

44. Gaga L. @LeadieGaga <https://www.facebook.com/pg/LeadieGaga/posts/>.

45. Rubin T. Lead Safe Mama. <https://tamararubin.com/>.

46. Anonymous letter from a parent, the University of Illinois Chicago Special Collections & University Archives, School of Public Health, "David E. Jacobs papers" the University of Illinois Chicago School of Public Health Library; April 19, 1999.

47. House of Representatives Committee on Appropriations, Department of Veterans Affairs and Housing and Urban Development and Independent Agencies, Report 107–159. 107th Congress; July 25, 2001:42.

48. CEHRC Discussion Draft. Respecting and Protecting the Rights of Families and Tenants (undated). <https://nchh.org/resource-library/CEHRC_Respecting_Rights_of_Families_and_Tenants.pdf>.

49. Respecting and Protecting the Rights of Families and Tenants. Ethical Issues Arising Out of Testing for Housing-Related Health Hazards as an Organizing and Advocacy Tool. National Center for Healthy Housing. (undated). <https://nchh.org/resource-library/CEHRC_Respecting_Rights_of_Families_and_Tenants.pdf>.

50. Scott R. Advocates for Healthy Housing. Shelterforce Magazine; March 1, 2005. <https://shelterforce.org/2005/03/01/advocates-for-healthy-housing/>.

51. CDC. Guidelines for The Identification and Management of Lead Exposure in Pregnant and Lactating Women; 2010. <https://www.cdc.gov/nceh/lead/docs/publications/leadandpregnancy2010.pdf>.

52. Lord P. The Providence Journal, 2001. Recounted in Kuffner A. Lead Paint Revisited. The Providence Journal: Oct 20, 2019 <https://www.providencejournal.com/news/20191020/pbs-series-looks-back-at-lead-paint-crisis-in-rhode-island>.

53. Children Poisoned by Lead as Cleanups Fail. Detroit Free Press; January 21, 2003:1. <https://www.newspapers.com/search/#lnd=1&ymd=2003-01-21&t=3676>.

54. Haner J. The Baltimore Experience: Despite a Century of Struggle by Heroic Doctors, the Lead Paint Scourge Continues to Strike Youngsters in Slum Housing. Baltimore Sun; October 22, 2000:1F.
55. Dissell R, Zeltner B. Toxic Neglect: Curing Cleveland's Legacy of Lead Poisoning. The Plain Dealer; Posted October 20, 2015. <https://www.cleveland.com/healthfit/2015/10/toxic_neglect_curing_cleveland.html#incart_m-rpt-1>.
56. Dissell R. The Plain Dealer and Brie Zeltner, The Plain Dealer Cleveland City Council legislation aims for "systemic change" in approach to lead poisoning; Updated June 03, 2019. Posted June 03, 2019. <https://www.cleveland.com/metro/2019/06/cleveland-city-council-legislation-aims-for-systemic-change-in-approach-to-lead-poisoning.html>.
57. Parents Resource Catalogue, UPAL of Michigan; November 27, 2002. The University of Illinois Chicago Special Collections & University Archives, School of Public Health, "David E. Jacobs papers" the University of Illinois Chicago School of Public Health Library.
58. Lead, the First Addiction. UPAL of Michigan, Lead Safe Northern Michigan Conference; October 10, 2000.
59. International, National, and Regional National Safe and Healthy Housing Coalition Member Organizations. Available from: https://nchh.org/build-the-movement/nshhc/coalition-members/.
60. NCHH. Health in All Policies (Childhood Lead Poisoning Prevention) Mini-Grants; 2019. <https://nchh.org/build-the-movement/grants-and-scholarships/2019-health-in-all-policies-mini-grants/>.
61. Green and Healthy Homes Initiative; 2020. <https://www.greenandhealthyhomes.org/>.
62. Equipping Communities for Action through the National Lead Poisoning Prevention Network. National Center for Healthy Housing; 2019. <https://nchh.org/build-the-movement/grants-and-scholarships/2019-lead-poisoning-prevention-grants_equipping-communities/>.
63. NCHH. Lead Poisoning Prevention Stories Case Studies; 2017. <https://nchh.org/who-we-are/nchh-publications/case-studies/lpp-stories-case-studies/>.
64. David P. Rall; 2018. Available from: https://www.niehs.nih.gov/about/history/pastdirectors/davidrall/index.cfm.
65. Alliance Alert. A Publication by the Alliance to End Childhood Lead Poisoning, November/December 2000:4. The University of Illinois Chicago Special Collections & University Archives, School of Public Health, "David E. Jacobs papers" the University of Illinois Chicago School of Public Health Library. Also see: David P. Rall Award for Advocacy, American Public Health Association. <https://www.apha.org/about-apha/apha-awards/david-p-rall-award-for-advocacy>.
66. Seelye KQ, Bingham E. Champion of Worker Safety, Dies at 90. NY Times; June 23, 2020. <https://www.nytimes.com/2020/06/23/us/eula-bingham-champion-of-worker-safety-dies-at-90.html>.
67. Jacobs DE. The OSHA Cancer Policy: Generic vs Substance-Specific Regulation in an Area of Scientific Uncertainty. Georgia Institute of Technology Master's Thesis; 1988. <https://smartech.gatech.edu/handle/1853/29880>.
68. Jacobs DE. Lead. In: Bingham E, Cohrssen B, eds. *Patty's Toxicology*. 6th ed. New York: John Wiley and Sons; 2012:381−426. [ISBN: 978-0-470-41081-3].
69. Jacobs DE. Responses to the National Association of Home Builders document titled "A Private Sector Strategy for Reducing Lead Exposure in Children"; April 15, 1992. (The University of Illinois Chicago Special Collections & University Archives, School of Public Health, "David E. Jacobs papers" the University of Illinois Chicago School of Public Health Library).
70. 29 CFR 1926.62.
71. OSHA. Regulatory Review Of 29 CFR 1926.62, Lead in Construction Pursuant to Section 610 of the Regulatory Flexibility Act and Section 5 of Executive Order 12866. Occupational Safety and Health Administration Directorate of Evaluation and Analysis, Office of Evaluations and Audit Analysis; August 2007. <https://www.osha.gov/dea/lookback/lead-construction-review.html#_Toc165376472>.
72. Kosnett MJ, Wedeen RP, Rothenberg SJ, et al. Recommendations for medical management of adult lead exposure. *Environ Health Perspect*. 2007;115:463−471. Available from: https://doi.org/10.1289/ehp.9784.
73. National Institute for Occupational Safety and Health. Lead Exposure Limits; June 18, 2018. <https://www.cdc.gov/niosh/topics/lead/limits.html#:~:text=The%20NIOSH%20Recommended%20Exposure%20Limit,over%20an%208%2Dhour%20period>.
74. Executive Order 13045: Protection of Children from Environmental Health Risks and Safety Risks. William J. Clinton. The White House; April 21, 1997. <https://www.epa.gov/children/executive-order-13045-protection-children-environmental-health-risks-and-safety-risks>.
75. The President's Task Force on Environmental Health Risks and Safety Risks to Children. Activities and Accomplishments; February 14, 2003. The University of Illinois Chicago Special Collections & University Archives, School of Public Health, "David E. Jacobs papers" the University of Illinois Chicago School of Public Health Library.

76. Levy M, Wang V. The framingham heart study and the epidemiology of cardiovascular disease: a historical perspective. *Lancet*. 2013;383(9921):999–1008. Available from: https://doi.org/10.1016/S01406736(13)61752-3. PMC 4159698. PMID 24084292.

77. National Children's Study (NCS) Archive. <https://www.nichd.nih.gov/research/supported/NCS>.

78. Schmidt C. The Death of a Study. An ambitious study of childhood disease became what critics call a $1.3 billion scientific humiliation. What happened? Undark magazine. May. 2020;25. https://undark.org/2016/05/25/the-death-of-a-study-national-childrens-study/.

79. President's Task Force. Eliminating Childhood Lead Poisoning: A Federal Strategy Targeting Lead Paint Hazards; February 2000. <https://www.cdc.gov/nceh/lead/about/fedstrategy2000.pdf>.

80. Nevin R. Understanding international crime trends: the legacy of preschool lead exposure. *Environ Res*. 2007;104:315–336.

81. Nevin R. How lead exposure relates to temporal changes in IQ, violent crime, and unwed pregnancy. *Environ Res Section A*. 2000;83(1):22. Available from: https://doi.org/10.1006/enrs.1999.4045.

82. Carpenter DO, Nevin R. Environmental causes of violence. *Physiol Behav*. 2010;99:260–268.

83. Nevin R. Lucifer Curves: The Legacy of Lead Poisoning. BookBaby; 1st edition (6 July 2016) <https://www.amazon.com.au/Lucifer-Curves-Legacy-Lead-Poisoning-ebook/dp/B01I3LTR4W>.

84. Alliance to End Childhood Lead Poisoning. Innovative Financing Sources for Lead Hazard Control; 1995. The University of Illinois Chicago Special Collections & University Archives, School of Public Health, "David E. Jacobs papers" the University of Illinois Chicago School of Public Health Library.

85. In 2003 Senator Clinton proposed "The Home Lead Safety Tax Credit Act of 2003." S. 1228. https://www.govtrack.us/congress/bills/108/s1228.

86. In 2016, Senator Whitehouse proposed S.2573 – "Home Lead Safety Tax Credit Act of 2016" https://www.congress.gov/bill/114th-congress/senate-bill/2573/text Senators Obama and Clinton and Congressman Charlie Rangle also introduced lead paint tax credit bills, but none passed.

87. A Lead Abatement Tax Credit: The Most Effective Means of Financing Home Lead Abatement. Discusses H.R. 2922, July 17; 1991. Representatives Cardin and Waxman. The University of Illinois Chicago Special Collections & University Archives, School of Public Health, "David E. Jacobs papers" the University of Illinois Chicago School of Public Health Library.

88. Press Release from Congressman Ben Cardin, July 17, 1991. Press Releases from Alliance to End Childhood Lead Poisoning and Environmental Defense Fund. HR 2922, 102nd Congress, July 16, 1991. The University of Illinois Chicago Special Collections & University Archives, School of Public Health, "David E. Jacobs papers" the University of Illinois Chicago School of Public Health Library.

89. Department of Energy, Weatherization Program Notice 02–6; July 12, 2002.

90. Letter to Gail McKinley, Director of Department of Energy Office of Building Technology Assistance from David Jacobs, HUD, June 3, 2002. The University of Illinois Chicago Special Collections & University Archives, School of Public Health, "David E. Jacobs papers" the University of Illinois Chicago School of Public Health Library.

91. Department of Energy. Weatherization Program Notice 22-7. Issued Date: December 15, 2021. Weatherization Health and Safety. <https://www.energy.gov/sites/default/files/2021-12/wpn-22-7.pdf>.

92. Nevin R, Jacobs DE. Windows of opportunity: lead poisoning prevention, housing affordability and energy conservation. *Housing Policy Debate*. 2006;17(1):185–207.

93. Viet SM, Rogers J, Marker D, et al. Lead, allergen, and pesticide levels in licensed child care centers in the United States. *J Environ Health*. 2013;76(5):8–14. Nov-Dec.

94. LIHEAP Information Memorandum IM 2001-15 on Lead Paint Hazard Control and Weatherization; 2001. <https://www.acf.hhs.gov/ocs/resource/lead-paint-hazard-control-and-weatherization>.

95. Letter to Christine Ferguson, RI Director of Health from Nancy-Ann Min DeParle, Administrator Health Financing Administration, US Department of Health and Human Services; December 9, 1999. The University of Illinois Chicago Special Collections & University Archives, School of Public Health, "David E. Jacobs papers" the University of Illinois Chicago School of Public Health Library.

96. Expenditures for window replacement for homes which are the primary residence of eligible children who are lead poisoned. <https://www.medicaid.gov/Medicaid-CHIP-Program-Information/By-Topics/Waivers/1115/downloads/ri/ri-global-consumer-choice-compact-ca.pdf>.

Part 2. Breaking the Barriers to Progress (1986–2001)

97. Eggert D. Michigan wins OK to spend federal funds on lead abatement. Crain's Detroit Business, Associated Press. November 14, 2016 <https://www.crainsdetroit.com/article/20161114/NEWS01/161119887/michigan-gets-federal-ok-to-spend-119-million-on-lead-abatement>.

98. Personal Communication. HUD Secretary Martinez and David Jacobs; 2001.

99. Notes on Shae Case from David Jacobs; November 15, 2002. The University of Illinois Chicago Special Collections & University Archives, School of Public Health, "David E. Jacobs papers" the University of Illinois Chicago School of Public Health Library.

100. Mitigation Notice Letter to Mary Beth Weishaar from Anthony Amato, Chicago Department of Public Health; November 15, 2002. The University of Illinois Chicago Special Collections & University Archives, School of Public Health, "David E. Jacobs papers" the University of Illinois Chicago School of Public Health Library.

101. Letter to Mary Beth Weishaar from David Jacobs; November 19, 2002. The University of Illinois Chicago Special Collections & University Archives, School of Public Health, "David E. Jacobs papers" the University of Illinois Chicago School of Public Health Library.

102. A Message from the Secretary; February 12, 2001. The University of Illinois Chicago Special Collections & University Archives, School of Public Health, "David E. Jacobs papers" the University of Illinois Chicago School of Public Health Library.

103. Congressional Testimony of David Jacobs before the US Senate Committee on Banking Housing and Urban Affairs, Subcommittee on Housing and Transportation; June 5, 2002. The University of Illinois Chicago Special Collections & University Archives, School of Public Health, "David E. Jacobs papers" the University of Illinois Chicago School of Public Health Library.

104. National Center for Healthy Housing. HUD lead paint and healthy homes appropriations. 2022 <https://nchh.org/resource/appropriations_fy20_labor-hhs/>.

105. Federal Action Plan to Reduce Lead Exposures and Associated Health Impacts. President's Task Force; 2018. <https://www.epa.gov/sites/production/files/2018-12/documents/fedactionplan_lead_final.pdf>

106. Letter from Antonio Diaz to David Jacobs, Re: A Partnership for Lead Hazard Control; October 3, 2001. The University of Illinois Chicago Special Collections & University Archives, School of Public Health, "David E. Jacobs papers" the University of Illinois Chicago School of Public Health Library.

107. Letter from Carl Minchew, Benjamin Moore Paint Company Director of Government Relations to David Jacobs, Director HUD Office of Healthy Housing and Lead Hazard Control; July 18, 2001. The University of Illinois Chicago Special Collections & University Archives, School of Public Health, "David E. Jacobs papers" the University of Illinois Chicago School of Public Health Library.

108. Memorandum for Rob Woodson HUD Deputy to the Chief of Staff for Policy and Programs from David Jacobs, Director, HUD Office of Healthy Homes and Lead Hazard Control, with concurrence from HUD office of general counsel. The University of Illinois Chicago Special Collections & University Archives, School of Public Health, "David E. Jacobs papers" the University of Illinois Chicago School of Public Health Library; October: 2001.

109. Email to Phillip A Musser, HUD Deputy Chief of Staff from David Jacobs; October 21, 2001. The University of Illinois Chicago Special Collections & University Archives, School of Public Health, "David E. Jacobs papers" the University of Illinois Chicago School of Public Health Library.

110. CLEARCorps Network News, Fall 2001, UMBC and the Shriver Center. The University of Illinois Chicago Special Collections & University Archives, School of Public Health, "David E. Jacobs papers" the University of Illinois Chicago School of Public Health Library.

111. National Center for Lead-Safe Housing, Management evaluation of the CLEARCorps program; May 1, 2001. The University of Illinois Chicago Special Collections & University Archives, School of Public Health, "David E. Jacobs papers" the University of Illinois Chicago School of Public Health Library.

112. CLEARCorps shows promise despite initial obstacles. Lead Detection and Abatement Contractor newspaper; April/May 1998:1. The University of Illinois Chicago Special Collections & University Archives, School of Public Health, "David E. Jacobs papers" the University of Illinois Chicago School of Public Health Library.

113. Environmental Evaluation of CLEARCorps Lead Hazard Control Interventions. Quantech, Inc.; May 1, 2001. The University of Illinois Chicago Special Collections & University Archives, School of Public Health, "David E. Jacobs papers" the University of Illinois Chicago School of Public Health Library.

114. State orders nonprofit to stop Metro Detroit lead paint work after child gets sick. By Christine MacDonald, The Detroit News. Published 10:38 p.m. ET Jan. 24, 2019. Updated 11:49 a.m. ET; January 25, 2019. <https://www.detroitnews.com/story/news/local/detroit-city/2019/01/24/state-orders-nonprofit-stop-lead-paint-work-after-child-gets-sick/2660103002/>.

115. Letter from Peter Beilenson, Baltimore Commissioner of Health to Freeman Hrabowski III, President, University of Maryland, Baltimore County; January 3, 2001. The University of Illinois Chicago Special Collections & University Archives, School of Public Health, "David E. Jacobs papers" the University of Illinois Chicago School of Public Health Library.

116. Jacobs, Keynote address, Second Annual Indoor Environmental Health and Technologies Conference; June 23–25, 2004. Orlando, FL and HUD/CDC Lead and Healthy Homes Grantee conference, Lead and Environmental Hazards Association, Environmental Solutions Association, Alliance for Healthy homes, American Indoor Air Quality Association, National Coalition for Lead-Safe Kids, National Environmental Health Association, Children's Environmental Health Network. The University of Illinois Chicago Special Collections & University Archives, School of Public Health, "David E. Jacobs papers" the University of Illinois Chicago School of Public Health Library.

117. The Mop-Up, By Atul Gawande, the New Yorker magazine; January 5, 2004.

118. Wheeler W, Brown MJ. Blood lead levels in children aged 1–5 years—United States, 1999–2010. *Morb Mortal Wkly Rep.* 2013;62(13):245.

119. American Healthy Housing Survey. Lead and Arsenic Findings. US Department of Housing and Urban Development; 2012. <https://www.hud.gov/sites/documents/AHHS_REPORT.PDF>.

120. Comprehensive and Workable Plan for the Abatement of Lead-Based Paint in Privately Owned Housing; December 1990. US Department of Housing and Urban Development. <https://www.huduser.gov/portal/publications/affhsg/comp_work_plan_1990.html>.

121. Day One Project and National Center for Healthy Housing. A Draft Executive Order to Ensure Healthy Homes: Eliminating Lead and Other Housing Hazards; December 2020. <https://9381c384-0c59-41d7-bbdf-62bbf54449a6.filesusr.com/ugd/14d834_63a5a86dc5a84eae86234319d9640c08.pdf>.

122. Pew Charitable Trusts and Robert Wood Johnson Foundation. 10 Policies to Prevent and Respond to Childhood Lead Exposure; 2017. <https://nchh.org/resource/report_10-policies-to-prevent-and-respond-to-childhood-lead-exposure_english/>.

123. HUD Lead Paint Budget Authority. Fiscal year 2021. <https://www.hud.gov/sites/dfiles/CFO/documents/2_FY21CJ_BudgetAuthorityTable.pdf>.

124. Economic analysis supporting Letter to Congress from the National Center for Healthy Housing and hundreds of others; May 7, 2021.

125. Jacobs DE, Clickner RL, Zhou JL, et al. The prevalence of lead-based paint hazards in U.S. housing. *Environ Health Perspect.* 2002;110:A599–A606.

126. Brody D, Brown MJ, Jones RL, et al. Blood lead levels- United States, 1999–2002, U.S. centers for disease control and prevention. *Morb Mortal Wkly Rep.* 2005;54(20):513–516.

127. Jacobs DE, Nevin R. Validation of a twenty-year forecast of U.S. childhood lead poisoning: updated prospects for 2010. *Environ Res.* 2006;102(3):352–364.

128. HUD Secretary Alphonso Jackson Resigns. By Christine Perez, Dallas Business Journal, Mar 31, 2008, 6:38am CDT Updated March 31, 2008, 12:28pm CDT, <https://www.bizjournals.com/dallas/stories/2008/03/31/daily1.html>.

129. HUD Scandals by Tad DeHaven; June 1, 2009. <https://www.downsizinggovernment.org/hud/scandals>.

130. HUD Office of Inspector General, Criminal Investigations Division, Investigation Number HQ-06–00015-I; September 18, 2006.

131. HUD Secretary. Jackson Kicks Off 'Healthy Homes For Healthy Kids Campaign' 30-City Blitz To Protect Children From Home Hazards Press Release HUD No. 06–045; Wednesday, April 19, 2006.

132. Email from Emma Thomas to David Jacobs; August 23, 2002 (the University of Illinois Chicago Special Collections & University Archives, School of Public Health, "David E. Jacobs papers" the University of Illinois Chicago School of Public Health Library).

133. UPAL Progress Report; 2002 (the University of Illinois Chicago Special Collections & University Archives, School of Public Health, "David E. Jacobs papers" the University of Illinois Chicago School of Public Health Library).

134. Note to George Weidenfeller, Deputy General Counsel at HUD from John P Kennedy, Associate General Counsel (undated) (the University of Illinois Chicago Special Collections & University Archives, School of Public Health, "David E. Jacobs papers" the University of Illinois Chicago School of Public Health Library).

135. Letter to Rachel Riley (HUD) from Margaret Sauser (UPAL of Michigan); December 3, 2003. The University of Illinois Chicago Special Collections & University Archives, School of Public Health, "David E. Jacobs papers" the University of Illinois Chicago School of Public Health Library.

136. Email from Michigan Department of Community Health, Lead Hazard Remediation program; October 13. The University of Illinois Chicago Special Collections & University Archives, School of Public Health, "David E. Jacobs papers" the University of Illinois Chicago School of Public Health Library; 2003.

137. Thomas removal memo. Secretary Martinez to David Jacobs (the University of Illinois Chicago Special Collections & University Archives, School of Public Health, "David E. Jacobs papers" the University of Illinois Chicago School of Public Health Library).

138. Emma Thomas performance September 9, 2002 to October 8, 2002. Memorandum from David Jacobs to Secretary Martinez (the University of Illinois Chicago Special Collections & University Archives, School of Public Health, "David E. Jacobs papers" the University of Illinois Chicago School of Public Health Library).

139. Program Assessment Rating Tool. Assessing Program Performance. White House OMB (office of Management and Budget). < https://georgewbush-whitehouse.archives.gov/omb/performance/index.html >.

140. Questions from Senator Jack Reed to Alphonso Jackson, April 12, 2004, before the Senate Banking Committee. the University of Illinois Chicago Special Collections & University Archives, School of Public Health, "David E. Jacobs papers" the University of Illinois Chicago School of Public Health Library.

141. Bush Budget would Cut Lead Funds. By Avram Goldstein, Washinton Post, April 11, 2004, P. C-1.

142. Memorandum for the Deputy Secretary Alphonso Jackson from David Jacobs, Quality Management Review Progress report; December 11, 2002. The University of Illinois Chicago Special Collections & University Archives, School of Public Health, "David E. Jacobs papers" the University of Illinois Chicago School of Public Health Library.

143. HUD IG Audit Report 2004-AO-0001. Award and Administration of Lead-Based Paint Hazard Reduction Grants; February 6, 2004.

144. Memorandum for Ronald Morony Office of Lead-based paint abatement from Edward Girovasi, HUD director. of Policy and Evaluation; May 20, 1993. The University of Illinois Chicago Special Collections & University Archives, School of Public Health, "David E. Jacobs papers" the University of Illinois Chicago School of Public Health Library.

145. Inspector General working papers, the University of Illinois Chicago Special Collections & University Archives, School of Public Health, "David E. Jacobs papers" the University of Illinois Chicago School of Public Health Library; 2004.

146. Memorandum to Saundra Elion HUD IG Office from David Jacobs; January 21, 2004. The University of Illinois Chicago Special Collections & University Archives, School of Public Health, "David E. Jacobs papers" the University of Illinois Chicago School of Public Health Library.

147. Memorandum to Saundra Elion HUD IG Office from David Jacobs, Lead Paint office; December 22, 2003.

148. OMB PART report, the University of Illinois Chicago Special Collections & University Archives, School of Public Health, "David E. Jacobs papers" the University of Illinois Chicago School of Public Health Library.

149. Memorandum for David Jacobs from Alphonso Jackson, Grant Authority; January 8, 2004. The University of Illinois Chicago Special Collections & University Archives, School of Public Health, "David E. Jacobs papers" the University of Illinois Chicago School of Public Health Library.

150. Memorandum for David Jacobs from Alphonso Jackson, Proposal to Remove; April 13, 2004. The University of Illinois Chicago Special Collections & University Archives, School of Public Health, "David E. Jacobs papers" the University of Illinois Chicago School of Public Health Library.

151. Letter from Debra Roth to HUD Secretary Jackson; June 14, 2004. The University of Illinois Chicago Special Collections & University Archives, School of Public Health, "David E. Jacobs papers" the University of Illinois Chicago School of Public Health Library.

152. Letter from Darlene Hall, HUD Freedom of Information Officer to Debra Roth, Esq; August 2, 2005. The University of Illinois Chicago Special Collections & University Archives, School of Public Health, "David E. Jacobs papers" the University of Illinois Chicago School of Public Health Library.

153. Memorandum for David E Jacobs from Alphonso Jackson; August 18, 2004. The University of Illinois Chicago Special Collections & University Archives, School of Public Health, "David E. Jacobs papers" the University of Illinois Chicago School of Public Health Library.

154. Letter from Dorr G Dearborn, PhD, MD, case Western Reserve University to Alphonso Jackson, HUD Secretary; May 24, 2004. The University of Illinois Chicago Special Collections & University Archives, School of Public Health, "David E. Jacobs papers" the University of Illinois Chicago School of Public Health Library.

155. Letter from Mercedes Marquez, General Manager Los Angeles Housing Department to Alphonso Jackson, HUD Secretary; May 26, 2004. The University of Illinois Chicago Special Collections & University Archives, School of Public Health, "David E. Jacobs papers" the University of Illinois Chicago School of Public Health Library.

156. Letter from Nancy Ibrahim, Esperanza Community Housing Corporation; May 23, 2004. The University of Illinois Chicago Special Collections & University Archives, School of Public Health, "David E. Jacobs papers" the University of Illinois Chicago School of Public Health Library.

157. Letter from Jim LaRue, the Home Mender, Inc. to HD Secretary Jackson; May 21, 2004. The University of Illinois Chicago Special Collections & University Archives, School of Public Health, "David E. Jacobs papers" the University of Illinois Chicago School of Public Health Library.

158. Letter from Harold Shultz, New York City Housing Preservation Division to HUD Secretary Jackson; June 1, 2004. The University of Illinois Chicago Special Collections & University Archives, School of Public Health, "David E. Jacobs papers" the University of Illinois Chicago School of Public Health Library.

159. Another letter came from a Rhode Island Health Department official who had briefly worked in the Lead Paint Office. He wrote, "I find ludicrous the claims that lax administrative practices resulted in awards of unsolicited contracts that benefited Mr. Jacobs personally" (Letter from Robert Vanderslice, PhD to Janice Simons (lawyer for David Jacobs), May 21, 2004. The University of Illinois Chicago Special Collections & University Archives, School of Public Health, "David E. Jacobs papers" the University of Illinois Chicago School of Public Health Library). Another came from Nick Farr, who was Director of the HUD Model Cities Administration in the 1960s and later was Assistant Secretary for HUD's Community Planning and Development Office and Director of the National Center for Healthy Housing. He noted that the Inspector General had not bothered to interview him or Steve Weitz, HUD's staffer in charge of the Evaluation of the HUD Grant Program, which the IG had criticized. He wrote, "The National Center contributed $177,212 of its own funds to the HUD Evaluation." (Letter from Walter G Farr, National Center for Healthy Housing to HUD Secretary Alphonso Jackson, May 20, 2004. The University of Illinois Chicago Special Collections & University Archives, School of Public Health, "David E. Jacobs papers" the University of Illinois Chicago School of Public Health Library). A leader in the Environmental Justice community, Max Weintraub, who served on the board of Community Toolbox, which had received HUD funding through the Tides Foundation, wrote, "the 60-odd community groups we worked with ... have become more successful in promoting lead poisoning prevention ... and additional funding has been obtained from private and public foundations," which had been Secretary Martinez' intent in launching Operation LEAP (Letter from Max Weintraub to HUD Secretary Alphonso Jackson, June 1, 2004. The University of Illinois Chicago Special Collections & University Archives, School of Public Health, "David E. Jacobs papers" the University of Illinois Chicago School of Public Health Library). The health commissioner of Cleveland wrote that Jacobs was: "universally respected and liked for his warm personality, his strong scientific approach, his unwavering professionalism, and over brimming integrity. He simply gets things done." (Letter from Terrence M. Allen, RS, MPH on behalf of the Cuyahoga County Board of Health to Alphonso Jackson HUD Secretary, May 24, 2004. The University of Illinois Chicago Special Collections & University Archives, School of Public Health, "David E. Jacobs papers" the University of Illinois Chicago School of Public Health Library). The director of Milwaukee's lead program, Amy Murphy, reflected on the trust that the lead paint office had achieved: "I have come to understand that leadership is the difference between adequate public sector programs and those that instill public trust in government." (Letter from Amy Murphy, Milwaukee Health Department to HUD Secretary Jackson, June 1, 2004. The University of Illinois Chicago Special Collections & University Archives, School of Public Health, "David E. Jacobs papers" the University of Illinois Chicago School of Public Health Library). Recognizing HUD's earlier inability to deal with lead poisoning, a medical doctor who directed California's lead program and who subsequently led the Food and Drug Administration's pediatric program, Susan Cummins wrote: "When one considers HUD's lead hazard control efforts had been at a virtual standstill for the decade preceding his arrival at HUD, the progress made

during Dr. Jacobs' tenure ... is even more striking." (Letter from Susan Cummins, to HUD Secretary Jackson, May 23, 2004. The University of Illinois Chicago Special Collections & University Archives, School of Public Health, "David E. Jacobs papers" the University of Illinois Chicago School of Public Health Library). A parent from North Carolina, Kristen Joyner, wrote that Jacobs: "has been one of the braver persons, because he was willing to fund some of the more non-traditional groups, and to help mentor them to achieve their potential." (Letter from Kristen Joyner, United Parents Against Lead North Carolina, to HUD Secretary Jackson May 22, 2004. The University of Illinois Chicago Special Collections & University Archives, School of Public Health, "David E. Jacobs papers" the University of Illinois Chicago School of Public Health Library). The director of the Philadelphia lead poisoning program wrote, "Bringing together housing and health people has been like getting carpenters and masons to talk to each other at the tower of Babel.... There is entropy in government that makes it difficult for us to change what we do." (Letter from Richard Tobin, Philadelphia Dept of Health to HUD Secretary Jackson, May 27, 2004. The University of Illinois Chicago Special Collections & University Archives, School of Public Health, "David E. Jacobs papers" the University of Illinois Chicago School of Public Health Library). Philanthropic organizations, such as the National Safety Council also wrote protest letters, as did those in the private sector, including lead inspectors, abatement contractors, laboratory directors, the National Conference of State Legislatures, academics, the Lead and Environmental Hazards Association, non-profits, lawyers, parent's groups, physicians, and even a lawyer from another federal agency (The University of Illinois Chicago Special Collections & University Archives, School of Public Health, "David E. Jacobs papers" the University of Illinois Chicago School of Public Health Library).

160. Letter from Don Ryan, Executive Director to Kenneth Donohue, Sr, HUD IG, March 11; 2004. The University of Illinois Chicago Special Collections & University Archives, School of Public Health, "David E. Jacobs papers" the University of Illinois Chicago School of Public Health Library.

161. Memorandum from Alphonso Jackson appointing Joseph Smith to Acting Deputy Director of the Office of Healthy Homes and Lead Hazard Control, October 28, 2002, stating "Mr. Smith will be responsible for the day-to-day operations and activities for the office until further notice." The University of Illinois Chicago Special Collections & University Archives, School of Public Health, "David E. Jacobs papers" the University of Illinois Chicago School of Public Health Library.

162. Alliance Calls on HUD Inspector General to Investigate Grant Making Procedures. E&H Solutions Magazine Vol 5, No. 3, October/November 2004.

163. Barr S. From HUD, A Cautionary Tale on Contracting, Washington Post; May 26, 2005:B2.

164. Letter to Alfsonso Jackson HUD Secretary from Don Ryan (undated). The University of Illinois Chicago Special Collections & University Archives, School of Public Health, "David E. Jacobs papers" the University of Illinois Chicago School of Public Health Library.

165. Letter to Roy Bernardi, HUD Deputy Secretary from Staff of HUD Office of Healthy Homes and Lead Hazard Control; April 24, 2004. The University of Illinois Chicago Special Collections & University Archives, School of Public Health, "David E. Jacobs papers" the University of Illinois Chicago School of Public Health Library.

166. Letter from David Cox, President of Quantech to Cherita Hammond, HUD Grant Officer; March 18, 2004. The University of Illinois Chicago Special Collections & University Archives, School of Public Health, "David E. Jacobs papers" the University of Illinois Chicago School of Public Health Library.

167. Jacobs. Note to Camille Pierce. The University of Illinois Chicago Special Collections & University Archives, School of Public Health, "David E. Jacobs papers" the University of Illinois Chicago School of Public Health Library.

168. Stephan Barr, From HUD, a Cautionary Tale on Contracting, Washington Post; May 26, 2005:B2.

169. Turning Lead into Gold: How the Bush Administration is Poisoning the CDC Lead Advisory Committee, A Report by the Staff of Rep. Edward Markey, Congress; October 8, 2002. <http://www.house.gov/markey/iss_environment_rpt021008.pdf>.

170. Union of Concerned Scientists. Lead Poisoning Prevention Panel Influenced by Industry. Published February 18, 2005. <https://www.ucsusa.org/resources/lead-poisoning-prevention-panel-influenced-industry>.

171. Deposition of William Banner Rhode Island v. Lead Industries Association (Sup Ct RI, No. 99-5226).

172. Anne Evens. Former Director of Chicago's Lead Poisoning Program, Personal Communication with David Jacobs, 2008.

173. Centers for Disease Control and Prevention. Winnable Battles Final Report; November 2016. <https://www.cdc.gov/winnablebattles/report/docs/winnable-battles-final-report.pdf>.

174. Minutes of the Annual Meeting of the Advisory Committee on Childhood Lead Poisoning Prevention, Atlanta, Georgia; November 14–16, 2012. <https://www.cdc.gov/nceh/lead/acclpp/meetings/minutes/2012novminutes.pdf>.

175. Thousands of Kids with Lead Poisoning Won't Get Help. Kaitlyn Ridel, USA TODAY; October 2, 2012. <https://www.usatoday.com/story/news/health/2012/10/02/lead-poisoning-children/1605533/>.

176. 2012 Declaration of National Lead Poisoning Prevention Delegates; May 4, 2012. The University of Illinois Chicago Special Collections & University Archives, School of Public Health, "David E. Jacobs papers" the University of Illinois Chicago School of Public Health Library.

177. Letter from EPA Children's Health Protection Advisory Committee to Lisa Jackson, EPA Administrator; March 29, 2012. <https://www.epa.gov/sites/production/files/2015-10/documents/chpac_lead_letter_2012_03_29.pdf>.

178. Testimony from David Jacobs, Chair, Lead and Healthy Homes Advisory Committee to the District of Columbia Department of Energy and Environment. DDOE Proposed FY 2013 Budget for the DC Childhood Lead Poisoning Prevention Program. Councilmember Mary M. Cheh Committee on the Environment, Public Works, and Transportation Council of the District of Columbia, April 25, 2012 (the University of Illinois Chicago Special Collections & University Archives, School of Public Health, "David E. Jacobs papers" the University of Illinois Chicago School of Public Health Library).

179. Position Statement, Funding for CDC's Healthy Homes and Lead Poisoning Prevention Program, National Safe and Healthy Housing Coalition; 2012. <http://nchharchive.org/Portals/0/Contents/Position%20Statement%20-%20CDC%20Healthy%20Homes%20and%20Lead%20Poisoning%20Prevention%20Final.pdf>.

180. Meyer PA, Pivetz T, Dignam TA, et al. Surveillance for elevated blood lead levels among children–United States, 1997–2001. *MMWR Surveill Summ.* 2003;52(10):1–21. PMID: 14532866.

181. Hanna-Attisha M, LaChance J, Sadler RC, Champney Schnepp A. Elevated blood lead levels in children associated with the flint drinking water crisis: a spatial analysis of risk and public health response. *Am J Public Health.* 2016;106(2):283–290. Available from: https://doi.org/10.2105/AJPH.2015.303003. Epub 2015 Dec 21. PMID: 26691115; PMCID: PMC4985856.

182. Kennedy C, Yard E, Dignam T, et al. Blood lead levels among children aged <6 years—Flint Michigan 2013–2016. MMWR, Early Release 65; June 24, 2016.

183. Why President Obama just drank the water in Flint, By Libby Nelson; May 4, 2016. <https://www.vox.com/2016/5/4/11591894/obama-flint-water>.

184. Advisory Committee on Childhood Lead Poisoning Prevention (ACCLPP). <https://www.cdc.gov/nceh/lead/advisory/acclpp.htm>. states: "The most recent charter...expired on October 31, 2013."

185. Perry Gottesfeld Personal Communication with. David Jacobs; 2013.

CHAPTER

6

Research ethics and the *Grimes* court case

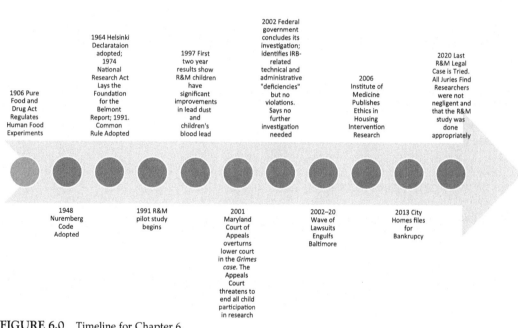

FIGURE 6.0   Timeline for Chapter 6.

6.1 The context: lead poisoning and the courts

Earlier chapters have shown that research and citizen action were both essential between 1970 and 2020 in establishing safe and proven methods of identifying and eliminating lead paint hazards to protect children and creating the political will to put those research findings into policy and action.

Most court cases around the country during this era were tort suits directed at landlords, with a handful awarding large damages to a few children. But most children received nothing (Chapter 3). And most property owners were fearful of their liability and remained perplexed and paralyzed on how to bring lead poisoning prevention into their ongoing routine mainte-nance and capital improvement operations. In Baltimore like elsewhere, most children and homes remained untested and uninspected. Lead poisoning remained at epidemic propor-tions in many neighborhoods, despite heroic efforts from an overwhelmed health department.

In response, applied research in Baltimore and elsewhere increasingly turned to finding effective and practical ways to the vexing problem of integrating lead hazard control into property management practices, instead of a highly specialized activity funded by small government grants aimed at already-poisoned children. Property management includes both maintenance and capital improvements.

One of the first research efforts along these lines was the Lead Paint Abatement and Repair and Maintenance study in Baltimore, known as the "R&M" study, which began as a pilot in 1991 and was carried out by Kennedy Krieger Institute (KKI), with collaborators at Johns Hopkins University (KKI is described in more detail later in this chapter). The study compared new ways to address lead paint and dust hazards and to stop the harm being done by traditional paint removal. The study aimed to contribute to the scientific basis for proactive and widespread housing-based policies and interventions to prevent lead poisoning. By 1997 preliminary results showed that both house dust and blood lead levels had improved considerably and were sustained over time.

Although most court cases were aimed at property owners (and a few more against the paint companies), one charged the R&M research team, alleging that two children had been harmed by their participation in the study. The Baltimore City Circuit Court in July 2000 summarily dismissed the charges, because it regarded the researchers as "Good Samaritans," who were protecting the children.[1] In dismissing the charges, the lower court wrote:

> KKI was sort of an institutional volunteer in the community. Coming in to collect dust and blood sam-ples, the next thing you know they get sued…

This "Good Samaritan" precedent[2] held that liability is limited for those who voluntar-ily perform care and rescue in emergency situations. This precedent was established because emergencies outside of the umbrella "medical setting" or "clinical environment" are common. The lower court thought that KKI researchers implementing lead paint reme-diation in homes was precisely such an emergency. The basic theory of Good Samaritan laws is that society is improved if the volunteer rescuers (i.e., the good Samaritans) help a person (including children in need) as opposed to worrying about the possible liability associated with assisting others. Parents voluntarily agreed to participate in the study and had signed an informed consent form (discussed later in this chapter).

But in August 2001, the Maryland Appeals Court[3] overturned the lower court's Good Samaritan-based ruling and shocked the nation by comparing Baltimore researchers to Nazis and the notorious Tuskegee study decades earlier. It said that researchers had per-formed their duties unethically, failing to uphold the widely recognized ethical principles of justice, respect (including full informed consent), and beneficence.[a] It questioned whether parents even had the right to allow their children to participate in such studies. The Appeals Court thought that researchers had a "special duty" to warn parents of risks

[a] Beneficence means doing good for others.

to children, even if those children were not actually in the studies. The Appeals Court thought it was creating new legal findings because it mistakenly thought there were no regulations governing research ethics.

It reached these and other conclusions without "facts in evidence," that is, as a matter of law. It ruled that the case be remanded back to the lower court to examine the facts.

This chapter examines those facts—both the scientific and the legal ones.

The R&M study, initially funded by the US Environmental Protection Agency (EPA) and later by the US Department of Housing and Urban Development (HUD), was carried out by the KKI and the Johns Hopkins University. It was led by Mark Farfel, who had completed many earlier important studies, and his mentor J. Julian Chisolm, who had been a leader in lead poisoning prevention for decades. The lead poisoning prevention community was dumbfounded by the Appeals Court ruling. They knew the study results, which had demonstrated reductions in both blood lead and dust lead. These early results had been presented at conferences over the years, showing the children had been protected, not harmed.

In one of his final actions before leaving HUD, Jacobs commissioned the Institute of Medicine (now part of the National Academies of Sciences, Engineering, and Medicine) in 2004 to examine this Baltimore study (and others like it) to help determine if indeed there had been something unethical or harmful in the way it was done. This was the first thorough examination of what came to be known as "housing intervention research,"[4] paving the way for larger examinations of housing conditions, remediation, and improved health.

If the Maryland Appeals Court ruling was simply the product of poor legal judgment and lack of scientific understanding, perhaps it would be a mere footnote in history. All of the subsequent Baltimore court cases in which juries heard the actual facts about those participating in the study found the researchers to not be negligent with regard to the study participants and how the study was done.[b] No other court anywhere made a similar ruling in the past 50 years.

Yet this particular case was important in this era's history of lead poisoning prevention. It was yet another example of why the facts are important, why failure to consider them leads to disaster, and why science must inform policy as well as legal precedent.

One other book, several other articles, and law school curricula tell this story incorrectly, suggesting that although the researchers had the "best of intentions," the study was flawed and unethical, harmed children, and somehow should have been done differently. But this book presents the ensuing debate through an environmental justice lens, setting the record straight.

The next chapter discusses how two other major court cases that also both began in the early 2000s (one in Rhode Island and the other in California) assessed accountability differently. These later cases determined that it was not the researchers who had placed children in harm's way; instead, the lead, paint and pigment industries were accountable.

But first, we turn to the Baltimore R&M legal case.

6.1.1 The Kennedy Krieger Institute and health research

KKI, a nonprofit community health entity, was loosely associated with Johns Hopkins University for decades and both played an immense role in the nation's lead poisoning

---

[b] One bizarre exception to this involved a child who was not even enrolled in the study.

prevention efforts. Hopkins was the first to diagnose lead paint as the cause of childhood seizures and fatalities in the United States in the 1920s.[5] They played a central role in the invention of the blood lead test, clinical treatments for lead poisoning, and recognition of the importance of controlling lead in paint and dust through the work of J. Julian Chisolm[6] and others.

In the early 1980s, KKI and Hopkins were the first to show that lead paint removal by open-flame burning and power sanding (the "traditional" methods of lead abatement at the time) caused more harm than good. Chisolm and his protégé, Mark Farfel, led the efforts to have those methods banned and new safer ones established.[7] Farfel developed the scientific foundation for the first health-based lead dust hazard and clearance standards in the late 1980s that were subsequently adopted by HUD in the 1990 Interim Public Housing Guidelines and in HUD and EPA regulations in 1999[8] and 2001.[9,10] His work was among the first to reveal the importance of the lead paint and lead dust exposure pathway in the 1980s that ultimately led to Congress' adoption of the scientific definition of what a lead-based paint hazard meant in Title X in 1992 (Chapter 3).

Farfel's graduate thesis in 1987 documented these findings. Among other things, his research showed that open-flame burning and power sanding (Figures 2.3 and 2.4) increased dust lead levels by 10–100 times and that one-third of children living in such homes had to be admitted to the hospital for chelation therapy because their blood lead levels had increased to dangerous levels.[11]

Johns Hopkins University awarded the John C. Hume award to Farfel because of the significance of his findings.[12] John Hume was descended from six generations of Congregational church missionaries and was known for asking his children and his students: "What are you going to do for humanity?"[13] Farfel also received a citizen citation award for his lead poisoning prevention work from the Baltimore Mayor at the time, Clarence H. "Du" Burns, who was that city's first Black Mayor.[14]

Hopkins was (and still is) one of 18 national educational centers that train environmental and occupational health scientists. They are funded by the National Institute for Occupational Safety and Health,[15] another reminder of the link between community and occupational health.

Farfel involved the community in the research efforts, founding and supervising "InterAction," the community outreach program of the Hopkins School of Public Health, which maintained relationships with dozens of Baltimore community organizations to engage students and faculty in meeting community needs.[16] Farfel said, "Nothing will happen if the community, parents, ministers, elected leaders, and property owners do not get on board."[17] He hired community members in his studies because he said they often knew more than the academics about community needs. In short, community involvement was a key ingredient in the way KKI and Hopkins conducted their research.

One account of Farfel said he wanted to be part of the:

> Twenty year history of working to merge policy and science for real-world change ... [Farfel said:] "Abatement shouldn't be a haphazard process.... Kids shouldn't get sick as a result of it." Farfel also knew how important the outreach workers were. He said, "They know the city—their rapport with the families is the key."[18]

The KKI outreach workers included Pat Tracy (who lived in the neighborhood where the study took place) and Don Cooper, who both played important roles in lead poisoning prevention for years. Other key figures included Peter Lees, Ruth Quinn, Susan and Michael Kleinhammer, Desmond Bannon, Susan Guyaux, and many others. The study also worked with the local Coalition to End Childhood Lead Poisoning, the Baltimore City Health Department, the Maryland Housing and Community Development office, the Maryland Department of the Environment, the Enterprise Foundation, City Homes (described later), and many others.

KKI was founded as the Children's Rehabilitation Institute in 1937. It was renamed the Kennedy Institute in 1968 in honor of President John Kennedy, who had passed the Medical Training Act during his administration to protect the rights and improve the lives of persons with disabilities. It later was renamed the Kennedy Krieger Institute to honor a long-time board member and supporter, Zanvyl Krieger.[19]

KKI believed that interdisciplinary care was required to enable children to benefit from various rehabilitation interventions. In the late 1980s KKI expanded its mission to include all brain disorders, including lead poisoning, not limited to developmental disabilities. Recognizing it as a key indicator of lead poisoning, KKI and Hopkins became more focused on housing, which had begun through Chisolm's work in the 1940s and 1950s, because lead poisoning from paint was recognized as a brain disorder.

One of the earliest studies in 1983 to focus on lead dust interventions was conducted by this team, which also included Evan Charney at Baltimore Sinai Hospital, Chisolm, and Farfel (then a graduate student). It showed that in homes that had professional wet mopping every 2 weeks over a 1-year period plus regular cleaning by the occupants, blood lead reductions averaged 6.9 μg/dL, compared to a control group that had only routine household cleaning. The authors concluded that "a focused dust control program can reduce blood lead levels more than standard lead [paint] removal in the home."[20] This was the beginning of many studies by KKI and Hopkins to examine which methods were effective and which were not. KKI had one of the premier childhood lead poisoning clinics and blood lead testing labs in the United States.

6.1.2 City Homes and affordable housing

City Homes was one of the Baltimore housing providers that collaborated in the KKI studies. Founded in 1986, it was a nonprofit organization providing housing to many low-income families. It enabled occupants in its homes to volunteer to participate in lead paint research projects if they chose to do so. City Homes voluntarily provided most of the homes in the R&M study, with the strong support of its residents.

At its peak, City Homes provided nearly 500 rental housing units to low-income Baltimore families and had a vacancy rate of under 1.5%, much lower than prevailing market conditions, likely because it was perceived to offer quality housing, unlike some other private sector landlords. One tenant remarked that City Homes was "the best landlord I ever had."[21]

In 1991 City Homes had an existing lead poisoning prevention program that included education about lead poisoning. It also made its housing units available to families with lead-poisoned children after the units had been improved, making it one of the few housing providers in the city to reach out proactively and help to address the problem of lead poisoning.[22]

Together with various rental subsidy programs, City Homes also received funding from the Enterprise Foundation (now Enterprise Community Partners). An Enterprise Foundation vice president, Nick Farr, served on the City Homes board in the late 1980s. City Homes was sued by a young child who had been lead-poisoned. Farr, a former Yale law Professor who later became the first executive director of the National Center for Lead Safe Housing, realized that there were no adequate standards and little technical assistance for nonprofit entities like City Homes. He became motivated to help find ways to assist such nonprofit low-income housing providers, remarking that "No one at City Homes ever wanted to see a child poisoned by lead."

But as a result of the Maryland Appeals Court ruling and the wave of litigation it produced, City Homes declared bankruptcy in 2013 and ended all of its operations.[21] No other nonprofit entity in Baltimore since then filled that void at the same scale. Quality low-income rental housing in Baltimore remained a dire need. In 2018 more than 3000 persons in the city were homeless; and over 20% of the city's population was spending more than half their income on housing.[23]

6.2 Legal and scientific evidence

Evidence has different meanings in the scientific and legal communities. Scientific evidence usually means empirical data collected and developed to prove or disprove a hypothesis, supported by repeated studies; it is decidedly *not* based on opinion.

On the other hand, legal evidence usually is presented at a trial to convince a judge or jury of alleged facts. For example, one jury instruction stated:

> [Legal] evidence can come in many forms. It can be testimony about what someone saw or heard or smelled. It can be an exhibit admitted into evidence. It can be someone's opinion. Direct evidence can prove a fact by itself.[24]

To simplify, scientific evidence is established by repeated studies and experimentation, legal evidence by argumentation.

Both methods are different ways of getting at the truth. Benjamin Franklin opined on a "useful Truth." He suffered from lead poisoning during his early years as a typesetter. Still prescient today, he noted in his letter to Benjamin Vaughan in 1786:

> You will see by it that the Opinion of this mischievous effect from Lead, is at least above Sixty years old; and you will observe with Concern how long a useful Truth may be known, and exist, before it is generally receiv'd and practic'd on.[25]

This section examines the state of the scientific evidence as it existed at the time of the R&M study and the Maryland Appeals Court decision. There was good reason, based on prior research, to think the R&M interventions, though still unproven, were likely to be effective. What was not fully known at the time of the study was the relative effectiveness and longevity of the different housing interventions.

By the 1980s and 1990s, substantial scientific evidence existed on interventions likely to protect and benefit children from lead paint hazards. Beginning in the early 1980s, researchers showed that simplistic removal of lead paint without dust controls and worker protection backfired badly. These paint removal methods resulted in *increased* (not improved) blood lead levels.[26–29] Two of the most important of these early 1980 scientific studies were done by Johns Hopkins University and KKI in Baltimore.[6,20,30]

But this scientific evidence did not stop a District of Columbia court from requiring under a consent decree that all lead paint be removed in 1982 in *Ashton v Pierce*.[31] That decision cited another earlier Philadelphia court decision where "the judge ordered the removal of all lead-based paint in 1200 federally owned residences in Philadelphia."[32] That city actually required open flame burning to remove lead paint at the time,[33] but the order was never fully carried out because of its disastrous consequences.[30] This method and others like it (power sanding, abrasive blasting, and dry scraping) were ultimately banned in both federal guidelines and regulations,[34] but it would take research to show how misguided this approach had been and to put a stop to it.

Court decisions calling for removal of lead paint in the 20 years between 1970 and 1990 had been a clear failure. With the hindsight of history, the *Ashton v Pierce* court case, despite "best intentions" by the lawyers who brought it played a major role in the policy paralysis that stymied the nation's attempts to address the problem during much of the 1980s.

During the 1980s, 1990s, and 2000s research found that new remediation methods worked, that is, both dust lead levels and blood lead levels in children who resided in the remediated homes could be reduced and that they were durable. This is why most studies measured both changes in dust lead as well as blood lead over time, an issue that would be litigated in the *Grimes* case. Following these and other studies, abatement came to mean remediation methods where lead paint was not simply removed.

Instead, newer methods of abatement meant building component replacement (e.g., new windows and doors), enclosure (e.g., new siding and drywall), encapsulation (a special long-lasting coating), and limited paint removal (through low-temperature heat guns, chemical strippers, and contained abrasive methods that did not produce much lead dust).

Interim controls included dust removal using specialized cleaning, paint stabilization, and soil covering to reduce lead exposure. Interim controls also addressed friction surfaces like windows where lead dust could be generated. Building component replacement, enclosure, encapsulation, and paint stabilization all meant that usually some lead paint remained after the intervention, but in a safe condition. These methods informed the R&M interventions, as did expert advice from the Baltimore City Health Department, CDC, the Maryland Department of Environment, and property managers from City Homes and elsewhere.

The largest and most comprehensive of these studies was mandated by Congress, the Evaluation of the HUD lead hazard control grant program (Chapter 4). In 1998, EPA reviewed this and other similar studies, including preliminary R&M findings, showing that declines in blood-lead concentration on the order of 25% had been reported for both abatement and interim controls (interim controls were sometimes called "partial abatement"), and that these methods showed statistically significant improvements in both dust lead and blood lead results over time.[35] The HUD evaluation study had also shown that dust lead improvements (66%–95%) were greater than blood lead level improvements (a 37% improvement),[36] because blood lead levels depend on other nonhousing sources of lead exposure, as well as internal exposures from bone lead.

There was also an important difference between the HUD Evaluation and the R&M study. The larger HUD evaluation was unable to randomize housing interventions because it involved the ongoing programmatic work of the HUD grantees. But the R&M study could include such randomization, which greatly improved the scientific strength of the study.

Dust lead proved to be a better measure of hazard control effectiveness because blood lead integrated all sources and pathways of lead exposure. Some of these other sources were not housing related. In other words, even if all exposures to housing-related lead paint, dust, and soil were completely eliminated, blood lead levels could not decline to zero because of other sources of lead in diet, food, secondary homes children might visit, schools, air, water, playgrounds, consumer goods, and so on. Internal bone lead from earlier exposures also contributed to blood lead levels, because the half-life of lead in bone is 10–20 years.[37]

But dust lead measurements alone could not prove a health benefit, which was why most of the studies also included blood lead measurements. Such measurements could be tied more directly to adverse health effects. Although they included this biomarker, these were essentially housing (not health) studies. The focus was on housing interventions to reduce lead exposure, with blood lead measurements as additional secondary measures of effectiveness. This proved to be important in the R&M legal case because lawyers later alleged that blood lead testing meant "experimenting" on children. In fact, the children benefitted from the blood lead testing, which many Baltimore children at the time did not actually receive, despite the CDC guidance at the time.

The EPA 1998 review showed that more focused remediation (not simply paint removal), together with dust removal, lowered both dust and blood-lead levels in at least six studies at the time (the Boston Retrospective,[38] Central Massachusetts Retrospective,[29] St. Louis Retrospective,[39] 1990 St. Louis Retrospective,[40] the HUD evaluation,[36] and the New York Chelation study).[41–43] Overall, these and other studies found large declines in dust lead levels and 18%–29% declines in blood lead. However, the EPA 1998 review concluded:

> The literature was limited in scope.… There was no definitive evidence in the literature that one of these categories of methods was more efficacious than the other and yet there are disadvantages to both.… Four critical data gaps remain: (1) no data exist regarding primary prevention effectiveness rather than secondary; (2) little documentation is available regarding the long-term effectiveness (i.e., greater than one year) of lead hazard intervention; (3) some promising intervention strategies, such as soil cover, encapsulation and enclosure of lead-based paint, have received relatively little study; and (4) the reasons for blood-lead levels not being brought below 10 µg/dL [the CDC level of concern at the time] are not well understood.

One study done in Baltimore in the early 1980s found that "regular, extensive dust-lead hazard management efforts by trained personnel produced an 18% decline in mean blood lead concentration."[20] Fig. 6.1 shows the time periods in which these and other studies had been undertaken during the 1980s and 1990s. Of these 30 studies, many (but not all)

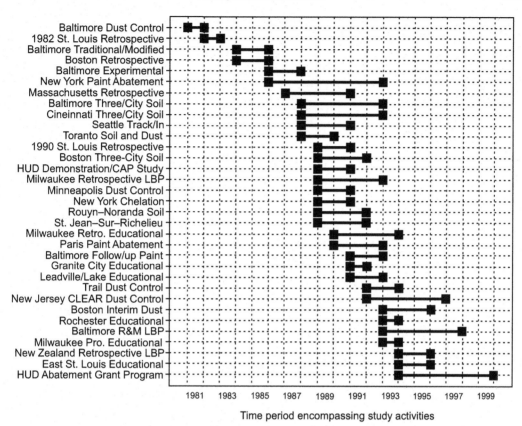

FIGURE 6.1 Lead intervention study time periods, 1980–99. *Source: From EPA. Review of Studies Addressing Lead Abatement Effectiveness: Updated Edition. EPA 747-B-98-001; December 1998.*

showed reductions in blood lead level, but usually only when beginning blood lead levels were very high and sometimes there was no significant effect at all. Some did not do paint abatement but were limited to education or soil interventions. The need for further research was clear.

In light of all this scientific evidence, why was it necessary to undertake more research when the R&M pilot study began in 1991? Although there was good reason to believe that both interim controls and abatement were effective, questions remained about how much disturbance of lead paint should be allowed, how long interventions would last, and how the methods could be brought into housing maintenance and capital improvement operations.

This is precisely why Congress included a requirement for additional research in Title X and why CDC, EPA, and HUD funded that research, including the Baltimore R&M study. Yet none of this scientific evidence would be considered in the case before the Maryland Appeals Court.

Lawyers in the Maryland court cases sometimes suggested the studies could have been done differently. What if these studies had not included blood lead measurements and none of the houses had children in them? What if the studies had only simulated the results or only used adults? What if they did not occur at all?

The answer: the results would not be generalizable to houses that *did* include children, and the studies would have been considered to be inconclusive and weak. They would not have been tied to improved health. They would not have generated the Congressional confidence to provide additional funding for widespread lead hazard control. Most important, children would not have benefitted from the research.

But of course, these benefits and the "greater good" would not have been worth it (and indeed would have been terribly wrong) if children had somehow been put in harm's way. Were they being experimented on? How did researchers ensure that participants in these studies were protected, respected, and benefitted? How should government and the courts ensure such research was done ethically? And did they in fact do so?

Answers to these questions requires an understanding of how research ethics are formulated and enforced.

6.3 Research ethics and protection of research study participants

The history of how research must be conducted in a manner that protects and respects those people who choose to participate in such studies has been described in greater detail elsewhere,[44] and is presented in a timeline (Fig. 6.2).

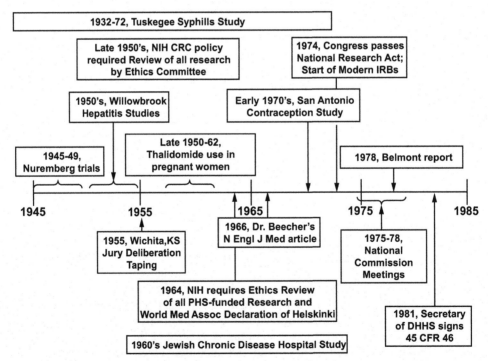

FIGURE 6.2 Timeline of human subjects research protection. *Reprinted with Permission: From Rice TW. The historical, ethical, and legal background of human-subjects research.* Respir Care 2008;53(10):1325–1329.

The first regulations related to the protection of human subjects in the United States predate the timeline above. In 1906 the Pure Food and Drug Act[45] was passed as a result of public outrage over the findings of unsanitary dangerous conditions in meatpacking industries by journalists like Upton Sinclair, who published a best-selling book the year before, *The Jungle*.[46] This early law was primitive by today's standards and there is no record of enforcement beyond better food and drug labeling.

The years leading up to its passage included experiments on humans, in which volunteers were fed food with common preservatives in use at the time (now known to be dangerous), including borax, benzoate, formaldehyde, sulfites, and salicylates. The dinner "table trials" and their human subjects came to be known as "The Poison Squad." The men who participated in the trials reportedly adopted the motto "Only the Brave dare eat the Fare."[47] But this law did not explicitly include provisions for research; it was a specific response to the dinner table trials.

The first significant effort to protect people participating in research occurred in 1946, when an international trial brought criminal charges against 23 Nazi physicians and administrators for their willing participation in war crimes and crimes against humanity. They had conducted medical experiments on thousands of concentration camp prisoners without their consent, often resulting in death or permanent damage. As a direct result of the trial, the Nuremberg Code was established in 1948, stating,

The voluntary consent of the human subject is absolutely essential.[48]

It also said the benefits of research must outweigh its risks. This was the first international document to provide for voluntary (not coercive) participation in research through informed consent, although the Nuremberg Code did not have the force of law.

The Tuskegee Syphilis Study was conducted between 1932 and 1972 by the US Public Health Service. It included 600 low-income African American men, 400 of whom had syphilis. Although free medical examinations were given, the participants were not told about their disease, even though a cure (penicillin) later became available in the 1950s. It continued without any medical treatment and in some cases the researchers actually stopped medical treatment offered by others outside the study; many died from the disease who would have lived had it not been for their participation in the study.[49] In 1997 President Clinton apologized to those who participated in the study and to their families[50] and the US government provided lifetime medical and other benefits.

These Nazi and Tuskegee despicable "studies" are briefly recounted here because the Maryland Appeals Court would later state that the lead paint R&M study in Baltimore in the early 1990s was similar to them (if the allegations were true).

In 1964, the Helsinki Declaration stated research with humans should be reviewed by an independent committee (later to become known as Institutional Review Boards) before any data collection began. In addition to this institutional review board (IRB), the Declaration also stated that informed consent is necessary, that research should be done only by scientifically qualified individuals and that risks should not exceed benefits.[51]

Neither the Helsinki Declaration nor the outrage over the Nazi studies (and others like them shown in the timeline above (Fig. 6.2) had produced anything that carried the force of law. But in 1974, in the wake of the Tuskegee study, Congress passed the National Research Act, which *could* be enforced.[52] Among other things, it created the National Commission for the Protection of Human Subjects of Biomedical and Behavioral Research, which produced the Belmont Report,[53] still one of the key documents governing research ethics that virtually all research entities engaged in studies involving humans adopted, including KKI and Johns Hopkins University.

The Belmont report articulated three essential ethical principles for research involving human participants (I choose to call them "participants" instead of the more common term "subjects" to underscore the importance of true collaboration between researchers and those who choose to participate). The three core Belmont principles are:

1. Respect for people, which requires informed consent (meaning that information should be complete, participants should be able to comprehend the research, and their participation is voluntary).
2. Beneficence, which requires that participants must not be harmed, that benefits should be maximized, and risks minimized.
3. Justice, meaning that risks and benefits should be distributed equally, with clear and fair research procedures.

The Belmont report resulted in another regulation in 1981, when the US Department of Health and Human Services (HHS) and the Food and Drug Administration issued a new regulation [45 Code of Federal Regulations Part 46, Subparts A, B, C, and D]. Together, these became known as the "Common Rule," which was formally adopted in 1991 by more than a dozen other federal departments and agencies conducting or funding research involving human subjects, including EPA and HUD.[54]

The Common Rule included requirements for assuring compliance by research institutions; requirements for researchers obtaining and documenting informed consent; requirements for Institutional Review Board membership, function, operations, review of research, and record keeping; and additional protections for certain vulnerable research participants (subjects), such as pregnant women, prisoners, and children. The historical record is unclear if the adoption of the Common Rule by these agencies included all its subparts, including subpart D, which pertains to research with children, or whether each subpart had to be formally adopted by each agency.

In any case, the Common Rule was enforceable, with an oversight office established within HHS. Virtually all research entities in the United States involved in human subjects pledged to comply with the Common Rule in its entirety, including Johns Hopkins University and KKI and there was oversight by the federal government to enforce it.

6.4 The Baltimore lead paint abatement and repair and maintenance study

The R&M study began in 1991 with a pilot study.[55] It was designed to examine three different combinations of interim controls and abatement, comparing them to two other groups of homes—ones that had been "fully abated" previously and another group of newer urban homes without lead paint.

Congress had recognized interim controls when it passed Title X at about the same time in 1992, defining it as

> a set of measures designed to reduce temporarily human exposure or likely exposure to lead-based paint hazards, including specialized cleaning, repairs, maintenance, painting, temporary containment, ongoing monitoring of lead-based paint hazards or potential hazards, and the establishment and operation of management and resident education programs.[56]

The study was originally funded by EPA and then by HUD to extend it over a longer time period. The funding was awarded through competitive processes in which government personnel scored the proposal and found it worthy of support. As a government-funded study, compliance with the Common Rule and IRB approval were required.

The study's main purpose was to fill a gap in knowledge about the "short- and long-term effectiveness of these approaches [interim controls] in terms of reducing lead in dust and in children's blood."[57] The research was intended to support science-based practices and policies for preventing lead poisoning through widespread residential lead remediation. In the research section of Title X, Congress had specifically mandated that research studies be conducted to "evaluate the long-term cost-effectiveness of interim control and abatement strategies."[56]

In mirroring the Title X language in its study protocols, KKI, Hopkins, and their government funders were in one sense merely complying with the Congressional mandates.

According to the EPA's summary:

> The [R&M] study was designed to characterize and compare the short-term (two months to six months) and longer-term (12 months to 24 months) effectiveness of three levels of interim control interventions (R&M I-III) in structurally sound housing where children were at risk of exposure to lead in settled house dust and paint. *At the time of this study, owners were not required to reduce lead exposure in their rental properties prior to children becoming poisoned. Thus, study houses received R&M interventions that they were not likely to have gotten otherwise.* Funds for R&M work provided by the Maryland Department of Housing and Community Development were capped at $1650 for R&M I, $3500 for R&M II, and $7000 for R&M III (emphasis added).[57]

All three of these interventions were designed to remove all deteriorated lead paint, and dust wipe testing was done after all the interventions were completed, even though Baltimore's local law did not require that at the time (it only required clearance testing after abatement that occurred after a child had already been poisoned).

The R&M study was the first to compare what were essentially five different hazard control strategies, each with varying degrees of lead paint disturbance and dust generation:

1. Homes built after 1978 where lead paint was unlikely (a type of control group where no paint was disturbed);
2. Homes that had already been abated previously (a second type of control group);
3. Homes with a wet scraping of peeling and flaking lead-based paint on interior surfaces; limited repainting of scraped surfaces; wet cleaning with a trisodium phosphate detergent and vacuuming with a high-efficiency particulate air vacuum; the provision of an entryway mat (to minimize track-in of lead-contaminated exterior dust); education of occupants on lead; and stabilization of lead-based paint on exterior surfaces (one of three intervention groups, R&M I);

4. In addition to all the items in number 3, homes with sealants and paints to make floors smoother and more easily cleanable and in-place window and door treatments to reduce abrasion of lead-painted surfaces (the second of three intervention groups, R&M II);
5. In addition to all the items in numbers 3 and 4 above, the last group of homes had window replacement and enclosure of exterior window trim with aluminum coverings as the primary window treatment, enclosure of exterior door trim with aluminum, and the use of coverings (e.g., vinyl tile) on some floors and stairs to make them smooth and more easily cleanable (the third of three intervention groups, R&M III).[58]

The last two categories were differing mixtures of interim controls and abatement. The 5th group was most like abatement but would also likely cause the greatest disturbance of lead paint. For this reason, the 4th and 5th group interventions were done while the homes were vacant. When the study was launched, which of the strategies would be most effective was not known, which was one of the reasons the study was undertaken. But all three were likely to result in reduced exposure, based on the earlier research cited above. An earlier pilot study of vacant homes also showed that the interim strategies were successful in reducing dust lead, with an assumption that the repaired homes would therefore also result in lower blood lead levels once the homes were occupied.[58]

All families in the five groups received cleaning kits and supplies for their own routine use. Although not required by law at the time, blood lead levels were measured at regular intervals and reported to the child's parent or legal guardian and the local health department in accordance with the existing standard of care and local case management if needed, consistent with local regulations. All homes had dust containment and professional cleaning, and all the work was done by workers trained in lead paint abatement. All groups received health education about lead poisoning as well. Families who chose to participate were temporarily relocated during the several days it took to complete the work, although some of the homes were vacant during the work and later occupied by families.

In short, the R&M study did not replace usual medical care and did not relieve housing owners of their duties to comply with the local Baltimore lead poisoning prevention regulations in effect at the time. But it did provide interventions that the families would not have otherwise received. Participation in the study was entirely voluntary.

Perhaps most importantly, especially in light of the later Court of Appeals ruling, the R&M study protocol explicitly stated that it would "not include a non-intervention control group of houses that contained lead-based paint hazards." In other words, no homes in the study had lead paint hazards that were not corrected, because that would have harmed children and been unethical.

The researchers reached out to both landlords and residents who might want to participate, using the community networks they had built over the decades, including such groups as the Baltimore Coalition to End Lead Poisoning, the Baltimore Property Owners Association, and City Homes. Like the HUD Evaluation study, the research was conducted mostly in homes and neighborhoods where the risks were greatest, where the communities stood to benefit the most and where participating owners had their properties. Because the houses would have hazards reduced, this was the best way to optimize benefits and minimize risks, as required in the Common Rule. Also similar to the HUD evaluation and other studies, the homes had children between the ages of 1 and 6 when blood lead levels were typically at their highest to maximize benefits.

The researchers and community residents conducted over 1100 home visits to over 650 modern, previously abated and older occupied dwellings during the spring and summer of 1992. Reflecting the respect KKI had in the community, over 90% of eligible households indicated an interest in participating in the study.[59]

Although not by design, the R&M study participants would end up being nearly all African American, reflecting the characteristics of high-risk housing neighborhoods in Baltimore. The exception to this was the group of modern urban homes. They were located in a few modern housing clusters built in high-risk neighborhoods after the 1978 ban on lead paint and were therefore presumably free of lead paint. The racial makeup of the participants figured prominently in the court's later allegation that this study "experimented on poor Black children."

The study had been approved by the Hopkins IRB, as well as by federal agency scientists to ensure that it met ethical standards and complied with the Common Rule. The informed consent form that caregivers signed stated:

> The repairs were not intended, or expected, to *completely* remove exposure to lead. We are now doing a study to learn how well different practices work for reducing exposure to lead in paint and dust (emphasis added).[60]

The benefits of the study were described in the informed consent form as free dust, soil, water, and blood lead testing, a summary of the house testing results and "steps you could take to reduce any risks of exposure." Another benefit included a $15 ($29 in year 2020 dollars) compensation for each time a questionnaire was answered every 6 months. Parents were also encouraged to do cleaning to continue to further reduce exposures, because lead dust could reaccumulate due to their location in leaded older housing neighborhoods. This was beyond the professional cleaning that was done after the work was completed, which is why the study provided cleaning supplies and health education to the participating families.

The risks associated with the study were described in the informed consent form as the minor discomfort or pain a child might experience from the blood lead testing. The letter sent to the families that described the results of the lead dust testing highlighted areas where dust lead levels were thought to be higher than those in modern urban homes, although there were no standards for the amount of lead dust remaining in a home after remediation using interim controls at the time. The letter explicitly said: "Remember, there is no rule for how much lead is allowed in dust from a house like yours."

Some criticized both this consent form and the results letter for somehow concealing information or implying that all lead had been removed. But the presence of reported lead dust together with the ongoing encouragement to continue cleaning with the provided cleaning supplies meant that no one should reasonably think that lead had been completely removed from the homes. This idea that lead was still present would figure prominently in the court cases to come.

The Appeals Court also appeared to be confused about dust lead levels reported to occupants, described in more detail later in this chapter. Because the best method of measuring lead in dust had not been settled at the time, the R&M study used wipe sampling immediately after the work had been done. This was not technically required by law, but the study went above and beyond the minimum requirements. In addition to the wipe sampling done immediately after the work was completed, the R&M study also used a newer vacuum dust sampling method that EPA had developed[61] at specific time intervals throughout the study. EPA wanted this method used, because the vacuum method could report lead concentration, lead dust loading, and total dust loading (Chapter 4). But because it was different than wipe sampling, the dust lead loadings measured were fundamentally different between the two methods. Some lawyers would attack this vacuum method as a "nonstandard" method, even though it was based on an ASTM voluntary consensus standard at the time.[62]

When this study was undertaken, there were in fact no legally binding "standard" lead dust sampling methods, either vacuum or wipe sampling, and there were no allowable limits for lead dust for each method beyond the guideline postabatement levels (Chapter 4). Specifically, there were no dust lead standards or guidance values for vacuum sampling. Regulatory standards for wipe sampling would not be promulgated nationally until the end of the 1990s, when HUD and EPA issued them for lead in dust via wipe sampling in 1999 and 2001, respectively. There was no valid way to convert wipe sampling results into those from the vacuum method, a basic scientific fact that eluded the Maryland Court of Appeals. Instead, the two methods were treated interchangeably, even though they clearly were not.

Figs. 6.3 and 6.4 show the results of the study, which turned out to be compelling and encouraging: all three levels of R&M intervention resulted in statistically significant reductions in house dust lead loadings that were sustained below preintervention levels during 2 years of follow-up. Note that the dust lead values are from vacuum sampling, not wipe sampling. Fig. 6.3 shows floors, window sills, and window troughs combined; Fig. 6.4 shows the results for floors.

Surprisingly, the most intensive R&M III group actually had *lower* dust lead levels than the fully abated houses that had been completed a few years earlier, suggesting that full

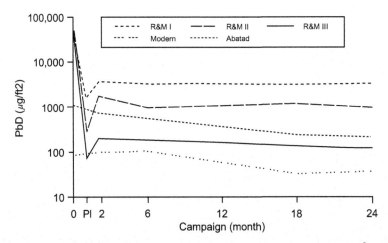

FIGURE 6.3 R&M median dust lead trends pre-remediation to 24 months. PbD (μg/ft^2) lead dust loading values are from vacuum sampling, not wipe sampling (vacuum sampling results cannot be compared to dust lead hazard standards). *Source: From Lead-Based Paint Abatement and Repair and Maintenance Study in Baltimore: Findings Based on Two Years of Followup. US EPA 747-R-97-005; November 1997.*

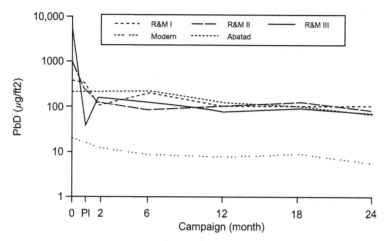

FIGURE 6.4 R&M median floor dust lead trends pre-remediation to 24 months. PbD ($\mu g/ft^2$) lead dust loading values are from vacuum sampling, not wipe sampling (vacuum sampling results cannot be compared to dust lead hazard standards). Source: *From Lead-Based Paint Abatement and Repair and Maintenance Study in Baltimore: Findings Based on Two Years of Followup. US EPA 747-R-97-005; November 1997.*

abatement might not be the optimal strategy. The Appeals Court decision alleged that anything short of full abatement should not have been permitted in the study, even though the actual data showed otherwise. Because the Appeals Court ruled without facts in evidence, these findings were not considered until juries eventually heard them years later.

The slight increase immediately after the interventions followed by the long-term sustained reduction in dust lead was most likely due to moving furniture back into the home and tracking of exterior dust into interiors by the normal living patterns of the residents. This initial "rebound" followed by long term declines had been observed in many other studies. In any case, dust lead levels remained far lower than where they were at the beginning in all three intervention groups, sometimes by very large amounts.

For floors, which is the surface that young children contact the most, the results were even more compelling. All 3 interventions had about the same effectiveness as the fully abated homes (Fig. 6.4).

As hypothesized at the start of the study, children with baseline blood lead concentrations both above and below 15/dL (which was the CDC environmental intervention blood lead level at the time) in each of the three R&M groups and in the previously abated group had statistically significant reductions, after controlling for age, gender, and season. The predicted blood lead concentration at 24 months was on average 20% lower than at baseline. Finally, all 3 R&M levels produced about the same decrease in blood lead levels, suggesting that the differing paint disturbance practices had been successfully mitigated by the dust containment and cleanup methods.

Despite the reductions in both lead dust and blood lead levels, some lawyers alleged that the children in the study had been harmed and that blood lead levels had somehow *increased*. In one case, for example, a child's blood lead level did in fact increase. But because the dust lead levels in that child's home had in fact *decreased*, a jury found that

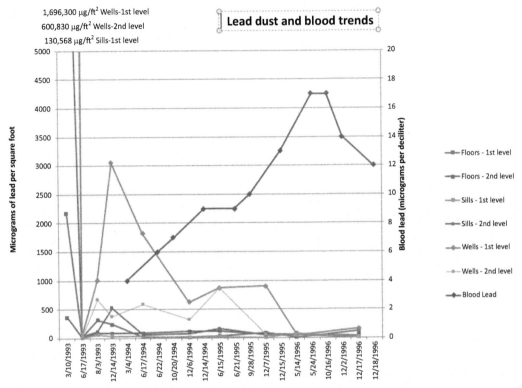

FIGURE 6.5   Lead dust and blood lead trends in one R&M study home. Source: *From The University of Illinois Chicago Special Collections & University Archives, School of Public Health, "David E. Jacobs papers" the University of Illinois Chicago School of Public Health Library.*

this child had been exposed to lead from some other source, not the R&M study home.[63] The facts proved to be compelling.

The chart in Fig. 6.5 convinced the jury that the R&M interventions could not be the cause of the child's increased blood lead level. The jury later stated that this chart was the key reason they found the researchers to not be negligent in the conduct of the study. Fig. 6.5 shows vacuum dust lead levels, not wipe dust lead results. It showed that dust lead levels decreased by a large amount, but the child's blood lead level increased. The obvious reason was that the child was being exposed to another source of lead outside the home, and the jury agreed.

6.5 The *Grimes* decision

A lower trial court initially dismissed allegations that KKI and Hopkins had harmed two children (Erika Grimes and Myron Higgins), whose families had each chosen to participate in the R&M study, because it found the children had been protected and the researchers had acted in the best interests of those children and the community through the "Good Samaritan" principle discussed earlier.[1]

But on August 16, 2001, the Maryland Court of Appeals issued a decision with no facts in evidence.[3] Earlier, the lower court had granted summary judgment, meaning that researchers had no legal duty, dismissing the case. There were not many facts about the study in the defendant's brief, so the Court of Appeals was left to "assume" facts that had been asserted by the Plaintiffs as uncontroverted. The Court of Appeals ordered a new trial using blistering language. Central to its ruling was a distinction between therapeutic and nontherapeutic research.

6.5.1 Therapeutic and nontherapeutic research

The Maryland Appeals Court ruled:

> In these present cases, a prestigious research institute, associated with Johns Hopkins University, based on this record, created a non-therapeutic research program whereby it required certain classes of homes to have only partial lead paint abatement modifications performed, and in at least some instances, including at least one of the cases at bar, arranged for the landlords to receive public funding by way of grants or loans to aid in the modifications. The research institute then encouraged, and in at least one of the cases at bar, required, the landlords to rent the premises to families with young children.

The Court defined so-called nontherapeutic research as:

> research [that] generally utilizes subjects who are not known to have the *condition* the objectives of the research are designed to address, and/or is not designed to directly benefit the subjects utilized in the research, but, rather, is designed to achieve beneficial results for the public at large (or, under some circumstances, for profit) (emphasis added).

This distinction between therapeutic and nontherapeutic research, as applied in the case was important. The former generally meant drug research, in which experiments typically randomize individuals into one of two groups: those that receive the experimental drug (the "therapy") and the second group receives a placebo. (Some drug trials now use a third group that receives neither the experimental drug nor the placebo to help control for the "placebo effect.") But as we have seen for lead poisoning, there was no drug intervention that had been shown to be effective.

The body of scientific evidence reviewed at the beginning of this chapter showed that remediation of lead-based paint hazards in housing with proper control of lead dust was likely to be the only effective *"therapy"* to benefit the participants. Much of that research was already known (or should have been known) to the Appeals Court. In short, if we regard the "therapy" for lead poisoning to be remediation of lead exposures in housing, then the court's distinction between the two types of research loses all logic and validity.

In some ways, the Appeals Court decision was a reflection of the false distinction between communicable and noncommunicable diseases (Chapter 9). The separation of research into these two categories historically served children suffering from environmental or housing diseases very poorly or not at all. For many children, the house that had lead paint hazards clearly "communicated" lead poisoning. But lead poisoning prevention was typically categorized as a "noncommunicable" disease.

The "condition" that the therapy in the R&M study was intended to address was elevated blood lead levels and homes with lead hazards. The Appeals Court's determination that the R&M study was "nontherapeutic" meant by definition that it could not "directly benefit the subjects." But this was at odds with the fact that both blood lead levels and dust lead levels were reduced across all levels of intervention, which was a benefit and thus therapeutic.

What clearly would have been a violation of the Common Rule and ethical norms would be to allow children to live in homes with known lead-based paint hazards, without performing any intervention. This did not occur in the R&M study. The only group of homes that did not have an intervention in this study was the comparison group of modern urban homes that did not have lead paint. In a drug trial (therapeutic research), this group would be regarded as the control group with the placebo (living in a home without lead paint).

But in housing intervention research, the R&M study was designed explicitly to not have a control group where lead-based paint hazards were found but left untreated (as in a placebo). Instead, the R&M study had two other comparison groups: one in which homes had been fully abated some years earlier and the second in homes without lead paint.

The definitions of therapeutic and nontherapeutic research were problematic in the housing context. If by therapeutic research one means, for example, a drug trial, the standard for protection is actually lower than for nontherapeutic research. In a drug trial, half of the randomized participants typically receive the experimental drug (a potential benefit) and the other half do not (no benefit). The Appeals Court held that because it classified the R&M study as nontherapeutic research, there could be no direct benefit to any of the participants. But in fact, there *was* a benefit, shown by the reduced dust lead and blood lead levels.

In its review of this and other housing studies (described in detail later in this chapter), the National Academy of Sciences, Institute of Medicine suggested that the distinction between therapeutic and nontherapeutic research made little sense in the housing context. Instead of a dichotomy between no-benefit (nontherapeutic research), versus some benefit (therapeutic research), it was more ethical and important to strike a balance between risks and benefits in all housing intervention studies and that this balance is in fact more consistent with research ethics.

The Appeals Court also alleged that the study included "nontherapeutic procedures that are potentially hazardous, dangerous, or deleterious to their [children's] health." By this the Court apparently meant that the three R&M study groups, because they did not use complete paint removal or full abatement, would put children into potentially hazardous conditions. But each of the three groups had eliminated all lead paint hazards.

This plainly did not increase risk. Rather, all three R&M interventions reduced the risk of lead poisoning.

Indeed, the *Grimes child lived in a "fully abated" home*, the standard that this court apparently considered to be the only acceptable treatment, much like the earlier court in *Ashton v Pierce* thought that only paint removal could be a solution.

The Maryland Appeals Court alleged that the Grimes child's family had been misled, because lead was still present, even though their home had been fully abated before the study began.

The Court of Appeals held that:

> children should not have been used for the purpose of measuring how much lead they would accumulate in their blood while living in partially abated houses to which they were recruited initially or encouraged to remain, because of the study.

But the record shows that in fact lead did *not* accumulate in their blood. The levels actually decreased. It was never shown how the children were encouraged to remain in their homes by the researchers and how they did not receive a benefit.

6.5.2 Protecting children or putting them in harm's way?

The Maryland Appeals Court decision in *Grimes* held that children were somehow being purposely exposed to risk:

> Apparently, it was anticipated that the children, who were the human subjects in the program, would, or at least might, accumulate lead in their blood from the dust, thus helping the researchers to determine the extent to which the various partial abatement methods worked. There was no complete and clear explanation in the consent agreements signed by the parents of the children that the research to be conducted was designed, at least in significant part, to measure the success of the abatement procedures by measuring the extent to which the children's blood was being contaminated. It can be argued that the researchers intended that the children be the canaries in the mines but never clearly told the parents. (It was a practice in earlier years, and perhaps even now, for subsurface miners to rely on canaries to determine whether dangerous levels of toxic gasses were accumulating in the mines. Canaries were particularly susceptible to such gasses. When the canaries began to die, the miners knew that dangerous levels of gasses were accumulating.).... *Nothing about the research was designed for treatment of the subject children* (emphasis added).

By "treatment," the Court apparently meant medical treatment, not housing treatment. But as we have seen, there is no medical treatment for lead poisoning, other than chelation (which is typically used only for children with extremely high blood lead levels above $45\,\mu g/dL$ and which has its own risks), well above anything seen in the R&M study. By eliminating the lead paint hazards in their homes, the children did in fact receive a housing treatment that was successful.

The question of whether blood lead levels were increasing or decreasing was a central concern of the Appeals Court. But because it ruled without a record of facts, it could not know the answer to the question of whether blood lead levels were rising or declining. As we shall see, in the 11 separate cases that were brought after the *Grimes* Appeals Court decision (one of which was a class action suit),[64] juries determined that the children in the study did indeed benefit from participation, because both dust lead and blood lead levels were reduced as a result of the interventions.

Getting those facts in front of the juries required enormous legal skill in order to overcome repeated attempts to exclude the evidence. The KKI legal team was led by Barry Goldstein, John Sly, Michael Brown, Michael Blumenfeld, and other lawyers, who were able to craft successful defenses or dismissals in all 11 cases of participating children that focused on the scientific facts of the case, not the Court of Appeals' incendiary allegations.

The result of the juries' deliberations was unanimous verdicts exonerating the researchers regarding claims made by the plaintiffs (the R&M study participants).

6.5.3 New ground

The Appeals Court also thought that it had broken new legal ground:

> As far as we are aware, or have been informed, there are no federal or state (Maryland) statutes that mandate that all research be subject to certain conditions. Certain international "codes" or "declarations" exist (one of which is supposedly binding but has never been so held) that, at least in theory, establish standards. We shall describe them, infra. Accordingly, *we write on a clean slate in this case.* We are guided, as we determine what is appropriate, by those international "codes" or "declarations," as well as by studies conducted by various governmental entities, by the treatises and other writings on the ethics of using children as research subjects, and by the duties, if any, arising out of the use of children as subjects of research (emphasis added).

But more than twenty years before, as explained earlier, Congress had passed the National Research Act in 1974, a federal statute that could be enforced, and regulations were promulgated that same year as 45 CFR part 46, with subsequent regulatory changes in 1978, 1981, 1983 (which established specific rules for children participating in the research), and 1991.[65]

There was no "clean slate," as imagined by the Appeals Court. Indeed, there was a substantial record of regulatory action to ensure children were protected during their participation in research.

6.5.4 Parents versus the court of appeals

The Appeals Court decision went so far as to suggest that parents did not even have the right to allow their own children to participate in "nontherapeutic" studies:

> Otherwise healthy children, in our view, should not be enticed into living in, or remaining in, potentially lead tainted housing and intentionally subjected to a research program, which contemplates the probability, or even the possibility, of lead poisoning or even the accumulation of lower levels of lead in blood, in order for the extent of the contamination of the children's blood to be used by scientific researchers to assess the success of lead paint or lead dust abatement measures. Moreover, in our view, *parents, whether improperly enticed by trinkets, food stamps, money or other items, have no more right to intentionally and unnecessarily place children in potentially hazardous nontherapeutic research surroundings, than do researchers. In such cases, parental consent, no matter how informed, is insufficient* (emphasis added).

In effect, the Appeals Court said it knew better than parents about what was good for their own children. It said that no child should be living in homes with lead paint, despite the fact that millions of children already did live in such housing (and continued to do so), whether they were in the study or not. The idea that the children were "otherwise healthy" seemed to contradict the widespread reality that most children in cities like Baltimore already had elevated blood lead levels.

The Appeals Court held that somehow children were brought into the study homes by the researchers, but in fact the parents found the homes through the normal leasing and advertising for rental properties done by the owners, including the Grimes child, whose family had moved into a "fully" abated and renovated home before the R&M study even began. In short, the researchers were not the owners, and it was the owners who rented the homes to occupants.

The Appeals Court did not explain where the children should go to live, and how they were "otherwise healthy."

The Appeals Court held that parental consent, no matter how well informed, was not sufficient to allow parents to enroll their children in the research if they chose to do so. The Court later backtracked, because its initial ruling threatened to end all child participation, effectively ending all research on vaccines, other drugs, and other health measures for children.[66]

In its retreat, the Appeals Court stated that it was making no finding of fact, acknowledging that it did not have all the facts. It said in a subsequent ruling that it did not mean "zero risk" for children, but any "articulable risk," which it never defined. The Appeals Court was at odds with the widely held view that parents generally knew what was best for their own children, not the courts, and that courts are not parents.

6.5.5 Institutional review boards

The Appeals Court decision also attacked IRBs as inherently biased in favor of researchers:

> It is clear to this Court that the scientific and medical communities cannot be permitted to assume sole authority to determine ultimately what is right and appropriate in respect to research projects involving young children free of the limitations and consequences of the application of Maryland law. The Institutional Review Boards, IRBs, are, primarily, in-house organs.

The Appeals Court did not recognize that others outside the scientific community, notably those government agencies approving the funding and other government offices conducting oversight, had also reviewed and approved the study, not only the IRB.

The Department of Health and Human Services (HHS) Secretary's Advisory Committee on Human Research Protection examined some calls to license or create more independent IRBs, but there was not a perceived need to do this. In any case, the researchers did *not* in fact have "sole authority" to do whatever they wanted in light of the many layers of review and oversight.

6.5.6 Special duty

The narrow legal question considered by the Court of Appeals was as follows:

> Was the [lower] trial court incorrect in ruling on a motion for summary judgment that as a matter of law a research entity conducting an ongoing non-therapeutic scientific study does not have a duty to warn a minor volunteer participant and/or his legal guardian regarding dangers present when the researcher has knowledge of the potential for harm to the subject and the subject is unaware of the danger?

The lower trial court had originally ruled that it was the landlord (not the researchers) whose duty it was to comply with local lead laws, because the landlord owned the property and had control of it, not the researchers. The lower court had ruled:

> KKI was sort of an institutional volunteer in the community. Coming in to collect dust and blood samples, the next thing you know they get sued and I think that there is absolutely no duty on the part of KKI simply because it came in to then assume a higher standard of ... [responsibility] in respect to these facts. KKI was not the owner of the property, not an agent for the owner, it didn't [accept] other properties from the landlord. It did not prefer the properties to the landlord.[1]

In other words, the Court of Appeals seemed to think that the researchers had a "special duty" to warn residents of risks that came from outside the study, not the risks associated with the study itself. Some suggested that this expansive interpretation would lead to "paralysis in similar public health studies and have negative consequences for … victims, the very same children the court seemed to be concerned about."[67]

In its argument before the Appeals Court, KKI's lawyers argued that:

> it was not the study that presented any risk of harm to Appellants, it was the circumstances in which Appellants found themselves completely independent of whether they participated in the study.[68]

This duty to somehow warn of risks not created by the study itself was never defined by the Appeals Court. It would be hard to know which risks outside the study researchers would be required to warn about and how far their obligations would go. For example, if a study of nutrition found dietary deficiencies, and the researchers told the parents to change the diet and see their physician, but the parents failed to do so, would the researchers be required to report the parents to law enforcement authorities for child neglect?[67]

6.5.7 Informed consent

The Appeals Court thought were deficiencies in the informed consent form that the IRB had previously approved:

> Nowhere in the consent form was it clearly disclosed to the mother that the researchers contemplated that, as a result of the experiment, the child might accumulate lead in her blood, and that in order for the experiment to succeed it was necessary that the child remain in the house as the lead in the child's blood increased or decreased, so that it could be measured.

Yet the informed consent form did say this:

> As you may know, lead poisoning in children is a problem in Baltimore City and other communities across the country. Lead in paint, house dust and outside soil are major sources of lead exposure for children. Children can also be exposed to lead in drinking water and other sources. We understand that your house is going to have special repairs done in order to reduce exposure to lead in paint and dust…. The repairs are not intended, or expected, *to completely remove exposure to lead*. (emphasis added).

In other words, the Appeals Court held that researchers should have informed the participants of risks associated not only with the study itself but also the risks of lead poisoning in general. The very first sentence in the informed consent form seemed to do exactly that, by stating lead poisoning was a problem in Baltimore. Yet the Appeals Court held that the researchers failed in their special duty, because it allowed the families to remain in homes with lead paint somehow without warning them. But clearly, the informed consent form *did* warn about the dangers of lead paint and that not all lead paint would be removed in their home. And the families were provided with cleaning supplies and lead paint educational materials, thereby acknowledging that some lead would remain in the home.

It was difficult to see how the informed consent form could have been clearer: it said that lead exposures would be "reduced" (not "eliminated") and that the repairs would not "completely" remove lead, because the researchers knew that there is no intervention that could completely remove all exposure.

6.5.8 The facts in the case

Indeed, the Appeals Court seemed to think that the actual facts were irrelevant:

> The evidence suggests and the parties appeared to agree during oral argument before this Court that the Monroe Street [Grimes] property was a member of Group 4 [the full abatement group]. Regardless, because we are reviewing this matter in the context of the granting of summary judgment based upon a trial court determination that no duty existed as a matter of law and, on remand, the facts of each case will, of necessity, need to be addressed, *we do not need to resolve to which group it was a member or whether there was, as a matter of fact, a breach of duty in that case*, or even damages for that matter (emphasis added).

The lower court had ruled that it was the property owner, not the researchers, who had the duty to maintain the fully abated home. For the Grimes child, the Baltimore health department investigated after KKI reported her blood lead results. After the health department went to the home and collected dust wipe samples, it found that there was no elevated lead dust in the Grimes home, with a notation that this property was "lead-free." In short, the health department found that Erika Grimes clearly had a lead exposure from someplace other than the fully abated home, although the alternative source was never identified. The Grimes child was born on May 30, 1992. Grimes' blood lead levels were 9 µg/dL on April 9, 1993, 32 µg/dL on September 15, 1993, and 22 µg/dL on March 25, 1994. Erika Grimes' mother moved into the fully abated home in the summer of 1990 and the unit was fully abated on October 15, 1990 (before the R&M study began).

The second child in the case, Myron Higgins, moved into a vacant unit that had dust lead levels below the clearance standards in effect at the time for abatement and in which all lead paint hazards had been remediated. The family moved into the home in May 1994. KKI measured his blood lead on June 8, 1994, July 29, 1994, and November 9, 1994, and notified the mother and the health department of the results, which were 17.5, 21, and 11 µg/dL, respectively. The child's overall blood lead declined while residing in the study home, consistent with the study findings.

In this time period, blood lead measurements typically had an analytical error rate of about plus or minus 3 µg/dL, so the difference between 17 and 21 µg/dL may have just been an artifact of measurement (not a true increase). In any case, KKI reported the blood lead levels to the appropriate local authorities for follow-up and the combination of that follow-up and the safer housing brought about by the study appears to have been successful in ultimately reducing the blood lead levels for both children. The other children in the study also showed declines in both blood and dust lead levels as described earlier.

6.5.9 Confusion on lead dust testing methods

As summarized earlier, the Appeals Court also was confused about the lead dust testing, because the R&M study used two different methods. Dust wipe sampling was performed after the intervention was completed. Lead dust testing by vacuum sampling was also completed before and after remediation and periodically after the child moved into the home. Due to the nature of vacuum sampling, it collected more lead dust than the traditional wipe sampling method. There were no generally accepted standards for lead safety by vacuum sampling.

Because the two methods reported results in the same units of measure ($\mu g/ft^2$) it is possible an Appeals Court without scientific capacity thought the sample collection method was irrelevant. But as we have seen, it was well known within the scientific community that the sampling method did indeed matter a great deal, because each had differing abilities to predict blood lead level, and each had differing dust collection efficiencies.

Existing clearance dust levels were based on wipe sampling and were applicable only to abatement projects, not the three R&M study groups. Although all the homes in the three R&M groups had both wipe sampling and vacuum testing, the *Grimes* decision erroneously asserted that the vacuum samples demonstrated a dust lead hazard.

But because there were no dust lead clearance or hazard levels for vacuum sampling, the Appeals Court demonstrated a basic ignorance of measurement science. In other words, the lack of lead dust standards for vacuum samples meant that a determination of whether a hazard was present or absent could not be made using that method.

Wipe samples were the only way to determine whether a dust lead hazard existed, and the data in the study showed that the remediation had successfully removed dust lead hazards. This confusion over the two methods would be manifested in the subsequent litigation. The decision to use a vacuum method was requested by EPA, which historically wanted dust lead levels to be reported in concentration (ppm), not loading ($\mu g/ft^2$) so that it could use its Integrated Exposure Uptake Biokinetic model (Chapter 4). EPA developed the vacuum method used in the R&M study.[61]

In any case, the study exceeded the existing standard of care, because it conducted wipe sampling even though the existing local law at the time did not require it for lead reduction work, only for lead abatement. Such clearance testing was the duty of the city in the normal course of its case management duties for children with elevated blood lead levels, not that of the researchers.

The Appeals Court did not consider any of this evidence, because it believed it was not necessary. In the conclusion to the Grimes decision, the Appeals Court admitted it had not considered the facts in the case:

> We hold that in every issue bearing on liability or damages remains *open for further factual development* (emphasis added).

Lawyers are taught that there are principles of law that are somehow independent of all contexts and that rulings can be made without that context, or in this case, "without facts in evidence." We shall return this problem at the end of this chapter.

6.5.10 Federal oversight

The Appeals Court also held that "The research did not comply with the regulations," and indicated that it alone had the authority to determine regulatory compliance, even though it also seemed to think there was a "clean slate," suggesting a basic ignorance of the Common Rule.

In fact, the Office for Human Research Protections (OHRP) at HHS, had performed its own investigation of the R&M study and did not find any violations of the regulations. Instead, they found there were three "deficiencies."

First, OHRP determined that the IRB should not have done an expedited review of a study protocol amendment involving blood testing. Second, the informed consent form should have provided more details on the various types of hazard controls that would be used in the study (the letter notes that this information was provided orally, but that there should have been a "script" used by researchers to inform the study participants; it was later determined that there was in fact such a script). And third, the IRB should have had a pediatrician on its committee. None of these deficiencies alleged that the study had placed children at any increased risk of lead poisoning or that the study should not have been conducted. Rather, the "deficiencies" were administrative or technical in nature, not ethical violations.

The letter detailing these deficiencies noted that all were addressed by KKI and Hopkins. It said, "JHUSOM [Johns Hopkins University School of Medicine] has developed and implemented a markedly enhanced system for protection of human subjects,"[69] and as a result, "there should be no need of further involvement of OHRP."[70]

6.5.11 The Appeals Court comparison to Nazi and Tuskegee research

Not all the Maryland Appeals Court justices agreed with its decision, particularly on the incendiary charge that the R&M study was similar to Tuskegee or Nazi research. One of the judges issued a "concurring in result only opinion" criticizing the other Court of Appeals judges:

> I cannot join in the majority's sweeping factual determinations that the risks associated with exposing children to lead-based paint were foreseeable and well known to appellees and that appellees contemplated lead contamination in participants' blood, *that the children's health was put at risk.… I cannot join the majority in holding that, in Maryland, a parent or guardian cannot consent to the participation of a minor child in a nontherapeutic research study* in which there is any risk of injury or damage to the health of the child without prior judicial approval and oversight. Nor can I join in the majority's holding that the research conducted in these cases was per se inappropriate, unethical, and illegal.

> *Such sweeping holdings are far beyond the question presented in these appeals,* and their resolution by the Court, at this time, is inappropriate. I also do not join in what I perceive as the majority's wholesale adoption of the Nuremberg Code into Maryland state tort law.

> Finally, I do not join in the majority's comparisons between the research at issue in this case and extreme historical abuses, such as those of the Nazis or the Tuskegee Syphilis Study.[3] (emphasis added)

The other Appeals Court judges wrote that the R&M researchers:

> intentionally exposed [study participants] to infectious or poisonous substances in the name of scientific research [similar to] the Tuskegee Syphilis Study, aforesaid, where patients infected with syphilis were not subsequently informed of the availability of penicillin for treatment of the illness.

The Appeals Court decision goes on at length to describe other such unethical studies, including Nazi research, and that the children in the R&M study were somehow similar to prisoners in concentration camps.

The Appeals Court never defined exactly what the "penicillin" treatment (the treatment that had been withheld from the Tuskegee men) would be in the lead paint context, or how the parents of the children were forced to participate (as the Nazi prisoners had been).

Throughout the decision, the Court seemed to think that full removal of lead paint was somehow the only effective treatment, even though the earlier research had shown that in fact such full removal did much more harm than good, and even though the Grimes child had lived in a fully abated home.

This comparison to the deplorable Nazi and Tuskegee research, although successful in capturing headlines, did real damage. This is the true legacy of the Grimes decision, examined at the end of this chapter.

6.6 Ethics in housing intervention research

In the wake of the confusion, HUD, CDC, and EPA decided that the National Academy of Sciences and its Institute of Medicine should examine the issues involved. Together with Susan Cummins, who at the time directed the Institute of Medicine's Board on Children, Youth, and Families, a uniquely qualified committee was assembled. Its members were composed of leading health, housing, environmental justice, ethics, and law professionals to produce a major report on the subject that remains the authoritative work in the housing intervention research field today.

In the charge to the committee, the problem was stated this way:

> On one hand, research is needed to better understand housing health hazards and to learn how best to ameliorate them. Children who are most at risk for these hazards may benefit the most from such research. On the other hand, such research presents ethical problems because children who are most at risk may also be vulnerable in multiple ways. The challenge is to carry out research that ultimately will lead to improvements in the health of children, while assuring that vulnerable children participating in research do not face inappropriate risks and that their parents are truly informed about the research.[4]

The Academy chose not to examine the specific legal issues presented in the *Grimes* case. Instead, it decided that it could not make a judgment because the case was resolved after the courts ruled on a motion for summary judgment and because no testimony was introduced in [the Appeals] court on crucial contested factual issues. But the Academy suggested two central ways to help prevent future confusion in both the legal and scientific communities.

First, it said there should be more systematic attempts to discuss a planned study within the community to better grasp that community's concerns and needs and to respond to them. This would likely produce more scientifically robust study protocols and ethics and it reinforced future efforts to develop community-based participatory research. The committee noted that the organization chart for the R&M study included community groups as part of the overall structure of the study, which was in keeping with KKI and Hopkins long-standing traditions in lead poisoning prevention research.

The second theme in the Academy's report was to improve the informed consent process in ways that enabled parents to better understand what their participation meant: Specifically, it said that it must be made clear that no intervention can ever be capable of eliminating all risks. Neither vaccines nor housing improvements can ever be 100% effective.

Unlike the Appeals Court, the committee put the need for housing intervention research squarely in the context of the reality of substandard housing that afflicts many communities at risk, especially for children in low-income families. Because they are at higher risk, the committee thought they were also likely to gain the most from such research.

Their report contained a series of recommendations, some of which were implemented in ensuring years. All their recommendations reflected a "systems approach" which they defined as not only researchers who design and implement studies, but also research institutions and IRBs, who approve and supervise research, the federal government, and other sponsors of research. The committee chose *not* to include the courts in its description of a "systems approach."

The committee also noted the limitations of randomized controlled trials in housing studies, the chief one being that community members who are in the "comparison" or "control" group may not receive a direct benefit from participating. Some housing studies have included ways of providing some benefits to such comparison groups so that all benefit. Failure to have such comparison or control groups is sometimes thought to result in a "lower" quality study.

For example, in 2019 a US Preventive Services Task Force report stated that, "There is insufficient evidence to assess the utility of screening for elevated blood lead levels in asymptomatic children," which is eerily similar to the initial Surgeon General's guidance in the 1970s. That early guidance required symptoms to be observed, even though most cases of lead poisoning were known (even then) to be mostly asymptomatic. Others thought the 2019 Preventive Services Task Force was wrong, and that there is in fact evidence showing the importance of blood lead testing for asymptomatic children.[71,72]

Finally, the Academy issued other recommendations about research on children, housing, communities, economically and educationally disadvantaged populations, other health hazards in children's homes, and IRB expertise. Specifically, the recommendations included more explicit adoption of Subpart D in the Common Rule covering children, who cannot truly undergo informed consent (but who can, at a certain age, provide assent) and better definition of terms used in that regulation. It suggested some ways in which informed consent might be done in ways that are more understandable to the participants.

It also recommended that communities be involved not only in data collection but in the design and implementation of research so that the studies could be more responsive to community needs. It regarded successful community engagement to mean:

> an assessment of economic conditions, history, norms, demographic trends, political structure, how residents maintain their communities, how community strengths can best be used, what skills and assets are present, existing organizations and institutions, such as places of worship and most importantly, equal relationship building, which could include membership on IRBs and employment on the study team.

The Academy noted that costs for these activities are often not included by funding agencies and foundations. Shortly after this Academy report appeared, CDC published a major paper on community-based participatory research in 2007, noting that despite progress, standards for such research remained fragmented.[73]

The Academy report defined "research equipoise" in the housing context to assess the genuine uncertainty that should exist about whether the intervention or control arm is better (equipoise means a balance of interests). This uncertainty met the equipoise requirement, because it was not known whether one of the three R&M study groups would be better and because it was not known if the previously abated homes would be better than any of the three study groups.

Because homes are private places, the committee recommended that researchers develop plans on what to do if they observe risks outside the study during the home visits, for example, child abuse or illegal drug use, including possibly notifying authorities when required by law or good practice. It also thought that compensation of participants in housing studies should be balanced, meaning that it should be adequate, but not so large that it could be coercive. The R&M study provided $15 ($28 in 2020 US dollars) to compensate the parents each time they answered a short questionnaire. Had it been larger, some might have construed this to be coercive; had it been smaller, it might be thought to insufficiently value the participants' time.

The Academy also recommended that studies should be designed in a way to provide benefits to both study and control groups. This is a higher standard than in drug trials, in which members of the control group (who receive the placebo) often do not get a benefit. Finally, the committee recommended that IRBs ensure that they have the necessary expertise to review research studies in housing pertaining to children, specifically pediatricians and community members.

Although the committee did not explicitly respond to the Maryland Appeals Court decision, it did reject "the suggestion [by the Appeals Court] that parents not be allowed to enroll children in research that does not offer the prospect of direct benefit ('nontherapeutic' research)." It noted that almost all (95%) of the private low-income housing in Baltimore at the time had lead paint in the home.

The committee's report stated:

> Johns Hopkins University, together with the Association of American Medical Colleges and the Association of American Universities, asked the Appeals Court to reconsider, since the ruling would have prohibited research otherwise allowable under federal regulations (Amici Curiae Association, 2000). The court denied the motion for reconsideration, but it issued a clarification of the opinion, stating that by "any risk" it meant, "any articulable risk beyond the minimal kind of risk that is inherent in any endeavor."

The committee also cited the National Commission for the Protection of Human Subjects of Biomedical and Behavioral Research, which was authorized by Congress in 1974, which it said had:

> recommended a *balancing* of risks and benefits and rejected the notion that non-therapeutic research with children is ethically unacceptable. (emphasis added).

Although the Appeals Court seemed to think it was writing on a "clean slate," this was yet another example that it was not.

In a drug trial, there is typically a researcher and the participant (the human subject). But in housing intervention research, it is far more complicated. In addition to the researcher and the subject, other parties may include the owner, renters, neighbors, a property management firm, building code inspection departments, other occupants, housing finance agencies, health departments (some of which have legal duties to investigate and order remediation of hazardous conditions), construction firms, architects, engineers, neighborhood associations, and others, many with specific legal duties.

The committee examined this landscape and suggested that the very distinction between therapeutic and nontherapeutic was simply not valid in the housing context. It cited another report from the National Commission[74] for the Protection of Human Subjects of Biomedical and Behavioral Research that was completed after the Belmont Report, in which the distinction between therapeutic and nontherapeutic research was rejected in favor of balancing risks and benefits for three reasons:

First, the report noted that the very term "therapeutic research" was logically inconsistent because research "is intended to develop general knowledge," while therapy is "for the benefit of an individual and therefore does not involve any generalizable component" (National Commission, 1977, p. 54). In other words, because therapy is by definition for the benefit of an individual, it cannot be research, because research is not individually based but intended to be generalizable.

Second, the report expressed concern that "simply because a benefit (therapy) is included in a "therapeutic" research protocol, all sorts of additional interventions not germane to the therapeutic intervention but useful for general knowledge can be regarded as justified." These additional interventions might be approved with a lower threshold of allowable risk.

Third, the National Commission in 1977 rejected earlier suggestions that all "nontherapeutic" research for children be rejected, because it neglects discussion of the low level of risk involved in most research involving children. The Commision's finding that children should not be prevented from participating was at odds with the Appeals Court initial decision that it was forced later to retract.

The committee stated:

> The guiding principle that appears to have been ignored by the Court of Appeals was that earlier work had determined certain conditions where it was ethically permissible to enroll children in research that offered them no prospect of direct benefit, provided that the risk was no more than a minor increase over minimal risk and other conditions applied.

This seemed to match the R&M study procedures and IRB approval, which represented a balancing of risks, not some underhanded attempt to misrepresent the study to parents or secretly entice them into participating.

The committee noted that operationalizing this principle was complex. It can be hampered sometimes by "therapeutic misconception," in which participants in both therapeutic (e.g., drug trials) and nontherapeutic (e.g., housing studies) may choose to be part of the study in the belief that the drug or housing intervention will improve their health, even though that is typically not fully known. This is the reason such studies are undertaken in the first place and why the informed consent process is so critical in explaining what will and will not occur. This is yet another example of how the dichotomy between therapeutic and nontherapeutic research is not on point when it comes to housing studies.

If EPA or HUD had formally adopted subpart D of the Common Rule, might the study design have somehow been changed? This seemed unlikely because Hopkins and KKI already complied with Subpart D, which covers research involving children in three categories:

1. Research that does not pose more than minimal risk (45 CFR 46.404);
2. Research that involves greater than minimal risk, but for which research participation provides a prospect of individual benefit sufficient to offset that risk (45 CFR 46.405);
3. Research that involves greater than minimal risk without individual benefit sufficient to offset the risk—but for which the risk is only a minor increase over minimal risk, the excess risk is commensurate with ordinary experiences of the children under study, and the knowledge to be gained from the research is "...of vital importance for the understanding or amelioration of the subjects' disorder or condition..." (45 CFR 46.406).

Which category might have been most relevant to the R&M study? The IRB at Hopkins seemed to think that the first category was appropriate because the only risk was discomfort from blood lead testing. If the study might have involved greater than minimal risk, participation may have provided the prospect of individual benefit, because all homes in the study (except the modern homes without lead paint) all received home improvements that reduced exposures. The third category did not seem to be relevant, because the children *did* appear to benefit as shown by the reduced blood lead and dust lead levels.

There is a fourth category in Subpart D (research that presents a greater than minor increase over minimal risk with no prospect of direct benefit), but this would have required direct review by a national panel and approval by the Secretary of DHHS (45 CFR 46.407). No one believed that the R&M study was in this fourth category.

The Academy committee wrestled with the value of housing research. Can research help to improve people's living conditions? If a drug trial is successful, then that finding can be translated into benefits for all of society through the availability of the new drug. But housing markets generally did not have a good mechanism to reap the benefits proven through healthy housing research, because the health benefits are not necessarily translated into better housing prices (Chapter 9). Therefore people who chose to participate in housing studies may be disappointed that their participation did not directly result in benefits, especially in low-income housing. In such communities, housing prices were often determined by factors other than health (e.g., vacancy rates, jobs, proximity to schools and transport, and others).

Yet without the evidence to show that housing interventions did improve health, it seemed unlikely that this market failure could ever be corrected. Promoting evidence-based housing codes may be one tool to help end the short-sighted cost-shifting between the housing and health economic sectors (Chapter 9).

Another criticism included in the *Grimes* case was that the investigators did not report testing results in a timely manner and did not properly interpret their meaning. The Academy's report stated that:

> The threshold criteria in determining whether to share test results are typically whether the results are valid.... There is no clinical benefit to reporting the results to individual parents if they cannot be meaningfully interpreted. In biomedical research, when the validity of experimental tests on biological specimens is not established, individual results generally are not reported to participants.

In the R&M study, dust lead testing was done by vacuum sampling, which could not be meaningfully interpreted by either the participants or even the researchers themselves, as there were no exposure limits for such sampling. The R&M researchers did attempt to provide a way to interpret the vacuum results by comparing the results to those found in previously fully abated homes where dust had been sampled using the same vacuum method in a pilot study. But this was only a few housing units, and no standards were ever developed to indicate if certain dust lead levels obtained by vacuum sampling really constituted a hazard.

Should the vacuum sampling results have been reported at all? If even the Appeals Court could not grasp the distinction between wipe and vacuum sampling, how could parents be expected to do so? There were two options: report the results and with limited knowledge attempt some interpretation, or not report them at all. Each had their own dilemma: Reporting results with poor ability to interpret them could lead to improper conclusions on the part of parents. Failure to disclose could lead to a perception that results were being concealed.

Today, all results are typically reported in housing studies, even if researchers are not able to fully understand what they mean. Such is the uncertainty inherent in research. In this way, perhaps participants were able to gain insight into the value of research in improving the ability to interpret findings.

In some of the R&M legal trials, there were allegations that the research study was somehow deficient because "nonstandard" methods were used, meaning vacuum sampling. Experts spent much time on the witness stand explaining what was and what was not known to the court, but this uncertainty initially fed courtroom theatrics by lawyers attempting to argue that uncertainty inherent to research had posed risks. Of course, scientific uncertainty is different from risk.

The Academy concluded its report:

> It is unrealistic and unfair to hold individual research investigators responsible for ameliorating the social circumstances that they study. Individuals can only be held responsible for actions, decisions, and conditions that are under their control. Researchers are not responsible for the housing of children in research, except in some intervention studies that assign subjects to housing conditions.

In the R&M study, the houses were randomized such that the researchers did not "assign subjects to housing conditions." All R&M study houses were required to be in structurally sound condition to be eligible and the Maryland Department of Housing and Community Development that issued loans to qualifying rental property owners also required that the work be done in properties occupied by low-income tenants, that is, people at high risk of childhood lead poisoning.

Juries agreed, finding the researchers were not negligent with regard to their interactions with study participants. The Academy also concluded that parents (not the courts) are generally in the best position to judge what is best for their own children.

6.7 The best of intentions or the best of community-based science?

The debate was not confined to the courtroom or to the Academy.

In the wake of the *Grimes* decision and the Academy's report, there were two camps of public opinion: The first thought the Appeals Court's language was incendiary and had reached conclusions not borne out by the facts. The second camp thought the researchers had committed grave errors by acting unethically and should have somehow designed the study differently, even if they had the "best intentions."

The second camp included a book *Lead Wars*,[75] which was summarized in a shorter article.[76] Its authors were two historians, Gerald Markowitz and David Rosner, who in an earlier book had documented the role of the lead paint industry in knowingly marketing its dangerous products. That first book mostly focused on the time before 1970,[77] and the second attempted to tell the *Grimes* lawsuit story. Markowitz and Rosner characterized the R&M interventions as dangerous "sellouts" because they did not use full abatement or paint removal.

A few years earlier in 1998, Herbert Needleman (who played a large role in devising the 1991 CDC Strategic Plan and proving the more subtle health effects of lower-level lead exposures) later said the 1991 CDC documents were a "high water mark." Wrongly, he came to regard everything to follow as a retreat, and that only complete paint removal should be allowed.[78]

Needleman resigned from the Alliance board over the question of whether the nation should pursue only "full abatement" like the house in which the Grimes child lived, or instead should pursue the Title X primary prevention approach that combined abatement and interim controls. These were the same misconceptions in *Lead Wars*.

The Alliance published a response to Needleman's 1998 assertions. It said that the 1991 CDC document did not in fact call for full abatement or paint removal. Instead, the text from the CDC document called for an "abatement program that will maximize the number of children benefited, given the fixed resources for abatement, using safe and effective methods." Indeed, CDC had called for "development of cheaper abatement methods, research on preventive maintenance and alternative abatement methods" and integration of safe lead work procedures into the building trades.

The Alliance also said Needleman's paper used old National Health and Nutrition Examination Survey blood lead data from the late 1980s instead of the more current data from the mid-1990s and that Needleman used old lead abatement cost-benefit data from the 1970s, instead of the more current estimates from HUD and CDC. It also said that Needleman was wrong about how property owners had stacked the Title X Task Force.[79]

In 1996 Needleman met with then-US Attorney Janet Reno to discuss how the Department of Justice might respond to his important findings on the association of early childhood lead exposure and criminal behavior in later life. HUD had arranged the meeting, thinking it would help bring the criminal justice community into the lead poisoning prevention effort. But she left the meeting confused, because she had just launched some of the most far-reaching lead enforcement cases in housing ever undertaken (Chapter 5). How could that be a "retreat" she wondered?[80]

Needleman's inability to understand the newer approach was likely influenced by the lead industry. Needleman had been the target of vicious industry attacks charging him with scientific misconduct, which have been described elsewhere.[81] He was eventually vindicated by the Department of Health and Human Services and the University of Pittsburgh, which wrote:

> The Office of Research Integrity found no scientific misconduct in Needleman's research, concurring with the University of Pittsburgh Hearing Board. . . . The scientific debate over lead's toxicity is over, and it is pointless and distracting to delay further action.[82]

But the attacks by the lead industry on Needleman did distract attention from how to involve the housing world. In a personal note, Don Ryan (the Executive Director of the Alliance) wrote to Needleman:

> I was struck by what a terrible waste it is that you are devoting so much of your time and energy to attacking those who are committed to the same goals as you. . . . The question now is whether you will focus your remaining advocacy energies on assigning blame (as you so confidently see it) for how the battle was lost or to joining in the battle that is still to be won.[83]

Markowitz and Rosner made the same mistake in thinking that the R&M study was somehow a retreat, instead of embodying the Title X primary prevention mandate. Markowitz and Rosner presented the question this way:

> Should we demand a massive program to eliminate lead paint on older structures, the major source of lead in the children's environment, or should we reduce exposures through various abatement procedures, knowing that low level lead exposures still threaten poor children living in decaying, older homes?... Although some researchers developed protocols aimed at eliminating lead as a widespread urban pollutant through its complete removal, others sought more pragmatic solutions.

But Markowitz and Rosner failed to cite *any* examples of researchers who had developed such complete removal protocols. In fact, such protocols did not exist anywhere in the nation at the time of the R&M study, because it had been proven that such complete removal had the effect of *increasing*, not decreasing exposures (Chapter 2). Moreover, HUD had estimated that such complete removal would cost at least $500 billion (Chapter 4), and no one was willing to pay for it. Markowitz and Rosner thought that somehow the move away from complete removal occurred because conservatives came to power in the Reagan Administration during the 1980s. As recounted previously, the move to block grants, the corruption at HUD, and the *Ashton v Pierce* case were all hallmarks of the Reagan years.

But the move away from lead paint removal was not one of them because the scientific evidence showing its danger was so overwhelming. The 1980s were marked by research on pathways that led to a more scientifically based redefinition of lead paint hazards. This was in fact a move away from the *Ashton v Pierce* definition that had led to the ill-fated attempts at complete removal. Markowitz and Rosner's "massive program to eliminate lead paint on older structures" had only led to more poisoned children and policy paralysis from 1970 to 1990. Perhaps not surprisingly, the enactment of Title X was strangely absent in their historical account.

Markowitz and Rosner erroneously presented the Hopkins legal defense argument this way:

> Because the study was merely observational, [KKI's lawyers] argued, they did not need to inform or warn parents or the children themselves about the potential danger that lurked in the dust that was in their homes.

Yet the facts presented earlier show the informed consent *did* in fact warn parents and the study enabled them to protect their children. The ongoing provision of cleaning supplies and education about the importance of dust control seemed to be at odds with the allegation that KKI somehow failed in its duty to inform parents about the dangers of lead dust.

Markowitz and Rosner's belief that the R&M study was "observational," was not supported by any of the R&M study documents, which never described the study that way. And no one thought that parents should not be informed. This was apparently an attempt to draw a distinction between the duties of the building owners, which were to comply with local lead poisoning laws, and the duties of the researchers, which were to carry out the study.

The consequence of this legal argument seemed to have led the first Hopkins and KKI lawyers in *Grimes* to suggest that the researchers had no duties to the participants in the study at all. Technically, it may have been true that it was the owner's duty to comply with local lead laws, not the researchers. In hindsight, this may have triggered the incendiary language from the Appeals Court. After all, how could it be that researchers had "no" responsibility at all?

Led by Barry Goldstein, a new team of lawyers for Hopkins and KKI abandoned this legal theory, opting instead to focus on the facts in the faith that juries would do the same. It was clear that researchers *did* have responsibility, but it was to carry out the study, not become building owners. Perhaps a more nuanced legal argument at the beginning, elucidating the duties of researchers, building owners, local health and housing departments, and others could have helped the Appeals Court grasp the landscape.

But this was a more complicated argument, one that escaped both the Appeals Court and Markowitz and Rosner. After examining the evidence the juries *did* understand what the researchers were (and were not) responsible for, unanimously finding them to be not negligent to the participants in the study. For a historical account, it was strange that *Lead Wars* omitted how the juries ultimately ruled in these cases.

Markowitz and Rosner, like the plaintiff lawyers who appealed the lower court's ruling, stated "KKI had a special obligation to make it absolutely clear that the families were not benefiting from the research and that their homes were still dangerous environments for young children ... sacrificing the health of the study children" and was "utilizing them as guinea pigs to determine cost-effective environmental treatment of lead-based paint."

As documented earlier, this statement was simply untrue, because all lead paint hazards were addressed and because the dust and blood lead levels declined. How could the homes still be "dangerous" if they had no remaining lead paint hazards? And how were children being "sacrificed" if their blood lead levels were actually improving? This seemed to appeal to the simplistic and dangerous (and now disproven position) that only complete paint removal could work and anything else somehow harmed children.

The lawyers for Grimes and Higgins (the plaintiffs) also alleged that "Kennedy [Krieger Institute] was well aware of the fact that the home contained unacceptable levels of lead in house dust ... [and] still contained high levels of lead in dust above the clearance criteria in Maryland for abated homes." This was an apparent reference to the vacuum dust lead results, but it was plainly false, because those results could not be compared to clearance criteria, which required the use of wipe sampling. Markowitz and Rosner, like the Appeals Court, failed to grasp the exposure science involved in vacuum versus wipe sampling methods. In short, *Lead Wars*, like the Appeals Court, failed to grasp the science.

Much was made of the costs of controlling lead hazards, suggesting that the real purpose of the R&M study was:

> ...not trying to find the safest way to prevent children from getting lead poisoning. Kennedy [Krieger Institute] already knows that the safest way... is to fully and completely abate homes or put children in homes that were built after 1980 that don't have any lead paint in them.[84]

This was at best disingenuous, because the Grimes child was in fact in a fully abated home, not one of the three groups of R&M study homes. Scientists who were in the lead poisoning prevention field had not in fact concluded that full abatement (paint removal) was always the best or only way to protect children (Chapter 5), because full paint removal had so badly backfired.

Lead Wars suggested that the real purpose of the study "was to see if there was anything cheaper that could be done [because] landlords say it is too expensive to fully abate these homes." Perhaps the idea that KKI was only interested in saving landlords money could have been better countered by not presenting maximum costs for each of the three categories of study homes, and simply to present them as differing intensities of paint disturbance and dust generation, which is what they really were. In any case, the Appeals Court, Markowitz and Rosner, and many others failed to grasp the critical issue of how much disturbance of lead paint should be allowed.

An account in the *New York Times* cited the view of Gary Goldstein, the CEO of KKI during this time period:

> [Goldstein said] "If you come into East Baltimore, it's not like one child in 100 is exposed. It's everyone, and we're trying to fix it." Goldstein denounced the court for suggesting that KKI and its doctors would purposely put children in harm's way. Rather, the intentions were noble and benevolent, he said. Baltimore's poor were stuck in neighborhoods "where 95 percent of the houses contain lead" and where "35 percent of the kids have lead poisoning." The research was aimed at finding a way to enable housing residents to move into safe homes and not get poisoned. In fact, that's exactly what happened, because for the majority of kids in the study, [blood] lead levels did go down.[85]

One parent accused the researchers of using children as "guinea pigs," but others defended the study as Goldstein had.[86]

A Washington Post editorial attacked the Appeals Court ruling:

> The [R&M] study [did not come] within a country mile of serious ethical lapses, let alone the horrendous abuses at Tuskegee or [the Nazi Concentration Camp] Buchenwald. Far more offensive is the social indifference that has let generations of children suffer the insidious menace of lead poisoning. If a judge's irresponsible comments impair helpful research in this area or mire it in opportunistic lawsuits, that will be the real moral outrage.[87]

Others worried about the legal implications of the court's decision, its impact on future research, the prospects for evidence-based policy, and widespread change. For example, Michael Weissman, the director of the American Academy of Pediatrics, Center for Child Health Research, said that the court's decision could make it more difficult to conduct similar types of research in the future:

> You never, ever want to hurt a child or put a child in harm's way. But for very large numbers of America's children, the only way that we can protect them from lead, at this particular point in history, is to do this type of research.[88]

6.8 Environmental justice and community participation in research

Although many like Markowitz and Rosner fell prey to the incendiary language of the Appeals Court's language comparing the R&M study to Tuskegee and Nazi research, many others refused to go along, citing environmental justice principles.

For example, Don Ryan wrote an op-ed in the Baltimore Sun under the headline: "Research on lead hazards is solution, not problem." The article said:

> The outrage expressed in news reports is rooted in the false premise that this study placed children in harm's way. The reality is that this research made homes safer, not only for the children in Baltimore but for hundreds of thousands of others across the nation. Children do not live in lead-burdened houses because researchers want to "experiment" on them but because so much of our housing is contaminated by lead.... In contrast to the picture portrayed in the news media and by the court, their study was purposefully designed to provide lead safety treatments to families' homes that far exceeded state and local requirements and made these homes safer than they had been and safer than typical neighboring properties.[89]

Janet Phoenix, a physician active in the Environmental Justice movement (and the mother of a lead poisoned child herself) wrote:

> It is unfortunate that the study participants were almost all from a single demographic grouping (African American). This was not by design, but instead occurred as a result of the fact that a limited number of property owners collaborated with this study. Nevertheless, when something like this occurs there is always suspicion that whites would have been treated differently. There is no evidence that I can see, however, that participants were uninformed or that the study subjected them to any harm. Quite the contrary, study participants seemed better off than other children living in Baltimore in the same neighborhoods.[90]

Environmental justice in the housing context has often been simplistically conflated with fair housing and housing segregation issues, as important as those are. Equally important, however, is the quality of housing and lead paint hazards, which are more concentrated in low-income communities of color. The research ethic in the KKI study, with its focus on involving the community from the beginning and identifying solutions to housing disparities (in this case the higher prevalence of lead paint hazards) appeared to be consistent with environmental justice principles. The study came to be seen as an exemplary example of housing-based environmental justice.

HUD, which provided additional funding for the R&M study after the initial EPA funding, highlighted this by citing lead poisoning prevention as one of its greatest environmental justice accomplishments in 1995 and again in 2012:

> HUD ... has made homes safer and healthier, especially for children. These [lead paint] programs focus their efforts on addressing high-risk communities, including economically disadvantaged and racially concentrated areas of poverty. The neighborhoods impacted by these [lead paint] grant programs prioritize resources based on the age of the housing stock, household income, and prevalence of health conditions due to poor housing conditions. The programs have a strong track record and have contributed to a significant reduction in lead poisoning cases.[91]

The history of housing policy in the United States is replete with both intended and unintended consequences of segregation, disinvestment, and associated adverse health outcomes. Lead paint policy in the time period covered here appeared to be an exception. Disparities in blood lead levels in high-risk communities of color were dramatically reduced, even as housing segregation remained pronounced in 2022 (described later in this book).

Lead poisoning was a central issue in the environmental justice movement since its beginning, reflected in Charles Lee's United Church of Christ report regarding toxic waste site location near communities of color (Chapter 5).

Environmental justice has been defined as the "fair treatment of people of all races, income, and cultures with respect to the development, implementation and enforcement of environmental laws, regulations, and policies."[92] The Natural Resources Defense Council went further, calling for improvement, not only fair treatment: "Environmental justice is an important part of the struggle to improve and maintain a clean and healthful environment, especially for those who have traditionally lived, worked and played closest to the sources of pollution."[93]

The R&M research to identify ways to ameliorate lead paint hazards in housing was consistent with these principles. The Maryland Court of Appeals language comparing the R&M study to the notorious Tuskegee and Nazi experiences may have at first glance given an initial appearance of enforcing environmental justice. But the verdicts of the Baltimore juries, who were able to examine the facts in the case, clearly found otherwise. With its long record of community involvement and its realization of an improved living environment, the R&M study appeared far more consistent with environmental justice principles.

6.9 The legacy of the Maryland Court of Appeals *Grimes* decision

What was the result of the *Grimes* decision and the resulting commentary?

The Appeals Court's characterization of the R&M study as somehow being similar to Nazi and Tuskegee research, its apparent inability to understand ethical research designs or basic exposure science, its ignorance of researchers' duties and how they were enforced by existing regulations, its dismissal of factual context, and its initial (but later retracted) attempt to deny families the right to participate in beneficial research studies involving their children were all rebuffed once citizens serving on juries had a chance to examine the facts and have their say in the subsequent cases.

In all the ensuing years, there was no other similar court decision in any other jurisdiction.

It was hard to identify anything of consequence resulting from the *Grimes* case, with one important exception: It led to a wave of expensive lawsuits, including a class action suit, all confined to Baltimore.[64] Ads appeared on late night TV in Baltimore, saying "If you were in this study, we can get money for you." But none of that actually happened.

What *did* happen was the Appeals Court decision forced City Homes, one of the non-profit housing providers that participated in the study, and that had led much of Baltimore's quality low-income housing world, into bankruptcy as a direct result of the *Grimes* decision. No other nonprofit housing provider filled its shoes in later years, and decent, low-income housing in Baltimore remained elusive for far too many families. The Appeals Court decision appeared to have only made the problem worse.

There were personal consequences as well. Most of these suits included as defendants not only the KKI and Hopkins institutions, but also their IRBs, and individual researchers, such as Mark Farfel, Peter Lees, and others. In all cases involving the research participants, KKI and Hopkins researchers were found by the juries not to be negligent.

Mark Farfel, the Principal Investigator of the R&M study who had championed community involvement in the R&M and earlier studies, who founded Hopkins' community engagement efforts, and had made some of the key findings on how best to protect children, left the lead poisoning prevention research field in the ensuing wave of litigation. He once remarked that as a Jew, he found the court's language comparing his research to that of the Nazis particularly painful.[17] Indeed, Farfel was the son of a Baltimore pediatrician who had courageously admitted the first black child to Baltimore's Sinai Hospital, which was against hospital policy at the time.[94] There is much irony that Farfel, a fierce advocate of community involvement and civil rights, was driven out of the lead poisoning prevention field by what were ultimately proven to be unfounded false allegations. A gifted researcher, he went on a lead some of the nation's most important health studies.

As a direct result of the many years of both research and advocacy by Farfel, Chisolm, KKI, and Hopkins, Maryland's lead poisoning rates plummeted by more than 95% between 1985 and 2017, and lead paint hazard control methods became standardized and validated. Maryland's lead law required remediation that was drawn from the R&M study in virtually all pre-1950 rental units in Maryland. These methods became widely used across the country and greatly reduced children's exposure, due to the R&M study, the HUD evaluation study, and many others like it. Maryland now has one of the nation's most successful lead laws.[95]

Perhaps more importantly for the future, law students today are sometimes taught (according to a popular "Case Brief Study Buddy") that "Under Maryland law, informed consent agreements in nontherapeutic research projects can constitute contracts and can constitute special relationships giving rise to duties, which could result in negligence actions if breached." But nowhere in the brief are the actual results of the study disclosed, nor are the

References 283

"special duties" ever defined (because the court never did either). Perhaps most importantly, the brief does not reveal the actual outcomes of the cases that went to trial.

Instead, there is this:

> The Kennedy Krieger Institute (KKI) (Defendant), a prestigious research institute associated with Johns Hopkins University, created a non-therapeutic research program which required particular classes of homes to have only partial lead paint abatement modifications completed. In at least some instances, including this case, the institute arranged for the landlords to receive public funding by way of loans or grants to aid in the modifications. The research institute then encouraged, and sometimes required, that the landlords rent the properties to families with young children. Children were encouraged to live in the houses where the possibility of lead dust was known to the researchers to be likely, so that lead dust and the lead content of their blood could be compared with the level of lead dust in the houses at periodic intervals over a period of two years.[96]

The fact that the children's health improved, as shown by lower blood and dust lead levels was omitted from this inaccurate account. It was not a "nontherapeutic program." It did not require that landlords rent to families with young children. It did not "encourage" children to live in their homes.

The lesson of the *Grimes* Appeals Court decision is that the most enduring legal decisions are those that are based not on incendiary headlines but rooted in scientific contextual evidence made understandable by skilled lawyers to enable it to be grasped by juries who can rule on the evidence.

In its view of the R&M study and others like it, *Lead Wars* stated: "the only thing the public health community did during the 1990s was to conduct more studies on partial abatement." This was rejected in a later rebuttal, published in 2013:

> In fact, the 1990s were marked by a broad and deep social movement to secure resources and force action to protect children at highest risk, in addition to high quality research. Sadly, [the Markowitz and Rosner] historical account overlooks how researchers and public health practitioners achieved meaningful progress by joining with parents of poisoned children, leaders of community-based organizations, all levels of government, inspection and abatement businesses, and advocates for affordable housing, tenants' rights, children's health, environmental justice, and social justice.[97]

Despite the shortcomings of their book, Markowitz and Rosner's earlier historical account of the lead and paint industries played a much more important role in two later court cases, one in Rhode Island and another in California.

Citizen's sustained rejection of the Maryland Court of Appeal decision revealed a new trend. We now turn to two landmark legal cases in Rhode Island and California against the lead, paint and pigment industries, and to an emerging consensus on the nature of housing, environment, and health.

References

1. Order. Ericka Grimes v Kennedy Krieger Institute, Inc. et al. Baltimore City Circuit Court Case No. 24-C-99–000925. Maryland, July 26; 2000. Judge Stuart R. Berger.
2. West B, Varacallo M. *Good Samaritan Laws*. *StatPearls*. Treasure Island (FL): StatPearls Publishing; 2020. Available from: https://www.ncbi.nlm.nih.gov/books/NBK542176/.

Part 2. Breaking the Barriers to Progress (1986–2001)

3. Grimes v. Kennedy Krieger Institute, Inc., 366 Md. 29, 782 A.2d 807 (Md. 2001), P. 53 August 16; 2001. Court of Appeals of Maryland, No. 128, 129 September Term. Opinion by Cathell, J, Raker J concurs in result only.

4. Institute of Medicine. *Ethical Considerations for Research on Housing-Related Health Hazards Involving Children.* Washington, DC: The National Academies Press; 2005. Available from: https://doi.org/10.17226/11450.

5. Blackfan Kenneth D. Lead poisoning in children with special reference to lead as a cause of convulsions. *Am J Med Sci.* 1917;153(June 1):377−380.

6. Chisolm JJ, 79, Dies; Lead-Poison Crusader by Carmel McCoubrey June 26; 2001 (NY Times). <https://www.nytimes.com/2001/06/26/us/j-j-chisolm-79-dies-lead-poison-crusader.html/>.

7. Farfel MR, Chisolm JJ. Health and environmental outcomes of traditional and modified practices for abatement of residential lead paint. *Am J Public Health.* 1990;80:1240−1245.

8. US Department of Housing and Urban Development. Lead based paint poisoning prevention in certain residential structures. 24 CFR Part 35. 1999.

9. Lanphear BP, Matte TD, Rogers J, et al. The contribution of lead-contaminated house dust and residential soil to children's blood lead levels: a pooled analysis of 12 epidemiologic studies. *Environ Res.* 1998;79 (1):51−68.

10. US EPA. Subpart D - Lead-Based Paint Hazards. *Code of Federal Regulations.* 2001;40(CFR Part 745).

11. Farfel M. Evaluation of Health and Environmental Effects of Two Methods for Residential Lead Paint Removal. Thesis; 1987. <https://catalyst.library.jhu.edu/catalog/bib_91837>.

12. Johns Hopkins University; Convocation. John C. Hume Award to Mark Farfel. 1987. (A doctoral graduate for significance in the Department of Health Policy and Management). The University of Illinois Chicago Special Collections & University Archives, School of Public Health, "David E. Jacobs papers" the University of Illinois Chicago School of Public Health Library.

13. Hume J. Medical pioneer, dies VD expert led JHU public health school. By Fred Rasmussen, The Baltimore Sun, February 21; 1998. <https://www.baltimoresun.com/news/bs-xpm-1998-02-21-1998052105-story.html/>.

14. Du Burns CH. First black mayor of Baltimore. Baltimore Sun, February 21; 2007. <https://www.baltimoresun.com/features/bal-blackhistory-burns-story.html/>.

15. Education and Research Centers. Portfolio (ERC). National Institute for Occupational Safety and Health. 2022. NIOSH. https://www.cdc.gov/niosh/oep/ercportfolio.html/.

16. Memorandum to Andrew Sorenson from Mark Farfel, July 17, 1989. Memorandum to Scott Zeger, Associate Dean for Academic Affairs from Mark Farfel, Student Interest in the InterAction Program, September 10; 1991. The University of Illinois Chicago Special Collections & University Archives, School of Public Health, "David E. Jacobs papers" the University of Illinois Chicago School of Public Health Library.

17. Personal Communication Mark Farfel and David Jacobs; 2002.

18. Chipping away at lead's legacy, by Mary Mashburn. Johns Hopkins Public Health. Fall; 2000:26−35.

19. Kennedy Krieger Institute. *Our Mission and Vision.* 2022. Available from: Our Mission and Vision. 2022.

20. Charney E, Kessler B, Farfel M, Jackson D. Childhood lead poisoning: a controlled trial of the effect of dust-control measures on blood lead levels. *N Engl J Med.* 1983;309:1089−1093.

21. City Homes Celebrating 25 years. 2012. Powerpoint at < http://cityhomesbalto.org/about.html >.

22. Pollack J. The Baltimore lead-based paint abatement and repair and maintenance study in Baltimore: historic framework and study design. *J Health Care Law Policy.* 2002;6(90−110).

23. Baltimore City Department of Housing & Community Development. Annual Expenditures for Affordable Rental Housing in Baltimore City; 2018. <https://dhcd.baltimorecity.gov/nd/affordable-housing-inventory/>.

24. Civil Jury Instruction 202, cited in The People of The State of California, Plaintiff, Vs. Atlantic Richfield Company, Conagra Grocery Products Company, E.I. Du Pont De Nemours And Company, Nl Industries, Inc., And The Sherwin-Williams Company, Defendants. Case No.: 1-00-Cv-788657, Statement of Decision, January 7; 2014.

25. The Famous Benjamin Franklin Letter on Lead Poisoning. Phila July 31, 1786 (To Benjamin Vaughan). <http://environmentaleducation.com/wp-content/uploads/userfiles/Ben%20Franklin%20Letter%20on%20EEA(1).pdf/>.

26. Rabinowitz M, Leviton A, Bellinger D. Home refinishing: lead paint and infant blood lead levels. *Am J Public Health.* 1985;75:403−404.

27. Shannon MW, Graef JW. Lead intoxication in infancy. *Pediatrics.* 1992;89(1):87−90.

28. Amitai Y, Graef JW, Brown MJ, et al. Hazards of deleading homes of children with lead poisoning. *Am J Dis Child.* 1987;141:758−760.

29. Swindell SL, Charney E, Brown MJ, Delaney J. Home abatement and blood lead changes in children with class III lead poisoning. *Clin Pediatr.* 1994;33:536–541.

30. Chisolm Jr. JJ, O'Hara DM, eds. *Lead Absorption in Children: Management, Clinical, and Environmental Aspects.* Baltimore/Munich: Urban & Schwarzenberg; 1982:229. Referenced in Farfel M., Reducing lead exposure in children. Annu Rev Public Health. 1985;6:333-60.

31. Ashton v Pierce. 541 F. Supp. 635 (D.D.C. 1982) US District Court for the District of Columbia - 541 F. Supp. 635 (D.D.C. 1982) June 22; 1982.

32. 119 Cong. Rec. 14882; 1973.

33. Ryan D. Available from: https://vimeo.com/user70809293/download/232587289/cda0386d3d/.

34. US Department of Housing and Urban Development. Guidelines for the Evaluation and Control of Lead-Based Paint Hazards in Housing. 1995.

35. Review of Studies Addressing Lead Abatement Effectiveness: Updated Edition EPA 747-B-98-001, December, Prepared by Battelle for Technical Programs Branch, Chemical Management Division, Office of Pollution Prevention and Toxics, U.S. Environmental Protection Agency, Washington, DC, 1998.

36. National Center for Healthy Housing and University of Cincinnati. Evaluation of the HUD Lead Hazard Control Grant Program; 2004. <https://www.hud.gov/sites/dfiles/HH/documents/Evaluation-of-the-HUD-Lead-Based-Paint-Hazard-Control-Grant-Program_Final-Report.pdf/>.

37. Rabinowitz MB. Toxicokinetics of bone lead. *Env Health Perspect.* 1991;91:33–37.

38. Amitai Y, Brown MJ, Graef JW, Cosgrove E. Residential deleading: effects on the blood lead levels of lead-poisoned children. *Pediatrics.* 1991;88(5):893–897.

39. Copley CG. The Effect of Lead Hazard Source Abatement and Clinic Appointment Compliance on the Mean Decrease of Blood Lead and Zinc Protoporphyrin Levels. Mimeo., City of St. Louis, Department of Health and Hospitals, Division of Health, Office of the Health Commissioner, St. Louis, MO; 1983. The University of Illinois Chicago Special Collections & University Archives, School of Public Health, "David E. Jacobs papers" the University of Illinois Chicago School of Public Health Library.

40. Staes C, Matte T, Copley G, Flanders D, Binder S. Retrospective study of the impact of lead-based paint hazard remediation on children's blood lead levels in St. Louis, Missouri. *Am J Epidemiol.* 1994;139(10):1016–1026.

41. Rosen J, Markowitz M, Bijur P, et al. Sequential measurements of bone lead content by L-X-ray fluorescence in CaNa2EDTA-treated lead-toxic children. *Environ Health Perspect.* 1991;91:57–62.

42. Markowitz ME, Bijur PE, Ruff HA, Rosen JF. Effects of calcium disodiumnversenate (CaNa2EDTA) chelation in moderate childhood lead poisoning. *Pediatrics.* 1993;92(2):265–271.

43. Ruff HA, Bijur PE, Markowitz ME, Ma Y, Rosen JF. Declining blood-lead levels and cognitive changes in moderately lead-poisoned children. *J Am Med Assoc.* 1993;269(13):1641–1646.

44. Rice TW. The historical, ethical, and legal background of human-subjects research. *Respir Care.* 2008;53(10):1325–1329.

45. Pure Food and. Drug Act of 1906 (PL. 59-384) 1906.

46. Sinclair published the book in serial form between February 25, 1905, and November 4, 1905, in Appeal to Reason, the socialist newspaper that had supported Sinclair's undercover investigation the previous year.

47. Pure Food and Drug Act. History. < https://en.wikipedia.org/wiki/Pure_Food_and_Drug_Act/ >. April 1, 2022

48. Trials of War Criminals before the Nuremberg Military Tribunals under Control Council Law No. 10. Vol. 2. Washington, DC: U.S. Government Printing Office; 1949:181–182.

49. U.S. Public Health Service Syphilis Study at Tuskegee. April 22, 2021 < https://www.cdc.gov/tuskegee/timeline.htm/ >.

50. Presidential Apology, The White House Office of the Press Secretary. For Immediate Release May 16; 1997. Remarks By the President In Apology For Study Done In Tuskegee. The East Room 2:26 P.M. EDT. <https://www.cdc.gov/tuskegee/clintonp.htm/>.

51. World Medical Association. Declaration of Helsinki 1975. Adopted by the 18th World Medical Assembly, Helsinki, Finland, June 1964 and as revised by the 29th World Medical Assembly, Tokyo, Japan, October 1975. <https://www.wma.net/what-we-do/medical-ethics/declaration-of-helsinki/doh-oct1975/>.

52. National Research Act, Public Law 93-348, July 2; 1974. <https://history.nih.gov/research/downloads/PL93-348.pdf/>.

53. Belmont Report. Office of the Secretary, Ethical Principles and Guidelines for the Protection of Human Subjects of Research, The National Commission for the Protection of Human Subjects of Biomedical and Behavioral Research, April 18; 1979. <https://www.hhs.gov/ohrp/sites/default/files/the-belmont-report-508c_FINAL.pdf/>.

54. Federal Policy for the Protection of Human Subjects; Notices and Rules. June 18; 1991 (56 FR 28003).

55. Kennedy Krieger Institute. Quality Assurance Project Plan for the Lead Paint Abatement and Repair and Maintenance Project – May 1, 1991, EPA Contract No. 68-00-0126.

56. Title X of the 1992 Housing and Community Development Act, Section 1004(13). Public Law 102-550. signed by President George H.W. Bush. October 1992.

57. US EPA. Lead-Based Paint Abatement and Repair and Maintenance Study in Baltimore: Findings Based on Two Years of Follow-Up No. 747-R-97-005; December 1997. <https://www.epa.gov/sites/production/files/documents/24folup.pdf/>.

58. Draft Final Report. Lead-Based Paint Abatement and Repair and Maintenance Pilot Study. EPA contract 68-DO-0126, May 5; 1992.

59. Farfel MR, Lim BS. The lead paint abatement and repair and maintenance study in Baltimore. ASTM STP 1226 In: Beard ME, Allen Iske SD, eds. *Lead in Paint, Soil, and Dust: Health Risks, Exposure Studies, Control Measures, Measurement Methods. and Quality Assurance.* Philadelphia: American Society for Testing and Materials; 1995:107–118.

60. Clinical Investigation Consent Form (revised May 1991), Johns Hopkins Hospital and Kennedy Krieger Research Institute, Lead Paint Abatement and Repair and Maintenance Study.

61. Roberts J, Ruby MG. Development of a High-Volume Surface Sampler. EPA/600/4-88/036; November 1988.

62. ASTM D5438-00. Standard Practice for Collection of Floor Dust for Chemical Analysis, ASTM International, West Conshohocken, PA; 2000. <http://www.astm.org/>.

63. Cecil Harris, III v City Homes et al. Baltimore City Circuit Court. Case NO. 24-C-12–006682.

64. Charnice Carpenter v Kennedy Krieger Institute (KKI) et al. Baltimore City Circuit Court Case Number 24-C-12-004854; Antwan Covington v KKI et al. Baltimore City Circuit Court Case Number 24-C-07-003562LP; Dontae Wallace v KKI et al. Baltimore City Circuit Court Case Number 1:07-CV-1140; Antwan Bryant v KKI et al. Baltimore City Circuit Court Case Number 24-C-08-003884LP; David Armstrong, Jr. v KKI et al. (Class Action) Baltimore City Cuit Court Case Number 24-C-11-005913; Ernest Bell v KKI et al. Baltimore City Circuit Court Case Number WDQ-12-0158; Tymesia Pugh v KKI et al. Baltimore City Circuit Court Case Number 24-C-12-002400LP; Jacquitta Sabb v KKI et al. Baltimore City Circuit Court Case Number 24-C-12-005322LP; Brandon Abrams v KKI et al. Baltimore City Cicuit Case Number 24-C-12-006682LP; Cecil Harris v KKI et al. Baltimore City Circuit Court Case Number 24-C-07-003562LP; Ashley Partlow v KKI et al. Baltimore City Circuit Court Case Number 24-C-09-008243.

65. History of the Common Rule Since 1974. <https://bioethicsarchive.georgetown.edu/achre/final/chap14_1.html/>.

66. Order denying Appellee's Motion of Partial Reconsideration and Modification of the Opinion. Grimes 782 A.2d 807 (Md. 2001).

67. Hoffman DE, Rothenberg KH. Whose duty is it anyway? The Kennedy Krieger opinion and its implications for public health research. *J Health Care Law Policy.* 2002;109:109–147.

68. Brief of Appellee at 24 Grimes (No. 129). October 2001.

69. Office of Human Research Protection, Department of Health and Human Services. Letter to Michael Klag (Hopkins), August 23; 2002.

70. Office of Human Research Protection, Department of Health and Human Services. Letter to Ch Van Dang (Hopkins), Michael Klag (Hopkins) and Gary Goldstein (Kennedy Krieger Institute) from Patrick McNeilly, Compliance Oversight Coordinator, Division of Compliance Oversight, August 19; 2002.

71. Lead screening update from the US Preventive Services Task Force, Cantor AG, Hendrickson R, et al. Screening for elevated blood lead levels in childhood and pregnancy: updated evidence report and systematic review for the US preventive services task force. *JAMA.* 2019;321:1510–1526.

72. Jacobs DE. Best evidence. Response to US preventive services task force. *J Pediatr.* 2019;212:243.

73. Faridi Z, Grunbaum JA, Gray BS, Franks A, Simoes E. Community-based participatory research: necessary next steps. *Prev Chronic Dis.* 2007. Available from <http://www.cdc.gov/pcd/issues/2007/jul/06_0182.htm>.

74. Research Involving Children. Report and Recommendations. The National Commission for the Protection of Human Subjects of Biomedical and Behavioral Research. Department of Health Education and Welfare. DHEW Publication no. (OS) 77-0004, September 6; 1977. <https://repository.library.georgetown.edu/bitstream/handle/10822/559373/Research_involving_children.pdf?sequence = 1&isAllowed = y/>.

75. Rosner and Markowitz. *Lead Wars: The Politics of Science and the Fate of America's Children.* University of California Press; 2014.

76. Rosner D, Markowitz G. With the best intentions: lead research and the challenge to public health. *Am J Public Health*. 2012;102:e19–e33. Available from: https://doi.org/10.2105/AJPH.2012.301004.
77. Markowitz and Rosner. *Deceit and Denial*. 1st ed. Univ of California Press; 2013.
78. Needleman H. Childhood lead poisoning: the promise and abandonment of primary prevention. *Am J Public Health*. 1998;88:1871–1877.
79. Alliance to End Childhood Lead Poisoning. *A Technical Analysis of Major Errors, Omissions and Misrepresentations in Needleman's article*. The University of Illinois Chicago Special Collections & University Archives, School of Public Health, "David E. Jacobs papers" the University of Illinois Chicago School of Public Health Library; 1998.
80. Personal Communication between Janet Reno and David Jacobs. 1996.
81. Lydia D. *Toxic Truth: A Scientist, a Doctor, and the Battle over Lead*. Beacon Press; 2008.
82. Summary of HHS Review of Needleman Study: Final report by the Office of Research Integrity (undated) "ORI found no scientific misconduct in Needleman's research, concurring with the University of Pittsburgh Hearing Board. The scientific debate over lead's toxicity is over, and it is pointless and distracting to delay further action."The University of Illinois Chicago Special Collections & University Archives, School of Public Health, "David E. Jacobs papers" the University of Illinois Chicago School of Public Health Library; 1998.
83. Don Ryan Letter to Herb Needleman, February 26; 1999. The University of Illinois Chicago Special Collections & University Archives, School of Public Health, "David E. Jacobs papers" the University of Illinois Chicago School of Public Health Library.
84. Oral Arguments Taken in Reference to 'In re: Myron Higgins,' State of Maryland, Court of Appeals, July 10; 2001:7–8.
85. Lewin, T. U.S. Investigating Johns Hopkins Study of Lead Paint Hazard. New York Times, August 24; 2001:A-11.
86. Roig-Franzia M. My Kids Were Used as Guinea Pigs. Washington Post, August 25, 2001, A-1; Tom Scocca, "Test Case: Hopkins Arm Sued Over Lead-Paint Study," Baltimore City Paper, August 8, 2001. Bor, "Kennedy Krieger Doctor Defends Lead Paint Study."
87. The Judge Went Too Far. Washington Post Editorial, September 29; 2001:A26.
88. Young G. Lead paint suit breaks new ground, sparks government investigation. *Natl Law J*. 2001;28.
89. Don R. Research on Lead Hazards is Solution, Not Problem. Baltimore Sun. August 28; 2001. <https://www.baltimoresun.com/news/bs-xpm-2001-08-28-0108280052-story.html/>.
90. Letter from Janet Phoenix to Grover Hankins, Thurgood Marshall School of Law, November 24; 1998. The University of Illinois Chicago Special Collections & University Archives, School of Public Health, "David E. Jacobs papers" the University of Illinois Chicago School of Public Health Library.
91. U.S. Department of Housing and Urban Development. 2012–2015 Environmental Justice Strategy, March 30; 2012. <https://www.hud.gov/sites/documents/ENVJUSTICE.PDF/>.
92. US Commission on Civil Rights. What Is Environmental Justice? Not in My Backyard: Executive Order 12,898 and Title VI as Tools for Achieving Environmental Justice, Chapter 2. (undated) <https://www.usccr.gov/pubs/envjust/ch2.htm/>.
93. Skelton R, Miller V. The Environmental Justice Movement. Natural Resources Defense Council, March 17; 2016. <https://www.nrdc.org/stories/environmental-justice-movement/>.
94. Farfel HS, by Gina Davis. The Baltimore Sun. October 14; 2007. <https://www.baltimoresun.com/news/bs-xpm-2007-10-14-0710140080-story.html/>.
95. National Center for Healthy Housing. Maryland Lead Law Case Study; 2017. <https://www.tfah.org/story/marylands-efforts-to-prevent-and-respond-to-childhood-lead-exposure/>.
96. Case Briefs. Grimes. v. Kennedy Krieger Institute, Inc. 2022 <https://www.casebriefs.com/blog/law/health-law/health-law-keyed-to-furrow/regulation-of-research-involving-human-subjects/grimes-v-kennedy-krieger-institute-inc/>.
97. Jacobs DE. Call for an accurate historical account of childhood lead poisoning prevention. Letter. *Am J Public Health*. 2013;103(5):e5.

PART 3

The new consensus (2001–22)

Despite the *Grimes* aberrant ruling accusing researchers of putting children in harm's way, other court decisions during this era focused on those who manufactured lead paint.

Part 1 told the story of missed opportunities and the abject failure of the medical approach. Part 2 presented the story of how successful Title X and citizen action had been in translating science into policy to drive new regulations, enforcement, and federal funding for privately owned housing.

Despite these successes, Title X did not establish accountability and it did not address housing and health beyond lead paint.

Beginning around 2001, a new accountability and a new consensus emerged.

Part 3 documents how that happened, not only in the United States but around the world.

If "you make a mess, you have to clean it up"- the Rhode Island and California court decisions

By the 1990s the legislative branch of government had passed Title X and appropriated significant funding for both public and private housing, responding to the new science and citizen action. The executive branch had reformed federal housing regulations and initiated serious enforcement, also reacting to citizens and science.

Where was the judicial branch? Would it also interpret the law in a way that paid attention to what its people wanted and what the science had proven? Or would it remain mired in the *Ashton v Pierce* and *Grimes* debacles?

7.1 Local jurisdiction lawsuits against the lead paint industry

More than 50 local jurisdictions filed lawsuits between 1987 and 2020[1] to require the lead, pigment, and paint companies to contribute to the cost of addressing residential lead paint hazards. The first suit against the Lead Industries Association and Sherwin-Williams paint company, and other paint and pigment manufacturers was filed in Boston by a lead-poisoned child in 1987.[2] New York City quickly followed in 1989 to "make 5 companies pay for all present and future costs incurred by New York City to eliminate the dangers of lead paint."[3] New York City Mayor Ed Koch said: "If we prevail, the bill for the industry's negligence will finally and rightfully come due." His remark centered on who should pay: the taxpayers or the industry?

These suits were often a response to pressure from community organizations, parents, and citizens. In Milwaukee, Wisconsin Citizen Action asked the City Council to authorize a lawsuit against the industry. Other community groups in New York City pressured the city and its housing authority to sue several lead pigment producers, and in the mid-to-late 1990s more were filed by St. Louis, Milwaukee, Chicago, and others.[4]

A landlord group also sued the paint companies in Wisconsin, arguing that the landlords had not put the lead paint in their housing units, but that the industry had contaminated the homes they rented out.[5]

Two US Senators, Jack Reed of Rhode Island and Robert Toricelli of New Jersey spon-sored the Lead Poisoning Expense Recovery Act of 1999, which authorized the federal government to recover damages from the lead industry. Their press release stated:

> To date an industry that has over $30 billion in assets has yet to make a significant contribution towards addressing the costly health and social problems associated with its product.... On the other hand the gov-ernment has spent over $1 billion to mitigate the effects of lead poisoning.... As cities and states stand up and say enough is enough, it is only appropriate for the federal government to join them to hold the indus-try accountable.[6]

In a legal opinion from the Congressional Research Office:

> It appears the Medical Care Recovery Act would permit the United States to sue lead-based paint man-ufacturers to recover funds it expends on medical care.... There appears to be federal law that would [also] permit the United States to recover funds it spends on lead-based paint abatement in housing.[7]

The federal government ultimately decided not to join the lawsuits and Reed's bill was never passed.

Although the lead paint industry charged that community groups were funded by trial lawyers, there was no evidence of that. As shown in Chapter 5, most community and parent groups remained woefully underfunded, relying largely on volunteers, and most of them were motivated by the realization that not enough was being done to protect children.

Some of these early lead lawsuits followed other class-action suits against manufacturers of tobacco, guns, asbestos, and other toxic substances, usually under product liability or other legal constructs. For example, the first tobacco cases were brought using the following legal theories:

- product liability—the tobacco companies made and marketed a product that was unfit to use;
- negligence—the tobacco companies failed to act with reasonable care in making and marketing cigarettes;
- false advertising—the tobacco companies failed to warn consumers of the risks of smoking cigarettes;
- fraud; and
- violation of state consumer protection statutes prohibiting unfair and deceptive business practices.[8]

Among these differing theories, product liability emerged as the main cause of action in the tobacco and asbestos cases. Product liability law is defined as:

> ...a manufacturer or seller being held liable for placing a defective product into the hands of a con-sumer. Responsibility for a product defect that causes injury lies with all sellers of the product who are in the distribution chain. In general terms, the law requires that a product meet the ordinary expectations of the consumer. When a product has an unexpected defect or danger, the product cannot be said to meet the ordinary expectations of the consumer.[9]

But product liability did not prove to be applicable to lead paint in this time period, because it would have required identifying which specific manufacturer had placed which lead paint into which house on which surface. And of course, one surface could conceivably contain more than one layer of lead paint so that the combined paint film could be from more than one manufacturer. Although there were valid analytical methods that potentially could make such determinations, they were very expensive and the prospect of applying them to millions of homes seemed daunting.

Product liability would also have required identifying all sellers, distributors, and others throughout the supply chain. Product liability claims would have required proving that the lead paint was "defective" in addition to being dangerous. Although some thought that market share could be used to allocate product liability for the paint and lead industries, the technical barriers proved to be insurmountable on a broad scale. This suggests why public nuisance emerged as the key legal theory in the lawsuits in Rhode Island and California, not product liability.

Public nuisance law succeeded in California and with a jury in Rhode Island (although a Rhode Island higher court later reversed the jury verdict). The reasons are explained later in this chapter.

Although some critics alleged that the tobacco and other earlier class-action lawsuits had only enriched trial lawyers, others argued that they promoted responsible corporate behavior and industry funds (at least in part) helped to correct some of the public health damage done.[10,11] The ensuing lead paint cases in Rhode Island and California both required that industry funds go to actual remediation, not only to pay for damages to children who had already been harmed, reflecting the increased attention to primary prevention (acting *before* children are harmed).

Many of the early lead paint lawsuits in the 1980s and 1990s failed for a variety of legal reasons, the main one being that most were brought under the legal theory of product liability. But two class-action lawsuits succeeded at trial using a public nuisance theory.

Legal theories were one thing, but the key lawyers were also important. Fidelma Fitzpatrick of Motley Rice played a leading role in both the Rhode Island and California cases, together with Bob and Jack McConnell, Peter Earle, Sheldon Whitehouse, Patrick Lynch, Neil Leifer (who had helped teach the Georgia Tech courses in the late 1980s Chapter 3), Danny Chou, Jenny Lam, Lorraine Van Kirk, Mary Alexander and many others too numerous to list here. A third successful suit in New Orleans public housing against the lead paint industry involved Jennifer Willis, a lawyer who had also participated in the Georgia Tech classes.[12] This integration of legal and technical expertise would prove to be essential in the cases.

The early class action lawsuits were limited to recovering the costs of lead poisoning for children, public health, and the government. But the Rhode Island and California cases, each starting around the year 2000, took a new approach. Although it never was a party to any of the lawsuits, the Alliance to End Childhood Lead Poisoning released a white paper on the new litigation, documenting the new approach:

> The new lawsuits fundamentally are aimed at achieving proactive lead hazard control on a community- or state-wide basis.... Current regulations and programs put the exclusive onus on property owners to control and eliminate lead-based paint and dust hazards.... The economics of affordable housing, principles of basic justice, and the urgent need for expanded resources for abatement require that the lead-based paint products industry contribute its fair share to solving the lead poisoning problem that it is ultimately responsible for creating.... Past lead poisoning lawsuits have not proved directly helpful in advancing broader prevention

initiatives because they primarily have consisted of private personal injury actions brought by those already poisoned seeking monetary compensation, typically from rental property owners and their insurers.[13]

For its part, the lead, paint, and pigment industries responded using two main strategies.

The first involved a legal approach arguing that as a matter of law, they could not be sued, so the facts were not relevant. This bore a striking resemblance to the *Grimes* decision, where the Maryland Appeals Court also ruled with no facts in evidence (Chapter 6). In a kind of legal "shell game," the industry argued that they were immune from lawsuits because the wrong legal theory was being used, or too much time had passed, or paint was not the real cause of lead poisoning, or the industry had already complied with public health and housing laws, or government had largely solved the problem already, or even that their Constitutional rights had been violated, along with many other defenses.

The second strategy was to launch a major public relations (PR) effort that used "...high-priced corporate defense lawyers and an army of politically well-connected lobbyists and public relations shops in an effort to fight, or kill, the suits."[14]

The PR strategy was led by none other than Robert H. Bork Jr., the son of President Reagan's nominee to the Supreme Court. Bork Sr. was not confirmed by the US Senate following a controversial hearing showing he had opposed civil rights, women's rights, evolution, and other matters. Bork Sr. had previously carried out President Richard Nixon's order to fire the special prosecutor in the Watergate scandal, an order that ultimately led to impeachment proceedings and the President's resignation.

Working with Bork Jr., Alan Wheat, a former Congressman turned lobbyist was also enlisted, as was Gail Norton, a Secretary of the Interior who had previously worked for the lead industry; she served under the second Bush Administration from 2001 to 2006.

In testimony before the Washington DC City Council, which was considering a suit against the lead paint industry, Wheat said the Europeans were wrong to ban lead paint in 1920, that his clients had acted responsibly and that landlords (not the industry) were to blame, and indeed that blame should not be contemplated at all:

> These companies never hid any information about lead paint risks from anyone.... In Europe, some nations had banned lead paint to protect painters. In this country on the other hand improved work practices were employed instead to prevent *potential* risks.... The District of Columbia should step up its efforts to require landlords of deteriorated housing to make their properties lead safe.... It is not good public health policy to seek scapegoats.... Good public policy seeks effective solutions ... not lawsuits (emphasis added).[15]

In some cases, the high-powered lobbying backfired. A state senator from Rhode Island, John Roney, said:

> I am struck that no one was here from the paint industry to tell us they are sorry, they made a mistake. Instead, they sent us two law professors, one former Congressman ... and the Rhode Island Banker's Association.[14]

In Milwaukee, a report from the local newspaper said:

> Another local official, Milwaukee Common Council President Melvin Pratt said: "I'll probably get a call from them [lead industry lobbyists] asking me out to dinner" ... he laughed.... It's been one of the most

heavily-lobbied issues in City Hall history.... Lead lobbyists asked for more time to persuade the federal government to put money into lead cleanup.... Parents Against Lead pressured Pratt to consider the facts... [Lawyer] Peter Earle is "feeling optimistic" about winning the case he started last year. "We are very confident we're going to win, and once that happens, many lawsuits will follow.... The basic decision of law will have been decided."[16]

The industry's PR strategy attempted to fund local groups. One account reported:

> In a move that has drawn the ire of activists ... the lobbyists are working on lead poisoning prevention programs with city groups and governments in New Haven, Conn, Newark, NJ and Rochester NY.[14]

Most community groups refused funding from the lead paint companies. One exception was CLEARCorps (Chapter 5).

In Houston, school districts were the first schools to also sue the industry. The school districts argued that paint companies were:

> still responsible for health hazards from lead paint, even if it was applied to walls 30 or 40 years ago.... "If we have applied somebody's product and it has to be removed at exorbitant cost, we don't feel the tax-payers should have to pay for the removal" said a school superintendent.[17]

Reflecting the new citizen action, State Attorneys General became more active in trying to find solutions to the lead paint problem and focused on paint can labels. On May 12, 2003, 45 of them reached an agreement between member companies of the National Paint and Coatings Association (NPCA), and 45 states, Washington DC, Puerto Rico, Guam, the Virgin Islands, and the Northern Mariana Islands. Major paint manufacturers in the NPCA included Sherwin-Williams, Glidden, Benjamin Moore, Thompson Miniwax, Rust-Oleum, Valspar, and Truserve. The agreement followed months of negotiations, which were led on the government side by California and six other states. The new label stated:

> Warning! If you scrape, sand, or remove old paint, you may release lead dust. LEAD IS TOXIC. EXPOSURE TO LEAD DUST CAN CAUSE SERIOUS ILLNESS, SUCH AS BRAIN DAMAGE, ESPECIALLY IN CHILDREN. PREGNANT WOMEN SHOULD ALSO AVOID EXPOSURE. Wear a NIOSH-approved respirator to control lead exposure. Clean up carefully with a HEPA vacuum and a wet mop. Before you start, find out how to protect yourself and your family by contacting the National Lead Information Hotline at 1–800–424–LEAD or log on to http://www.epa.gov/lead.[18]

Although the labels were important, some thought this was unlikely to do much good, because labels were frequently not read. Attention turned to actions attempting to force the industry to contribute to remediation.

7.2 The Rhode Island court decision, 1999–2008

The first significant class action case in which a jury found against the lead and paint industries occurred in Rhode Island in 2006.[19] The state's Attorney General, Sheldon Whitehouse (later to become a US Senator) and his assistant Attorney General Patrick Lynch (later to become Rhode Island Attorney General himself) filed suit in 1999. The case initially charged the paint, pigment and lead industries with creating a public nuisance, wrongful indemnity, unjust enrichment, and conspiracy. It was later limited to public nuisance.

The initial defendants were NL Industries (formerly the National Lead Company, which had marketed the "Dutch Boy" logo), Sherwin-Williams, Millennium Holdings (the parent company of Glidden paints), Atlantic Richfield (which had acquired the International Smelting and Refining Co.), and Du Pont. The Lead Industries Association, first incorporated in 1928 and which was also named in the suit, filed for bankruptcy and was dismissed from the case.

The New York Times reported:

> Rhode Island was the first state to take on paint manufacturers when it filed a lawsuit against the companies and the Lead Industries Association, a trade group, in October 1999. The state claimed that lead paint constituted a public nuisance in Rhode Island, where more than 43 percent of the houses were built before 1950.[20]

The State of Rhode Island hired outside lawyers because it knew it was up against a much larger and better-funded army of industry lawyers. Some of the industry lawyers were employed by the Jones Day law firm, the fifth largest law firm in the United States and the 13th highest grossing law firm in the world, with a long record of industry defense. Some of the Jones-Day alumni included former US Supreme Court Justice Antonin Scalia and Donald McGahn, a lawyer who served in President Trump's White House in 2018.[21]

The jury reached its guilty verdict seven years later, on February 22, 2006, following a 4-month civil trial, the longest in the state's history.[22]

The trial court summarized the case this way:

> Here, the [Rhode Island] Attorney General, as guardian of the public, asserts that the defendants, as manufacturers, promoters, and suppliers, are responsible for the presence of lead, a substance alleged to be a health hazard to members of the public, in public and private buildings throughout the State. Further, the State also contends that the defendants' misconduct, by causing a public health crisis, has caused the State to incur substantial damages. In its expansive request for relief, the State, in part, seeks an order for the abatement of lead.[19]

The first jury to hear the case could not reach a verdict, resulting in a hung jury (mistrial). But a second one did render a verdict against the industries. In that second case, the State amassed the vast evidence of toxicity, extent of harm, and historical evidence. The companies offered no defense at all, saying that "as a matter of law" they could not be found guilty.

In its deliberations, the judge instructed the jury to answer three questions: (1) did the presence of lead paint in approximately 250,000 of Rhode Island's homes constitute a public nuisance; (2) if so, were the defendant former lead. pigment and paint producers responsible for that nuisance; and (3) if they were, should they be required to pay to abate that nuisance?[4] The second jury unanimously said "yes" to all three.

The Rhode Island trial court ordered "abatement of the public nuisance" pursuant to the jury's verdict against Millennium Holdings, NL (formerly National Lead) Industries, and the Sherwin-Williams Company. Two other defendants, Atlantic Richfield, and Con

Agra Groceries, both of which had acquired companies that previously made lead paint were released. Another, DuPont, settled out of court for $12 million before the verdict, which was used for abatement, public education, and other lead poisoning prevention work.

The final trial court judgment said that the jury found both lead pigment and lead paint were a "public nuisance," that the three defendants "caused or substantially contributed to" the public nuisance, and they were ordered to abate it.[19] Because this trial did not determine exactly what constituted "abatement of the nuisance," the court ordered a Special Master to determine what that would mean. In response to the jury's verdict in Rhode Island, the companies asked for a new trial and to overturn the jury's verdict. Both requests were denied by the trial court.

The Rhode Island case did not present a detailed abatement plan at trial. After the trial, an abatement plan was drafted, and the Court ordered two special masters (Susan Klitzman from New York University and Rhode Island Superior Court Judge Michael Silverstein) to review it and take testimony from its author (Jacobs). The plan was deemed workable, but because it came after the trial and was not part of it, some thought this may have contributed to the industry's view that remediation was too difficult and not workable. The Rhode Island Supreme court's final decision would hinge on this very issue of a "remedy." A subsequent case in California presented a similar plan at trial, which the Court there mostly adopted (described later in this chapter).

In 2008 the Rhode Island Supreme Court overturned the jury verdict on appeal, ruling that the case had been brought under the wrong legal theory (public nuisance vs product liability law). In its opinion, the State Supreme court ruled that "public nuisance law simply does not provide a remedy for this harm." Perhaps with some sense of remorse and its own abdication in failing to confront the problem, the Rhode Island Supreme Court stated:

> Our hearts go out to those children whose lives forever have been changed by the poisonous presence of lead.[23]

7.2.1 Public nuisance law

Why did the legal meaning of "public nuisance" fail in Rhode Island, yet ultimately succeed in a later case in California? Before turning to that later case, the theory of public nuisance law is noteworthy.

Public nuisance law dates back to at least the 16th century and had long been recognized in the United States, Rhode Island, and many other states.[24] The concept was also codified in numerous environmental laws, all flowing from the "shared commons" concept (Chapter 9).

Decades before the lead paint case, the Rhode Island Supreme Court had defined a "public nuisance" to mean:

> An unreasonable interference with a right common to the general public: it is behavior that unreasonably interferes with the health, safety, peace, comfort or convenience of the general community.[25]

With so many homes contaminated with lead paint, the State thought it was apparent that lead paint interfered with the "health of the community." But instead the Rhode Island Supreme Court stated:

> a necessary element of public nuisance is an interference with a public right — those *indivisible* resources shared by the public at large, such as air, water, or public rights of way (emphasis added).[26]

Because most homes were privately owned, the Rhode Island Supreme Court thought they were not shared by the public and therefore were not "indivisible." Yet homes were bought and sold and the public had an interest in preserving the viability of the housing market, as demonstrated in the 2008 financial crisis, which began as a housing mortgage finance issue.

The rejection of any public interest in housing was basic to the Rhode Island Supreme Court's narrow interpretation. This idea that housing was unlike other shared common resources like air and water would later resurface (Chapters 8 and 9).

The Rhode Island Supreme Court decision essentially conflated private nuisances with public ones, defining public nuisance out of existence, even though earlier Rhode Island court decisions had made clear the differences between public and private nuisances. Most other courts had relied on a precedent from 1978, which held that:

> When private nuisances [such as those found *within private homes*] become so great and extensive as to constitute a public annoyance and inconvenience, and a wrong against the community, [it can] properly be the subject of a public prosecution (emphasis added).[27]

But the Rhode Island Supreme court ignored precedent and instead relied on commentary from a law Professor, who wrote:

> the exposure to lead-based paint usually occurs within the most private and intimate of surroundings, his or her own home. Injuries occurring in this context do not resemble the rights traditionally understood as public rights for public nuisance purposes — obstruction of highways and waterways, or pollution of air or navigable streams.[24]

Fidelma Fitzpatrick noted in her law review of the Rhode Island decision that there are many instances in which hazards that occur in the private home are clearly public nuisances. These include, for example:

> Keeping diseased animals on private property; interfering with public safety by storing explosives in the midst of a city; or interfering with the public morals by operating houses of prostitution. . . . The reality is that the Rhode Island Supreme Court failed to consider or acknowledge the public harm standard when it overturned the prior legal rulings and jury verdict in State of Rhode Island v. Lead Indus. Association. The Court fundamentally misconstrued the law of nuisance, which concerns itself with interference with public rights. It is the *nature* of the right at issue that determines whether a particular nuisance is public or not, not the *location* of the condition (emphasis added).[24]

Like the Maryland Appeals Court decision that thought it could create some kind of new but undefined "special duty" for researchers, the Rhode Island Supreme court decision marked a radical departure from prior legal conceptions of public nuisance and tried to create a new legal construct. And like the Maryland Appeals *Grimes* court decision, no other court later confirmed the Rhode Island Supreme Court's interpretation that a public nuisance cannot exist in a home.

Indeed, the Rhode Island Supreme Court had reversed itself on lead paint, because in an earlier case in 1998, it had ruled:

> the persistence of the continuing hazard of lead paint presents immediate and irreparable *harm to the public* so long as that hazard remains unabated (emphasis added).[28]

When it passed the Rhode Island Lead Poisoning Prevention Law in 1991, the state legislature had also recognized the public nature of lead paint hazards:

> Childhood lead poisoning is dangerous to the *public* health, safety, and general welfare of the people (emphasis added).[29]

The Rhode Island Supreme Court decision also held that the defendants "must have control over the instrumentality causing the alleged nuisance at the time the damage occurs,"[30] which meant that because the paint, pigment and lead industry defendants did not own the homes, they did not have control over them, and therefore could not be liable.

But this also was at odds with established public nuisance law. A leading legal textbook stated:

> Generally, one who *creates* a nuisance is liable for the resulting damages, and ordinarily such person's liability *continues as long as the nuisance continues.* Liability for nuisance may be imposed upon one who sets in motion the forces which eventually cause [it] (emphasis added).[31]

The Rhode Island Supreme Court's position was rejected by the later, successful public nuisance case against the lead and paint companies in California, discussed in the next section. The California court ruled:

> liability for nuisance does not hinge on whether the defendant *owns, possesses or controls the property,* nor on whether he is in a position to abate the nuisance; the critical question is whether the defendant *created or assisted in the creation of the nuisance* (emphasis added).[32]

Although the Rhode Island Supreme Court over-ruled the jury verdict, it laid the foundation for the California case. Liz Colon, the mother of a lead poisoned child in Rhode Island said, "The ruling in California wouldn't have come without Rhode Island's lawsuit."[33]

7.3 The California court decision, 2000–2022

Unlike the Rhode Island case, California's suit against the lead, pigment and paint industries[34] survived all higher court appeals, including an attempt to go all the way to the US Supreme Court. The California decision marked the first time that the industries were held accountable and were forced to contribute financially to the abatement remedy across many jurisdictions. At the time this book is being written, the companies have made the first payments to the 10 California jurisdictions that brought the case.

But none of it came easily or quickly.

Initially filed in 2000, the case took nearly 20 years, one of the longest trials in the nation's history. The judge's initial ruling occurred on January 7, 2014, following a bench trial that lasted 23 trial days with 450 exhibits introduced into evidence. The bench trial occurred on July 15, 2013, and ended on September 23, 2013. This time, the companies *did* attempt a defense (unlike the Rhode Island case). And this time, an abatement plan was presented to the court and became part of the trial itself, instead of being developed after the trial had concluded, as had been the case in Rhode Island.

Before the actual trial began, there had been at least 60 earlier hearings over more than a decade. But among its many appeals, Sherwin-Williams still argued that the discovery time had been too short. The defendants finally settled on July 10, 2019. The Court entered a new final judgment approving the settlement agreement and dismissing the action with prejudice, described later in this chapter.

The initial judgment required abatement in the interiors of most pre-1978 homes in the 10 California counties and cities that brought the suit, which the court estimated would cost the defendants $1.15 billion, an estimate from testimony in the trial. This was later reduced to $400 million, because an Appeals Court decided to limit the abatement work to only pre-1950 homes (not pre-1978 homes), in the belief that the companies did not market lead paint after 1950.

The final settlement was $305 million,[35] which was more than the Housing and Urban Development (HUD) lead paint appropriation for the entire nation in 2020 ($290 million).

Unlike many common lawsuits against landlords, the California and Rhode Island cases did not seek damages on behalf of poisoned children. Instead, they required abatement of the public nuisance, meaning the elimination of lead paint hazards in homes, regardless of whether a child had been poisoned. This made it the nation's largest primary prevention program ever, taking action *before* a child was harmed, the foundational principle behind Title X.

The abatement plans for California and Rhode Island were similar. The lawyers who tried both cases felt that the absence of an abatement plan in the Rhode Island court proceedings may have contributed to the state supreme court's reversal of the jury verdict, because the abatement remedy had not been defined at the trial. Therefore an abatement plan was presented to the California court during the trial to prove that a feasible and scientifically validated remedy to the public nuisance did in fact exist.

The final California ruling found that three lead paint companies had created a public nuisance by concealing the dangers of lead and that they had pursued a campaign against the regulation of lead while actively promoting it for use in homes, despite knowing that lead paint was highly toxic. Three defendants (ConAgra Grocery Products Company, LLC, NL Industries, Inc., and Sherwin-Williams Company) were held liable under the state's public nuisance law, the same legal theory that the Rhode Island Supreme Court had rejected.

Two other initial defendants (Atlantic Richfield Company (ARCO) and DuPont) were found not to be liable. ARCO avoided liability because the court thought that it had only advertised lead paint to the painting trade (not to the public). DuPont avoided liability because the court thought that its interior paint did not contain "white lead pigments."

The plaintiffs in the California case were the Counties of Santa Clara (where the trial was conducted), Alameda, Los Angeles, Monterey, San Mateo, Solano, and Ventura; three cities were also plaintiffs (Oakland, San Diego, and San Francisco). Together, these 10 jurisdictions included about half of the older housing stock in the state.

The initial charges in California included fraud, strict liability, negligence, unfair business practices, and public nuisance. This was later limited to public nuisance, with abatement as the remedy.

Importantly, California's definition of a public nuisance applied to both the community and to individuals:

> Anything which is injurious to health … or is indecent or offensive to the senses, or an obstruction to the free use of property, so as to interfere with the comfortable enjoyment of life or property … is a nuisance…. A public nuisance is one which affects at the same time an entire community or neighborhood, or any considerable number of persons, although the extent of the annoyance or damage inflicted upon individuals may be unequal.[36]

And with the experience of the Rhode Island case, the California court made it clear that product liability law did *not* apply and that public nuisance law *did*. The California court ruled that:

> A public nuisance cause of action is not premised on a defect in a product or a failure to warn but on affirmative conduct that assisted in the creation of a hazardous condition. Here, the alleged basis for defendants' liability for the public nuisance created by lead paint is their affirmative promotion of lead paint for interior use, not their mere manufacture and distribution of lead paint or their failure to warn of its hazards.[34]

The California Superior Court trial judge (James Kleinberg) was focused on the facts, not only legal theories. After he was thanked for listening to complicated scientific testimony, he said with a bemused surprised look, "Well, it *is* my job."

Other witnesses in the California case included the author of the Agency for Toxic Substances and Disease Registry 1988 report (Paul Mushak, who passed away before the final verdict); the historians Markowitz and Rosner; leading toxicologists and medical experts, such as Bruce Lanphear and Michael Kosnett; an exposure scientist and inspection and remediation expert (Jacobs); and many other experts from the local jurisdictions, some of whom had participated in the earlier HUD evaluation study (Chapter 4).

This blend of legal talent and national and local experts proved to be essential because part of the companies' defense was that lead poisoning in children came from something other than lead paint, such as bad landlords, bad parents, soil, air, food, and others, the same argument they had been using for decades. For example, one local health official was asked by an industry lawyer at the trial:

> Isn't it true that in your county, the main source of lead exposure is from grasshoppers?

The room fell silent. How could lead poisoning be caused by grasshoppers?

But a local health official calmly explained that there were indeed a few cases where fried grasshoppers had been stored in bowls from Mexico with lead glazing (with the lead leaching into the food). But he said that the main source of exposure was clearly from paint, not grasshoppers, and he provided the evidence to back it up. But the industry lawyer persisted in blaming grasshopers, until Judge Kleinberg finally intervened and said: "I get it; I urge you to move on!"

In his decision, Judge Kleinberg ruled on 10 main questions:

1. Is white lead carbonate (the pigment in most lead paint) harmful to children?
2. Is there still a present danger?
3. Did the companies promote and sell the paint in each of the jurisdictions that brought the suit?
4. Did the companies know that their product was harmful, but promote and sell it anyway?
5. Do other sources of lead mean that the companies can be held not accountable?
6. Is intact lead paint a hazard?
7. Is the problem still present in light of the long-term decline in blood lead levels?
8. Is it the job of private companies to solve the problem when there are existing government programs?
9. Is the proposed abatement solution cost-effective, timely, manageable, and likely to be successful?
10. Have the companies been afforded their Constitutional right to due process?

The companies tried relentlessly to show the answer to each of these was "no," and that therefore they could not be held liable and were not required to contribute to the cost of abatement.

But in the end, all the lower and upper California courts agreed the answer to each question was "yes." The last question was posed by the defendants in an attempt to move the case to the federal court system, but the US Supreme Court refused to hear the case.

The first two questions were rather straightforward yet required careful legal analysis and assembling mountains of scientific evidence. The vast majority of lead in old paint was in fact white lead carbonate, as demonstrated by Mushak's painstaking analysis of old mineral use yearbooks (Fig. 7.1).[37]

The second question was answered by an extensive analysis of the state's blood lead surveillance system, which showed that millions of children in the ten jurisdictions had been harmed by lead paint and the contaminated dust and soil it generated.

For the second and fifth questions, Sherwin-Williams argued that it was necessary to know exactly which properties in California had lead paint before liability could be assessed, but the court ruled that this was the purpose of lead paint inspections and was not needed to assess liability. Particularly noteworthy was the Court's determination that paint was the main source of lead exposure:

> alternate sources of lead such as water and air contain only trace amounts of lead, and neither appreciably contributes to lead poisoning in the Jurisdictions.... But the existence of other sources of lead exposure has no bearing on whether lead paint constitutes a public nuisance. It does not change the fact that lead paint is the primary source of lead poisoning for children in the Jurisdictions who live in pre-1978 housing.[38]

The third and fourth questions were answered by the paint company's own internal documents and by the historians Markowitz and Rosner. One example of how the Court

Uses of white lead 1929–74

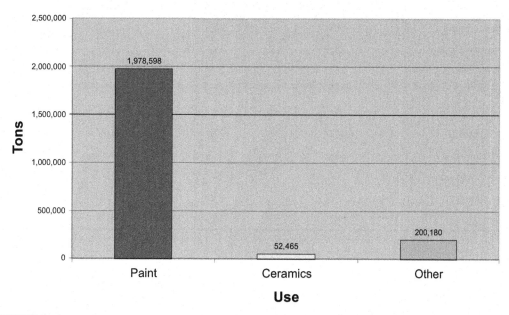

FIGURE 7.1   Uses of white lead carbonate, 1929−74. From: *Paul Mushak Court Affidavit; 2010.*

determined the companies knew that their product was hazardous came from an internal Sherwin-Williams document from 1900, cited by the Court in its final decision, in which the company stated:

> It is also familiarly known that white lead is a deadly cumulative poison, while zinc white is innocuous. It is true, therefore, that any paint is poisonous in proportion to the percentage of lead contained in it.[39]

An internal "inter-office letter" from Sherwin-Williams in 1969 reported a building research meeting at Rockefeller University in New York City. The company had sold lead-based paint for decades, and the memo showed how much they knew, the need for secrecy, and how alarming the costs of remediation were:

> A memo of the meeting will be prepared by the University, but no individual comments will be quoted.... When acute cases are detected, they nearly always can show by x-ray that the child has ingested flakes of paint. The child is hospitalized and treated ... for about 10−20 days at $100 per day.... However, this de-leading of the blood is ... temporary.... The child may be in trouble again, particularly if he is returned to a home that is not properly repaired and ingests more lead materials. If these individuals in the adult age, because of the mental impairment, become wards of the State, it is estimated their total cost to the State during their lifetime could range between $200−300,000.... On a national basis the total number of children afflicted is very high indeed.... As to a solution to the problem, a very simple statement, but very difficult to carry out, would be to remove the source of lead or put it behind barriers so that the children could not get to it.... The entire population is becoming more and more exposed to lead.... *The entire problem is certainly depressing and the outlook for an economical practical solution is not too optimistic.* (emphasis added)[40]

TABLE 7.1   Number of lead paint advertisements in the ten jurisdictions, 1900–72.

Entity	DuPont	Fuller	NL	SW
Alameda County	269	233	240	401
Los Angeles County	28	131	81	350
Monterey County	167	328	162	704
Oakland	162	143	168	221
City of San Diego	63	269	98	685
San Francisco	127	272	126	229
San Mateo County	111	183	219	149
Santa Clara County	207	347	444	305
Solano County	137	152	260	301
Ventura County	14	28	127	229

Gerald Markowitz and David Rosner. Reproduced from The People of the State of California, Plaintiff, vs. Atlantic Richfield, et al., Defendants. Amended Statement of Decision, Superior Coury of California, County of Santa Clara, Case No. 1–00-CV-788657, March 26, 2014. P. 32.

Proving that the companies had promoted and sold paint in each of the ten jurisdictions bringing the suit required enormous historical research. The historians Gerald Markowitz and David Rosner (and their students) compiled a listing of the number of advertisements appearing in each of the 10 jurisdictions between 1900–1972, which the court cited in its final ruling. It showed the magnitude of the historical research and the extent of the companies' promotion of its products (Table 7.1). (NL is National Lead Industries and SW is Sherwin-Williams)

Other historical data would be assembled to document the number of stores in each jurisdiction that sold lead paint, how many advertising campaigns there were, and when each of the companies came to know that their product was dangerous. The historical evidence showed that the industries marketed their products that they knew to be dangerous but did so anyway. The historians showed that each of the companies had advertised its products in each of the 10 jurisdictions decades ago, combing through thousands of old magazine and other ads.

It was not only about advertisements, however. The Court found that Sherwin-Williams:

> had two plants in the jurisdictions, as well as stores and dealers selling lead paint [and] transported millions of pounds of lead pigment to its warehouses and factories during the first four decades of the 20th century. [The company] knew at an early date of the *occupational* risks to factory workers from lead dust exposure and it is a reasonable conclusion that it knew or should have known of the *hazards in the home* (emphasis added).

Again, the link between occupational health and housing health was apparent.

For the sixth question, the confusion over intact and deteriorated paint continued in this case. The California court addressed this issue directly, stating:

> Furthermore, lead paint that is currently intact poses a substantial risk of future harm because it will inevitably degrade and be disturbed by normal residential activities, such as renovations.

For the seventh and eighth questions, although the defendants had argued that abatement was already happening under existing government programs, the court ruled:

> Childhood Lead Poisoning Prevention Programs operated by the Public Entities have largely reached their limits. The Public Entities lack the resources...

On the ninth question (the abatement plan), the court ruled that:

> The People's abatement plan, it is argued, can abate the public nuisance in this case at a reasonable cost and by reasonable means (p. 86) [and] and in a reasonable amount of time ... the abatement experts on both sides agree that abatement is needed.

The abatement plan considered by the court stated:

> The total cost of the Plan as proposed at trial by the People's abatement expert, Dr. David Jacobs, is $1.618 billion if implemented by the Public Entities. P. 88. The Court is persuaded by Dr. Jacobs' experience and expertise which greatly eclipse that of the Defendants' expert in these matters. P. 98

The court noted how ambitious the abatement plan was:

> Of course, by any measure, the remedy sought by the People is of substantial, even massive proportions. Seeking the abatement of lead by inspections and rehabilitation of tens of thousands of homes – at a minimum – is a daunting decision. But the Court is convinced that although great strides in reducing lead exposure have been made, and the incidence of exposure with correlative blood lead levels has declined to a low level, thousands of children in the jurisdictions are still presently and potentially victimized by this chemical. (p. 95)

The court found the California abatement plan was valid and authoritative, unlike the Rhode Island case, where no abatement plan was presented at trial.

The abatement plan was targeted to those at greatest risk, which meant that certain types of properties could be excluded. This was mostly consistent with the reformed federal lead paint regulations that had been promulgated previously in 1999. The exclusions included: Institutional group quarters, including correctional facilities, nursing homes, dormitories, nonfamily military housing (e.g., barracks), mental health psychiatric rehabilitation residences, alcohol/detox living facilities, supervised apartment living quarters for youths over 16, schools, and nonhome-based daycare centers not otherwise included; housing designated exclusively for the elderly (unless children are present); houses not occupied by a young child, houses constructed after 1980; and properties documented by an inspection to not contain any lead-based paint.

The California abatement plan also included specific types of validated lead hazard control, which were drawn almost entirely from the 1995 and 2012 HUD Guidelines. The plan also included other important matters that were needed, many of which had been developed by the HUD lead paint programs, such as public education, priorities (which included a "worst-first" approach), eligibility, workforce development, bidding and payment procedures, contracting, clearance testing, and evaluations to ensure the abatement was proceeding as intended. The court included specific X-Ray Fluorescence (XRF) testing procedures in its court order and prohibited other testing procedures (such as spot test procedures) that had not been validated. It also included specific hazard control procedures, including specific options for specific building components, all consistent with HUD guidelines. The court used the science in its decision and to ensure it was actually implemented, it even included specific staffing requirements in each of the jurisdictions.

Ultimately, Judge James P. Kleinberg ordered:

> The Defendants against whom judgment is entered, jointly and severally, shall pay... $1,150,000,000 (One Billion One Hundred Fifty Million Dollars) into a specifically designated, dedicated, and restricted abatement fund (the "Fund"). P. 109

One odd feature of the California ruling involved a debate over interior versus exterior lead paint. Both the trial court and the California Appeals Court believed that there were other sources of exterior lead, such as soil residues from previous use of lead gasoline, lead smelting and mining, and exterior lead paint. The courts felt the jurisdictions had not proven the companies had marketed their product for exterior use, although they did prove that they had marketed it for interior use. At the same time, the court ruled:

> Lead contamination in soil and dust in older homes is almost always due to lead in paint rather than other environmental contaminates.

Although this answered question 5, no existing program had ever abated only *interior* lead paint. For example, the HUD program addressed both interior and exterior hazards, including contmainated bare soil. This eventually became a moot point, because the companies finally settled, and the funds could be used for both interiors and exteriors. But this was another example of how legal evidence can collide with scientific evidence.

Another oddity involved an Appeals Court ruling that abatement should only occur in pre-1951 housing, because it thought that the companies had not promoted its lead paint after that date.[41] Although there was evidence that lead hazards were most prevalent and severe in older housing, all other abatement programs had used pre-1978 as the cut-off date, because that was when lead paint was finally banned for residential use in the United States (Chapter 1).

In recalculating the costs of abatement however, the court decided to use the same cost estimate for all pre-1978 housing, even though the costs for each housing unit would obviously be greater in the older pre-1951 housing, which had more lead paint. Because the number of houses to be remediated also declined, the expected economies of scale also needed to be reduced, but this was not included in the final cost calculation by the court. All this ultimately also became a moot point in light of the final settlement.

7.4 The industry fights back

Even though they had lost in both the trial court and all Appeals Courts, the companies continued their resistance by appealing to the court of public opinion. For example, they proposed a ballot initiative that would have over-ruled the court's decision if the voters passed it. Known as the "The Healthy Homes and Schools Bond Act of 2018," the ballot initiative said it would provide $1.5 billion in bond funding for the remediation of structural and environmental hazards occurring in homes and schools.[42]

At first, this seemed like a worthwhile effort to many of California's healthy homes programs and advocates. The language contained specific money for remediation of mold, asbestos, radon, water, pests, ventilation, and lead hazards for children, seniors, and others.

But buried within the proposed statute's language was this "poison pill":

> Notwithstanding any other law, lead-based paint on or in private or public residential properties, whether considered individually, collectively, or in the aggregate, *is not a public nuisance* (emphasis added).

Danny Y. Chou, who at the time was a Santa Clara County lawyer playing a key role in the litigation and later became a judge, said:

> the [ballot] initiative declares that lead-based paint in residential properties is not a public nuisance, which cuts the heart out of the cities' and counties' legal victory.[43]

But Tiffany Moffatt, a public relations consultant for the paint and lead companies said in response:

> This initiative provides an important public benefit of addressing the state's housing crisis by increasing the supply of safe and affordable homes that would otherwise be unlivable.[44]

California's legislators reacted to the paint companies' subterfuge in fury, because they had already placed another $4 billion affordable housing bond on the same ballot. At hearings, they demanded that the paint companies withdraw their ballot initiative.

For example, California Assemblywoman Wendy Carrillo called the paint and lead industry's ballot initiative the "Let Us Get Off The Hook And You Pay Millions Of Dollars For Our Mistake Act." She said, "The companies knew their paint had lead in it and they sold it anyway." California Assemblyman David Chiu stated:

> The ballot measure also says that these three companies will be completely absolved of their liability, so that they will not have to pay for what they did. . . . [the companies] are proposing a ballot measure so that California homeowners and California taxpayers will pay $2 billion for lead paint clean up.[45]

The industry ballot initiative was led by the "Californians for Safe and Affordable Housing," a political action committee that had raised $8 million from only three sources: Conagra Grocery Products Company (which had acquired a lead paint company), NL Industries, Inc., and Sherwin-Williams, all defendants in the California case. The ballot initiative was finally withdrawn by the companies, but only after the State Attorney General decided to issue this text of how it would appear on the ballot:

> [The ballot initiative authorizes] $2 billion in general obligation bonds for the remediation of structural and environmental hazards in homes, schools, and assisted-living and senior housing and declares that *lead-based paint is not a public nuisance and eliminates the liability of lead-paint manufacturers* (emphasis added).[45]

Had the Attorney General's summary failed to include the elimination of the industry's liability, the companies' publicity campaign may have succeeded in transferring the companies' liability to the taxpayers.

Their publicity campaign also included other attacks. For example, Daniel Fisher, a columnist for the pro-business magazine Forbes wrote:

> And [the judge] wants the job supervised by a plaintiff expert and former Housing and Urban Development official [David Jacobs] who left his job as head of HUD's lead abatement program after government auditors found $90 million in improper grants to outside organizations.[46]

But in fact, Jacobs had not left HUD due to "improper grants." He was found to be innocent of these charges (Chapter 5) and he left HUD two years after the audit.

The Forbes columnist also stated that the judge's ruling was somehow a gift to slumlords, repeating the companies' claim that it was only poor housing maintenance that caused lead poisoning. Yet the money could not go to slumlords; instead, it could only go to inspectors and abatement contractors.

Fidelma Fitzpatrick summed up the California case this way:

> As a result of Judge Kleinberg's thoughtful consideration of both sides of this issue and resulting decision, the companies that willingly used and promoted lead pigment, despite knowledge of the harm it can cause, will finally be forced to take responsibility for the damage they have caused, instead of putting the burden on public entities.... The abatement plan is a proactive solution to preventing lead poisoning waiting to happen due to existing paint – a positive example of prevention instead of treating the negative effects after the fact.[47]

7.5 The new consensus

The California court decision established a legal precedent, reflecting a new consensus, because (like the earlier Rhode Island jury) it found the lead paint industry accountable.

One indication from Milwaukee suggested that the Rhode Island and California decisions had indeed established a new precedent. In 2019 a jury awarded $6 million to three children who had been poisoned years earlier, to be paid by the lead, pigment and paint companies (not landlords).[48]

The California ruling was a watershed moment. One account said:

> The ruling ... rewards scofflaw landlords who are responsible for the risk to children from poorly maintained lead paint" [said a spokeswoman for the industry defendants].... But Fidelma L. Fitzpatrick [who] ... represented the cities and counties, said she was "thrilled for all the kids who are actually going to have a shot to have a safe home environment.... It's a great lesson in responsibility and accountability," Fitzpatrick said. "You make a mess, you have to clean it up.[49]

Fitzpatrick's simple summation crystalized accountability and the California court agreed. "Blaming the well-worn stereotypes of 'slum landlords,' 'bad parents,' 'the poor,' and 'the government' does not relieve Defendants of liability," wrote Judge Kleinberg in his California decision.

Although the prosecuting jurisdictions finally forced the companies to settle, there were important limitations. Instead of $1.1 billion, the final settlement was $305 million. Even at this reduced sum, it still made the abatement work in California the largest ever in a single state. Yet, thousands of homes with lead hazards will remain. And the 20-year time period required in California to actually begin payments from the industry was daunting.

Despite the many legal setbacks between 1970 and 2020, it appeared that accountability was finally determined. The public did not support bringing legal action against scientists seeking solutions, as in the Baltimore *Grimes* cases. Instead, the jury in Rhode Island, the Courts in California, and a jury in Milwaukee all believed the ones who caused the problem in the first place should be required to help pay to correct it.

This emerging consensus was also manifested in a new appreciation of housing-related diseases and injuries beyond lead paint poisoning. In the United States, the experience with lead poisoning prevention became a springboard to a broader healthy homes movement, an effort to apply the lessons from both Title X and the court cases to asthma, mold, injuries, and other housing-related health problems (Chapter 8).

References

1. Drum K. How the Paint Industry Escapes Responsibility for Lead Poisoning. Mother Jones Magazine; August 8, 2013. <https://www.motherjones.com/kevin-drum/2013/08/how-paint-industry-escapes-responsibility-lead-poisoning/>.
2. Santiago v. Lead Industries Association; 1987 Santiago v. Sherwin-Williams Co., 782 F. Supp. 186 (D. Mass. 1992).
3. Trial Lawyers for Public Justice, Press Release, June 8; 1989. The University of Illinois Chicago Special Collections & University Archives, School of Public Health, "David E. Jacobs papers" the University of Illinois Chicago School of Public Health library.
4. Rabin R. The Rhode Island lead paint lawsuit: where do we go from here? N Solut. 2006;16(4):353–363.
5. Landlord group to sue paint companies, by Pete Millard, the Business Journal; July 23, 1999, p. 3.
6. Make Polluters Pay for Childhood Lead Poisoning. Press release from the Senate for the Lead Poisoning Expense Recovery Act of 1999 (S. 1821); March 16, 2000.
7. Memorandum to Honorable Jack Reed from Henry Cohen, American Law Division, Congressional Research Service; June 23, 1999. The University of Illinois Chicago Special Collections & University Archives, School of Public Health, "David E. Jacobs papers" the University of Illinois Chicago School of Public Health library.
8. Michon K. Tobacco Litigation: History & Recent Developments. Undated. <https://www.nolo.com/legal-encyclopedia/tobacco-litigation-history-and-development-32202.html#:~:text=product%20liability%20%2D%20the%20tobacco%20companies,the%20risks%20of%20smoking%20cigarettes&text=violation%20of%20state%20consumer%20protection,unfair%20and%20deceptive%20business%20practices>.
9. What is Product Liability. Find Law; July 2, 2019. <https://www.findlaw.com/injury/product-liability/what-is-product-liability.html>.
10. Phend C. Tobacco Master Settlement at 20 Years—Big Tobacco's payouts squandered from prevention standpoint, MedPage Today; November 24, 2018. <https://www.medpagetoday.com/primarycare/smoking/76496>.
11. Class Actions: A Powerful Consumer Tool By Ruth Susswein. <https://www.consumer-action.org/news/articles/the-benefits-of-class-actions-winter-2019-2020>.
12. The third case was in New Orleans public housing and was settled out of court, because the housing authority reportedly had records of exactly who sold them lead paint.
13. Memorandum from Alliance to End Childhood Lead Poisoning. Lawsuits Against the Lead Industry: Rationale and Remedies; August 7, 2000. The University of Illinois Chicago Special Collections & University Archives, School of Public Health, "David E. Jacobs papers" the University of Illinois Chicago School of Public Health Library.
14. Zeller S. Lead paint lobbyists. Natl J. 2001;2172–2174.
15. Testimony of Allan Wheat on Bill 13-721, Before the Human Services Committee DC City Council; September 25, 2000. The University of Illinois Chicago Special Collections & University Archives, School of Public Health, "David E. Jacobs papers" the University of Illinois Chicago School of Public Health library.

16. Hissom, D. Lead lawsuit picture clearer. Shepard Express Metro; October 12, 2000.
17. Texas schools sue over lead paint, Associated Press; July 7, 2000. https://apnews.com/article/937c41a8e2e66a4889a28511ef602391
18. Lead Exposure Warnings and Education and Training Programs Agreement between State Attorneys General and the National Paint and Coatings Association, Inc.; May 12, 2003 <http://www.ossh.com/newsalerts/2003/03-055.pdf>.
19. State of Rhode Island v. Lead Industry Association, et al. No. 99—5226, 2001 R.I. Super. LEXIS 37, at *19—28 (R.I. Super.; April 2, 2001).
20. Bhattarai A. *Rhode Island Court Throws Out Jury Finding in Lead Case*. New York Times; July 2, 2008. <https://www.nytimes.com/2008/07/02/business/02paint.html>.
21. Jones Day.Law.com. <https://www.law.com/law-firm-profile/?id = 163&name = Jones-Day&slreturn = 20200327164949>.
22. Providence Journal, 3 Companies Found Liable in Lead-paint Nuisance Suit; February 23, 2006.
23. Supreme Court of Rhode Island. State of Rhode Island v. Lead Industries Association, Inc., et al. Nos. 2004-63-M.P., 2006-158-Appeal, 2007-121-Appeal. Decided; July 01, 2008. <https://caselaw.findlaw.com/ri-supreme-court/1142902.html>.
24. Professor Donald Gifford, cited in Rhode Island Supreme Court decision, referenced in Fitzpatrick F. Painting over long-standing precedent: how the Rhode Island supreme court misapplied nuisance in state v. lead industries association. *Roger Williams Univ Law Rev*. 2010;15(2):437—471.
25. Citizens for Pres. of Waterman Lake v. Davis, 420 A.2d 53, 59 (R.I. 1980) (quoting Copart Indus., Inc. v. Consolidated Edison Co., 362 N.E.2d 968, 971 (N.Y. 1977)).
26. State of Rhode Island v. Lead Indus. Ass'n, 951 A.2d 428, 453 (R.I. 2008).
27. New York v. Waterloo Stock Car Raceway, Inc., 409 N.Y.S.2d 40, 43 (N.Y. Sup. Ct.); 1978.
28. Pine v. Kalian, No. PC 96—2673, 1998 WL 34090599 (R.I. Super. Feb. 2, 1998), affd, Pine v. Kalian, 723 A.2d 804 (R.I. 1998).
29. R.I. General Laws Section 23—24.6-3; 1991.
30. State v. Lead Indus. Ass'n., 951 A.2d 428, 449 (R.I. 2008).
31. Melley AE. Nuisances, 58 AM. JUR. 2D Nuisances Section 112 (2010).
32. 40 Cal. Rptr. 3d 313, 325 (Cal. Ct. App. 2006).
33. Alex K. PBS series looks back at the lead paint crisis in Rhode Island. Providence Journal; October 21, 2019. <https://www.providencejournal.com/news/20191020/pbs-series-looks-back-at-lead-paint-crisis-in-rhode-island>.
34. The People of The State of California, Plaintiff, Vs. Atlantic Richfield Company, Conagra Grocery Products Company, E.I. Du Pont De Nemours and Company, NL Industries, Inc., And the Sherwin-Williams Company, Defendants. Case No.: 1-00-Cv-788657, Statement of Decision; January 7, 2014.
35. Final Agreement and Full and Complete Release, Joint Motion for Judgment of Dismissal with Prejudice; July 16, 2019. Honorable Judge Thomas E. Kuhnle.
36. California Civ. Code, Section 3479 and 3480.
37. Affidavit of Paul Mushak, Ph.D., Burton v. American Cyanamid et al., Case No. 07-C-0303, U.S.D.C., Eastern District of Wisconsin; September 29, 2010.
38. Kleinberg. The People of The State of California, Plaintiff, Vs. Atlantic Richfield Company, Conagra Grocery Products Company, E.I. Du Pont De Nemours and Company, NL Industries, Inc., And the Sherwin-Williams Company, Defendants. Case No.: 1-00-Cv-788657, Statement of Decision; January 7, 2014.
39. Williams S. Chameleon internal newsletter; 1900. Cited in Case No.: 1-00-CV-788657 Statement of Decision. Superior Court of California County of Santa Clara p. 30.
40. Baldwin EC. *Building Research Advisory Board Meeting, New York City. Inter-Office Letter for Cleveland Executive Office; June 2, 1969*. The University of Illinois Chicago Special Collections & University Archives, School of Public Health, "David E. Jacobs papers" the University of Illinois Chicago School of Public Health library.
41. Case No. 1-00-CV-788657. The People's Post-Remand Brief Re Recalculation of Abatement Fund for Pre-1951 Housing; May 24, 2018.
42. Request for Title and Summary for Proposed Initiative Constitutional Amendment, Pursuant to Article II, Section 10(d) of the California Constitution. Letter from Randy A. Perry (Nielsen, Merksamer, Parrinello, Gross & Leoni, LLP) to Hon. Xavier Becerra (California Attorney General); November 22, 2017.

43. Michael Hiltzik. Fearing a Rebuff in Court, Lead Paint Companies Are Trying to Stick Taxpayers With Their Cleanup Bill. Los Angles Times. January 18, 2018.

44. Lead poisons kids. Who should pay to clean it up: You or the companies that profited? By Dan Morain Sacramento Bee.

45. California Home and School Remediation Bond and Remove Status of Lead Paint as Public Nuisance Initiative (2018). Ballotpedia. <https://ballotpedia.org/California_Home_and_School_Remediation_Bond_and_Remove_Status_of_Lead_Paint_as_Public_Nuisance_Initiative_(2018)>.

46. Fisher D. Slumlords Are the Big Winners In California Judge's $1 Billion Lead-Paint Ruling, Forbes, 12/22/2013. <https://www.forbes.com/sites/danielfisher/2013/12/19/slumlords-are-the-big-winners-in-california-judges-1-billion-lead-paint-ruling/#4a679e5f2870>.

47. January 8, 2014, Lead Paint Ruling: Billion Dollar Public Health Decision in California. <https://www.motleyrice.com/article/lead-paint-ruling-public-health-decision>.

48. Vielmetti B. Jury awards $6 million in lead paint case, finds three firms responsible for lead poisoning of three kids in Milwaukee, Milwaukee Journal Sentinel. Published 12:27 p.m. CT May 30, 2019, Updated 5:06 p.m. CT; May 31, 2019. <https://www.jsonline.com/story/news/2019/05/30/lead-paint-jury-gets-lead-paint-poisoning-case-milwaukee/1284444001/>.

49. Stunning Loss for Lead Paint Makers in Lawsuit by California Cities and Counties, By Lilly Fowler and Myron Levin on December 16, 2013. <https://www.fairwarning.org/2013/12/stunning-loss-for-lead-paint-makers-in-lawsuit-by-california-cities-and-counties/>.

8

The US and international healthy homes movement

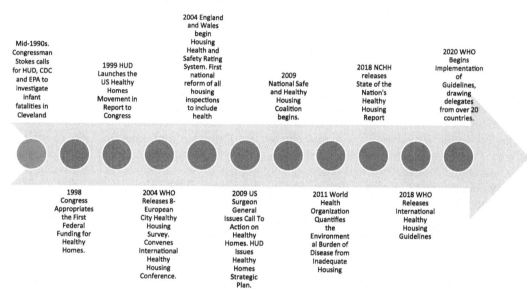

FIGURE 8.0   Timeline for Chapter 8.

8.1 The detective scientists who solved the Cleveland mold mystery

At a Congressional appropriations subcommittee on the US Department of Housing and Urban Development (HUD) in 1996, Congressman Louis Stokes from Cleveland asked for an update on how the Office of Lead Hazard Control had been implementing Title X's lead paint requirements, a routine part of such hearings. But toward the end, he unexpectedly asked what was being done about mold in homes. He reported that there were infants in Cleveland who had died from mold in their homes, and he asked HUD to look into the matter.

Fifty Years of Peeling Away the Lead Paint Problem
DOI: https://doi.org/10.1016/B978-0-443-18736-0.00010-8

HUD proceeded to investigate the Cleveland tragedy, even though it technically had statutory authority mainly for lead paint. There were several key figures there: Dorr Dearborn, a pulmonologist at a local hospital and a Professor at the Case Western Reserve Medical School, who had performed examinations on infants with pulmonary hemorrhage, made the initial contact with the Centers for Disease Control and Prevention (CDC). Ruth Etzel, a branch chief at CDC, led the initial investigation in 1994/1995 and uncovered the apparent involvement of toxigenic mold in the infants' homes. Subsequently, further investigation also involved Terry Allan (the county health department commissioner), John Sobolewski (also at the county health department), and Stu Greenberg, who ran a local nonprofit, Environmental Health Watch. All had years of experience in lead paint poisoning prevention.

Dearborn testified before the House Subcommittee on HUD Appropriations in 1998, where he reported:

> A previously rare disorder, acute pulmonary hemorrhage has been diagnosed in 41 infants in the Greater Cleveland area in the past 5 years. This serious disorder causes infants to cough blood and usually requires intensive care measures to save them. Fifteen of these infants have died including nine originally thought to have had Sudden Infant Death Syndrome.... An investigation of this outbreak was led by the CDC and has linked the disease to the exposure of a toxic mold called Stachybotrys, which was found in the infants' homes ... Stachybotrys, while not a common mold, is known to have a wide distribution. We are aware of a total of 124 cases of acute pulmonary hemorrhage in infants nationwide over the past 5 years. The rapidly growing lungs of young infants appear to be especially vulnerable to the toxins made by this mold.... We urge you to help us attack this newly recognized environmental hazard that is killing young infants in our community.[1]

The 124 cases were based on a mail survey by Dearborn and Etzel in 1996 of all the Pediatric Pulmonary centers in the United States, although no mold studies had been done outside Cleveland.[2] As Dearborn had reported, some of the deceased infants had received a preliminary diagnosis of Sudden Infant Death Syndrome (SIDS), which "is the sudden unexplained death of a child of less than one year of age. Diagnosis requires that the death remained unexplained even after a thorough autopsy and detailed death scene investigation."[3]

But Dearborn stated that the Cuyahoga County Medical Examiner's re-evaluations of 117 infant deaths found seven had "lesions in the lungs that are not consistent with SIDS but rather had evidence of alveolar hemorrhage prior to death. There must be something else going on here."[2] He ultimately settled on a diagnosis of idiopathic acute pulmonary hemorrhage, thought to be a relatively rare disease characterized by repeated episodes of bleeding into the lungs.[4]

What could be causing it? After pursuing some other possibilities, Etzel, Allan, Dearborn, and others thought "maybe we should go look at the houses where these infants lived." They discovered that most of the houses had a badly designed ventilation system, in which basement air was used to supply the furnace/air handler fan to ventilate the house. Normally, ventilation systems do not use basement air but instead have a combination of outdoor air and return conditioned air from living spaces. This bad design later came to be nicknamed the "Cleveland Drop," in which the breathing air in a home from top to bottom comes from the basement (Fig. 8.1).

Not only did the houses all have the same type of bad ventilation, but they also had experienced recent flooding and/or water damage. The basements were found to contain

FIGURE 8.1 Ventilation in Cleveland drawing in unhealthy basement air. *From U.S. Department of Housing and Urban Development. Healthy Homes Program Guidance Manual. Courtesy of Environmental Health Watch/Cuyahoga County Board of Health, Cleveland, Ohio; September 2011.*

FIGURE 8.2 Repaired Cleveland ventilation system. *From U.S. Department of Housing and Urban Development. Healthy Homes Program Guidance Manual. Courtesy of Environmental Health Watch/Cuyahoga County Board of Health, Cleveland, Ohio; September 2011.*

the mold *Stachybotrys chartarum*, which some media reports labeled "black toxic mold." CDC described it as

> a greenish-black mold. It can grow on material with a high cellulose and low nitrogen content, such as fiberboard, gypsum board, paper, dust, and lint. Growth occurs when there is moisture from water damage, excessive humidity, water leaks, condensation, water infiltration, or flooding.[5]

The solution proved to be rather straightforward: correct the basement moisture problem, safely remove the mold, and instead of using basement air for ventilation throughout the house, attach a duct so that air came from a combination of the outdoors and recirculated from the interior living space, not the basement (Figs. 8.1 and 8.2). But that turned out not to be the end of the story.

Kathryn Mahaffey later said the mold had the "Yuck" factor. Anyone who looked at it could see that it was bad news (Fig. 8.3).

Etzel and Dearborn decided to report their findings, thinking that perhaps there were others who had been misdiagnosed elsewhere. Idiopathic pulmonary hemorrhage is not

FIGURE 8.3   Stachybotrys mold. Reproduced with permission, Shutterstock.

tracked by CDC and remains difficult to diagnose. Together with colleagues they published the findings,[6] which Dearborn reported to Congress in 1998.[7]

Instead of thanks, Etzel, Dearborn, and Allan became the targets of a few high-ranking officials within CDC who believed that other factors might have contributed to the deaths and that the Cleveland cases were inconclusive. In a reversal of their previous findings, CDC stated:

> A review within CDC and by outside experts of an investigation of acute pulmonary hemorrhage/ hemosiderosis in infants has identified shortcomings in the implementation and reporting of the [earlier CDC] investigation ... a possible association between acute pulmonary hemorrhage/hemosiderosis in infants and exposure to molds, specifically Stachybotrys chartarum ... was not proven.[8]

The CDC's initial assessment that had been issued just a few months earlier stated:

> An investigation led by the Centers for Disease Control and Prevention has found an association with household exposure to a toxigenic mold, Stachybotrys chartarum.... Sufficient association and rationale exist to institute public health prevention measures...[9]

A bitter battle ensued within CDC over whether the Cleveland data were reliable, with attacks on Dearborn's professional and scientific integrity. Not everyone in CDC went along, notably Ruth Etzel, who believed that there was sufficient proof and found the evidence compelling. Because she advocated for a more productive response, she was sidelined by CDC and eventually forced out of the agency. In 2010 Dearborn was invited to give a research seminar at CDC on their subsequent studies and afterward received an informal apology from CDC.

HUD's reaction to what had happened in Cleveland was quite different from CDC's. HUD believed it was obvious that housing and its inadequate ventilation had played a key role in infant deaths. Instead of waiting for more conclusive clinical proof, HUD thought it was more important to prevent these and other tragedies by implementing housing remediation and ventilation improvements. HUD's examination of the data found it to be compelling and sound. HUD announced its first request for healthy homes funding in Feb 1998 for the FY 1999 budget.[10] In that budget request, HUD reported the preliminary findings of its own investigation of the Cleveland situation:

> An outbreak of "bleeding lungs" (pulmonary hemosiderosis) in 1996 killed at least 10 children in Ohio and another 60 infants nationwide before it was traced to a toxic mold caused by inadequate home ventilation and humidity controls.[11]

Congressman Stokes had been trying to get some action on the mold problem for years from both CDC and EPA. In an earlier hearing held in 1995, Congressman Stokes asked David Satcher, then the director of CDC, what was being done about mold:

> Mr. Stokes: There have been over 25 cases of infant pulmonary hemorrhaging in Ohio, about 10 in Illinois, and 1 or 2 throughout the rest of the country. What can you tell us about this disease, and are we just seeing the tip of the iceberg?

> Dr. Satcher: As you may know, CDC is reviewing this problem in Cleveland, Ohio. Acute pulmonary hemorrhage, typically manifested as blood from the nose or mouth, is uncommon in infants. Some cases of infant pulmonary hemorrhage are associated with cardiac or vascular malformations, infectious processes, or trauma; however, for most cases, *the cause is not identified*. Pulmonary hemorrhage results in the presence of iron in the lungs resulting from the breakdown of red blood cells. In cases of acute pulmonary hemorrhage, these hemosiderin - laden macrophages (iron in the lungs) can be detected beginning approximately two weeks after the bleeding episode. Chronic pulmonary hemorrhage is known as pulmonary hemosiderosis. Cases of pulmonary hemosiderosis for which the cause is undetermined traditionally have been classified as idiopathic (unknown cause) pulmonary hemosiderosis and account for most of the cases of pulmonary hemorrhage during infancy. Twenty - six cases of idiopathic pulmonary hemorrhage have been reported in Cleveland since 1993. Relying on voluntary case reporting from physicians and other health care providers, twelve cases have been reported from Illinois and thirty - three cases from various other states throughout the United States. The actual number of idiopathic pulmonary hemorrhage cases that occur among infants in the U.S. is unknown, since this is not a reportable condition.

> Mr. Stokes: Since November of 1994, the Centers for Disease Control has been investigating an outbreak of pulmonary hemorrhage in infants in my district. The disease appears to be due to a toxic fungus found in the home of the infants. What is the CDC doing about prevention of this environmental fungal problem, is there a prevention effort underway or planned?

> Dr. Satcher: Since November 1994, the Centers for Disease Control and Prevention has been investigating the cause of the outbreak of pulmonary hemorrhage in Cleveland. Based on this investigation, several lines of evidence indicate a possible causal link between this condition and a fungus called Stachybotrys atra. It is unknown if this link exists in conjunction with cases being reported in the other parts of the U. S. The planning of prevention efforts has begun in the Cleveland area. These efforts will be implemented as a cooperative effort by Rainbow Babies and Children's Hospital, the Cuyahoga County Board of Health, the Cleveland Department of Public Health, and the Cleveland Metropolitan Housing Authority. The prevention efforts will focus on 1) distribution of information to mothers of infants, 2) detection of mold, and 3) abatement / clean - up of molds. Upon completion of a planned National Case - Control Study of Idiopathic Pulmonary Hemorrhage in Infants, a determination can be made whether or not this is a problem unique to Cleveland or is spread throughout the U. S. If it is determined this problem exists in other parts of the US, a prevention plan will be formulated and implemented. As a part of the case - control study, prevention recommendations, informational materials, and technical assistance as requested will be provided to local health departments that identify and report cases (emphasis added).[12]

Stokes found this answer frustrating and wanted more done to prevent the problem, not just study it further to see if Cleveland was somehow unique. Although Satcher's statement seemed to indicate a cleanup of mold might be undertaken, no cleanup was undertaken by CDC and there was no mention of ventilation.

The answer from the EPA Administrator in 1998 also demonstrated the lack of cooperation among different government agencies, similar to what had stymied the lead poisoning prevention efforts in the 1970s and 1980s. Congressman Stokes put the same mold question to EPA Administrator Carol Browner and was told there was no clear answer and that there had been no coordination with HUD:

> Cong Stokes: As you know, there is an outbreak of pulmonary hemorrhage in infants in this cluster in my Congressional district. There have been 38 cases diagnosed in the Cleveland area, including 14 deaths and 122 cases nationwide in the past five years. Do we know why a third of the cases of pulmonary hemorrhage in infants have occurred in the Cleveland area and why half of the total number of cases are clustered around the Great Lakes area?

> Ms. Browner. We are looking at this issue. We are working with the Centers for Disease Control and the National Institute for Environmental Health Sciences. We all want to understand what is happening here. At this point in time, *we do not have a clear answer*, but we are continuing to look at the issue...

> Mr. Stokes: The issue crosses the interests and jurisdictions of the Department of Health and Human Services, HUD, and EPA. Can you tell me whether or not there are any collaborative efforts going on between those agencies relative to finding out the causes of this problem?

> Ms. Browner. Yes, we are working with HHS. *I am not sure that we are working with HUD* and that is an interesting point that you raised. We will contact them and see what would be appropriate in terms of their involvement (emphasis added).[13]

The first report of the problem in Cleveland was in 1994.[14] By 1997 Stokes had had enough of the delay from CDC and EPA. Recognizing this was fundamentally a housing problem, he turned to HUD. He convinced his fellow Congressional members on the House Appropriations Committee to provide the first federal money for a healthy homes effort at HUD, beginning in the fiscal year 1999.

The initial HUD request for the new healthy homes program was $25 million, which was reduced to $20 million in the House and $10 million at the House/Senate conference committee that made the final decision. This marked an important expansion of HUD's responsibilities beyond lead paint. The final appropriations bill for HUD said:

> For the Lead Hazard Reduction Program, as authorized by sections 1011 and 1053 of the Residential Lead-Based Hazard Reduction Act of 1992, $80,000,000 to remain available until expended, of which ... $10,000,000 shall be for a Healthy Homes Initiative, which shall be a program pursuant to sections 501 and 502 of the Housing and Urban Development Act of 1970 that shall include research, studies, testing, and demonstration efforts, including education and outreach concerning lead-based paint poisoning and other housing-related environmental diseases and hazards.[15]

> [Congress] Provides that $70,000,000 of the total appropriation shall be for lead hazard reduction ... and $10,000,000 shall be for the Healthy Homes Initiative, instead of $20,000,000 as proposed by the House.[16]

In the Committee report accompanying this first federal healthy homes appropriation, there were further details about what Congress wanted, with the first articulation of a key healthy homes concept:

> A central goal of the Healthy Homes Initiative is to develop and implement a program of research and demonstration projects that would address *multiple housing-related problems* affecting the health of children.... The Committee requests HUD to submit a plan by January 1, 1999 that inventories the problems to be addressed, describes their intersections, identifies key technical questions, and provides a spending plan. In developing this plan, HUD should seek input and advice from experts and researchers, other federal agencies, and experienced local practitioners. Within the [$10 million] Healthy Homes Initiative, the Committee directs that *a minimum of $4 million be devoted to preventive measures to correct moisture and mold problems* in inner-city housing occupied by families with infants in communities where toxic mold exposure has been linked to acute pulmonary hemorrhage and infant death. In addition, as part of the initiative, the Committee also expects HUD to undertake research on moisture and mold prevention *through proper ventilation* and other means, and to develop and disseminate model standards appropriate to residential housing (emphasis added).[17]

8.2 The Department of Housing and Urban Development healthy homes report to Congress

HUD released the mandated healthy homes plan in a report to Congress in 1999 written primarily by Warren Friedman, Peter Ashley, Tom Matte, Molly McNairy and David Jacobs.[18] The office would later be renamed as the HUD Office of Lead Hazard Control and Healthy Housing to reflect its broader mission.

In preparing the Report to Congress on Healthy Housing, HUD convened a large panel of experts to help conceptualize a broader approach to healthy housing beyond its lead poisoning program, focusing initially on mold as Congress had directed. CDC's Jerry Hershovitz, then the director of its lead poisoning program, cochaired the panel with HUD. He had vast experience in how substandard housing affected health, working in various pest control, lead poisoning prevention, and other health programs across the country. Tom Matte at CDC also helped in this new effort. He and Jacobs later published a peer-reviewed version of the Report to Congress.[19]

In addition to Dearborn and Allan from Cleveland, the experts included some who had been running housing and health programs for years, such as Dick Svenson's Healthy Neighborhoods program in New York State. Steve Schwartzberg in Alameda County had already begun to move his program beyond a sole focus on lead poisoning prevention. The effort also included Bruce Lanphear and other long-time researchers in the lead poisoning prevention field such as Michael Weitzman and Scott Clark. It was also informed by pediatricians, two of whom established the Doc4Kids Project in Boston.[20,21] One of the two (Megan Sandel) would later champion the idea of "housing as a vaccine."

Other experts included building scientists and engineers who understood ventilation, moisture, and dust control, like Terry Brennan and Phil Morey, an occupational health scientist who was perhaps the nation's leading mold expert at the time, together with J. David Miller in Canada. Brennan helped create a healthy homes training curriculum for a new nationwide training center and network, which would be housed at the National Center for Healthy Housing. The Center had changed its name from the National Center for Lead Safe Housing in 2001 to reflect the broader mission. The training effort would eventually coordinate about 20 universities and other training organizations across the country.

Nick Farr and Rebecca Morley from the National Center for Healthy Housing, Don Ryan and others also played important roles in expanding the mission from lead paint to other housing-related diseases and injuries. The Alliance to End Childhood Lead Poisoning was renamed the Alliance for Healthy Housing in 2002.

HUD's report to Congress[18] that launched the national healthy homes movement in the United States explained the deliberations of the experts and was the first to state its key principle: instead of having separate programs, campaigns, home visits, and professions for lead, mold, injuries, radon, asbestos, asthma, and other housing problems, it made much more sense to adopt a more integrated holistic approach.

In practice, that meant a departure from the traditional approach of attempting to correct one housing danger at a time, because a limited number of building deficiencies contributed to many hazards at the same time. Substantial advances in health and cost savings were possible using this new approach because each separate visit to a home by an inspector, public health nurse, outreach worker, or remediation contractor came at a significant cost.

An intense debate followed. What exactly should those "limited number of building deficiencies" be? And what did an "integrated holistic approach" really mean?

Initially, the healthy homes report to Congress identified four key factors: excess moisture; dust; ventilation to control toxic substances; and education, although the last one is not really a building deficiency. Over the ensuing years, this list expanded as the program matured.

By 2020 the key principles still included moisture and ventilation, but the four key factors had expanded to 10:

- pest control,
- maintenance,
- cleaning,
- heating and cooling,
- toxic substance control,
- safety from injuries,
- ventilation,
- moisture control,
- affordability, and
- accessibility.[22]

This was essentially a "systems" approach to housing. For example, controlling moisture not only addressed lead paint deterioration (moisture is the main cause of premature paint failure); it also helped to prevent injuries from falls due to rot and structural deterioration, pests, and asthma triggers (such as mold). The many ripple effects of water leaks to diseases, injuries, and numerous health and social outcomes are shown in Fig. 8.4 (starting with water leaks in the middle).

Housing problems cause a ripple effect of impacts.

Here is an example of how a single housing problem—water—can lead to multiple health effects and economic impacts:

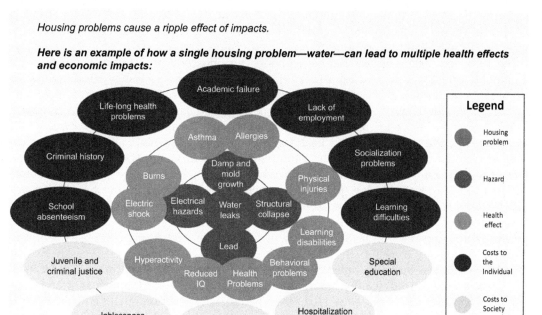

FIGURE 8.4   Effect of water leaks on houses, disease, and associated outcomes. *From US Department of Housing and Urban Development Office of Lead Hazard Control and Healthy Homes. The Impact of Housing Quality on Health, February 28; 2017:13.*

The main factors a healthy homes effort might address were potentially much longer than the initial four, which posed a dilemma: how could healthy homes be both focused and comprehensive? In preparing the Report to Congress, the experts initially came up with a much longer list: Lead, Asthma/allergens, Mold, Insect pests, Rodents, Pesticides, Other toxic chemicals, Environmental tobacco smoke, Combustion byproducts, Radon, Asbestos, Take-home contaminants (from work), Unintentional injuries (including fires), Uncontrolled moisture, Inadequate ventilation, Soil gases (other than radon), Hazardous building materials, Drinking water contamination, Sewage backup, Swimming pools, Noise and Vibration, Crowding, Firearms, Faulty Construction, Appliances (ozone generators, humidifiers, unvented clothing dryers), Food handling and sanitation, and even pets (their dander had been linked to allergens and asthma).

When first shown this list, Kathryn Mahaffey, a dog lover, jokingly suggested that the healthy homes program would become a pet eradication program. Of course, that longer list was unworkable (even with the removal of pets from the list).

This dilemma between comprehensiveness and focus remained part of the debate on how to implement the healthy homes concept in the years to come.

The HUD healthy homes report to Congress also called for renewed attention to housing codes and housing inspection methods. A Healthy Housing Reference Manual was released by CDC and HUD in 2006,[23] followed 2 years later by a Healthy Housing

Inspection Manual.[24] Both were updated versions of the 1938 and 1986 Basic Housing Inspection Manual (which were essentially model housing codes), issued by CDC and the American Public Health Association.[25]

The 1938 and 1986 manuals became a model housing code in 2014 when the National Healthy Housing Standard was released by the National Center for Healthy Housing and the American Public Health Association. Two former HUD Secretaries (Henry Cisneros and Shaun Donovan) spoke at the press conference, stating how much the new model code was needed.[26]

But other housing codes remained problematic when it came to health. Between 1976 and the 1990s, the early integrated health and housing codes from CDC and APHA had been supplanted by more fragmented plumbing, electrical, ventilation, maintenance, and other model codes from the International Code Council, which largely abandoned health concerns like lead paint, radon, pest management, and other key healthy housing principles. Most local housing codes were enforced only by complaints, although some jurisdictions became much more proactive in the 2010s.[27,28] Chapter 9 examines this disconnect between housing codes and healthy homes, and how a new emerging consensus promises reform.

In 1999 Congress had also called for a series of demonstration and research projects to prove that a modernized healthy homes approach could really work, much as it had done in Title X for an evaluation of the HUD lead paint grant program. HUD launched major new healthy housing research efforts, beginning with mold. By 2019 HUD had completed more than 90 healthy housing studies.[29] HUD's healthy housing research efforts were led by Peter Ashley, Warren Friedman, Ellen Roznowski-Taylor, Kofi Berko, Eugene Pinzer, Brenda Reyes, David Jacobs, and others.

HUD's 1999 report to Congress cited other healthy homes issues beyond mold: 80% of all fire deaths occurred in the home, with residential fire fatalities at a rate of 26 children under 6 years old per million; yet only 52% of households had working smoke alarms. It presented evidence that thousands of children died each year from exposure to pesticides, solvents, and other chemicals improperly stored and used in the home.

The 1999 Report to Congress presented an initial attempt to show the many links between housing and health (Table 8.1), and later attempts produced an even more complex "causal web" developed by the World Health Organization, discussed later in this chapter.[30]

The 1999 HUD Report to Congress examined how the new program could overcome both scientific and implementation barriers:

1. Scientific issues
   a. What is the causal relation of the housing hazard to health?
   b. What is the estimated prevalence of the hazard and burden of associated illness or injury?
   c. Are practical, valid, and reliable methods and protocols for assessing the hazard available?
   d. Is there scientific evidence to support practical, safe interventions that significantly reduce or eliminate the hazard?
   e. Which of the hazards are high-priority public health problems?
   f. For which hazards can action be taken without the use of specialized personnel, elaborate testing, or laboratory analysis?

TABLE 8.1 Housing deficiencies and associated health outcomes

	Inadequate ventilation	Dust traps	Moisture intrusion/ humidification systems	Broken Windows/other structural defects	Inadequate early warning alarms or prevention devices	Hobbies	Egress	Improper renovation	Deferred or inadequate maintenance
Lead poisoning	X	X	X	X		X		X	X
Allergens/ asthma	X	X	X	X		X		X	X
Toxigenic mold	X	X	X	X				X	X
Insect pests		X	X						X
Rodent pests		X	X						X
Pesticide exposure	X	X				X		X	X
Environmental tobacco smoke	X					X			
Carbon monoxide poisoning	X			X	X		X	X	X
Respiratory irritants	X		X				X	X	X
Radon exposure	X			X				X	X
Asbestos								X	X
Paraoccupational exposures	X								
Injury	X		X	X	X	X	X	X	X
Fire	X				X	X	X	X	X

From Report to Congress on the Healthy Homes Initiative, US Department of Housing and Urban Development; 1999.

g. Which corrective measures may introduce new hazards into the home environment or work at odds with interventions intended to control other hazards?

2. Implementation issues
   a. How should healthy homes activities address community-level hazards in relation to individual-home hazards?
   b. How can demonstration/evaluation projects be designed to test implementation models, and what are key elements to include in such demonstration projects?
   c. How can evaluation and assessment of housing-related health and safety problems be accomplished most efficiently?
   d. How can basic research on the specific causes and pathways of housing-related illnesses and injuries and building conditions be combined with demonstration projects?
   e. What are the best evaluation markers of healthy homes activities? Rates of illness, injury, or other biological markers? Economic data (e.g., healthcare costs, housing value, energy consumption)? Knowledge/behavior of stakeholders (e.g., occupants, landlords, property professionals, public health professionals)? Changes in environmental or housing conditions (vs the actual conditions themselves)?
   f. How should target groups (e.g., populations, housing, communities) be selected?

Finally, the HUD 1999 report to Congress included a detailed spending plan for the initial $10 million Congress had appropriated:

- $4 million to solve mold and moisture problems in inner-city housing by demonstrating the effectiveness of remediation methods developed in the research setting, but not yet implemented in large numbers of urban houses;
- $4 million to demonstrate and evaluate housing repairs that simultaneously prevent asthma, lead poisoning, pulmonary hemorrhage, injuries, and other health and safety threats to children in several hundred homes in several cities;
- $500,000 for a healthy homes public education campaign;
- $500,000 to develop a housing-based surveillance system; and
- $1 million to establish epidemiological baseline data, develop the capacity of code enforcement agencies to take on certain health and safety housing issues, research methods of measuring mold and allergens in the housing environment, examine home energy design methodologies, and other projects.

Congressman Stokes warmly greeted the plan. But others in Congress were not so sure. For example, Senator Mikulski, the long-time supporter of HUD's lead paint program and responsible for creating the HUD lead paint office, initially responded:

> I don't want snappy slogans! I want programs that actually help people.[31]

She later became convinced that the healthy homes concept was a good one, but initially she did not support it because it encompassed too many issues and initially appeared inadequately focused. Others worried that it would distract local governments' attention away from lead paint or duplicate other programs.

But the superiority of the healthy homes approach gained steam after overcoming these initial barriers.

8.3 The Surgeon General's Call to Action

In 2009 the US Surgeon General issued a "Call to Action to Promote Healthy Homes."[32] Surgeon General reports carried a great deal of weight among health professionals. For example, the large-scale smoking cessation public health campaign began in the 1960s from such a report.[33]

Largely written by Mary Jean Brown at CDC, the Surgeon General's Call to Action on Healthy Housing offered this more prevention-oriented definition:

> A healthy home is sited, designed, built, maintained, and renovated in ways that *support* the health of its residents (emphasis added).

This emphasized not only correcting housing deficiencies after they had already caused disease or injury. It was a move away from the medical model of treating people only after they become sick instead of preventing it in the first place. The Call to Action stressed the need for cross-discipline and interagency coordination. Title X had made clear assignments to different agencies and the emerging healthy homes consensus applied that lesson to other housing related disease and injuries.

A few in Congress questioned why HUD, CDC, and EPA all needed their own healthy homes programs. In 2010 the three agencies responded with a clearly delineated healthy homes framework for each agency (Table 8.2).[34] HUD, CDC, and EPA also collaborated to support infrastructure development for healthy homes, meaning trained and certified workers who could reliably identify and remediate other housing hazards. Some of these workers were also cross trained in related fields such as home weatherization. This had also been a key feature of Title X for lead paint inspectors, risk assessors, and abatement contractors.

TABLE 8.2 Healthy homes frameworks at HUD, CDC, and EPA, 2010.

HUD	CDC	EPA
• Evaluates housing interventions • Trains the workforce • Reduces lead paint violations in federally assisted housing • Initiates model housing codes/industry standards • Grants to state and local housing programs for lead hazard control	• Blood lead surveillance to identify populations and areas with high risk for exposure • Education of professionals and the general public on the hazards of lead and ways to reduce exposure. • Cooperative agreements with state and local health departments to identify and provide services to affected children.	• Authorizes states to license lead paint professionals • Maintains the "Lead Hotline" with HUD and CDC • Enforces the Disclosure Rule with HUD and DOJ • Addresses lead at superfund sites with CDC • Establishes standards for lead in environmental media.

From Mary Jean Brown, Matthew Ammon, Peter Grevatt. Federal agency support for healthy homes. J Public Health Manage Pract. 2010;16(5) E-Suppl:S90–S93.

But the lack of Congressional authority for healthy homes beyond HUD's appropriation had consequences for both CDC and EPA. CDC had separate asthma and lead programs. Avoiding duplication meant that the CDC healthy homes effort would focus on residential asthma triggers, and the asthma program would focus on medical surveillance, pharmaceutical interventions, and education. This division of labor was supported by virtually all CDC staff.

But Tom Friedan, the CDC director in 2012, distorted the healthy homes concept that CDC had previously supported to attack its lead poisoning prevention program. The CDC lead program was virtually eliminated by pitting the lead program against the asthma program and repurposing lead poisoning prevention funding into so-called "winnable battles" (Chapter 5). Perhaps the most recent example of the importance of CDC within the housing world has been its ban on housing evictions during the Covid-19 pandemic.[35]

At EPA, healthy homes activities were scattered among many separate offices, such as children's health, indoor air quality, pesticides, research and development, and others. Despite the staff efforts within CDC and EPA, those agencies never established a formal healthy homes program.

This is a reflection of Congress' failure to enact a "Title X for Healthy Housing." As we have seen, Title X had mandated clear lead poisoning prevention duties for each agency (discussed further at the end of this chapter).

8.4 Assembling the evidence

Initially focused on research, the HUD healthy homes program grew to include demonstration projects, as the field expanded into more widespread practice. By 2022, the demonstration projects had proven so successful that HUD expanded the effort into a "healthy home production grant program," funded at $105 million.[36] Together, they showed not only the connections between housing and health, but also the effectiveness of various interventions.

The seed planted by Congressman Stokes had sprouted.

By 2007 the need to systematically examine what worked (and what did not) had become obvious. If standards and broad implementation to reflect the emerging healthy housing consensus were to occur, the evidence base had to be assembled. Decisions would need to be made about which were ready to implement and which needed more research and development.

In addition to the Surgeon General's Call to Action described earlier, two other important developments occurred between 2007 and 2010: a new HUD healthy homes strategic plan and a systematic review of scientific evidence by CDC and the National Center for Healthy Housing.

After the report to Congress in 1999, HUD produced a new healthy homes strategic plan in 2009, approved by none other than Shaun Donovan, who became HUD Secretary that same year.[37] Donovan helped save the HUD lead paint regulation reform effort a decade earlier in 1999 when he was the deputy commissioner of the Federal Housing Administration (Chapter 4). He later supported HUD funding for the World Health Organization (WHO), together with other nations to enable its *Housing and Health Guidelines* at the international level (discussed later in this chapter).

The initial focus on the mold outbreak in Cleveland led to a HUD-funded randomized controlled trial there, published in 2006. It showed significant improvement in asthma symptoms (including reduced acute care trips to the doctor) among children following mold and moisture remediation. During the 12 months of follow-up, the control (nonintervention) group saw an almost 20% higher rate of emergency department visits or hospital in-patient visits compared to the intervention group. The difference between the two groups was 30%, a huge improvement.[38] This study, in part, ultimately led to a large mold remediation industry, with one estimate placing its overall market value at $2.5 billion in 2020,[39] evidence that the healthy homes effort was moving beyond research to practice.

In Seattle, researchers funded with a HUD healthy homes grant studied green-built public housing units nicknamed "Breathe Easy Homes," with construction beginning in 2004 (Fig. 8.5).[40] They had special features to improve indoor air quality and reduce indoor asthma triggers. Children with asthma who moved into these homes experienced 12.4 asthma symptom-free days per two-week period after 1 year, compared with 8.6 asthma symptom-free days in the control group who did not live in such homes. Urgent

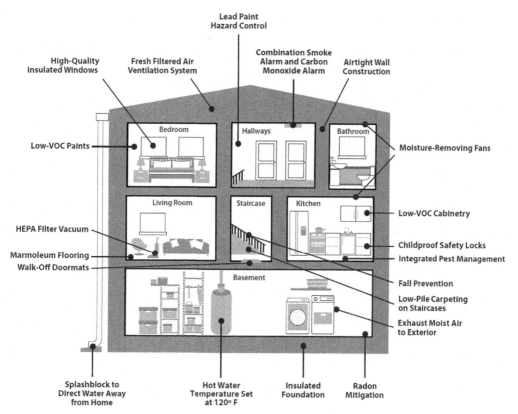

FIGURE 8.5 Key features of a healthy home. *From: Public Health/Seattle and King County, Washington State and the National Center for Healthy Housing.*

asthma-related clinical visits in the previous 3 months decreased from 62% to 21% and the caretakers' quality of life increased significantly. The financial benefits were enormous, because a 40% improvement in expensive emergency room visits was realized through the housing improvements in this study. Significant reductions in exposures to mold, rodents, and moisture were also reported in the Breathe Easy Homes.[41–43]

An evaluation of the New York State Healthy Neighborhoods Program, which had started in the early 1990s also showed significant benefits. Dick Svenson ran the program for years and served on the initial HUD/CDC healthy homes expert panel in 1999. By the mid-2010s, the program provided healthy homes services to over 36,000 residents in 13,120 dwellings in 12 counties across the state. There were significant improvements in tobacco control, fire safety, lead poisoning prevention, indoor air quality, and general environmental health and safety (e.g., pests, mold). For residents with asthma, there were significant improvements in environmental triggers, self-management, and short-term morbidity outcomes, including up to 3.5 fewer days with worsening asthma in a three-month period.[44,45] A cost-benefit analysis showed that the per person savings for those with asthma who received healthy homes services had a net benefit of $781 per home visit and a benefit to program cost ratio of 3.5 (for every dollar spent, there was $3.50 in benefits).[46]

Health foundations also began to fund healthy housing studies and activities, yet another reflection of the emerging consensus. The National Center for Healthy Housing teamed with the Enterprise Community Partners (which had helped to start the Center in 1992) to examine if "green building standards" could improve not only energy use and sustainability but also health. Enterprise created the Green Community Standards for use in low-income housing rehab and new construction in 2004. These and other green building standards were becoming more broadly used, with one 2020 estimate valuing the green building industry at $81 billion in the United States alone.[47]

The Blue Cross Foundation (a philanthropic institution associated with a health insurance company) in Minnesota wanted to know what the health benefits really were from those standards. It funded one of the first healthy green housing studies in 2008 in a low-income multifamily housing development known as Viking Terrace in Worthington Minnesota. The mayor had become alarmed about housing conditions after a young child nearly died from carbon monoxide from a poorly ventilated gas hot water heater. The results of the study showed there were statistically significant improvements in overall health, asthma, and nonasthma respiratory problems following green healthy housing improvements. Adults also reported that their children's overall health improved. New mechanical ventilation was installed (compared with no ventilation other than opening windows previously) and energy use was reduced by 45% over the 1-year postrenovation period, in part because residents no longer had to use fans to move air from one room to another. Because about 40% of energy use went to heat and cool buildings, the idea that both health and energy use could both improve was an important new finding,[48] especially important in the effort to combat climate change.

One of the Viking Terrace residents (Abang Ojullu) was a refugee from Somalia. She had a terrible journey, traveling through multiple refugee camps and her husband had been killed in the war; miraculously, she eventually arrived in Minnesota to make a home with her children in a housing development that was about to undergo green healthy housing improvements. Published by Robin Jacobs, the story described the difficulty her daughter faced in their new home:[49]

> Abang Ojullu remembers all too vividly the day she put her eldest daughter on a small ambulance jet bound for Sioux Falls. The child's asthma attack was too severe for doctors in rural Worthington, Minnesota, to treat. Her daughter Ananaya had experienced a long litany of maladies in her short life. Some winters the girl's asthma was compounded by pneumonia. Frequent hospitalizations and doctor visits meant missed school days. For two years, Abang made the hour-long drive to Sioux Falls once a month so that Ananaya could see a specialist. But last winter, six months after moving into the newly renovated Viking Terrace Apartments, Ananaya did not get sick once, her mother says. Neither did any of her five other children, although in the past each had bouts of asthma that often required nebulizer treatments.

8.4.1 Does it work? A systematic review of housing interventions and health

A systematic review of all the scientific evidence to determine which healthy housing interventions worked, which needed more research and which ones failed was completed by National Center for Healthy Housing (NCHH) and CDC in 2009. It would take more than 2 years to complete. (Another systematic review was completed a decade later from the World Health Organization, discussed later.)

The 2009 systematic review used a method that had been in place for many years at CDC, known as the "Community Guide."[50] The results of these systematic reviews were published in several peer-reviewed papers.[51-57] The interventions were grouped by biological, chemical, injury, and community-based solutions. There were some surprises.

For example, the group studying biological interventions found that for the asthma studies, just doing one thing, like putting on allergen-resistant mattress covers was not very effective. But using a more holistic multifaceted approach tailored to the individual, home-based asthma interventions clearly worked. Also, the initial assessment suggested that mold and moisture control only had weak evidence of health improvement, but further analysis found they were effective.[52]

The experts working on chemical contaminants in the home[53] found that certain radon mitigation methods were effective, although initially they could find only one published study on it. The group wondered why all the EPA studies over the years could not be found in the peer-reviewed literature. They did in fact exist and had been peer-reviewed by EPA but had mostly only been published in paper EPA reports at the time. So, they were scanned and placed on the internet so that future literature reviews would be able to find them.[54] When the additional evidence was found, so was the effectiveness of radon mitigation. A frequent criticism of systematic reviews is that they fail to examine all the evidence that is actually available.

The injury group was also presented with a dilemma. Many injury interventions had never been studied one-by-one; instead, they had been studied as a "package," posing a challenge for the systematic review process, which preferred the one-by-one approach (to help determine which of the individual interventions in the package had been responsible for which health improvement). Some wondered if separate studies to show that smoke alarms work, or hand railings prevented falls was really needed. The evidence of injury intervention "packages" (a reflection of the holistic healthy homes approach) did emerge from several studies,[55] including a randomized controlled trial in

Cincinnati. That study showed there was a significant reduction in the rate of modifiable medically attended injuries for the children compared with controls (2.3 injuries vs 7.7 injuries per 100 child-years).[56]

Finally, the community intervention group found that housing rental vouchers for low-income renters provided by HUD through its "section 8 housing choice voucher" program had produced significant health improvements.[57]

These systematic reviews helped to standardize health homes interventions, because they established the proof that they worked.

HUD, CDC, and EPA also collaborated to support infrastructure development for healthy homes. A wide variety of health care and housing workers in state and local agencies were trained and certified. Some of these workers were also cross trained in related fields such as home weatherization.

Additional healthy housing research efforts in later years continued to reveal new understanding of how building materials can influence health. Ohio would continue to be a focal point for such research, as it had been for mold. For example, in 2019 Ohio State University conducted a national symposium on health effects associated with carpets and how mold and other substances could influence genetic components of house dust embedded in such carpets. It highlighted how new genetic sequencing methods could be employed to better understand indoor chemistry and how chemical reactions in the home environment might influence health.[58]

These developments marked a significant improvement and more widespread implementation in other sectors, such as energy. Although the first attempts at energy conservation in the 1970s had produced "sick building syndrome" due primarily to inadequate ventilation,[59,60] the more recent studies of green housing and health showed health *benefits* from more energy-efficient homes with smarter approaches to ventilation.[61–65] These and many other studies were reviewed by the Department of Energy and NCHH.[66]

8.5 The formation of the National Safe and Healthy Housing Coalition

Launched in 2009, by 2022, many local parent and community groups were members of the National Safe and Healthy Housing Coalition, which had over 650 members representing over 400 organizations with representation in all 50 states and some territories.[67] A broad umbrella group organized through the efforts of the NCHH and others, the Coalition emerged as the nation's leading healthy homes entity through which advocates, local governments and scientists could coordinate. The Kresge Foundation provided support in its early years. NCHH also provided limited funding and technical assistance to local groups around the country,[68] as did the Green and Healthy Housing Initiative.[69] But as in earlier years the local groups remained far too poorly funded to meet the need. The experiences of these groups are detailed elsewhere.[70]

The Coalition organized "Hill Days" to meet with members of Congress and their staffs. Over a hundred delegates met with 140 different Congressional offices in 2018.[67] The typical visit included parents of lead poisoned children, local community groups, local political leaders, medical care providers and businesspeople engaged in fixing homes.

The presence of an active, organized coalition of citizens and scientists made all the difference, as it did in earlier years with the formation of the Alliance to End Childhood Lead Poisoning, the National Center for Lead Safe Housing, United Parents Against Lead and other parent and community groups. Philanthropy had helped to launch these and other groups in the 1990s and again in the 2010s when the National Safe and Healthy Housing Coalition was formed. As a result, appropriations for lead and healthy homes activities for HUD, CDC and EPA all increased substantially after 2017 (see Conclusion to this book).

8.6 The World Health Organization Healthy Homes Movement

The 2007 systematic review of healthy housing effectiveness in the United States benefitted from the involvement of David Ormandy in England. His healthy homes work there had started much earlier, in the 1970s, marking the first modern efforts to integrate housing and health.[71] (Chapter 9 describes the first intersection of housing and health through the sanitation movement starting in the late 1800s.)

In England and Wales, he and others succeeded in changing the way in which all housing inspections were carried out. He shifted the focus from one on the building to one on potential threats to health, which he called "the effects of defects." Together with many other collaborators, he helped to create a Housing Health and Safety Rating System,[72] which first appeared in the late 1990s. The legislation for it was passed in 2004[73] and nationwide training and implementation occurred 2 years later. All housing inspectors in England and Wales implemented it starting in 2006.

The system was evidence-based. It merged health data from hospitals and other sources with housing data to estimate severity of injury and disease from building defects, aided by England's National Health Service (a single payer health care model). The British housing inspectors were highly educated and well-trained in the new inspection system.

When asked why he had succeeded, Ormandy simply smiled and said it was because he had been at it longer.[74] A Professor at the University of Warwick, many admired his no-nonsense attitude and ability to implement broad changes. Other key collaborators in England that made possible their reform of housing inspections there included Richard Moore, Stephen Battersby at the University of Warwick, Simon Nichol from Building Research Establishment, and Paul Wilkinson of the London School of Hygiene and Tropical Medicine.

Ormandy also organized what was likely to have been the first international healthy homes conferences, which was held at the University of Warwick in 1986 [titled "(Un)healthy Housing: A Diagnosis"]. This was followed by four more: "Unhealthy Housing: Prevention and remedies (1987)," "Unhealthy Housing: The Public Health response (1991); "Healthy Housing: promoting good health (2003)" and the 5th Warwick Healthy Housing Conference (2008).[75]

Because it was largely focused on US cities, HUD historically was not very active internationally. One exception was when HUD Secretary Henry Cisneros led the United States delegation to the UN Habitat II conference in 1996.[76] That conference issued this general statement on housing:

> Member States are committed to endorse the universal goals on ensuring adequate shelter for all and making human settlements safer, healthier, and more livable, equitable, sustainable, and productive.[76]

The Alliance to End Childhood Lead Poisoning held a major international conference in 1994,[77] focused mostly on ending global use of lead gasoline and new lead paint. There had also been international lead poisoning conferences in Australia (1994),[78] Mexico (1996),[79] and other countries, but none of these were focused on broader housing and health issues.

Internationally, many lead poisoning prevention community groups operated under the International Pollutant Elimination Network (IPEN) umbrella, which reported over 500 nongovernmental organizations in 116 developing countries working to ban lead-based paint and raise awareness by 2018.[80] The last week in October was designated as "International Lead Poisoning Prevention Week of Action," which may have started in the United States with a declaration from Senator Jack Reed in 2000.[81,82] This resolution was conceived by Rebecca Morley, who was then a Presidential Management Intern at the HUD lead paint office detailed to Reed's office. She later launched the National Safe and Healthy Housing Coalition as director of NCHH in 2009. One of the longest-running international lead poisoning community groups was run by Elizabeth O'Brian in Australia called "The Lead Education and Abatement Design Group" (The LEAD Group).[83]

Most of the early international healthy homes activities came from the World Health Organization (WHO) in Europe and started before the Title X regulations had been implemented in the United States. Xavier Bonnefoy from France led many of those initial efforts in the 1990s; Nathalie Roebbel, Matthias Braubach, Carlos Dora, and others at WHO carried on his work. David Ormandy, with his experience in reforming housing inspections in England, emerged as one of many key WHO advisors.

Bonnefoy landed in some hot water with his superiors. He issued a WHO study on noise and health. The US government had attempted to prevent its issuance, because the report showed that adverse health effects were associated with short frequent episodes of high noise, not only average noise levels. Some within the US State Department apparently thought that this might be used to prevent aircraft takeoffs and landings during the wars in Iraq and sought to prevent its release behind the scenes. Bonnefoy was outraged by the attempted political interference to censor a technical health report and demanded the report come out. He eventually succeeded, but at some cost to his career.[84]

WHO held its first healthy homes conference in 2002 in Forli, Italy with limited attendance. But only 2 years later, Bonnefoy and his colleagues launched the second international healthy homes conference, occurring in Lithuania in 2004, which was attended by over 250 people from 24 countries, including the United States for the first time and where HUD delivered a keynote address.[85] The conference passed a unanimous statement (see Appendix for full text) that came to be known as the Vilnius Declaration (Fig. 8.6):

> A decent home for all citizens is to be a priority target for all national and local governments.... Development policies need to address the rehabilitation of unhealthy neighborhoods and ensure that health is at the center of all plans and programs for new housing, reconstruction, rehabilitation, and urban planning....[86]

VILNIUS DECLARATION

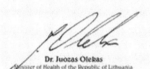

2ⁿᵈ **WHO** International

Housing & Health

S y m p o s i u m

We, Dr Juozas Olekas, Minister of Health, Lithuania,
Mr Arturas Zuokas, Mayor of Vilnius, Lithuania,
Mr Xavier Bonnefoy, Regional Adviser (WHO Regional Office for Europe Bonn Office),
together with 250 decision makers and scientists from 24 countries have gathered
in Vilnius 29-30 September and 1 October 2004 for a symposium on "Housing and Health", and

- Agree upon the following declaration. It has been designed for supporting local authorities and
national governments to develop detailed action plans that include goals, specific and measurable
objectives, and detailed sets of activities in the housing and health sector for which each authority
will be able to claim accountability.
- Acknowledge the role of social conditions as a major health determinant. Non-decent housing
conditions not only are associated with poor social conditions but also contribute to and strengthen
social inequities of health.

The signed Declaration, Fourth Ministerial Conference on Environment and Health, Budapest, Hungary, 23-25 June 2004 (Para 14b)

"We are therefore committed, within the limits of our national mandates, to taking action to ensure that health and environmental dimensions are placed at the core of all housing policies (from housing construction and rehabilitation plans, programmes and policies to the use of adequate building materials) and that healthy conditions are ensured and maintained in the existing housing stock. We commit ourselves to contributing to the development and strengthening of housing policies that address the specific needs of the poor and the disadvantaged, especially regarding children."

to help local authorities assess the impact of the existing housing stock on the health of the population. These tools can help to establish priorities for developing housing policies. The WHO, Bonn office, is requested to provide, upon request, help and support to those member states and local authorities willing to develop guidelines, housing policies and strategies designed to ensure health provision and health protection.

We the undersigned, on behalf of all the participants, reaffirm the commitments undertaken by previous conferences and pledge to continue to support the initiatives outlined above. We hereby fully adopt the commitments made in this Declaration.

Dr. Juozas Olekas
Minister of Health of the Republic of Lithuania

Mr. Artūras Zuokas
Vilnius City Mayor

Mr. Xavier Bonnefoy
Regional Adviser World Health Organization
Regional Office for Europe

FIGURE 8.6 Vilnius Declaration on Housing and Health, 2004. (See Appendix for full text of the Vilnius Declaration). Source: *From The University of Illinois Chicago Special Collections & University Archives, School of Public Health, "David E. Jacobs papers" the University of Illinois Chicago School of Public Health Library.*

It also confirmed the WHO Declaration of the Fourth Ministerial Conference on Environment and Health, Budapest, Hungary, June 23–25, 2004, held a few months earlier:

> We are therefore committed, within the limits of our national mandates, to taking action to ensure that health and environmental dimensions are placed at the core of all housing policies.[87]

To help prepare for the 2002 and 2004 international conferences, WHO conducted the world's largest representative multinational healthy housing study (starting around the year 2000). It marked the first modern study documenting the many links between the health of inhabitants and their housing conditions throughout Europe.[88] Known as the "Large Analysis and Review of European Housing and Health Status" (LARES), it identified indoor environment conditions (including dampness, mold, indoor emissions, infestations, etc.), thermal comfort, noise, and home accessibility as major risk factors.

The project was an ambitious one, consisting of eight cities (Angers, Bonn, Bratislava, Budapest, Ferreira do Alentejo, Forli, Geneva, and Vilnius). Data were gathered from over

8500 occupants in 3373 buildings and remains perhaps the largest healthy homes survey ever accomplished.

Results showed that across Europe, housing-related risks affected the health of large populations. Some risk factors were present in 20%−25% of the population. The data showed that unhealthy housing conditions were not equally distributed, with certain low-income neighborhoods and population groups bearing a much higher risk.

It also helped to show how a single housing condition could result in multiple adverse health effects. For example, one study using the LARES data showed that poor lighting was associated with trips and falls and also depression.[89] Another paper showed that those who were strongly annoyed by general neighborhood noise had twice the odds of a doctor-diagnosed asthma attack. Drainage problems at a housing unit were associated with 54% higher odds of experiencing respiratory symptoms, building structural problems with 27% higher odds, and a leaky roof with 35% higher odds.[90]

LARES resulted in many other peer-reviewed papers and a major book, edited by Ormandy.[88] It also resulted in many new friendships, including a marriage between two of the investigators.

WHO also published a series of short pamphlets on mold, injuries, and other specific housing issues, but it was clear to Bonnefoy that healthy homes needed better cohesion and the health risks from inadequate housing needed to be quantified. He decided to launch a WHO project in 2005 with support from the German Ministry of Environment to identify which housing-health relationships had sufficient scientific evidence (and which ones did not).[91] The ones with sufficient evidence would be included in the first assessment of the quantified burden of disease from inadequate housing. The initiative produced the first "causal web" of health and housing conditions and how they are related to immediate (proximate) and long-term (distal) outcomes.

This causal web served as a starting point in understanding and ranking the adequacy of the global evidence. It was also one of the first attempts to provide the healthy homes field with both breadth and focus by identifying causes and effects and tracing them through proximate and distal pathways. This was similar to the pathway studies in the late 1980s from lead paint to housedust pathways in the US that proved to be so critical in advancing policy.

This first attempt in 2005 failed to achieve Bonnefoy's vision of quantifying the Environmental Burden of Disease from Inadequate Housing. But that would come a few years after his untimely passing from a rare blood disease in 2007. Beginning as a French Sanitary Engineer, he later came to know heads of state and laid the foundation for WHO's healthy homes work in later years.[92]

Matthias Braubach took on the task of leading the quantification of the disease burden of inadequate housing that Bonnefoy had started. Working with Ormandy and Jacobs as coeditors with dozens of other experts from across the globe, WHO completed a massive literature review and used it to estimate, for the first time, how many premature deaths and disability-adjusted life years (DALYs) were due to adverse housing conditions.[93] It used a well-validated method of calculating the population attributable fraction of disease from relative risk ratios in the scientific literature. Not surprisingly, the estimates (mostly from European data) turned out to be enormous (Fig. 8.7).

Summary of exposure, population-attributable fraction (PAF) and EBD from inadequate housing conditions

Exposure	Health outcome	Exposure–risk relationship	PAF	EBD from housing per year
Mould	Asthma deaths and DALYs in children (0–14 years)	RR = 2.4	12.3%	45 countries of the European Region: 83 deaths (0.06 per 100 000) 55 842 DALYs (40 per 100 000)
Dampness	Asthma deaths and DALYs in children (0–14 years)	RR = 2.2	15.3%	45 countries of the European Region: 103 deaths (0.07 per 100 000) 69 462 DALYs (50 per 100 000)
Lack of window guards	Injury deaths and DALYs (0–14 years)	RR = 2.0	33–47%	European Region: ~10 deaths (0.007 per 100 000) ~3310 DALYs (2.0 per 100 000)
Lack of smoke detectors	Injury deaths and DALYs (all ages)	RR = 2.0	2–50%	European Region: 7523 deaths (0.9 per 100 000) 197 565 DALYs (22.4 per 100 000)
Crowding	Tuberculosis	RR = 1.5	4.8%	EURO B and EURO C subregions:[a] 15 351 cases (3.3 per 100 000) 3518 deaths (0.8 per 100 000) 81 210 DALYs (17.6 per 100 000)
Indoor cold	Excess winter mortality	0.15% increased mortality per °C	30%	11 European countries: 38 203 excess winter deaths (12.8 per 100 000)
Traffic noise	Ischaemic heart disease including myocardial infarction	RR = 1.17 per 10 dB(A)	2.9%	Germany only: 3900 myocardial infarcts (4.8 per 100 000) 24 700 ischaemic heart disease cases (30.1 per 100 000) 25 300 DALYs (30.8 per 100 000)
Radon	Lung cancer	RR = 1.08 per 100 Bq/m^3	2–12%	Three western European countries: France: 1234 deaths (2.1 per 100 000) Germany: 1896 deaths (2.3 per 100 000) Switzerland: 231 deaths (3.2 per 100 000)
Residential SHS	Lower respiratory infections, asthma, heart disease and lung cancer	Risk estimates range from 1.2 to 2.0 OR = 4.4	PAF estimates range from 0.6% to 23%	European Region: 64 700 deaths (7.3 per 100 000) 713 000 DALYs (80.7 per 100 000)
Lead	Mental retardation, cardiovascular disease, behavioural problems	Case fatality rate 3%;	66%	European Region: 694 980 DALYs (79.2 per 100 000)
Indoor carbon monoxide	Headache, nausea, cardiovascular ischaemia/insufficiency, seizures, coma, loss of consciousness, death	DNS/PNS incidence 3–40%	50–64%	EURO A subregion:[a] 114–1545 persons with DNS/PNS (0.03–0.4 per 100 000) 114 ± 97 deaths (0.03 ± 0.02 per 100 000)
Formaldehyde	Lower respiratory symptoms in children	OR = 1.4	3.7%	EURO A subregion:[a] 0.3–0.6% of wheezing in children
Indoor solid fuel use	COPD, ALRI, lung cancer	RR = 1.5–3.2	6–15%	European Region: 8490 ALRI deaths in children < 5 years (16.7 per 100 000) 293 600 ALRI DALYs in children < 5 years (577 per 100 000) 5800 COPD deaths in adults ≥ 30 years (1.1 per 100 000) 100 700 COPD DALYs in adults ≥ 30 years (19.3 per 100 000)

Note: OR = odds ratio; RR = relative risk; DALYs = disability-adjusted life years; N/A = not available; COPD = chronic obstructive pulmonary disease; ALRI = acute lower respiratory infections; DNS/PNS = delayed or persistent neurological sequelae.
[a] The lists of countries in the WHO European subregions are provided on page 4.

FIGURE 8.7 Environmental burden of disease from inadequate housing. Source: *From Braubach M, Jacobs DE, Ormandy D, eds.* Environmental Burden of Disease Associated with Inadequate Housing: A Method Guide to the Quantification of Health Impacts of Selected Housing Risks in the WHO European Region. *World Health Organization (Europe); June 2011.*

For example, lead exposure was shown to cause over 640,000 DALYs annually in Europe alone. Mold was responsible for 83 deaths and over 55,000 DALYs annually. Radon caused over 3000 deaths annually in France, Germany, and Switzerland alone. The report covered everything from crowding, dampness, noise, cold, formaldehyde, indoor use of solid fuels for cooking, lack of window guards, and many other housing deficiencies.

During the 2010s, Nathalie Roebbel at WHO took on the task of producing truly global housing and health guidelines, which took over 7 years to complete.[94] For the first time, WHO completed evidence-based recommendations on housing and health that were accompanied by systematic literature reviews. In later years, Roebbel led the WHO efforts that recognized housing as a key social determinant of health (Chapter 9).

Producing global guidelines proved to be a difficult challenge because housing varied greatly around the world. The guidelines development group confronted an ethical dilemma at the very beginning: Should the standards be less rigorous in low-income countries because of the cost and other priorities? For example, one physician from a low-income country thought that structural instability of buildings was far more important than installing smoke alarms in homes, because building collapse was more frequent than fires in her country. Another thought that the pros and cons of housing density would differ in areas with severe overcrowding (prevalent in low-income countries) compared to areas where urban sprawl caused health problems (prevalent in high-income countries). After much deliberation and debate, the group unanimously concluded that healthy housing was a basic right for all people, and it would not be ethical to provide separate recommendations for low- and high-income countries. The guidelines stated:

> While the guidelines development group deems the access to safe and healthy housing a right for populations across all Member States, it also acknowledges that implementing these recommendations will be challenging and vary according to a country's context.

The WHO Guidelines may never have been finalized without the skilled leadership of both Roebbel and the group's chair, Philippa Howden-Chapman from New Zealand. Howden-Chapman had conducted path-breaking research on how insulation can dramatically improve health, and she also helped to implement New Zealand's Warrant of Fitness law for housing.[95,96] She later won New Zealand's highest scientific award in 2021, the Rutherford Medal for her work on healthy housing.[97] For her part, Roebbel was the unsung hero at WHO who skillfully and patiently responded to the seemingly endless peer review process and more than anyone was responsible for releasing the guidelines.

When the Guidelines were launched at an international conference in Uganda, the Mayor of Kampala attended and said, "This is exactly what we need, we have barely begun to consider how our houses should be built to support health."[98] They were also launched at a conference in New Zealand[99] and another in the United States.[100] HUD had funded much of the WHO guidelines work, a departure from its earlier sole focus on internal United States matters. These WHO Guidelines produced new recommendations on crowding, cold and heat indoors, home safety and injuries, and accessibility. They also summarized previous WHO work on other healthy homes subjects such as lead and radon. WHO released an infographic summarizing key healthy housing opportunities (Fig. 8.8).

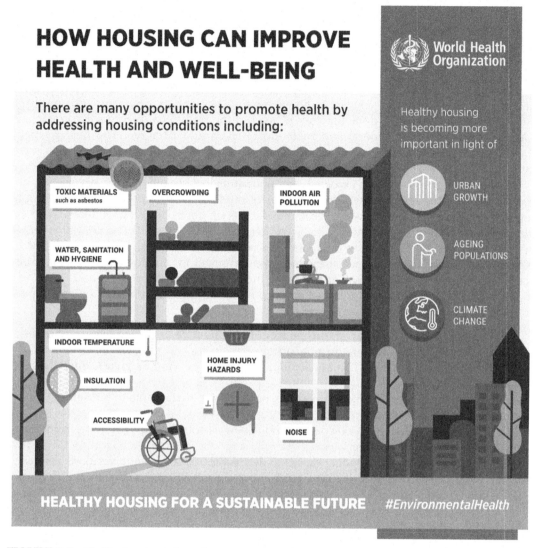

FIGURE 8.8 Healthy housing infographic. Source: *From World Health Organization; 2018. <https://www.who.int/publications/i/item/9789241550376/>.*

One reason it had taken so long was that there was disagreement about the strength of the evidence found in some of the systematic reviews. The experts who drafted the guidelines felt the evidence was quite strong, but some reviewers within WHO felt it was weaker, mainly because they had used a review procedure intended not for housing research, but instead for clinical trials, known as "grading of recommendations, assessment, development, and evaluation" (GRADE). Fundamentally, this disagreement was because individual interventions were often bundled with other interventions (the "package problem" described earlier). This made it sometimes difficult to tease out exactly which interventions had produced which specific health outcomes.

This is because housing improvements typically occur in packages, not one at a time. Clinical trials on the other hand typically have a single intervention such as an experimental drug. This was an ongoing problem for those who examine research designs (known as "methodologists") and those who design and carry out housing and health studies. Because the healthy homes concept is about treating housing as a system, with multiple interventions that could be done together instead of the traditional piecemeal approach, it was not surprising that experts and methodologists would disagree. The disagreement was eventually solved, with calls for additional research to fill the gaps; these gaps were articulated in the WHO guidelines.

For example, one such gap involved the association between heat and adverse health. Because most of the literature was from studies of outdoor temperature (not indoors), some believed this meant that only weak recommendations on controlling heat indoors could be made. But others thought the human body will react to heat in much the same way whether the heat is experienced indoors *or* outdoors. This remained an important avenue of research, especially in light of the need for housing to be more resilient to climate change and the likelihood that indoor temperatures will also increase in the coming years.

The WHO Guidelines also produced a series of startling global statistics on the importance of the global healthy housing problem:

- 110 000 people die every year in Europe from home injuries or during leisure activities, and another 32 million require hospital admission.
- 7500 deaths and 200 000 DALYs are attributable to lack of window guards and smoke detectors in Europe.
- Approximately 10% of hospital admissions per year are attributable to household crowding in New Zealand.
- In 2012 India recorded over 2600 deaths and 850 of various injuries resulting from the collapse of over 2700 buildings.
- In Kyrgyzstan, household crowding causes 13 deaths per 100 000 from tuberculosis per year.
- Worldwide exposure to lead is estimated to have caused 853 000 deaths in 2013.
- 3.8 million deaths globally were attributable to household air pollution from solid fuels for cooking, mostly in low- and middle-income countries.
- About 15% of new childhood asthma in Europe was attributable to indoor dampness (over 69 000 DALYs and 103 deaths every year).

The systematic reviews were important because they showed that many of the adverse housing conditions are associated with health outcomes in a statistically significant way:

- Structural deficiencies in homes are associated with slips or falls.
- Poor accessibility for disabled and elderly people is associated with injury, stress and isolation.
- Insecure or nonaffordable housing is associated with stress.
- Poorly insulated homes are associated with respiratory and cardiovascular problems.
- High indoor temperatures are related to cardiovascular mortality.

- Indoor air pollution is related to disease, respiratory and cardiovascular problems, and allergic and irritant reactions.
- Crowded housing is related to infectious diseases.
- Inadequate water supply and sanitation facilities are related to food safety, personal hygiene and associated diseases.

The WHO guidelines also articulated the many cobenefits of healthy housing, such as:

- Improving thermal insulation, weatherization and ventilation, and installing energy-efficient heating improves indoor temperatures that support health, but also lowers expenditure on energy and reducing carbon emissions.
- Reducing crowding contributes to improved educational outcomes, as children are able to study more effectively.
- Improving thermal comfort through installing insulation and heating reduces days off school and work.
- Improving housing can also create jobs and stimulate investment.

In 2020 WHO began implementing the guidelines by helping nations around the globe.[101]

8.7 "A Kid Who Grew Up in Public Housing"

First triggered in the United States by the mold disaster in Cleveland and with a foundation built from the lead paint experience, the healthy homes idea emerged as a national movement by 2010 in the United States. Great Britain, New Zealand, and other nations had already started decades earlier, as had the World Health Organization. Guidelines, strategic plans, coordination among agencies, research, and active citizen advocacy had all accelerated.

For policymakers in the United States, the focus on lead was clear. So was the mold issue. On the other hand, the healthy homes concept was logical and was more comprehensive but required more focus.

For example, the HUD office name was changed from the Office of Lead Hazard Control to the Office of *Healthy Homes* and Lead Hazard Control in 2000. This earned it the dubious honor of having the longest acronym within HUD. But in 2019, Congress changed the name back again to reflect its desire that lead paint remain the priority: The Office of *Lead Hazard Control and Healthy Housing*. Most of the funding appropriated by Congress remained targeted to lead paint, not healthy homes.

Despite the name changes and funding disparity, Congress became increasingly committed to the healthy housing idea, reflecting the new consensus. Funding for the healthy homes program remained relatively constant between $10 and $20 million per year from 1999 until 2017. But then it increased to $30 million in 2018,[102] and in 2020 it increased to $45 million,[103] reflecting a maturing of the program and increased confidence by members of Congress. In 2021 funding increased even further to $50-$70 million. Far from taking money away from lead paint, funding for lead paint also received increased record appropriations during this same time. These were all reflections of a new, emerging consensus that health and housing could no longer remain separated (Chapter 9).

In 2017 Congress authorized HUD to enable its lead hazard control grantees to spend a small part of their lead paint grants on other healthy homes repairs as well, an initiative championed by Matt Ammon and his deputy Michelle Miller at HUD and others.[104] Part of the increased confidence in Congress came from an evaluation of the healthy homes program carried out by the same entity that had evaluated the HUD lead paint grant program a decade earlier—the National Center for Healthy Housing. HUD also published guidelines in the form of a program manual, an effort led by Amy Murphy and Carol Kawecki and others.[105] Increasingly, the consensus was that it made little sense to correct lead paint hazards, but not fix a broken smoke alarm or a mold problem at the same time.

There were still important barriers, however. By 2022 there was still no healthy homes authorizing legislation similar to the 1992 Title X lead paint law, although several such laws were proposed, led by Senators Jack Reed of Rhode Island and Sherrod Brown of Ohio (and others).[106] HUD's healthy homes authority flowed from its statutes on research and demonstrations and its appropriations statutes.

There were several attempts to pass a comprehensive healthy housing law in the United States. On March 7, 2008, Senators Jack Reed and Chuck Hagel introduced a bipartisan "Healthy Housing Council Act of 2008" that would accomplish many of the same things Title X did. Sen. Reed stated:

> This legislation would establish an independent interagency Council on Healthy Housing in the executive branch. The bill would improve the coordination of existing but fragmented programs, so that families can access Government programs and services in a more efficient and effective manner. While there are many programs in place to address housing-related health hazards, these programs are fragmented and spread across many agencies, making it difficult for at-risk families to access assistance or to receive the comprehensive information they need. It is time for better coordination. This bill authorizes $750,000 for each of fiscal years 2009 to 2013 for an independent Council on Healthy Housing, which would bring Federal, State, and local government representatives, as well as industry and nonprofit representatives, to the table at least once a year. The council would review, monitor, and evaluate existing housing, health, energy, and environmental programs. The council would then make recommendations to reduce duplication, ensure collaboration, identify best practices, and develop a comprehensive healthy housing research agenda.[107,108]

This bill was reintroduced in 2012 and 2013 but failed to pass Congress on both occasions.

Although increasing, funding for both lead and healthy homes remained far below what was needed, given the condition of housing across the country. This condition was documented by the first report on the State of the Nation's Healthy Housing in 2018, ranking the nation's largest cities on the healthfulness of the housing stock.[109]

The evidence, the research and demonstrations and increased appropriations, together with increased collaboration at WHO and around the world all showed that healthy homes had emerged as much more than a "snappy slogan"; it had emerged as a significant effort to link housing and health to produce change.

And none of it might have happened in the United States without the persistence of Congressman Stokes and the scientists in Cleveland.

FIGURE 8.9 Healthy housing awards ceremony honoring Congressman Stokes, 2012. From left to right: Dr. Dorr Dearborn, Congressman Louis Stokes, Dr. David Jacobs, and Cuyahoga County Health Commissioner Terry Allan. Source: *David Jacobs.*

In 2012 Congressman Stokes received a national healthy homes award for his vision in creating the US Healthy Homes program (Fig. 8.9). Senator Sherrod Brown, also from Ohio, was in attendance and in future years he continued Stokes' work, supporting HUD lead and healthy homes funding requests.

At that ceremony, Congressman Stokes was asked what he thought of the seed he had planted more than a decade earlier. With characteristic modesty, he answered:

> I don't know what all this fuss is about; I'm just a kid who grew up in public housing and wanted to do the right thing for our children.[110,111]

Congressman Stokes had planted an important seed from a good idea. The question remained whether a good idea could stimulate a more widespread consensus among the housing, health, and environmental communities (Chapter 9).

Appendix: Vilnius Declaration

Second World Health Organization International Housing and Health Symposium
We, Dr. Junzas Olekas, Minister of Health Lithuania
Mr. Arturas Zuokas, Mayor of Vilnius, Lithuania
Mr. Xavier Bonnefoy, Regional Advisor (WHO Regional Office for Europe (Bonn Office)
Together with 250 decisionmakers and scientists from 24 countries have gathered in Vilnius 29—30 September and 1 October 2004 for a symposium on "Housing and Health," and

— Agree upon the following declaration. It has been designed for supporting local authorities and national governments to develop detailed action plans that include goals, specific and measurable objectives, and detailed sets of activities in the housing and health sector for which each authority will be able to claim accountability.

— Acknowledge the role of social conditions as a major health determinant. Non-decent housing conditions not only are associated with poor social conditions but also contributed to and strengthen social inequities of health.

The signed Declaration, Fourth Ministerial Conference on Environment and Health, Budapest, Hungary 23–25 June 2004 (Para 14b).

We are therefore committed with the limits of our national mandates, in taking action to ensure that health and environmental dimensions are placed at the core of all housing policies (from housing construction and rehabilitation plans, programmes and policies). In the use of adequate building materials and that health conditions are ensured and maintained in the existing housing stock. We commit ourselves to contributing to the development and strengthening of housing policies that address the specific needs of the poor and the disadvantaged, especially regarding children.

We encourage all national and local governments to give a public affirmation of the above Ministerial statement. They should adopt the Budapest principle that all local and national housing policies and programmes are designed to provide a shelter, a living and social environment that supports health, safety and quality of life for all citizens regardless of economic status.

We recognize that:

— Housing has to be considered as a whole, include the home, the household, the neighbourhood and the immediate environment: all impact on physical and mental health as well as on the wellbeing of inhabitants.

— A decent home for all citizens is to be a priority target for all national and local governments.

— Living conditions in certain areas are simply unacceptable. Development policies need to address the rehabilitation of existing unhealthy neighbourhoods in respect of territorial equity and sustainability, strengthening socialization. As important is the need to ensure that the promotion and protection of health is at the centre of all plans and programmes for new housing, construction, rehabilitation and urban planning developments.

— Healthy housing policies, programmes and action plans will only be effective if they involve all the actors. This includes all authorities and departments with an interest in housing, health and planning as well as citizens. The policies have to address the needs of those most vulnerable to the threats from unhealthy housing: children, the elderly, the disabled, the unemployed, the migrants and those living at or below poverty levels. They should encourage citizens to participate with governments and other interested parties in finding original solutions to concrete social situations; these policies should reconcile the role of the market and of the state in sharing the necessary responsibility and comply with the principle of sustainable development.

— Economic analysis can express costs in explicit terms and hence encourage decision-makers to take them into account. We further recognize that economic analysis helps with setting priorities with regard to risk reduction, by assessing the cost-effectiveness of such measures. We affirm the potential of economic instruments as policy tools that contribute effectively in improving health and the housing conditions, and we recognize that it is possible to make far more use of them in order to reach full internalization of health costs.

- There can be no standardized approach or solution. It is necessary to take the local context into account. Every country and every city has its own dynamics, its own actors and its own mechanisms that need to be recognized and appreciated. Routine inspections of housing stocks are the basis of local policies.
- Synergies can develop through education programmes, the simple exchange of information, coordination, or real partnership. However, to facilitate this, the right conditions for dialogue need to be in place especially at the local or community level. The concept of a healthy housing "label" can be recommended. Such a label would advise inhabitants about the quality of their home, its furnishings and its maintenance standards.
- To properly inform policies, decisions and actions, there needs to be comprehensive strategies, including both qualitative and quantitative research and good models for health impact assessment. Such systems will allow social inequalities to be identified and responded to, and the effect of actions monitored.
- Research capacities in housing and health sectors need to be strengthen. Explicit work plans need to be adopted, which include public health interventions, and housing (dwellings and neighborhoods) changes that should be the focus of longitudinal studies.

We affirm that the impact of housing on health occurs through a large variety of pathways and translates into different forms.

- Access to basic amenities: All dwellings should be connected to electricity and to a safe drinking water network, have a kitchen, a private toilet and a shower and the benefit of a fixed system offering adequate thermal comfort
- Housing accidents: Also referred to as unintentional injuries, which are the leading case of death among 1–45 year olds in Europe
- Residential noise: Neighborhood and traffic noise, which are the two most important noise sources to mitigate from a public health perspective
- Indoor air quality: carbon monoxide, environmental tobacco smoke, indoor allergens, moulds, moisture, dust mites, particles, VOCs, and all pollutants lead to respiratory diseases and allergies
- Mental health consequences of poor housing and livability conditions: All conditions that hinder privacy and make the shelter reachable from outside aggressions (such as no tight windows, bad lighting, insufficient thermal and acoustic insulations of dwellings, lack of maintenance and respect of common spaces, pests infestations and those conditions that reduce the capacities of the dwelling to be a bridge with the outside would (lack of security of access to dwellings, fear of crime, difficulty to access to public services and facilities: shops, Multiactivity spaces, green sparces , no play areas, no visitable vegetation, bad view...) all lead to poor mental health.
- Ensure conditions for sufficient sleep for all: Noise temperature and light are among the environmental triggers that affect sleep quality
- Energy housing and health: With rising costs in energy an increasing number of households will have difficulties covering costs that ensure healthy thermal comfort of

their dwelling through the year. Local and national authorities should, through all possible avenues, address the issues of energy savings, the use of renewable energy sources, and access to minimum thermal comfort for all. Devices used to heat and cool housings should not pollute the indoor and outdoor environment. The number of deaths caused by excessively hot and cold housings should not increase in Europe, and more importantly should decrease during the coming years. In this respect, policies including targets and describing choices of sustainable techniques for energy supplies and strategies are most needed.

- The house as a whole, its building materials: Including recycling materials, its furnishings health and ventilation, systems, and its contents should not cause adverse health effects; they should be safe and not release any toxic substance. New products or systems used for construction or rehabilitation of housing should be evaluated prior to their use to determine their potential positive or negative impact on the health status of occupants.
- Pest control in urban environment: Emergence of new diseases as well as existing threats from many prevailing pests requires action.
- Housing design: Residential buildings and dwellings should be designed in a flexible was to allow the normal use of housing whatever the age, the physical limitations, the social economic status of the inhabitants is.
- The healthy choice for housing should be the easy choice: It should be designed and maintained to ensure a healthy lifestyle. Car dependent, sprawling, urban design should be avoided. Mixed-use pedestrian-oriented design that encourages social engagement must be the standard for future developments. More attention should be paid to facilitate exercise, walking and the use of bikes, rather than motorized transport.
- Housing maintenance and health lifestyle: Both housing owners and residents are responsible for providing health conditions in dwellings. This can be achieved through a regular maintenance of housing stock and buildings, and through health lifestyle and adequate residential behaviours of the residents.

The WHO tools developed for the LARES (Large Analysis and Review of European Housing and Health Status) project are recognized as useful instruments to help local authorities assess the impact of the existing housing stock on the health of the population. These tools can help to establish priorities for developing housing policies. The WHO Bonn office is requested to provide, upon request, help and support to those member states and local authorities willing to develop guidelines, housing policies and strategies designed to ensure health provision and health protections.

We the undersigned, on behalf of all the participants, reaffirm the commitments undertaken by previous conferences and pledge to continue to support the initiatives outlined above. We hereby fully adopt the commitments made in this Declaration.

Signed:

Dr. Juozas Olekas ((Minister of Health of the Republic of Lithuania, Mr. Arturas Zuokas, Vilnius City Mayor, and Mr. Xavier Bonnefoy, Regional Advisor, World Health Organization Regional Office for Europe.

References

1. Dorr Dearborn Testimony. Departments of Veterans' Affairs and Housing and Urban Development and Independent Agencies Appropriations for 1999, Tuesday, April 21; 1998. Testimony of Members of Congress and Other Interested Individuals and Organizations.
2. Personal Communication. Dorr Dearborn and David Jacobs; 2004.
3. Sudden Infant Death Syndrome (SIDS). 1998–2022 <https://www.mayoclinic.org/diseases-conditions/sudden-infant-death-syndrome/symptoms-causes/syc-20352800/>.
4. National Institute of Health. Idiopathic pulmonary hemosiderosis. <https://rarediseases.info.nih.gov/diseases/6763/idiopathic-pulmonary-hemosiderosis>.
5. Centers for Disease Control and Prevention. *Facts about Stachybotrys chartarum*. 2019. <https://www.cdc.gov/mold/stachy.htm/>.
6. Sorenson WG, Montana E, Etzel RA, et al. Clinical profile of 30 infants with acute pulmonary hemorrhage in Cleveland. *Pediatrics*. 2002;110:627–637.
7. Hearings before the House Appropriations Committee, Labor, Health & Human Services, & Education subcommittee, 104th Cong, 2nd Sess, 1998 testimony of DG Dearborn, M.D.
8. Update: Pulmonary Hemorrhage/Hemosiderosis Among Infants—Cleveland, Ohio, 1993–1996. Morbidity and Mortality Weekly Report. March 10, 2000/49(09);180-4
9. Dearborn DG, Yike I, Sorenson WG, Miller MJ, Etzel RA. Overview of investigations into pulmonary hemorrhage among infants in Cleveland, Ohio. *Environ Health Perspect*. 1999;107(suppl 3):4950–4959.
10. HUD Press Release, HUD Budget would boost funding for lead hazard control by 40%, HUD No. 98-64; February 10, 1998.
11. Departments of Veterans Affairs and Housing and Urban Development, and Independent Agencies Appropriations for 1999. Hearings Before a Subcommittee of The Committee on Appropriations. House of Representatives One Hundred Fifth Congress. Second Session. Subcommittee on Va, Hud, and Independent Agencies. Part 6; March 25, 1998:466.
12. United States. Congress. House. Committee on Appropriations. Subcommittee on the Departments of Labor, H. and Human Services. Departments of Labor, Health and Human Services, Education, and related agencies appropriations for 1997: hearings before a subcommittee of the Committee on Appropriations, House of Representatives, One Hundred Fourth Congress, second session. Washington: U.S. G.P.O.; 1996.
13. Departments of Veterans Affairs and Housing and Urban Development, and Independent Agencies Appropriations for 1999. Hearings Before a Subcommittee of The Committee on Appropriations. House of Representatives. One Hundred Fifth Congress. Second Session. Subcommittee on VA, HUD, And Independent Agencies. Part 7, March 10; 1998. <https://www.govinfo.gov/content/pkg/CHRG-105hhrg48508/html/CHRG-105hhrg48508.htm\>.
14. Epidemiologic Notes and Reports: Acute Pulmonary Hemorrhage/Hemosiderosis Among Infants -- Cleveland, January 1993–November 1994. Morbidity and Mortality Weekly Report; December 09, 1994;43 (48):881–883. <https://www.cdc.gov/mmwr/preview/mmwrhtml/00033843.htm/>.
15. Departments of Veterans Affairs and Housing and Urban Development, and Independent Agencies Appropriations Act, 1999, 105th Congress (1997–1998). <https://www.congress.gov/bill/105th-congress/house-bill/4194/text/> [Page 112 STAT. 2482 H.R.4194].
16. H. Rept. 105-769 - Making Appropriations for The Department of Veterans Affairs and Housing and Urban Development, and for Sundry Independent Agencies, Boards, Commission, Corporations, and Offices for The Fiscal Year Ending September 30, 1999, and for Other Purposes 105th Congress (1997–1998). <https://www.congress.gov/congressional-report/105th-congress/house-report/769/1?q = %7B%22search%22%3A%5B%22cite%3APL105-276%22%5D%7D&r = 1&overview = closed/>.
17. Committee reports and appropriations are quoted in The Healthy Homes Initiative: A Preliminary Plan (Full Report). U.S. Department of Housing and Urban Development Office of Lead Hazard Control, April 19; 1999. <https://nchh.org/information-and-evidence/healthy-housing-policy/archive/archived-national-policy-projects/report-to-congress-1999/>.
18. The Healthy Homes Initiative: A Preliminary Plan (Full Report). U.S. Department of Housing and Urban Development Office of Lead Hazard Control, April 19; 1999. <https://nchh.org/information-and-evidence/healthy-housing-policy/archive/archived-national-policy-projects/report-to-congress-1999/>.

19. Matte TD, Jacobs DE. Housing and health: current issues and implications for research and programs. *J Urban Health: Bull N Y Acad Med.* 2000;77(1):7–25.

20. Not Safe at Home, How America's Housing Crisis Threatens the Health of Its Children, Boston Medical Center. Doc4Kids Project. Editors Joshua Sharfstein and Megan Sandel; February 1998. The University of Illinois Chicago Special Collections & University Archives, School of Public Health, "David E. Jacobs papers" the University of Illinois Chicago School of Public Health Library.

21. Sandel M, Sharfstein J, Shaw R. There's no place like home. How America's Housing Crisis Threatens Our children. Housing America and Doc4Kids Project; March 1999. The University of Illinois Chicago Special Collections & University Archives, School of Public Health, "David E. Jacobs papers" the University of Illinois Chicago School of Public Health Library.

22. National Center for Healthy Housing. The Principles of a Healthy Home; 2020. <https://nchh.org/information-and-evidence/learn-about-healthy-housing/healthy-homes-principles/>.

23. Centers for Disease Control and Prevention and U.S. Department of Housing and Urban Development Atlanta. *Healthy Housing Reference Manual.* US Department of Health and Human Services; 2006. Available from: https://www.cdc.gov/nceh/publications/books/housing/summary.htm.

24. Centers for Disease Control and Prevention and U.S. Department of Housing and Urban Development. *Healthy Housing Inspection Manual.* Atlanta: US Department of Health and Human Services; 2008. Available from: https://www.cdc.gov/nceh/publications/books/inspectionmanual/healthy_housing_inspection_manual.pdf.

25. Committee on the Hygiene of Housing American Public Health Association. Basic Principles of Healthful Housing. Preliminary report. *Am J Public Health.* 1938;351–372. Available from: https://ajph.aphapublications.org/doi/pdfplus/10.2105/AJPH.28.3.351/.

26. National Center for Healthy Housing and American Public Health Association. National Healthy Housing Standard, May 16; 2014. <https://nchh.org/tools-and-data/housing-code-tools/national-healthy-housing-standard/>.

27. Change Lab Solutions. Up to Code: Code Enforcement Strategies for Healthy Housing. 2015. < https://www.changelabsolutions.org/product/code/ >.

28. Building Codes. National Center for Healthy Housing. <https://nchh.org/tools-and-data/standards-and-assessments/building-codes/>. 2020. Also see: Technical Assistance for Code Transformation Innovation Collaborative (TACTIC). <https://nchh.org/tools-and-data/technical-assistance/tactic/>. 2020

29. HHTS and LTS Grant Program Abstracts, 2006–2019. Healthy Homes Technical Studies and Lead Technical Studies. US Department of Housing and Urban Development. 2020. < https://www.hud.gov/program_offices/healthy_homes/hhi/hhts/ >.

30. World Health Organization Europe. Report on the WHO Technical Meeting on Quantifying Disease from Inadequate Housing. World Health Organization. Bonn Germany; November 2005. 28–30. <https://www.euro.who.int/data/assets/pdf_file/0007/98674/EBD_Bonn_Report.pdf/>.

31. Personal Communication. Senator Barbara Mikulski and David Jacobs. 2001.

32. U.S. Department of Health and Human Services. The Surgeon General's Call to Action to Promote Healthy Homes. U.S. Department of Health and Human Services, Office of the Surgeon General; 2009. <https://www.ncbi.nlm.nih.gov/books/NBK44192/>.

33. Centers for Disease Control and Prevention (US). The Health Consequences of Smoking—50 Years of Progress: A Report of the Surgeon General. National Center for Chronic Disease Prevention and Health Promotion (US) Office on Smoking and Health. Atlanta (GA); 2014. <https://www.ncbi.nlm.nih.gov/books/NBK294310/>.

34. Brown MJ, Ammon M, Grevatt P. Federal agency support for healthy homes. *Public Health Manag Pract.* 2010;16(5):S90–S93. E-Supp.

35. CDC. Temporary Halt in Residential Evictions to Prevent the Further Spread of COVID-19. September 4; 2020. 85 Federal Register 55292. <https://www.federalregister.gov/documents/2020/09/04/2020-19654/temporary-halt-in-residential-evictions-to-prevent-the-further-spread-of-covid-19/>.

36. US Department of Housing and Urban Development Press Release No. 22–004, January 12, 2022. <https://www.hud.gov/press/press_releases_media_advisories/HUD_No_22_004>

37. HUD. *Leading Our Nation to Healthier Homes: The Healthy Homes Strategic Plan.* 2009. Available from: <https://www.hud.gov/sites/documents/DOC_13701.PDF/>.

38. Kercsmar CM, Dearborn DG, Schlucter M, et al. Reduction in asthma morbidity in children as a result of home remediation aimed at moisture sources. *Environ Health Perspect.* 2006;114:1574–1580.

39. Commercial and residential water damage statistics: United States, August 6; 2020. <https://titanrebuild.com/commercial-residential-water-damage-statistics-united-states/>.

40. Phillips T. *High Point: The Inside Story of Seattle's First Green Mixed-Income Neighborhood*. Splash Block Publishing; 2020. Available from: https://highpointbook.com/.

41. Takaro TK, Krieger J, Song L, Sharify D, Beaudet N. The breathe-easy home: the impact of asthma-friendly home construction on clinical outcomes and trigger exposure. *Am J Public Health*. 2011;101:55–62.

42. Krieger J. Home is where the triggers are: increasing asthma control by improving the home environment. *Pediatr Allergy, Immunol, Pulmonol*. 2010;23(2).

43. Krieger J, Jacobs DE. Healthy housing (Book Chapter) Aug. In: Dannenberg AL, Frumkin H, Jackson RJ, eds. *Making Healthy Places: Designing and Building for Health, Well-being, and Sustainability*. Island Press; 2011.

44. Reddy A, Gomez M, Dixon S. The New York state healthy neighborhoods program: findings from an evaluation of a large-scale, multisite, state-funded healthy homes program. *J Public Health Manag Pract*. 2017;23(2):210–218.

45. Reddy A, Gomez M, Dixon S. An evaluation of a state-funded healthy homes intervention on asthma outcomes in adults and children. *J Public Health Manag Pract*. 2017;23(2):219–228.

46. Marta Gomez M, Reddy AL, Dixon SL, Wilson J, Jacobs DE. A cost-benefit analysis of a state-funded healthy homes program for residents with asthma: findings from the New York state healthy neighborhoods program. *J Public Health Manag Prot*. 2017;23(2):229–238. March/April.

47. Statista.com. Green Buildings in the U.S. - Statistics & Facts Published by Ian Tiseo, December 11; 2020. <https://www.statista.com/topics/1169/green-buildings-in-the-us/>.

48. Breysse JV, Jacobs DE, Weber W, et al. Health outcomes and green renovation of affordable housing. *Public Health Rep*. 2011;126(suppl 1):64–75. Available from: http://www.publichealthreports.org/issueopen.cfm?articleID=2647/.

49. Jacobs RL, Jacobs DE, Breysse J. Home is where the health is: bringing healthy sustainable housing to low-income populations. *Clgh Rev J Poverty Law Policy*. 2010;44(5–6):249–256.

50. The Community Guide to Preventive Health Services, US Preventive Services Task Force. Atlanta Georgia, 2000.

51. Jacobs DE, Brown MJ, Baeder A, et al. A systematic review of housing interventions and health: introduction, methods, and summary findings. *J Public Health Manag Pract*. 2010;(suppl)S3–S8. September.

52. Krieger J, Jacobs DE, Ashley PJ, et al. Housing interventions and control of asthma-related indoor biologic agents: a review of the evidence. *J Public Health Manag Pract*. 2010;(suppl)S9–S18. September.

53. Sandel M, Baeder A, Bradman A, et al. Housing interventions and control of health-related chemical agents: a review of the evidence. *J Public Health Manag Pract*. 2010;(suppl)S19–S28. September.

54. Radon Research. Available from: http://nchharchive.org/Search.aspx?Source = 3&SearchIn = 5&Name = radon&Status = 0,1&Audience = &Principle = &Keywords = random&Author = &Journal = &From = &To =), 2014.

55. DiGuiseppi C, Jacobs DE, Phelan KJ, Mickalide AD, Ormandy D. Housing interventions and control of injury-related structural deficiencies: a review of the evidence. *J Public Health Manag Pract*. 2010;(suppl)S32–S41. September.

56. Phelan KJ, Khoury J, Xu Y, Liddy S, Hornung R, Lanphear BP. A randomized controlled trial of home injury hazard reduction: the HOME injury study. *Arch Pediatr Adolscent Med*. 2011;165(4):339–445.

57. Lindberg RA, Shenassa ED, Acevedo-Garcia D, Popkin SJ, Villaveces A, Morley RL. Housing interventions at the neighborhood level and health: a review of the evidence. *J Public Health Manag Pract*. 2010; 16(5 suppl):S44–S52. Available from: https://doi.org/10.1097/PHH.0b013e3181dfbb72. Available from: https://journals.lww.com/jphmp/Fulltext/2010/09001/Housing_Interventions_at_the_Neighborhood_Level.8.aspx.

58. Haines S, Dannemiller K, et al. Ten questions concerning the implications of carpet on indoor chemistry and microbiology. *Build Environ*. 2020;170:106589. March.

59. Sundell J, Levin H, Nazaroff WW, et al. Ventilation rates and health: multidisciplinary review of the scientific literature. *Indoor Air*. 2011;21:191–204.

60. Grimsrud DT. Indoor air: the first 10 years. *Indoor Air*. 2011;21:179–181.

61. Jacobs DE, Ahonen E, Dixon SL, et al. Moving into green healthy housing. *J Public Health Manag Pract*. 2015;21(4):345–354.

62. Breysse J, Dixon SL, Jacobs DE, Lopez J, Weber W. Self-reported health outcomes associated with green-renovated public housing among primarily elderly residents. *J Public Health Manag Pract*. 2015;21(4):355–367.

63. Jacobs DE, Breysse J, Dixon SL, et al. Health and housing outcomes from Green Renovation of Low-Income Housing in Washington, DC. *J Environ Health*. 2014;76(7):8−16.

64. Garland E, Steenburgh ET, Sanchez SH, et al. Impact of LEED-certified affordable housing on asthma in the South Bronx. *Prog Comm Health Partnerships: Res, Educ, Action*. 2013;7(1). Available from: https://doi.org/10.1353/cpr.2013.0008.

65. Jacobs DE, Wilson J, Dixon S, et al. Studying the optimal ventilation for environmental indoor air quality (The STOVE Study). *Int J Env Res Public Health*. 2022.

66. Wilson J, Jacobs DE, Reddy AL, Tohn E, Cohen J, Jacobsohn E. Home RX: The Health Benefits of Home Performance − A Review of the Current Evidence. US Dept of Energy; December 2016. <https://betterbuildingssolutioncenter.energy.gov/sites/default/files/attachments/Home%20Rx%20The%20Health%20Benefits%20of%20Home%20Performance%20-%20A%20Review%20of%20the%20Current%20Evidence.pdf/>.

67. Goodwin S. Ten Years of the National Safe and Healthy Housing Coalition: A Decade of Appropriations Advocacy. September 27, 2019 <https://nchh.org/2019/09/nshhc-at-10_2/>.

68. National Center for Healthy Housing. Grant, Scholarship Opportunities, 2022. Available from: https://nchh.org/build-the-movement/grants-and-scholarships/.

69. Green, Healthy Homes Initiative. Technical Assistance, 2021. Available from: https://www.greenandhealthyhomes.org/.

70. National Center for Healthy Housing, Lead Poisoning Prevention Stories Case Studies, 2017.

71. Ormandy D. Historical development of housing hygiene policy. *J R Soc Health*. 1987;107:39−42. Available from: https://doi.org/10.1177/146642408710700201.

72. The Housing Health and Safety Rating System (HHSRS), Published Friday, July 12; 2019. United Kingdom Parliament. <https://researchbriefings.parliament.uk/ResearchBriefing/Summary/SN01917/>.

73. 2004 Housing Act authorized the 2005 regulations: Statutory Instruments No. 3208. HOUSING, ENGLAND The Housing Health and Safety Rating System (England) Regulations. Also see: Housing Health and Safety Rating System. Operating Guidance. Housing Act 2004. Guidance about inspections and assessment of hazards given under Section 9; February 2006. Office of the Deputy Prime Minister: London.

74. Personal Communication. David Ormandy and David Jacobs. 2019

75. Professor David Ormandy. (undated). Available from: https://warwick.ac.uk/fac/sci/med/staff/dormandy/.

76. United Nations Conference on Human Settlements (HABITAT II). Istanbul (Turkey) 3−14 June 1996. <https://www.un.org/ruleoflaw/wp-content/uploads/2015/10/istanbul-declaration.pdf/>.

77. Alliance to End Childhood Lead Poisoning, Global Dimensions of Lead Poisoning: The First International Prevention Conference. Final Report 64; 1994. The University of Illinois Chicago Special Collections & University Archives, School of Public Health, "David E. Jacobs papers" the University of Illinois Chicago School of Public Health Library.

78. International Conference on Lead Abatement and Remediation, Newcastle Australia. In: *The University of Illinois Chicago Special Collections* & University Archives, School of Public Health, "David E. Jacobs papers" the University of Illinois Chicago School of Public Health Library; 1994.

79. Mahaffey K, Jacobs DC. Lead in the Americas: A Call to Action. Working Group Two: Paint. Edited by Christopher Howson, Mauricio Hernandez-Avila and David P. Rall. Committee to Reduce Lead Exposure in the Americas. Board on International Health. Institute of Medicine, Washington, DC, USA, In Collaboration with The National Institute of Public Health, Cuernavaca, Morelos, Mexico; 1996. The University of Illinois Chicago Special Collections & University Archives, School of Public Health, "David E. Jacobs papers" the University of Illinois Chicago School of Public Health Library.

80. IPEN 2018. International Lead Poisoning Prevention Week of Action, October 21−27; 2018. <https://ipen.org/sites/default/files/documents/ipen-lead-week-of-action-2018-v1_4.pdf/>.

81. Senator Jack Reed Honored by National and Local Groups. March 21, 2014. Press Release. National Center for Healthy Housing. <https://nchh.org/2014/03/senator-jack-reed-honored-by-national-and-local-groups/>.

82. S.Res.199—106th Congress (1999−2000) A resolution designating the week 24, 1999, through October 30, 1999, and the week of October 22, 2000, through October 28, 2000, as "National Childhood Lead Poisoning Prevention Week." October 19, 1999. <https://www.congress.gov/bill/106th-congress/senate-resolution/199/text?r = 57&s = 1/>.

83. The LEAD Group, Inc. May 2022. https://lead.org.au/

84. Personal Communication. Xavier Bonnefoy and David Jacobs; 2004.

85. Jacobs DE. Housing and Health: Challenges and Opportunities, Keynote Address, Proceedings of the 2nd WHO International Housing and Health Symposium, WHO European Centre for Environment and Health (Bonn Office), Noise and Housing Unit, Bonn Germany, September 29 - October 1, 2004, Vilnius Lithuania, October 20; 2005:35−50. The University of Illinois Chicago Special Collections & University Archives, School of Public Health, "David E. Jacobs papers" the University of Illinois Chicago School of Public Health Library.

86. World Health Organization. Second WHO International Housing and Health Symposium. Vilnius, Lithuania, October 1; 2004. The University of Illinois Chicago Special Collections & University Archives, School of Public Health, "David E. Jacobs papers" the University of Illinois Chicago School of Public Health Library.

87. World Health Organization Europe. Fourth Ministerial Conference on Environment and Health Budapest, Hungary, June 23−25; 2004. <https://www.euro.who.int/data/assets/pdf_file/0008/88577/E83335.pdf/>.

88. Housing and Health in Europe: The WHO LARES project. Ed D Ormandy. Routledge; 2009. ISBN 9781138972001.

89. Brown MJ, Jacobs DE. Residential lighting and risk for depression and falls: results from eight European cities. *Public Health Rep.* 2011;126(suppl 1):131−140.

90. Miles R, Jacobs DE. Future directions in housing and public health: findings from Europe and implications for planners. *J Am Plan Assoc.* 2007;74(1):77−89.

91. World Health Organization Regional Office for Europe. Report on the WHO Technical Meeting on Quantifying Disease from Inadequate Housing, Bonn Germany, November 28−30, 2005; April 2006. <https://www.euro.who.int/__data/assets/pdf_file/0007/98674/EBD_Bonn_Report.pdf/>.

92. Bonnefoy X. Inadequate housing and health: an overview. *Int J Env and Pollution.* 2007;30(3/4):411−429. Available from: https://www.euro.who.int/__data/assets/pdf_file/0017/121832/E90676.pdf.

93. Braubach M, Jacobs DE, Ormandy D, eds. *Environmental Burden of Disease Associated With Inadequate Housing: A Method Guide to the Quantification of Health Impacts of Selected Housing Risks in the WHO European Region (book).* World Health Organization Europe; 2011. Available from: https://www.euro.who.int/_data/assets/pdf_file/0003/142077/e95004.pdf/.

94. World Health Organization. Housing and health guidelines. Recommendations to Promote Healthy Housing for a Sustainable and Equitable Future. 2018. November 23. <https://www.who.int/publications/i/item/9789241550376>.

95. New Zealand Rental Warrant of Fitness; Housing Law. NZ Rental WOF Limited. 2020. <https://www.nzrentalwof.co.nz/>.

96. Barnard LT, Bennett J, Howden-Chapman P, et al. Measuring the effect of housing quality interventions: the case of the New Zealand 'Rental Warrant of Fitness'. *Int J Env Res Public Health.* 2017;14:1352. Available from: https://doi.org/10.3390/ijerph14111352.

97. 2021 Rutherford Medal: Impact of Housing on Health. <https://www.royalsociety.org.nz/what-we-do/medals-and-awards/research-honours/2021-research-honours-aotearoa/2021-rutherford-medal/>.

98. Personal Communication. Nathalie Roebbel and David Jacobs; 2019.

99. Presentations from WHO International Housing and Health Guidelines Southern Hemisphere Launch. University of Otago Wellington, New Zealand; February 25, 2019. <otago.ac.nz/wellington/departments/publichealth/summerschool/otago712469.html/>.

100. Lead and Healthy Homes Conference. Lead and Environmental Hazards Association. Washington DC; March 2019.

101. World Health Organization. Report on Meeting to Implement WHO Housing and Health Guidelines; January 2020, <https://nchh.org/resource-library/who_promoting-healthy-housing-for-all_towards-an-implementation-strategy-for-the-who-housing-and-health-guidelines.pdf>.

102. FY 2018 Department of Transportation and HUD Appropriations, Department of Housing and Urban Development (HUD), Office of Lead Hazard Control and Healthy Homes (OLHCHH). <https://nchh.org/resource-library/Appropriations_FY18_T-HUD.pdf/>.

103. FY 2020 Department of Transportation and HUD Appropriations, Department of Housing and Urban Development (HUD), Office of Lead Hazard Control and Healthy Homes (OLHCHH). <https://nchh.org/resource-library/appropriations_fy20_t-hud.pdf/>.

104. H.R.244 - Consolidated Appropriations Act, 2017. 115th Congress (2017−2018) 2017 HUD Appropriations Act. Page 131 STAT. 778. <https://www.congress.gov/bill/115th-congress/house-bill/244/text/>.

105. Murphy A, Jacobs D, Kawecki C et al. Healthy Homes Program Manual. Office of Lead Hazard Control and Healthy Homes, July 19; 2012. <https://www.hud.gov/program_offices/healthy_homes/HHPGM/>.

106. US Congress. *Healthy Housing Council Act of 2013. Senators.* Reed, Johanns, Boxer, and Franken; 2013. Available from: https://www.congress.gov/bill/113th-congress/senate-bill/291/text.

107. Reed SJ. Floor Statement Introducing the Healthy Housing Council Act of 2008; March 07, 2008. <https://www.reed.senate.gov/news/speeches/floor-statement-introducing-the-healthy-housing-act/>.

108. S.2735 - Healthy Housing Council Act of 2008. March 7, 2008. <https://www.congress.gov/bill/110th-congress/senate-bill/2735?s = 2&r = 30>.

109. State of Healthy Housing. National Center for Healthy Housing; 2020. <https://nchh.org/2020/12/nchh-unveils-new-state-of-healthy-housing-report/>.

110. Personal Communication. Congressman Louis Stokes and David Jacobs; 2012.

111. Lou Stokes put health impacts of substandard housing on the national agenda, and in Cleveland: Terry Allan, Dorr Dearborn and Dave Jacobs (Opinion). Cleveland Plain Dealer. Posted August 30; 2015. <https://www.cleveland.com/opinion/2015/08/how_lou_stokes_put_substandard.html/>.

Reframing health, environment, and housing

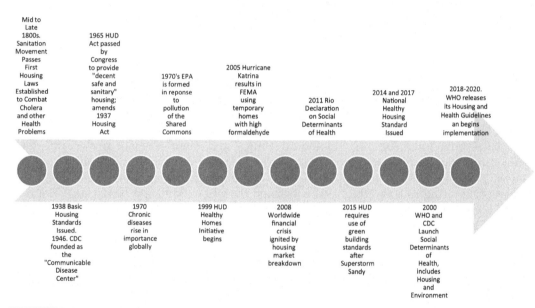

FIGURE 9.0 Timeline.

9.1 Health: reframing communicable and noncommunicable disease

In 1996 the Department of Housing and Urban Development (HUD) was invited to speak at a Centers for Disease Control and Prevention (CDC) Grand Rounds session. Traditionally marked by a dialog to impart new information on disease and injury, such sessions were intended to be a departure from the day-to-day routine rounds that physicians conduct with their individual patients. At their best, Grand Rounds presented the larger picture beyond the immediate clinical setting.[1]

The expectation was that there would simply be a salute to the progress on lead poisoning made possible by Title X and better relations between HUD, CDC, and EPA, which as we have seen were substantial. But in keeping with the larger picture of such sessions, there was much more than a salute.

Instead, the very definition of disease was examined. The long-standing organizing principle in the medical world between communicable and noncommunicable diseases was said to be artificial and had not served the nation well in the case of lead poisoning prevention. It was artificial because communicable diseases could not always be addressed solely with better medicine. It had not served the nation well because it created needless barriers among public health, medical, environmental, housing, and other allied professions.

For housing-related diseases like lead poisoning and certain asthma cases and injuries, HUD argued that the house "communicated" the health problem, much like a virus "communicated" the flu. Lead paint was like a contagion, spreading poisoning among millions of children over many years in millions of homes.[2] Traditionally, the perceived solution was better medicine for communicable diseases and better behavior or other nonmedical interventions for noncommunicable diseases, such as more exercise and better diet.

But the idea that housing "communicated" certain health problems was rejected by the CDC audience, because it did not fit into the existing medical models. Much like the Maryland Court of Appeals decision in the *Grimes* case, the perceived scientific distinction between "therapeutic" and "nontherapeutic" interventions (which was essentially the difference between communicable and noncommunicable diseases) resulted in confusion and wrong policy directions.

The physicians and other CDC professionals attending the grand rounds session thought lead poisoning was clearly a "noncommunicable" disease. It was not caused by a virus, bacterium, fungi, or other biological "bug," and there was not a medicine that would cure it. But more than a few scratched their heads and wondered if healthy housing could be thought of as a "cure" or "medicine" for some health conditions. Years later, the concept of "housing as a vaccine," was advanced by Megan Sandel, a physician.[3] The traditional communicable disease paradigm had started to give way to something new.

The division into communicable and noncommunicable diseases was reflected in how CDC was organized. Of the 12 "centers" that made up the Centers for Disease Control and Prevention in 2020, only 4 were devoted to "noncommunicable" diseases:

- Birth defects (which included developmental disabilities);
- Chronic diseases;
- Environmental health (where the lead poisoning prevention program was located); and
- Injuries.

Indeed, CDC was founded on July 1, 1946, as the "Communicable Disease Center" (the original name for the "CDC"), which was later changed to the "Centers for Disease Control and Prevention." Its primary first mission was to prevent malaria from spreading across the nation.[4]

Over many years CDC rightfully developed a worldwide reputation as a leader in combatting microbe-related disease outbreaks and pandemics both in the United States and around the world. This included drug-resistant tuberculosis, smallpox, Legionnaire's disease, swine flu, sexually transmitted diseases, measles, SARS (severe acute respiratory syndrome), Ebola, and others.

All these efforts were marked by a high degree of professionalism and were mostly free of political interference. CDC remained headquartered in Atlanta, not Washington DC to avoid any perception of political interference. One notable exception was CDC's response to the Covid-19 pandemic, when, by many accounts, CDC's reputation was tarnished by political interference. Over a thousand current and former CDC employees signed an open letter protesting that interference in 2020.[5]

It became increasingly clear that vaccines and other medicines, by themselves, could not be the only response. Although the power of better medicines such as vaccines has been undeniable for polio and smallpox, for other housing-related diseases, better medicine clearly could not be the only response. The Covid-19 pandemic was perhaps the most recent case in point. So-called "nontherapeutic" interventions, like wearing masks, social distancing, and staying home (assuming the home was a healthy one) emerged as critically important solutions, on par with the search for a vaccine. CDC even went so far as to order housing evictions to end temporarily during the Covid-19 pandemic,[6] perhaps the first time the agency has ever issued a housing order. Clearly, communicable disease control meant using interventions that had traditionally been used for so-called noncommunicable diseases.

Communicable disease is defined as follows:

> An infectious disease (such as cholera, hepatitis, influenza, malaria, measles, or tuberculosis) that is transmissible by contact with infected individuals or their bodily discharges or fluids (such as respiratory droplets, blood, or semen), *by contact with contaminated surfaces or objects*, by ingestion of contaminated food or water, or by direct or indirect contact with disease vectors (such as mosquitoes, fleas, or mice) (emphasis added).[7]

On the other hand, noncommunicable disease, sometimes called "chronic disease," is defined by WHO as follows:

> Diseases of long duration [which are] the result of a combination of genetic, physiological, *environmental* and behavioral factors. The main types of non-communicable diseases are cardiovascular diseases (like heart attacks and stroke), cancers, chronic respiratory diseases (such as chronic obstructive pulmonary disease and asthma) and diabetes (emphasis added).[8]

Both definitions show how the two categories overlap. "Contact with contaminated surfaces or objects" and "environmental factors" are both important in lead poisoning. Presumably, housing is one of the "environmental" factors that has "contaminated surfaces." Between 1970 and 2020, there was an increasing recognition that so-called "noncommunicable" diseases like asthma, obesity, lead poisoning, heart disease, and other chronic diseases were growing in importance. In 2018 WHO reported that noncommunicable diseases killed 41 million people each year, equivalent to 71% of all deaths globally.[8]

The continuing focus within CDC on communicable diseases was manifested in the labels of the two categories themselves. Noncommunicable disease was only defined by something it was not. More importantly, communicable disease did not adequately describe the linkages to its causes, because it was not only exposure to a "bug" that caused it.

Communicable diseases often have multiple contributing factors. Perhaps the best example comes from a century earlier, during the "sanitation" movement. The installation of indoor plumbing, improved ventilation, and other housing improvements together with better medicine helped to virtually eradicate cholera from the United States. In that case, it was *both* a bacterium (*Vibrio cholerae*) and the lack of potable water that caused the problem. Similarly, the Covid-19 pandemic showed that those with certain "noncommunicable" diseases, such as obesity, diabetes, or respiratory diseases like COPD, were predisposed to Covid-related deaths.

Of course, noncommunicable diseases also have multiple causes. Re-establishing the housing and health collaboration ultimately led to a realization that there is little to be gained by separating diseases into these two categories.

By the 2010 decade, there was an increasing consensus that a more integrated risk-based multi-intervention response was needed. This was similar to the recommendation from the National Academy of Sciences Institute of Medicine report on housing intervention research, which appeared in the wake of the *Grimes* decision (Chapter 6). The Academy said the distinction between therapeutic and nontherapeutic research held little meaning for housing-related diseases and the distinction between communicable and noncommunicable diseases was wanting.

The holistic healthy homes approach was one example of multiple interventions being bundled together. Another was the social determinants of health approach. Both are examined later in this chapter, and both marked a new emerging consensus and reframing of how we think of health.

9.2 Environment: reframing the shared commons

Just as the traditional concept of communicable and noncommunicable disease came to be seen as inadequate in the case of lead paint poisoning, so too did the traditional concept of the "Shared Commons" in the environmental realm. Historically, the legal structure for the environmental movement in the United States stood on two fundamental principles of English common law: "Shared Commons"[9] and "The Polluter Pays."[10]

The principle of the Shared Commons was derived from medieval times: Although everyone's cattle could graze on the common green, nobody's cattle should be permitted to overgraze the resource and deprive others of its use. This meant that the community had the authority to protect its common interest if private activity deprived the public of its right and reliance on a shared resource. This "commons" eventually came to include most (but not all) shared resources such as exterior air or water, not just grazing rights. And the Shared Commons underpins what constitutes a "public nuisance," which was the key legal theory in both the Rhode Island and California lead paint cases (Chapter 7).

The Polluter Pays principle held it was the originator of the pollution, not the injured public, that was responsible for the cost of its remediation. This principle enabled courts in California and a jury in Rhode Island to require the lead, pigment and paint industries to help pay for remediation.

Implementing the Shared Commons and Polluter Pays principles in the United States, the Clean Air[11] and Clean Water[12] Acts authorized an intricate system of regulation and oversight to support outdoor air and water quality, starting mostly in the 1970s. National legislation mandated standard setting and extended to states, counties, and localities for execution and monitoring. This system had broad national support for clean outdoor air and water and followed weaker earlier systems.

For example, "Fresh Air Funds" had a great deal of support. Started by a Congregational minister (one precursor to the United Church of Christ) in the late 1800s, these private charities sent urban children to camp for a week, allowing some to escape, if only briefly, urban exterior air pollution. But they were clearly inadequate in preventing the problem, which is why the Clean Air Act was ultimately passed.[13] Similarly, clean water regulation in the United States began in 1899 with the Rivers and Harbors Act, but this law was also weak.[14]

These earlier efforts failed to keep pace with increasing pollution. By the late 1960s, there was growing public outrage against the increasingly obvious and worsening outdoor air, water, and land pollution. For instance, in the 1950s and 1960s the Cuyahoga River in Cleveland, Ohio burst into flames several times because of pollution (Fig. 9.1).[15] As a boy, Jacobs had gone to junior high and senior high school in an Ohio town through which the river ran, spending many hours on the riverbanks collecting fossils as a "rock-hound" hobbyist, so the river's condition hit close to home for him.

One 1969 account stated:

> The Cuyahoga River was one of the most polluted rivers in the United States. The reach from Akron to Cleveland was devoid of fish throughout the 1950s and 60 s. There were at least 13 fires on the Cuyahoga River, the first occurring in 1868. The largest river fire, in 1952, caused more than $1 million in damage to boats and a riverfront office building. Fires erupted on the river several more times before June 22, 1969; on that date a river fire was covered in Time magazine, which described the Cuyahoga as the river that "oozes rather than flows" and in which a person "does not drown but decays."[16]

Cleveland's then-new mayor, Carl Stokes (the brother of Congressman Louis Stokes, who had played such an important role in launching the US healthy homes initiative

FIGURE 9.1   Cuyahoga River, November 3, 1952. From: *Cleveland Press Collection at Cleveland State University.* *<https://ohiohistorycentral.org/w/Cuyahoga_River_Fire>.*

described in Chapter 8), responded to the river fires to rally voters, who approved a $100 million bond to clean it up.[17]

The Cuyahoga River fires and other similar events mobilized public concern across the nation and helped initiate many water and air pollution control activities resulting in the Great Lakes Water Quality Agreement and ultimately the creation of the federal Environmental Protection Agency. Congress passed the Clean Air Act in 1970 and Clean Water Act in 1972.[18] Congress passed the National Environmental Policy Act in 1969, signed into law in 1970.[19] The first Earth Day occurred in 1970, the same year that President Richard Nixon established the US EPA by Executive Order in December 1970.[20] This was also around the same time that Congress passed the first (but ill-fated) Lead Based Paint Poisoning Prevention Act in 1971.

Despite the far-reaching impact of such laws, they were limited when it came to housing, which was not thought to be part of the Shared Commons. The commons was traditionally defined this way:

> The commons is everywhere. It is the air we breathe, the words we speak, the traditions we respect. It is tangible and intangible, ancient and modern, local and global. It is everything we inherit together, as part of a community, as distinct from things we inherit individually. It is everything that is *not privately or state-owned* (emphasis added).[21]

Therefore because houses were privately or state-owned, they were *not* part of the shared commons. As a result, there were no sweeping national housing standards like there were for outdoor air and water. Instead, housing codes were fragmented;[22] most jurisdictions established their own separate housing codes (discussed later in this chapter). Indeed, much of the 1971 lead-based paint poisoning prevention act was focused on public housing, not private housing, perhaps because public housing could be construed to be more of a "public" resource (even though it was "state-owned").

But for the other 99% of the privately owned housing stock, no national standards were passed, with the notable exception of Title X in 1992, which covered both public and private housing and set a standard for existing lead paint; Title X also authorized EPA to establish standards for lead dust and bare lead-contaminated soil in housing. But Title X was the exception. The lack of national housing standards was a direct result of the narrow definition of the shared commons.

The disconnect was evident: there were national environmental standards for outdoor air and water, and there were national standards of care for medical practice and the public expected and supported them.[23] For example, if one had gall bladder surgery, it was generally done the same way from Seattle to New York and across the nation. National standards for air and water included National Ambient Air Quality Standards and the National Primary Drinking Water Regulations.

With only a few exceptions, the only *indoor* environment that had national regulations was the workplace, regulated primarily by the Occupational Safety and Health Administration, which was also established in 1971 around the same time as the environmental laws. When Congress passed the occupational safety and health act in 1970, it stated the goal was:

> to assure so far as possible *every* working man and woman in the nation safe and healthful working conditions and to preserve *our human resources* (emphasis added).[24]

In other words, the OSHA Act embodied a shared commons approach for all workers.

Within the environmental community, the Shared Commons did not extend to the quality of air *indoors*, other than workplaces. Although homes and other nonoccupational places were where people did more of their breathing, legally enforceable indoor air quality standards were not established.[25] Instead, national, state, and local legislatures sometimes authorized public agencies to conduct voluntary programs intended to raise the public's awareness about indoor air problems and appropriate actions, but as a poor cousin to outdoor air regulation. In fact, EPA stated explicitly: "EPA does not regulate indoor air."[26]

For housing, there was not (yet) a perceived Shared Commons for which the public felt a communal benefit and responsibility. Instead, housing was thought to be mainly a private matter. It was left to each owner to determine what to do about lead paint and other health conditions in housing. Furthermore, housing was thought to be primarily an economic matter, a means to build individual wealth, described later in this chapter. HUD's Policy Development and Research office was primarily composed of economists, not health professionals, perhaps explaining why so little had been done on unhealthy housing and lead paint from 1970 to 1990.

In the housing world, individuals often acted for themselves or in small groups, rather than an inflamed public reacting to images of rivers on fire acting through the political process. Owners of private residences or tenants of buildings typically sued in the courts for redress of specific injury or health problems individually, rather than appeal to the legislature for the enactment of broad standards applicable across all occupied spaces.

Exactly who was the "polluter" in the housing context? In many cases, they could not be easily identified and tasked with payment for remediation, explaining in part why the courts did not recognize product liability law for existing lead paint. For housing, there was not a dramatic moment of recognition such as a river on fire to galvanize public action, despite a series of well-publicized serious housing problems: lead paint, asbestos, formaldehyde-contaminated insulation and flooring, and many others. These products were typically allowed into housing commerce with little or no prior testing, which occurred usually only after health problems became apparent and harm had been done.

Furthermore, responsibility for housing was shared by many, including architects, maintenance personnel, designers, planners, employers, owners, property managers, code and building inspectors, occupants, and others. In the absence of a villain on whom the burden of correction could be laid, there was historically less public support for mandates on healthy housing, private paint companies, lumber manufacturers, designers of unvented gas heaters and others.

This diffuse housing responsibility was different from both the medical and the environmental communities. In medicine, the responsibility rested mostly between patients and their physicians and sometimes with pharmaceutical and medical device manufacturers. In the environment, the responsibility rested mostly between polluting industries and their regulators. Unlike housing, the types and numbers of actors were fewer in both the health and environmental context.

But by 2008, it had become clear that the narrow definition of Shared Commons excluding most housing had failed, and a new concept began to emerge. The economic crisis that year had its origins in the housing market. It had become clear if that market failed, so did the larger economy, affecting virtually everyone. Although homes will always be a private space, the idea that all owners and tenants were entirely on their own was increasingly rejected as a failure that threatened everyone.

Instead, housing increasingly came to be seen as part of the Shared Commons. Although the Rhode Island Supreme Court thought that private homes were "indivisible" and therefore not a public concern, the later California court decision showed how much the ground had shifted when it found lead paint in homes was a "public nuisance." Housing financing, quality, and healthfulness all came to be as much a public concern as exterior air and water, as part of our shared commons and infrastructure.

This expanded concept was manifested in many ways. For example, green housing standards started to emerge within the environmental field during the 2000s. Enterprise Green Community Standards appeared in 2005. By 2022 27 states and Washington, DC, required that affordable housing developments receiving public funds comply with those standards.[27] These standards incorporated many healthy homes concepts and research had shown their importance (Chapter 8).

A few years later, the Green Building Council launched its "Leadership for Environmental and Energy Design" standards for homes.[28] Groups like the Healthy Building Network were beginning to enable builders to reject the use of toxic building materials, such as urea formaldehyde foam insulation, asbestos, and other similar substances. It assembled the first and most comprehensive database of chemical hazards in building products through its Pharos Project.[29]

These materials were all products that cost far more to remove or control than they ever saved in reduced construction or maintenance costs, improved health or improved housing durability. As we have seen, lead paint had originally been marketed as a way to preserve homes and keep them in a "sanitary" condition. The use of toxic building materials increasingly came to be viewed by the public as a "river on fire" demanding a public response.

Health-based improvements in housing weatherization and energy conservation also started to appear. Although the earlier attempts at energy conservation in homes during the 1970s had led to "sick building syndrome," later standards provided smarter energy conservation, better sustainability, improved ventilation and moisture control, and improved health; they were increasingly adopted within the environmental and healthy housing fields.[30,31] By 2005 EPA and the federal government had launched a series of green building initiatives,[32] marking a clear move to include housing within the meaning of the Shared Commons.

The Healthy Homes movement after 1999 fundamentally moved away from a substance-by-substance, categorical, trial and error approach (i.e., the incremental approach that allowed one dangerous product after another into widespread use followed by a separate remediation campaign). Instead, a smarter, more preventive, holistic, and comprehensive response took hold. Title X was one of the earliest laws in the US to recognize that both public and private housing were indeed part of the Shared Commons. For lead paint (and for other harmful housing conditions), it was no longer acceptable to just hope that individual owners and occupants would somehow figure out how to ensure their homes supported their health. Proposals to include lead and healthy homes in large infrastructure spending packages emerged.[33] Indeed, a bill introduced in Congress in 2021 was titled the "Housing is Infrastructure Act,"[34] a clear recognition of the new shared commons.

FIGURE 9.2 Cuyahoga River Park, Cleveland, 50 years after the 1969 river fire. *Reproduced with Permission: Ken Busch. kenbuschphotography.smugmug.com.*

In Cleveland, the new concept of an expanded shared commons that included housing bore fruit. In 2020 the city council passed a new law that required virtually all rental property owners to register and make the homes free of lead paint hazards before they could be occupied.[35] Gone was the burning river. Instead, the Cuyahoga river area was redeveloped into a beautiful park shared by all (Fig. 9.2).

9.3 Housing: reframing wealth, affordability, and equity

For most people, the investment in a home is the largest they will ever make, including both renters and owner-occupants. For much of the last half of the 20th century, the role of housing was perceived to be mostly limited to economic issues. Government policies promoted homeownership as a way to promote economic stability. America's homeownership rate rose from around 45% to 70% from 1940 to 2000.[36]

But such ownership came at a price. Globally, housing prices reached an all-time high in 2012 (Fig. 9.3). The trend in higher housing prices was also reflected in rental housing, where inflation-adjusted rents increased by 64% from 1960 to 2014, but household income only rose by 18%. As a result, the percent of cost-burdened renters doubled from 24% to 49%.[37] Increasingly, the traditional view that investment in housing resulted in more wealth became untenable for most of the population.

The historic failure to include housing as part of disease prevention within the medical sphere and as part of the Shared Commons in the environmental disciplines meant that instead of a stabilizing force, housing policies contributed to the exact opposite for nearly everyone. The effect on the highest risk communities was particularly pronounced.

The failures fed inequities in health, wealth, opportunity, and community vitality through segregation, unaffordability, substandard housing quality, homelessness, and housing shortages. Racist and discriminatory housing practices prevented communities of color and low-income households from building wealth through homeownership[38] (Fig. 9.4) and accessing healthy housing, worsening disparities in health and environmental outcomes.

Global average house prices, real terms
1990=100

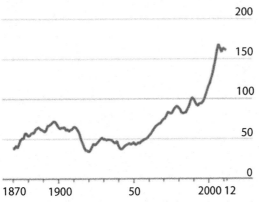

FIGURE 9.3 Global average house price 1870–2012. *Reproduced with permission: The Economist Magazine. How Housing Became the World's Biggest Asset Class, Special Report, January 16; 2020. <https://www.economist.com/special-report/2020/01/16/how-housing-became-the-worlds-biggest-asset-class>.*

Source: "No price like home: global house prices, 1870–2012" by K.Knoll, M, Schularick and T. Steger, *American Economic Review* 2017

The Economist

People of color have experienced lower homeownership rates for decades

Homeownership rate by race/ethnicity, 1940—2017

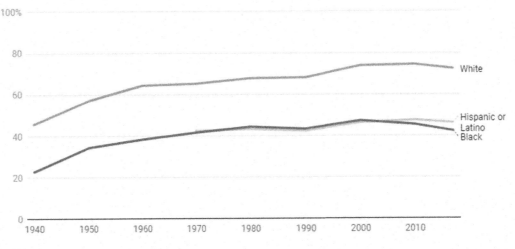

Hover over or click to see values.

Chart: Center for American Progress • Source: F. John Devaney, "Tracking the American Dream: 50 Years of Housing History from the Census Bureau: 1940 to 1990" (Washington: U.S. Department of Commerce, 1994), available at https://www.huduser.gov/portal/Publications/pdf/HUD-7775.pdf; U.S. Census Bureau, "Table 22. Homeownership Rates by Race and Ethnicity of Householder: 1994 to 2017," available at https://www.census.gov/housing/hvs/files/annual17/ann17t_22.xlsx (last accessed June 2019).

FIGURE 9.4 Homeownership rates by race and ethnicity. *From: Solomon D, Maxwell C, Castro A. Systemic Inequality: Displacement, Exclusion, and Segregation: How America's Housing System Undermines Wealth Building in Communities of Color. Center for American Progress. August 7; 2019. <https://www.americanprogress.org/issues/race/reports/2019/08/07/472617/systemic-inequality-displacement-exclusion-segregation/ > .*

High housing cost burdens harmed both physical and mental wellbeing. In the early to mid-2000s, severely cost-burdened families faced with high housing costs spent 75% less on healthcare and 40% less on food. Low-income seniors with high housing cost burdens were forced to spend 62% less on healthcare. These households often had less healthy food or delayed healthcare or medications to pay their high housing costs.[39–41]

By 2020 the burden of housing costs had become even starker, with low-income households increasingly unable to afford healthcare. The State of the Nation's Housing report for that year stated:

> Among households in the bottom expenditure quartile that included children under age 18, those with moderate cost burdens spent *57 percent less on healthcare* (including insurance premiums and out-of-pocket expenses) and *17 percent less on food* than unburdened households. Those with severe burdens spent *93 percent less on healthcare and 37 percent less on food*. Older adults with moderate [housing] cost burdens spent 31 percent less on healthcare and 21 percent less on food than same-age households without burdens, while those with severe burdens spent nearly 50 percent less on both healthcare and food (emphasis added).[42]

Communities of color had persistently lower quality housing and correspondingly poorer health outcomes. The disparities in both housing quality and health have persisted (Figs. 9.5, 9.6, and 9.7).[39,43]

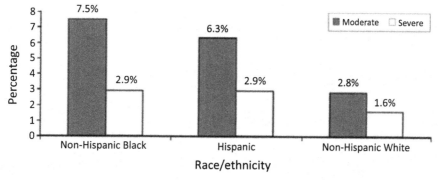

FIGURE 9.5 Prevalence of moderate and severe substandard housing by race and ethnicity. *Reproduced with permission: Jacobs DE. Environmental health disparities in housing.* J Am Public Health Assoc. 2011;101(suppl 1):S115–122.

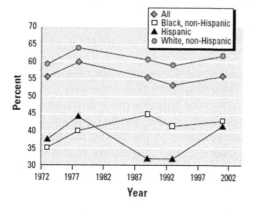

FIGURE 9.6 General health status by race and ethnicity percent of those reporting excellent or very good health. *From: Jacobs DE, Dixon SL, Wilson JW, Smith J, Evens A. 2009. The relationship of housing and public health: A 30 Year Retrospective Analysis in the US.* Environ Health Perspect. 2009:117;597–604.

Households of Color Are More Likely to Have Fallen
Behind on Housing Payments

Share of Households Behind on Rent/Mortgage in September 2020 (Percent)

FIGURE 9.7 Housing cost burden by income and race/ethnicity. *Reproduced with permission: Joint Center for Housing Studies. State of the Nation's Housing; 2020.*

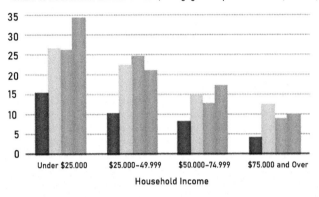

Notes: Households behind on rent or mortgage reported that they were not caught up at the time of survey. White, Black, and Asian households are non-Hispanic. Hispanic households may be of any race.
Source: JCHS tabulations of US Census Bureau Household Pulse Survey, Week 15

In 2020 the racial and ethnic disparities in housing cost burden accelerated:

> Among renters, Black and Hispanic households are particularly likely to have cost burdens. Black renters have the highest share at 53.7 percent, followed closely by Hispanic renters at 51.9 percent. By comparison, 41.9 percent of white renters were cost burdened last year, along with 42.2 percent of Asian renters and 46.6 percent of renter households identifying as multiracial or another race. Across most income groups, households of color are more likely to be cost burdened than white households.[42]

These health disparities and increasing housing costs for virtually the entire population stood in stark contrast to the nation's housing policy. The original 1937 Housing Act had stated the goal of "decent, safe and sanitary housing for all Americans." By including safe and sanitary, both communicable and noncommunicable disease prevention were included. And because the goal included all Americans, it had elements of the Shared Commons as well.[44]

But with high housing costs, achieving health investments in housing collided with the realities of the housing market financing system. Why were health investments in housing unlike other building improvements?

As the healthy homes movement attempted to integrate its small-scale research and demonstration findings into normal housing finance systems, there was a growing realization that health-related investments, whether for lead paint abatement, asthma or mold remediation or injury prevention suffered from a problem in the housing market itself. Simply put, such investments were not reflected in the value of the home, unlike other housing improvements. For example, a new roof or heating system typically resulted in a higher market value for that home. This enabled the owner to (at least partially) recover that investment when it came time to sell or rent the home.

But for health investments in housing, this did not happen. The financial benefit of reduced health care costs from healthier homes was not quantified by those who establish the market value of homes (appraisers). Furthermore, the benefits were not apparent to those who determine how much debt or rent a household could take on (mortgage lenders and landlords).

This meant that health investments were unlike other home improvements. From an owner's perspective, it made little financial sense to make a health investment in housing, because the savings went to the medical system in reduced health care costs, or for lead poisoning, the savings went to the school system in the form of reduced need for special education. But the financial benefit of health investments in housing did not go to the owner.

At the same time, what better time could there be to finance healthy homes improvements than at the time of sale or turnover? The loan (mortgage) or lease was often adjusted to finance corrections for other housing defects and in theory the same could happen for healthy housing issues. Lead paint, mold, or injury hazards would then become just like any other housing defect. But because these investments were not reflected in home value, the healthy homes investments were an "externality," in which the costs of unhealthy housing were borne by the rest of society.

This classic "market failure" was a key reason why many healthy housing improvements were difficult to implement broadly through the normal housing finance system. The alternative was to rely on subsidies and/or enforcement, which could be too little and too late.

Operationalizing housing market forces for health purposes were challenging. In 2004 HUD and CDC personnel spoke at the conference of the Window and Door Manufacturers Association to encourage them to advertise and market the health and energy benefits of new lead-free energy-efficient windows. The manufacturers and builders were intrigued but thought people primarily bought new windows for esthetic, not health reasons.[45] Other efforts were taken to make these costs and benefits more transparent. For example, window replacement was shown to increase the value of a home significantly, primarily because people are willing to pay more for a home with new windows. But old windows were also a major source of lead paint hazards and lead dust, so there are additional benefits for lead poisoning prevention (and likely asthma from avoided leaks and mold).[46-48]

If housing market forces and allied programs like weatherization did not include health investments, that left regulation, enforcement, and subsidy. Indeed, housing regulations began with health and social well-being at their core. The world's earliest known housing law is the Code of Hammurabi, dating back to 3000 BCE. It contained a drastic penalty by today's standards for inadequate housing:

> If a builder has built a house for a man and his work is not strong, and if the house he has built falls in and kills the householder, that builder shall be slain.[49]

More recent housing laws and codes had their origins in the sanitation movement that began in England around the 1830s and spread to the United States in the latter part of the 1800s with the 1890 publication of Jacob Riis' book How the Other Half Lives.[50] The book documented the deplorable unhealthy crowded housing conditions in tenement slums.

The sanitation movement is credited with improved housing ventilation, reduced crowding, and other conditions related to tuberculosis, typhoid, and cholera.[51] The sanitation movement was marked by the understanding that communicable and noncommunicable disease solutions were linked and that housing quality was a shared responsibility.

But the dawning of increased specialization ended the success of the sanitation movement. For its part, the public health world and medicine adopted the position that treatment occurred mostly after illness, in keeping with its communicable disease/therapeutic intervention focus. The environmental world focused on the traditional Shared Common concept that excluded housing, failing to regulate indoor air for example. And the housing world adopted the economic wealth-building approach, using housing codes to preserve the financial asset but not address housing-related diseases and injuries. The multidisciplinary emphasis on prevention and the connection between housing and health was lost, but not forgotten.

With the atrophy of the sanitation movement partnership also came the disappearance of health professionals in the housing code-making process. The first housing codes in the United States were developed principally by health entities, including the American Public Health Association's 1938 Basic Principles of Healthful Housing[52] and the 1986 CDC's Recommended Minimum Housing Standards.[53]

But by 1990 housing codes became governed by so-called "model" code authorities, such as the International Code Council (despite its name the ICC was a mostly US affair). Separate mechanical, electrical, structural, ventilation, plumbing, fire, energy, fuel gas, sewage, maintenance, and other model codes appeared.[54] Each of these were typically modified by each jurisdiction across the country.

The predictable result was fragmentation.[22] Virtually every jurisdiction came to have its own housing codes. Builders were often faced with a bewildering array of requirements that varied from one locality to the next. And virtually all codes either ignored health issues such as lead paint, or relegated it to an optional appendix, as was the case for radon. For example, the International Residential Code of 2018 stated: "The [radon] provisions contained in this Appendix [F] are not mandatory."[55]

Despite its fragmented nature, the main housing enforcement mechanism was through local housing codes. For the most part, housing code officials remained largely ill-trained and ill-equipped to carry out health-related enforcement. Instead, they tended to refer healthy homes matters to local health departments, who often did not have the required housing expertise. As we have seen, this stood in stark contrast to England, where housing inspectors enforced a code that explicitly included health considerations (Chapter 8). In many cases in the United States, however, code enforcement was driven mostly by complaints from occupants. Like blood lead screening, this reactive approach usually meant that conditions had to become very serious before action was taken. Litigation often ensued, but most received no compensation, and property managers were faced with unpredictable outcomes.

By 2000 it became increasingly clear that viewing housing solely as an economic matter and using fragmented housing codes was unsustainable. A new framework began to emerge. But like an environmental picture of a river on fire, it took a disaster to realize that a new housing framework was needed.

In 2005 in the wake of Hurricane Katrina in New Orleans, there was a desperate need for temporary housing for people who had been displaced from their homes by the storm.

The Federal Emergency Management Agency (FEMA) decided it would allow "travel trailers" to be used.

But these trailers were built with wood products that released high levels of formaldehyde. FEMA thought that because they would only be occupied for a few days, the higher exposures could be tolerated. But in fact, people would be forced to remain in them for much longer, sometimes years.

CDC was ordered by Congress to investigate, and the results created public outrage. Eleven different reports showed that the exposures were far too high and contributed to many illnesses.[56] People had seemingly been forced to trade one disaster for another. FEMA finally issued new standards to ensure its temporary housing had low formaldehyde,[57] but only after the harm had been proven scientifically. Mobile homes (also sometimes called "manufactured housing") were regulated by HUD, but travel trailers at the time were not covered. The costs of the illnesses and the wave of lawsuits were far greater than the costs of providing healthier temporary housing.

The walls between housing, health, and environment had started to crack. By 2019 one survey reported that nearly 72% of Americans believed a home inspection helped prevent potential problems when moving into a new residence.[58] Many people purchasing older homes increasingly sought home inspections before they completed the purchase so they could negotiate the cost of repairs with the seller. This was also the original concept behind the 1992 lead paint disclosure rule in Title X. In 2017 France required homes to have a lead paint inspection prior to sale.[59] As we have seen in Chapter 8, England had implemented its Healthy Housing Rating System for its home inspections. New Zealand had implemented its "Rental Housing Warrant of Fitness" requirements.[60]

This new trend was not confined to other countries, however. In a few cities such as Rochester New York, lead paint and code enforcement were successfully integrated into a new local ordinance in 2005, with very positive results. Code officials were trained to collect lead dust samples. For most rental homes, there were regularly scheduled code inspections every 3 years (with rapid responses to complaints). Owners came to include the costs of lead safety as part of their business models. Instead of causing housing abandonment as some had feared at its beginning, the housing market stabilized, because owners knew what to expect.

The results in Rochester were compelling. Blood lead levels in Rochester declined twice as fast as in the rest of the state. Part of the reason for this code's success was due to citizen action, the involvement of Katrina Korfmacher (a local university Professor), engaged citizens, and a committed housing commissioner, Gary Kirkmire, and a well-trained code inspectorate. The Rochester experience is recounted in a book about overcoming the "siloed" approach between housing codes, environment, and health.[61]

There were other successes in integrating health into housing codes.[62] Many cities began to adopt pro-active code enforcement programs that included healthy housing elements.[63] A new healthy housing model code appeared in 2017 when two former HUD Secretaries, the National Center for Healthy Housing (NCHH) and the American Public Health Association released the National Safe and Healthy Housing Standard.[64] NCHH also developed a "code comparison tool" that enabled better integration of health into housing codes in 2019.[65]

Another example of an emerging new healthy housing framework came in 2015. In its rebuilding efforts after Superstorm Sandy in New York, HUD required that builders comply with "green standards," although it did not specify exactly which green standard was to be used. The lessons from the Hurricane Katrina debacle had been learned. A major study from the National Academy of Sciences on integrating health into long-term disaster recovery operations highlighted this new requirement.[66] This report was the first to review the many studies that had been undertaken to determine if green building standards, which had mostly been adopted for environmental sustainability purposes, really did improve health, and found that they did. The Enterprise Green Community standards were the ones most extensively studied for their health impacts.

The widespread adoption of new green healthy housing standards marked a new understanding on how housing, health, and the environment were inseparable and pointed the way to a new consensus on healthy housing.

9.4 Toward a healthy housing consensus

The three biggest parts of the economy that caused shock waves across the United States and the rest of the world after 2000 were the high cost of healthcare, the housing crisis, and energy. In the United States, the second leading cause of failure to make a housing payment on time in 2014 was a high medical bill (the first was loss of a job) and those with insecure unhealthy housing had higher health care costs, accelerating the housing crisis in a repeating vicious cycle.[67]

Climate change, driven in part by energy use in buildings, caused thousands of homes to be destroyed through wildfires, flooding, landslides, and storms, exacerbating the existing housing shortage and creating its own healthcare crisis. In 2020 during the fires, California's governor declared, "This is a climate damn emergency" after dozens had died, hundreds of homes had been destroyed, and air quality deteriorated.[68] Australia and Siberia were also the scenes of major wildfires.

In 2020 a new pandemic that forced many to retreat to their homes for safety made the need for a healthy home clearer than ever before.

A new healthy housing framework meant that solutions needed to address all three spheres: housing, health, and environment. Each could no longer remain confined to different worlds, implemented by different professions with their own specialized training, with their own terminology, studied by different parts of the academy, with differing policies and financing systems, with different citizen advocacy groups, and with budgets far too small and disjointed to effect real change. Relatively tiny offices at HUD and at WHO could not be sufficient to meet the larger challenges brought on by these three related crises. Neither EPA nor the Department of Health and Human Services had formal healthy housing programs as late as 2022, even though Title X had demonstrated the strength of the HUD, EPA, and CDC collaboration for lead paint poisoning prevention.

A new healthy housing framework increasingly recognized the importance of racial and economic disparities by including affordability and equity in its 10 key principles. This meant recognizing the widespread nature of housing deficiencies, and that the problem could not be solved by restricting action to a few "hot spots." Instead, the entire nation and the entire world had a stake.

For example, the water lead problem in Flint Michigan rose to the public's attention in 2014 because unhealthy drinking water delivered into people's homes was perceived to be part of the Shared Commons. The problem affected everyone and was worse in low-income communities of color such as Flint. A case in which a home "communicated" a disease through an environmental cause in the form of a housing deficiency from lead water service lines and environmentally corrosive water could not be clearer.

As the new healthy housing concept matured, the emergence of the social determinants of health framework appeared around the same time. The WHO launched some of the earliest work on social determinants of health in 2000 through the work of an international commission.[69] In 2013 WHO identified three main scientifically proven interventions to combat social determinants of health: education (such as decreased classroom sizes), social protection (such as nutrition programs), and urban development (such as healthy affordable housing and transportation).[70] Nathalie Roebbel, who had led the WHO healthy homes effort, also played a key role in the new WHO social determinants of health initiative.

In 2008 CDC defined social determinants of health as follows:

> life-enhancing resources, such as food supply, *housing*, economic and social relationships, transportation, education, and health care, whose distribution across populations effectively determines length and quality of life (emphasis added).[71]

By 2017 housing had emerged as a key social determinant of health; much of the research focused on housing quality, affordability, stability, and neighborhood-level factors (including housing racial segregation).[72] These research efforts moved into larger scale implementation of proven interventions, reflected in international proceedings. For example, the 2011 Rio Declaration on Social Determinants of Health resulted in pledges from many nations, and WHO began to monitor how the multidisciplinary interventions on the social determinants of health improved health outcomes.[73,74]

This era saw a reframed healthy housing movement that extended its beginnings from a multidisciplinary holistic (yet still abstract) good "policy idea" within a few small government offices in the United States, a few other countries, and WHO into a much larger global consensus that impacted five large systems in many countries:

- a reformed housing finance system that reflected health investments in the value of a home to stimulate more public and private investment, principally through green healthy housing standards (a market-based solution);
- a reformed healthy housing model code that stimulated the housing inspectorate to enforce healthy housing codes through housing inspections and remediation (a local government solution);
- larger lead paint and healthy housing subsidies that provided investment where the private market failed (a national government solution);
- a reformed health system that adopted a risk-based social determinants of health model, largely replacing an artificial separation between communicable and noncommunicable disease (a primary prevention public health solution); and
- a reformed environmental system that recognized housing as part of the Shared Commons, enforced by court decisions enforcing the "polluter pays" principle that in this case required paint, pigment, and lead industries to contribute to remediation (a legal solution).

This new consensus resulted in new policies, record funding, and other actions to implement them at scale. Most importantly, it promised to end a strange but still powerful policy paralysis paradox, described next in the conclusion to this book.

References

1. Sandal S, Iannuzzi MC, Knohl SJ. Can we make grand rounds "grand" again? *J Grad Med Educ*. 2013;5 (4):560−563. Available from: https://doi.org/10.4300/JGME-D-12-00355.1.
2. Jacobs DE. The Effectiveness of Lead-Based Paint Hazard Control: A Review of the Scientific Evidence, presented at the Centers for Disease Control and Prevention Grand Rounds, Atlanta, Georgia, June 14; 1996.
3. Sandel M. Housing as a Vaccine. Pediatrician Sees Housing as a Vaccine. Housing Matters. Urban Institute. January 12; 2015. < https://housingmatters.urban.org/articles/pediatrician-sees-housing-vaccine > .
4. CDC. Our History, Our Story; 2020. <https://www.cdc.gov/about/history/index.html>.
5. Over 1,000 Current and Ex-CDC Officers Decry the "Politicization" of the Agency. Axios, October 17; 2020. <https://www.axios.com/1000-cdc-officers-politicization-covid19-5f80712b-2a99-445c-9336-28906a2a99fd.html>.
6. CDC. Federal Register Notice: Temporary Halt in Residential Evictions to Prevent the Further Spread of COVID-19. The Centers for Disease Control and Prevention (CDC), located within the Department of Health and Human Services (HHS) announces the issuance of an Order under Section 361 of the Public Health Service Act to temporarily halt residential evictions to prevent the further spread of COVID-19. 85 FR 55292. Updated April 2022. < https://www.cdc.gov/coronavirus/2019-ncov/covid-eviction-declaration.html > .
7. Definition of Communicable Disease. Miriam Webster Dictionary; 2020. <https://www.merriam-webster.com/dictionary/communicable%20disease>.
8. WHO. Non-Communicable Diseases; 2018. <https://www.who.int/news-room/fact-sheets/detail/noncommunicable-diseases>.
9. Hardin G. The tragedy of the commons. *Science*. 1968;162(3859):1243−1248.
10. United Nations. Organization for Economic and Cooperation and Development. The Polluter-Pays Principle. OECD Analyses and Recommendations, Paris; 1992. <https://www.oecd.org/officialdocuments/publicdisplaydocumentpdf/?cote = OCDE/GD(92)81&docLanguage = En>.
11. Clean Air Act. 42 U.S.C. Section7401 et seq.; 1970. <https://www.epa.gov/laws-regulations/summary-clean-air-act>.
12. Clean Water Act. 33 U.S.C. Section1251 et seq.; 1972. <https://www.epa.gov/laws-regulations/summary-clean-water-act>.
13. Guarneri J. Changing strategies for child welfare, enduring beliefs about childhood: the fresh air fund, 1877−1926. *J Gilded Age Progressive Era*. 2012;11(1):27−70. Available from: http://www.jstor.org/stable/23249057.
14. History of the Clean Water Act. Environmental Works, April 18; 2018. <https://www.google.com/search?q = history + of + clean + water + act&rlz = 1C1GCEU_enUS923US923&oq = history + of + clean + w&aqs = c-hrome.0.0i457j0j69i57j0j0i22i30l2j0i10i22i30j0i22i30.3217j0j7&sourceid = chrome&ie = UTF-8>.
15. Introduction to the Clean Water Act. Watershed Academy Web. <https://cfpub.epa.gov/watertrain/moduleFrame.cfm?parent_object_id = 2571>.
16. The Cities: The Price of Optimism. Time Magazine., August 1; 1969.
17. Adler JH. Fables of the Cuyahoga: reconstructing a history of environmental protection. *Fordham Environ Law J*. 2003;14(95−98):103−104. Case Western Reserve University.
18. Water and Air Pollution. Updated: March 30, 2020. Original: November 6, 2009. <https://www.history.com/topics/natural-disasters-and-environment/water-and-air-pollution>.
19. U.S. Code §. 1970. Congressional Declaration of National Environmental Policy; January.
20. Office of the President. Reorganization Plan No. 3 of 1970, July 9, 1970. Available: <http://www.epa.gov/history/org/origins/reorg.htm>.
21. The Demand for the Common Good by Jonathan Rowe (YES! Magazine. Summer 2004). <https://pov-tc.pbs.org/pov/downloads/2004/pov-thirst-lesson-plan.pdf>.
22. Jacobs DE. *Healthy housing standards: fragmentation or harmonization? Keynote address. Proceedings of the 5th Warwick Healthy Housing Conference*. Coventry, UK: University of Warwick, March 17; 2008.
23. Kinney ED. The origins and promise of medical standards of care. *AMA J Ethics*. 2004;6(12):574−576. Available from: https://doi.org/10.1001/virtualmentor.2004.6.12.mhst1-0412.

24. Occupational Safety and Health Act of 1970. Public Law 91-596. 84 STAT. 1590. 91st Congress, S.2193, December 29; 1970.
25. Jacobs DE, Kelly T, Sobolweski J. Linking public health, housing and indoor environmental policy: successes and challenges at local and federal agencies in the U.S. *Environ Health Perspect*. 2007;115:976–982.
26. EPA Indoor Air Program. Updated May 2021. <https://www.epa.gov/regulatory-information-topic/regulatory-information-topic-air#indoorair>.
27. Enterprise Green Community Standards. 2020. <https://www.enterprisecommunity.org/solutions-and-innovation/green-communities>.
28. US Green Building Council. LEED for Homes. Available from: https://www.usgbc.org/resources/leed-homes.
29. Healthy Building Network; 2020. <https://healthybuilding.net/>.
30. Engvall K, Norrby C, Norback D. Ocular, nasal, dermal and respiratory symptoms in relation to heating, ventilation, energy conservation, and reconstruction of older multi-family houses. *Indoor Air*. 2003;13: 206–211.
31. Hirsch T, Hering M, Burkner K, et al. House-dust mite allergen concentrations and mold spores in apartment bedrooms before and after installation of insulated windows and central heating systems. *Allergy*. 2000;55:79–83.
32. EPA. Green Building History; 2016. <https://archive.epa.gov/greenbuilding/web/html/about.html>.
33. Jacobs DE, Weinberg A. Infrastructure and Mortgages: What About the Kids? Blog for National Center for Healthy Housing; February 2017. <http://nchh.org/Resources/Blog/InfrastructureandMortgages.aspx>.
34. House of Representatives, Maxine Waters. H.R.4497 – Housing is Infrastructure Act of 2021 117th Congress (2021–2022). <https://www.congress.gov/bill/117th-congress/house-bill/4497>.
35. Dissell R. Cleveland City Council passes historic lead poisoning prevention law. The Plain Dealer and Brie Zeltner, The Cleveland Plain Dealer. Posted July 24; 2019. <https://www.cleveland.com/metro/2019/07/cleveland-city-council-passes-historic-lead-poisoning-prevention-law.html>.
36. The Economist. How Housing Became the World's Biggest Asset Class, January 16; 2020. <https://www.economist.com/special-report/2020/01/16/how-housing-became-the-worlds-biggest-asset-class>.
37. Woo A. How Have Rents Changed Since 1960? Apartment List, June 13; 2016. <https://www.apartmentlist.com/research/rent-growth-since-1960>.
38. Solomon D, Maxwell C, Castro A. Systemic Inequality: Displacement, Exclusion, and Segregation: How America's Housing System Undermines Wealth Building in Communities of Color. Center for American Progress, August 7; 2019. <https://www.americanprogress.org/issues/race/reports/2019/08/07/472617/systemic-inequality-displacement-exclusion-segregation/>.
39. Jacobs DE. Environmental health disparities in housing. *J Am Public Health Assoc*. 2011;101(suppl 1):S115–S122.
40. Lubell J, Crain R, Cohen R. *The Positive Impacts of Affordable Housing on Health*. Washington, D.C.: Center for Housing Policy and Enterprise Community Partners; 2007.
41. Reddy A. Housing and Health White Paper. National Center for Healthy Housing, with support from Academy Health, American Public Health Association, Kaiser Permanente; 2020.
42. Joint Center for Housing Studies of Harvard University. The State of The Nation's Housing 2020. <https://www.jchs.harvard.edu/state-nations-housing-2020>.
43. Jacobs DE, Dixon SL, Wilson JW, Smith J, Evens A. The relationship of housing and public health: a 30 year retrospective analysis in the US. *Environ Health Perspect*. 2009;117:597–604.
44. Public Law 75-412, Housing Act of 1937, September 1, 1937. Also 25 CFR Section 700.55 – Decent, Safe, and Sanitary Dwelling.
45. Jacobs DE, Brown MJ. Windows of Opportunity: Linking Energy Conservation, Property Value, Childhood Lead Poisoning Prevention and Window Replacement, Window and Door Manufacturer's Association, Albuquerque, NM, August 10, 2004. Window and Door Manufacturers accociation presentation.
46. Nevin R, Jacobs DE, Berg M, Cohen J. Monetary benefits of preventing childhood lead poisoning with lead-safe window replacement. *Environ Res*. 2008;106:410–419.
47. Nevin R, Jacobs DE. Windows of opportunity: lead poisoning prevention, housing affordability and energy conservation. *Hous Policy Debate*. 2006;17(1):185–207.
48. Jacobs DE, Tobin M, Targos L, et al. Replacing windows reduces childhood lead exposure: results from a state-funded program. *J Public Health Manag Pract*. 2016;22(5):482–491.
49. Code of Hammurabi. Circa 3000 BC (Section 229). <https://legacy.fordham.edu/halsall/ancient/hamcode.asp>.

50. How The Other Half Lives. Studies Among the Tenements of New York by Jacob A. Riis. With Illustrations Chiefly from Photographs Taken by the Author. New York. Charles Scribner's Sons. Trow's Printing and Bookbinding Company, New York; 1890. <https://www.gutenberg.org/files/45502/45502-h/45502-h.htm>.

51. Stein L. A study of respiratory tuberculosis in relation to housing conditions in Edinburgh; the pre-war period. *Br J Soc Med*. 1950;4:143–169.

52. Basic Principles of Healthful Housing, Preliminary Report. Committee on the hygiene of housing. American public health association. *Am J Public Health*. 1938;28(3):351–372.

53. American Public Health Association and Centers for Disease Control and Prevention. Recommended minimum housing standards. 1986. <http://nchharchive.org/LinkClick.aspx?fileticket = uGqGpbBc2h4%3D&tabid = 550>.

54. International Code Council 2021 I-Codes. <https://shop.iccsafe.org/2021-complete-14-collection.html>.

55. Appendix F Radon Control Methods. International Residential Code 2018. <https://codes.iccsafe.org/content/IRC2018/appendix-f-radon-control-methods>.

56. CDC issued 11 separate reports on the Katrina travel trailer formaldehyde problem. updated August 23, 2018. Available at < https://www.cdc.gov/air/trailerstudy/default.htm >.

57. Federal Emergency Management Agency. Improved Housing Ready to Deploy For 2009 Hurricane Season, Release date: July 9, 2009. Release Number: HQ-09-081. The policy states in part: "…construction specifications that reduce formaldehyde emission levels to less than 16 parts per billion [0.016 ppm] within all travel trailers, park models, mobile homes, and alternative housing units…" Before this policy, formaldehyde was not regulated in travel trailers.

58. Home Inspection Trends; 2019. <https://inspectorchecklist.com/home-inspection-trends-2019/>.

59. Lead in Your French Home, By FrenchEntrée, September 15, 2017. The Anti-exclusion Act 98-657 dated July 29 1998, Article 123 modifying the Code of Public Health. Ministerial Order n°2006-474 dated April 25 2006 concerning the control of risks of lead poisoning, modifying Articles R1334-1 to R1334- 13 of the Code of Public Health (regulatory rules). <https://www.frenchentree.com/french-property/lead-in-your-french-home/#: ~ : text = Throughout%20France%2C%20an%20inspection%20for,fluorescence%20apparatus%20(see%20photo)>.

60. Rental Housing Warrant of Fitness, New Zealand; 2019. <https://nzrentalwof.co.nz/>.

61. Korfmacher K. *Bridging Silos: Collaborating for Environmental Health and Justice in Urban Communities*. MIT Press; 2019.

62. National Center for Healthy Housing. Technical Assistance for Code Transformation Innovation Collaborative (TACTIC). 2019. < https://nchh.org/tools-and-data/technical-assistance/tactic/ >.

63. Change Lab Solutions. Healthy Housing Through Proactive Rental Inspection. 2014. < https://www.changelabsolutions.org/product/healthy-housing-through-proactive-rental-inspection >.

64. National Center for Healthy Housing. 2018 update. National Healthy Housing Standard. <https://nchh.org/tools-and-data/housing-code-tools/national-healthy-housing-standard/>.

65. National Center for Healthy Housing. Code Comparison Tool; 2019. <https://nchh.org/tools-and-data/housing-code-tools/cct/>.

66. IOM (Institute of Medicine), National Academy of Sciences. *Healthy, resilient, and sustainable communities after disasters*. Washington, DC: The National Academies Press; 2015. Available from: http://www.iom.edu/postdisaster.

67. Pollitz K, Cox C, Lucia K, Keith K. Medical Debt Among People with Health Insurance, January 07; 2014. <https://www.kff.org/report-section/medical-debt-among-people-with-health-insurance-consequences-of-medical-debt/>.

68. Becker R. 'Debate is over,' California's governor says. 'This is a climate damn emergency.' Cal Matters, September 11; 2020. Available from: https://calmatters.org/environment/2020/09/california-governor-climate-emergency/.

69. Wilkinson R, Marmot M, eds. *The Social Determinants of Health: The Solid Facts (PDF)*. 2nd ed. World Health Organization Europe; 2003. ISBN 978-92-890–1371-0.

70. World Health Organization. *The Economics of Social Determinants of Health and Health Inequalities: A Resource Book (PDF)*. World Health Organization; 2013, ISBN 978-92-4-154862-5. Available from: https://apps.who.int/iris/bitstream/handle/10665/84213/9789241548625_eng.pdf;jsessionid = 97F39CFE3967A7FE93BA46E18B4418E9?sequence = 1.

71. Ramirez B, Laura K, Baker EA, Metzler M. *Promoting Health Equity: A Resource to Help Communities Address Social Determinants of Health. United States Centers for Disease Control and Prevention*. Atlanta: U.S. Department

of Health and Human Services, Centers for Disease Control and Prevention; 2008. Available from: https://www.cdc.gov/nccdphp/dch/programs/healthycommunitiesprogram/tools/pdf/SDOH-workbook.pdf.

72. Housing and Health: An Overview of The Literature. Health Affairs Health Policy Brief, June 7; 2018. Available from: https://doi.org/10.1377/hpb20180313.396577. <https://www.healthaffairs.org/do/10.1377/hpb20180313.396577/full/>.

73. World Conference on Social Determinants of Health. *Rio Political Declaration on Social Determinants of Health (PDF)*. World Health Organization; 2011. Available from: https://www.who.int/sdhconference/declaration/en/.

74. Pega F, Valentine N, Rasanathan K, Hosseinpoor AR, Neira M. The need to monitor actions on the social determinants of health. *Bull World Health Organ*. 2017;95(11):784–787. Available from: https://doi.org/10.2471/BLT0.16.184622.

10

Conclusion: the triumph of science and citizen action over policy paralysis

10.1 Knowing and doing

The philosopher Johannes Goethe believed "It is not enough to know, we must also apply; it is not enough to will, we must also do."[1] This book has documented how knowledge was created and applied to advance political will and to implement solutions for lead paint poisoning.

Those solutions produced enormous progress. From 1976−80 to 2015−16, the blood lead level of the US population aged 1−74 years declined 93.6%, from 12.8 to $0.82 \mu g/dL$. In 2015−16, 0.2% of children aged 1−5 years had blood lead levels of $10 \mu g/dL$ or higher, and 1.3% were $5 \mu g/dL$ or higher (these were the two blood lead trigger levels used between 1991 and 2021)[2] (see Fig. 10.1A−D). The Centers for Disease Control and Prevention (CDC) stated that childhood lead poisoning prevention was 1 of 10 great US public health achievements.[3]

The facts demonstrating progress also showed that it was not shared equally. CDC data published in 2013 showed that the average (geometric mean) blood lead levels for Black children were 38% higher than for White children ($1.8 \mu g/dL$ and $1.3 \mu g/dL$, respectively). For low-income children versus other children, the difference was 33% ($1.6 \mu g/dL$ and $1.2 \mu g/dL$), respectively.[4] The difference in blood lead levels by race, ethnicity, and income remained statistically (and stubbornly) significant.[5] Fig. 10.1 shows more recent data from CDC's National Health and Nutrition Examination Survey from 1977 to 2015 (Fig. 10.1A shows that the difference between non-Hispanic Black and White children remained statistically significant in 2015; Fig. 10.1B−D shows the improvement in the percentage of children above the 2021 CDC reference blood lead value of $3.5 \mu g/dL$ and the CDC 1991 level of concern of $10 \mu g/dL$).

These differences were far smaller than they once were, and the rate of improvement in blood lead levels for non-Hispanic Black children was far greater. In other words, there was a significant improvement in reducing *both* disparities and blood lead levels in all children. In short, progress was greatest for those at higher risk and substantial for everyone else—both at the same time, because public education was widespread and remediation funding was carefully targeted.

The trend in lead-contaminated housing was similar to these trends in blood lead. In 2019, Department of Housing and Urban Development (HUD) data showed that

Fifty Years of Peeling Away the Lead Paint Problem
DOI: https://doi.org/10.1016/B978-0-443-18736-0.00007-8

34.6 million (29%) of homes had lead paint somewhere in the building, an improvement from the estimate of 37.9 million (40.1%) of homes with lead paint in 1998. Homes receiving government support had significantly lower prevalence of lead paint than those not receiving such support, and there were no significant differences in lead paint prevalence by tenure, urbanization, income, poverty status, or ethnicity, suggesting the targeted high-risk approach yielded progress. There was also a significant decrease in lead paint in housing occupied by Black families, from 41.2% in 1998 to 25.2% in 2019. Dust lead levels were also lower. In 2019 the arithmetic mean dust lead loading on floors nationwide was 73% lower ($3.68 \mu g/ft^2$, down from $13.6 \mu g/ft^2$ in 1998). On windowsills, mean dust lead levels in 2019 were $54 \mu g/ft^2$, much lower than the $195 \mu g/ft^2$ in 1998 (a 72% improvement).

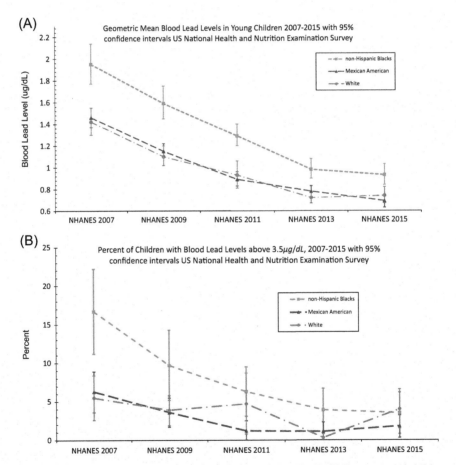

FIGURE 10.1   (A) Children's geometric mean blood lead levels, 2007–15. (B) Percentage of children with elevated blood lead ($3.5 \mu g/dL$), 2007–15. (C) Percentage of children with elevated blood lead ($10 \mu g/dL$), 1988–2002. (D) Percentage of children with elevated blood lead ($10 \mu g/dL$), 1976–2002. Data from: *National Health and Nutrition Examination Survey.*

(C)

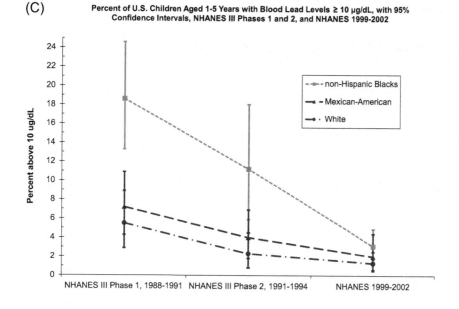

(D)

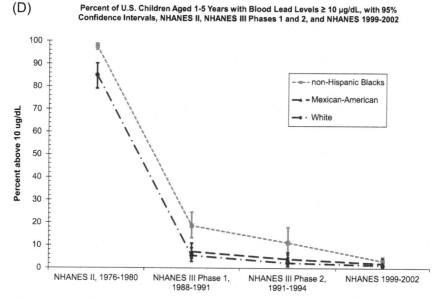

FIGURE 10.1   (Continued).

Despite these large improvements, the most recent housing survey also showed that the number of homes with deteriorated lead paint actually *increased* by 4.6 million homes in the past 8 years, likely because the housing stock continues to age and as it does there is more paint deterioration.[6,7]

There were several main reasons why progress happened:

- Science informed policy, which in turn drove even more scientific innovation.
- People skilled at creating and applying that knowledge occupied key positions both inside and outside government.
- Political will was created through mobilized and informed citizens demanding both policy change and the resources to put those policies into practice.
- Most importantly, houses were remediated.

Despite the ringing success of science-based lead paint policies and practice, "alternative facts" sometimes challenged real facts and the science behind them. In the case of lead paint, this generated ill-informed policies focused solely on the presence of lead paint instead of exposure, blocking progress during the 1970s and 1980s.

Centuries earlier, President John Adams stated:

> Facts are stubborn things; and whatever may be our wishes, our inclinations, or the dictates of our passion, they cannot alter the state of facts and evidence.[8]

The Enlightenment that heralded both the end of dictatorial rule by an unelected aristocracy and the advance of science to guide democratic decisions is only several hundred years old, a mere blink of the eye in human history. Earlier decisions were driven by autocratic dictate or "magic."[9] Although they have always been present in one form or another, attacks on science and denial of facts gathered momentum again in the late 2010s in the United States and around the world. In response, citizens and scientists both joined together to insist on science-based policies. The March for Science on April 22, 2017, featured thousands of citizens and scientists chanting, "What do you want? Good science! When do you want it? After peer view!" came the odd and nerdy (yet prescient) response (Fig. 10.2).

The history of lead paint poisoning prevention is understood best as one in which science triumphed by driving new evidence-based policies that were supported, indeed

FIGURE 10.2 The March for Science, 2017. Reproduced with Permission from Shutterstock.

demanded by, the public and then put into practice. This book has presented the complex facts and the debate that swirled around them. Is it a true triumph?

Internationally, in 2019 lead exposure annually accounted for 540,000 deaths and 13.9 million years of healthy life lost in disability-adjusted life years. The highest burden was in low- and middle-income countries. World-wide lead exposure accounted for 63.8% of the global burden of idiopathic developmental intellectual disability, 3% of the global burden of ischemic heart disease, and 3.1% of the global burden of stroke.[10] Lead poisoning rates remained quite high among immigrant children arriving in the United States; one study in 2019 showed nearly 42% of such children had an elevated blood lead level above 5 µg/dL.[11] Yet more and more countries adopted model laws that prohibited new lead paint production across the globe. The UN reports that as of December 31, 2020, 79 countries have legally binding controls to limit the production, import, and sale of lead paints (41% of all countries).[12]

To eliminate lead poisoning as a major public health problem both in the US and internationally, the triumphs and progress of the past must guide future actions.

10.2 Two steps forward, one step back

Although not a straight line, there were many more advances than retreats between 1970 and 2022. The early 1970s marked steps forward with a recognition of the magnitude of the problem when Congress passed the first Lead Based Paint Poisoning Prevention Act. But because health professionals were the only ones tasked with solving it in the late 1970s and early 1980s, it marked a step back and the problem remained largely hidden until the Agency for Toxic Substances and Disease Registry's bombshell report in the late 1980s awakened both Congress and the public. When the housing world finally came on board in the late 1980s and 1990s, there were steps forward.

Getting the science right on pathways of lead exposure and implementing them first in public housing and other housing later marked steps forward. But when the *Ashton v Pierce* court case and the 1987 Housing Act in Congress in the late 1980s erroneously declared lead paint and lead paint hazards to be the same thing, it was a step back.

When dangerous methods of lead paint removal were recognized by researchers in the early 1980s and subsequently banned, there were steps forward. But when some persisted in calls for simplistic and dangerous lead paint removal, it marked a step back. Yet when newer evidence-based remediation methods were broadly implemented and when scientific evaluations proved they worked, they marked steps forward.

When modern evidence-based lead abatement failed to become part of normal housing financing, it marked a step back. But when Title X was enacted in 1992 using an enlightened scientific definition of "lead paint hazard," it marked a major step forward. The law's implementation over the succeeding years through authorized federal funding, enforcement, standards, clear agency responsibilities, and expanded local capacity all marked major steps forward.

When national surveys counted the number of children harmed and the number of homes contaminated, they marked important steps forward by focusing the nation's attention. But when the counting of children with high blood lead levels temporarily stopped in the 1980s, it appeared to some that the problem was out of sight (and out of mind), marking a step back. When the surveys were restored in the 1990s, it marked a step forward.

When one HUD Secretary, Henry Cisneros, brought science to HUD, it marked more steps forward. When another HUD Secretary, Andrew Cuomo, stopped the Campaign for a Lead-Safe America, it marked a step back. When yet another HUD Secretary, Mel Martinez, made lead paint a priority, it marked steps forward. But when the next HUD Secretary, Alphonso Jackson, attempted to eliminate support for parent and community groups, forced out key scientists, substituted incompetent managers, and tried to bury the lead paint office by moving it out of the office of the Secretary, it marked retreats. But more steps forward occurred when those attempts were rebuffed by leaders across the country, and new HUD Secretaries (Shaun Donovan and Julian Castro) restored science and competence at HUD.

When X-ray fluorescence lead paint analyzers failed to measure lead paint accurately in the 1970s, it marked a step backward. But when new analyzers and new laboratory and field quality control procedures appeared as a result of federal standards in the 1990s, the greater reliability marked steps forward, spurring more scientific innovation, with faster and more inexpensive lead paint inspections.

When reports to Congress, strategic plans, task forces, and other bodies completed their works, important steps forward were made. But when such plans were not fully implemented, it marked a step back. More steps forward occurred when a national 10-year strategic plan was launched by the President's cabinet in 2000. But then a step backward when it was not fully funded, and private funding failed to appear. Steps forward happened when Congress provided record funding by the end of the 2010 decade.

The 2018 strategic plan issued by the President's Task Force on Lead was one of these retreats—a retreat from the previous one issued in 2000. The 2018 plan began by saying "This is *not* a budget document" (emphasis added). Previous calls for "ending" childhood lead poisoning became "reducing" it. Gone were the Cabinet Secretaries who had adopted the previous one in 2000. The goal of eliminating lead paint hazards as a major public health problem became instead "identify lead exposed children and improve their health outcomes," a throwback to the reactive failed approach two decades earlier. Instead of an ambitious campaign for a Lead-Safe America, there was an ambiguous goal to "communicate more effectively."[13] Steps forward occurred with the release of the Find It Fix It Fund It campaign (described later in this concluding chapter). Further steps forward occurred when Congress rejected the 2018 plan by providing record-high funding for lead paint and healthy homes.

Attacking, falsely accusing, and attempting to remove skilled leaders in government, research institutions, parent's groups, and advisory committees all marked steps backward. These attacks were often later rebuffed, but at great cost to their livelihood and reputations. Those who were wrongly attacked included Xavier Bonnefoy at WHO, Mark Farfel at Johns Hopkins University, Mary Jean Brown at CDC, Paul Mushak at the Public Health Service, Margaret Sauser at UPAL, Dorr Dearborn at Case Western University, Ruth Etzel at CDC, Barry Mankowitz at City Homes, Herb Needleman at the University of Pittsburgh, David Jacobs at HUD, and others. All were later exonerated.

Another retreat occurred when the CDC lead poisoning prevention advisory committee was disbanded in 2016, robbing the nation of key scientific input. This advisory committee had been in place for decades. But in 2020, more steps forward occurred when the committee was re-established through Congressional action to force CDC to do so.[14]

Losing court battles to hold the lead, pigment, and paint industries accountable marked steps backward. Yet a Rhode Island jury verdict and a California judge's ruling requiring the industries to help pay for remediation marked steps forward. The loss of the Rhode

Island jury's decision by the state's supreme court marked a step back. But newer cases against the lead paint industry also prevailed, making even more steps forward.

Other steps forward included applying the lessons from lead paint to other healthy housing problems not only in the United States but around the world. A step backward happened when Congress failed to pass a "Title X for Healthy Housing," with clear roles for different agencies authorized, funded, and defined, as it had done for lead paint. Yet another step forward occurred when the WHO released its first international housing and health guidelines in 2018.

More steps forward occurred as local jurisdictions increasingly adopted primary prevention approaches, instead of waiting for children to become lead poisoned. For example, the state of New York implemented its primary prevention program in 2007 in over 10 cities. By 2018 it had inspected over 40,000 homes, protecting tens of thousands of children. With a total of over $70 million invested since the program began, estimated to be about $4,700 per child, it was only a fraction of the cost of special education for a child with lead exposure, which was about $38,000 over 3 years.[15] New local lead paint laws were passed in Cleveland, Philadelphia, Pittsburgh, and other cities, marking steps forward.

The high estimated cost of lead abatement in the 1990 HUD Comprehensive and Workable Plan produced paralysis, marking a step backward. But later cost and benefit estimates, most recently in 2017 showed the costs were far less and the benefits far greater, marking steps forward. More steps forward occurred when local private and public funding appeared, such as more than $120 million in Cleveland alone.

The abdication of the housing code world in addressing lead paint in the early 1990s was a step backward. But new proactive lead and healthy housing codes appeared in the 2000s, led by Rochester NY and others, marking more steps forward.

In an ideal world, one can imagine a more straightforward history without all the temporary retreats—as new evidence emerged, new policies could be smoothly implemented to match the new understanding. In the real world, the course was more uneven. The links between science and enlightened policy and putting it into practice required continuous efforts that succeeded in overcoming temporary setbacks.

10.3 The Find It, Fix It, Fund It campaign

Over the past 50 years, there were three main phases of lead paint poisoning prevention:

- Blood lead screening and the medical model (1970–92)
- Emergence of a primary prevention (but relatively small) housing model (1992–2015)
- Bringing lead paint into mainstream housing and financing systems.

During the 2010 decade, a campaign to bring the third phase into being involved hundreds of groups that banded together to articulate specific lead poisoning prevention recommendations under the umbrella of a "Find It, Fix It, Fund It" (FFF) campaign.[16,17] It identified new efforts to identify which houses had lead problems ("Find It"). Once found, those problems would be corrected using proven methods ("Fix It"). Finally, funding was to be obtained to execute the first two steps at scale ("Fund It"). The name of the new national campaign was borrowed from an earlier citizen's effort in Rochester, NY.[18] A few of the hundreds of grassroots organizations in the FFF campaign are shown in Fig. 10.3 and were a direct result of the many groups that sprang up around the country in earlier decades, consisting of parents of

Find It, Fix It, Fund It
Action Drive Members

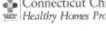
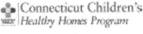

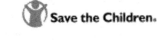

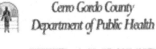

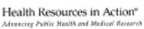

For more information, contact jkruse@nchh.org | For details on the action drive, see bit.ly/FindFixLEAD

FIGURE 10.3   Find It Fix It Fund It graphic. From: *National Center for Healthy Housing https://nchh.org/build-the-movement/find-fix-fund/.*

poisoned children, local health and housing professionals, scientists, concerned citizens, churches, environmental groups, energy and home weatherization professionals, and many others. The FFF campaign demonstrated again that the most effective way to present both problems and solutions was through the combined voices of parent and community groups, housing, environmental, and health professionals and scientists.

Internationally, many community groups operated under the IPEN umbrella, which reported over 500 nongovernmental organizations in 116 developing countries working to ban lead-based paint.[19] The Australians (where it all began at the end of the 19th century) had robust community-based organizations devoted to solving lead paint and other housing problems.[20] The last week in October, designated as "International Lead Poisoning Prevention Week of Action," likely began in the United States with a declaration from Senator Jack Reed in 2000, written by Rebecca Morley when she served as a Presidential Management Intern in his office.[21,22]

The FFF campaign documented the economic impact of including lead in an infrastructure Congressional bill. Every dollar invested in infrastructure created at least $1.75 in related economic benefits. Correcting lead hazards in the most at-risk homes would create an estimated 52,000–75,000 jobs, with tens of thousands more for moderate risk homes.[16]

In many ways, the FFF campaign's recommendations distilled the knowledge that had been gained over the previous 5 decades. Specifically, the campaign called for Congress to ensure that its appropriations were large enough to cover all housing, not only affordable, public, or other federally assisted housing. This marked a turn from a relatively small program into a much larger one, based on the new consensus shown in Chapter 9.

There were four main areas in the FFF campaign:

- Policy recommendations to locate and eliminate lead exposure, modernized regulations, and updated exposure limits;
- Improvements for surveillance of blood lead levels and follow-up for children already exposed to lead;
- Financing strategies; and
- Infrastructure and workforce development.

Congress reacted to the FFF campaign by increasing appropriations for HUD's lead and healthy homes program to its highest level ever: $360 million in 2021, with increases starting in 2017. There was also a funding increase at CDC (Fig. 10.4).

The FFF campaign also advanced other policy recommendations beyond appropriations, including reform of the disclosure rule, updating the antiquated HUD single-family mortgage insurance and section 8 housing choice voucher regulations, and funding for community groups to enable "citizen science" projects to identify where lead hazards were located.

The campaign also called for increased enforcement, especially at EPA for its renovation, repair, and painting rule. In December 2020, EPA and the Department of Justice announced the biggest fine ever levied for violations of that regulation against one of the largest hardware store chains in the country, Home Depot, for over $20 million.[23]

The FFF campaign did not limit its focus to only government, however. It specifically called on US companies that still made lead paint in other countries, notably Sherwin Williams, to stop doing so, consistent with recommendations from the World Health Organization and the United Nations Environment Program.[24] The industry's actions had contaminated tens of millions of homes in the United States before the 1978 ban and continued to do so in other

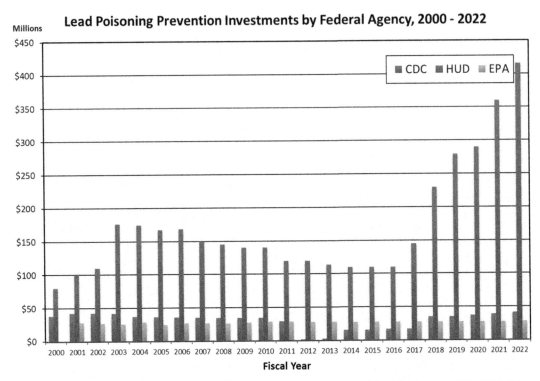

Millions

Lead Poisoning Prevention Investments by Federal Agency, 2000 - 2022

FIGURE 10.4 Federal appropriations for lead poisoning prevention and healthy homes, 2000–2022. *From: National Center for Healthy Housing.*

countries. Through its World Coatings Council, the industry continued to falsely claim that substitutes for lead paint did not exist or were somehow not viable, even though the world's largest paint company, Akzol Nobel, stated, "There is no good reason to put lead in paint."[25]

In 2022 a few states in the United States still did not have blood lead surveillance systems for children who were already exposed to lead. The FFF campaign advocated for reporting by physicians and by all states, ending the patchwork incomplete voluntary reporting system.

The FFF campaign also had an ambitious research agenda. For example, it suggested that the National Institutes of Health should determine if there were effective interventions *after* children had already been exposed. Because lead poisoning is a brain injury, CDC had determined that there could be some therapies that held some promise in at least partially overcoming the injury although this remained a largely unfulfilled research area.[26]

When tax reform was proposed in 2017, the FFF campaign recommended a federal income tax credit to enable lower- and moderate-income families to finance lead hazard control in their homes, as Congress had attempted earlier. The campaign thought it unfair that landlords were allowed to deduct lead abatement as a business expense, but owner-occupants could not.

This financing issue remained an important policy gap. There was some funding available for very low-income families through the HUD lead and healthy homes program and high-income families did not require financial assistance. But for the much larger number of lower- and moderate-income families, there remained little federal, state, local, philanthropic, or private financial support to eliminate lead hazards. This posed challenges to bringing lead hazard control into mainstream housing systems.

Another financing method advanced by the campaign included requiring paint companies to pay into a special fund to be used for correcting lead hazards in homes, which is what the California court had ruled. The idea was also similar to hazardous waste site cleanup funds, where polluting industries were required to contribute to cleanup.

The FFF campaign called for the replacement of old lead-contaminated windows (most of which only had single panes) with new energy-efficient ones. Windows had the highest levels of lead paint and contaminated dust compared to other building components and were also one of the important sources of energy inefficiency in housing. Modeled after a successful pilot program in Illinois known as "CLEAR-WIN" (Comprehensive Lead Education and Reduction through Window replacement), such an effort reduced fuel costs, a clear link to climate change mitigation. It also provided jobs for both window installers and manufacturers, improved property values, and eliminated a major source of lead exposure—all through a single focused effort that the public rated highly.[27]

Still another "mainstreaming recommendation" included the expansion of Medicaid/Children's Health Insurance Program waivers to not only pay for medical testing of lead-poisoned children, but also actual remediation of their homes as a reimbursable medical expense. Such waivers were obtained in Michigan, Ohio, Rhode Island, Indiana, Maryland, and a few other states by 2022.[28] This marked an important example of the new consensus on housing and health described in the previous chapter and in some cases became large enough in a few states to rival the HUD lead paint grants.

10.4 Getting the housing market to work

Another way to bring lead paint poisoning prevention to scale involved the disclosure rule. The lead paint disclosure law[29] was an example of how the failure to consider housing as part of the "shared commons" played out in the 1990s (because housing was not thought to be part of the commons, it was up to each homeowner to figure out what to do about lead paint). The initial version of the law would have worked to protect *all* children, not only a relative few served by small government programs. Introduced by Congressman Henry Waxman in the early 1990s, that initial version would have required an actual lead inspection to determine exactly where lead paint was in each home, not just leave it up to the whims of negotiations between buyers and sellers.[30]

We have seen how the real estate industry and other business interests succeeded in watering it down so that only "known" lead paint had to be disclosed and inspections would not be required. The result was that there was typically nothing "known" to disclose because most houses remained uninspected. This seemed to reward ignorance. The only homes that had "known" lead paint hazards were those that were inspected, usually because a poisoned child lived there—in other words after the exposure already occurred, the very thing the disclosure law was meant to prevent. Although vigorous enforcement of this rather weak "education only" law produced substantial numbers of remediated homes (Chapter 5), the fact remained that most homes (except public housing) remained uninspected and disclosure largely existed on paper only.

Joe Ventrone (who served as deputy chief of staff at HUD in 2001—04 during the time when HUD Secretary Mel Martinez made lead paint a HUD priority) later became a senior official at the National Association of Realtors. To thank him upon his departure from HUD for making lead a priority and for obtaining increased Congressional funding, Jacobs gave Ventrone a small lead ingot, probably one of the more unusual going-away presents a federal official ever received. He told Jacobs in 2018 that he proudly still displayed it on his desk. Perhaps realtors will one day support home lead inspections as part of the increasingly normal home inspection process, making lead more than an odd desk ornament.

When reliable information existed to guide investments, the housing market had at least a chance to stimulate more private funding on lead paint. Although most mortgage financing institutions had underwriting standards that required correction of certain housing defects, they failed to include lead paint. Such underwriting standards had been recommended by the Title X task force, but they were adopted only in the multifamily FHA mortgage insurance program at HUD in this time period. In France, and a few other local US jurisdictions like Cleveland, Rochester, NY, Baltimore MD, lead paint inspections and/or lead dust testing at the time of sale became routine, but in most areas disclosure and widespread lead paint inspections remained the exception. Reforming the disclosure law is needed to permit the housing market to work correctly.

10.5 Getting government to work

With a few exceptions, federal agencies largely carried out their Title X mandates, reflecting the Congressional clarity of their respective assignments and the on-going Congressional oversight. In 1992 that Act charged HUD with administering the main source of funding to remediate privately owned low-income housing with grants to local governments, as well as regulating federally assisted housing, conducting public education, enforcement, and undertaking certain technical studies. CDC was charged with estimating the number of lead-poisoned children through its National Health and Nutrition Examination population survey, surveillance of local data, ensuring blood lead testing was done properly to help identify highest risk children, providing guidance to clinicians, and implementing quality assurance for blood testing laboratories (many of these requirements predated Title X). EPA was to ensure that licensed professionals were trained, certified, and did inspection and remediation work correctly. EPA was also to set exposure standards for house dust and bare soil and establish laboratory quality controls for environmental paint, dust, and soil sample analysis. HUD and EPA shared disclosure enforcement responsibilities.

Although there were the inevitable bureaucratic turf struggles, this book showed that, for the most part, the three-legged stool of HUD, CDC, and EPA worked reasonably well. But a chair needs four legs. Private sector investment was that fourth leg, and there were new signs it was increasing as part of the new consensus on healthy housing.

HUD, CDC, and EPA played the main roles in this time period. However, the Occupational Safety and Health Administration also made progress. OSHA was mandated by Title X to issue a lead exposure standard for construction workers who would be doing the lead hazard control work under the HUD program and in other construction. The very first comprehensive occupational health standard (which focused on lead) passed by OSHA in 1978 was restricted to industrial workers, such as those in battery plants, smelters, and the like. How odd that it took a housing law to mandate OSHA to establish a construction worker standard for lead.

As many homes increasingly also become workplaces, there would be new opportunities for collaboration between the healthy housing and occupational health fields.[31]

Other agencies with housing programs, such as the Departments of Defense, Agriculture, and Energy also played important roles. This interagency coordination depended on Congressional oversight and the President's cabinet.

In short, the legislative and executive branches of government contributed significantly to solutions in this time period. However, progress was more uneven in the judicial branch of government. Many courts undermined progress on lead paint poisoning prevention by issuing flawed decisions that ignored or misunderstood scientific facts, failed to assign accountability, or were unresponsive to the plight faced by millions of children (most of whom never had their day in court). Examples include a DC court's decision in the *Ashton v. Pierce* case in the 1980s, which attempted to mandate paint removal using dangerous methods because it believed that the presence of lead paint was the same as a hazard. Without facts in evidence, a Maryland Appeals Court wrongly accused researchers of not protecting children, when the facts showed that they did, as later juries ruled. The Rhode Island Supreme Court overruled a unanimous jury verdict, in which the jury found the lead, paint, and pigment industries were accountable for the harm caused by their products and should pay for remediation. These and other unfortunate decisions were ultimately rejected in the court of public opinion, and by the legislative and executive branches. However, other more recent court decisions in California and in Milwaukee where scientific evidence was considered created new precedents, because they held the industry accountable and ordered them to pay for remediation and the damage their product had on children's health.

10.6 Getting the procedures right and recruiting the necessary expertise

Solutions to the lead paint problem required accurate and precise measurements. Limited measurement technology emerged as a significant barrier in the 1980s, when XRF portable lead paint analyzers clearly had high error rates. Furthermore, the initial failure to standardize safe work practices in the 1970s and 1980s meant that attempts to do the "right" thing by quickly removing lead paint using dangerous methods backfired, increasing exposures for both children who lived in the homes and the workers who attempted to remediate those homes.

There were numerous examples throughout the 1970s and 1980s of mistaken and in some cases downright fraudulent lead paint testing. This is why Title X mandated standardized training and licensing. The problem was not limited to lead paint measurements. Although an early study in the 1970s quantified lead dust measurements, it was not until the late 1980s and 1990s that lead dust measurement technology was standardized, correlated with children's blood lead levels, and promulgated in federal HUD and EPA regulation.[32]

These barriers were overcome in the 1990s and later by the standardization of both measurements and hazard control practices, embodied in the 1995 HUD Guidelines for the Evaluation and Control of Lead Paint Hazards in Housing that was written by the National Center for Healthy Housing. These Guidelines were later incorporated into standard field operating procedures, and local, state, and federal regulations.[33]

It was not only a question of technology and standardization, however. Who would carry out those procedures reliably and how could the public be ensured they were done reliably? It was occupational and environmental health professionals who largely played that key role. The lead paint ban for residential paint was first passed by occupational health professionals in 1920 at the International Labor Organization. It was yet another example of how harmful workplace conditions often led to the recognition of broader community health problems. And it was largely occupational and public environmental health professionals who brought scientific expertise to HUD in the 1990s, grounding the policies in practice.

Yet the connection between housing and occupational health remained tenuous; many environmental and occupational health professionals received no housing training, and the converse was also true.[31] With the notable exception of England and Wales, a trained, credentialed healthy homes inspectorate remained relatively small in the United States and in most other countries, especially in low-income developing nations, where housing inspections were often more the exception than the rule. Better training programs and curricula in both academia and professional continuing education to create a more robust, educated, and credentialed housing and health workforce remained a clear need.

Where exactly could the expertise be found? First embodied in the sanitation movement in the 1800s, the historic collaboration between housing, health, and environment was reborn in the United States between 1970 and 2000 with the lead paint experience. The needed multidisciplinary approach started in housing and health departments and became nationwide with the public housing insurance program and other efforts in the late 1980s. It later became institutionalized with the emergence of the National Center for Healthy Housing, the Alliance to End Childhood Lead Poisoning, the World Health Organization's healthy housing efforts, the National Safe and Healthy Housing Coalition, the Lead and Environmental Hazards Association and high-level offices at HUD, EPA, and CDC and many other entities.

10.7 Strategic plans

What are we to make of the various strategic plans on lead paint that were produced over the past 50 years? Were they just paper or did they change policy and practice?

In 1988 a report to Congress was the first to scientifically show the magnitude of the problem, describe toxic effects, and rank the importance of different sources and pathways of exposure. This report focused the nation's attention in the late 1980s due in no small part to the courage of Paul Mushak its principal author in forcing the government to release it. Yet its recommendations on how to fix the problem were poorly defined, reflecting the state of knowledge at the time. It called for:

Comprehensive approaches to controlling lead exposure in high-risk areas of the United States, establish and maintain effective and efficient screening programs, develop environmental measurement techniques for field use, and conduct research on childhood lead poisoning and develop effective legal sanctions.[34]

In 1990 under Congressional orders, HUD produced a report titled a Comprehensive and Workable Plan for the Abatement of Lead-Based Paint in Privately Owned Housing,[35] which turned out to be anything but "workable." It was the first to attempt to quantify both the extent and costs (but not the benefits) of lead paint remediation. This HUD estimate was that 64 million housing units contained lead paint and that the cost would be at least \$500 billion to remove it all. That price tag produced policy paralysis and again demonstrated the lack of scientific understanding of exposure pathways. But it did point to the direction the nation needed to take, that is, the development of a national strategy.

In 1991 CDC produced its own strategic plan, but it mainly focused on the role of clinicians and pediatric healthcare providers, blood lead screening, diagnosis, and guidance on chelation and other medical procedures. Although it included a short section on housing remediation at the very end, it was chiefly focused on the medical community.[36] An earlier CDC statement was issued in 1985, which contained the first government statement that "the precise threshold for the harmful effects of lead on the central nervous system is not known," but again there was little focus on solutions other than medical care, which was targeted to those with very high exposures in need of hospitalization and chelation to prevent death and acute health effects.[37]

Despite these limitations, the 1991 CDC plan was the first to estimate the financial benefits of lead paint remediation. The declaration issued that same year by the Department of Health and Human Services Secretary Louis Sullivan stated that lead poisoning was the nation's number 1 environmental disease in children and played a large role in focusing the nation's attention despite White House objections.

Congress responded by holding extensive hearings and passing Title X the very next year, in 1992. Among many other things, that act mandated the first serious housing-based strategic plan to address lead paint through a Congressionally mandated task force convened by HUD Secretary Cisneros, which produced its report in 1995.[38] For the first time, representatives of the broad array of interests affected by lead hazards in housing, including parents, rental property owners and managers, lenders and insurers, real estate agents, physicians and public health experts, housing and urban planning experts, contractors, state and local government officials, advocates for tenants' rights, environmental protection, and affordable housing providers were brought together. The recommendations in this plan became much more specific to housing:

1. Adopt benchmark lead-based paint maintenance and hazard control standards for rental housing (including essential maintenance practices and standard treatments);
2. Provide public and private financing of lead paint hazard control in economically distressed housing;
3. Modify the liability and insurance systems to ensure that they both compensate poisoned children and clearly set standards for preventive measures that property owners should take;
4. Increase public awareness;
5. Follow strategies that match families with young children to lead-safe housing; and
6. Promote more research on cost-effective strategies.

However, that plan did not include a timeline or a cost/benefit analysis. But it did lay the foundation for both short-term and long-term strategies that housing professionals understood and could implement.

President Clinton's Cabinet-Level Task Force plan in 2000 was the most robust government plan to date, including a forecast, cost-benefit analysis, the first-ever interagency budget to implement it, and a timeline. It presented the evidence that both interim controls and abatement were effective and the cost and benefits of each. Its interagency budget[39] resulted in a record high appropriation at the time, and its forecasting methodology was validated using empirical data.[40] Although it was never fully funded, which meant that Task Force's 2010 goal to eliminate childhood lead poisoning as a major public health problem was not achieved, it did provide a much clearer roadmap.

10.7.1 A new forecast to eliminate lead paint poisoning by 2027

What could a new roadmap look like in 2022? Using new data from the American Community Survey, the Residential Energy Consumption Survey (to estimate the rate of window replacement), ongoing rates of demolition and substantial housing renovation (both of which tend to reduce the number of surfaces with lead paint), and ongoing lead remediation, the new plan suggested here shows that much of the lead paint problem could be eliminated in affordable housing occupied by children by early 2027. This would require an annual spending level of at least $2.5 billion over 5 years (a total cost of nearly $13 billion).[41] These beginning estimates are based on updated costs of inspection and remediation from the HUD lead hazard control grant program.

The US House of Representatives in 2022 approved $5 billion for lead paint in an infrastructure law, but the Senate failed to include lead paint. The debate in the Senate reflected the antiquated definition of infrastructure, one that excludes housing from the "shared commons" and the new consensus described in the previous chapter. In any case, this marked an important milestone in "mainstreaming" lead into housing. The *net* benefit would be at least $15 billion annually for the group of children born in 2022 alone (compared to the cost of $2.5 billion), with additional benefits accruing each year as more children are born or move into homes without lead paint, dust, or soil hazards; such estimates are likely underestimates because they do not include certain intangible but very real benefits (such as stress, litigation, property management and others).[42] In 2019 the appearance of a "cost-benefit calculator" made such estimates at the local level feasible for the first time.[43] Based on the experience of the 2000 Cabinet-level task force, a new task force can be created in 2022 to implement a new strategy, which is the subject of a draft Presidential Executive Order. This book documents the power and influence of this type of task force.[44]

10.8 The influence of industry

Lead, paint, pigment, real estate, and construction industries all attempted to thwart solutions. During 1970–2022, the lead, pigment and paint industries resisted calls to contribute to the solution, but by 2020 they finally lost in several legal courts and in the court of public opinion. There was now consensus that these industries should contribute to the solution. But the consequences of hollowing out the environmental public health workforce as well as ongoing disinvestment in housing and the housing profession itself meant that responses were often possible only after the damage had already occurred instead of preventing it. Some referred to childhood lead poisoning as an on-going "silent" and insidious pandemic due to the inability to fully implement primary prevention.[45]

Lead paint was only one of many toxic materials that various industries placed into housing and housing commerce. The list became a long one: asbestos, poly-cholorinated biphenols in lighting and electrical materials, Chinese dry wall, insulation and flooring with formalde-hyde, carpets with stain-resistant dangerous chemicals, and many more. Even vinyl wallpa-per, initially marketed as an attractive building material, had been abandoned because of its moisture trapping quality, leading to extensive mold contamination. Whatever cost savings the industries claimed for lead paint and these other materials were dwarfed by their remedi-ation costs, often borne by private homeowners or taxpayers.

In 2018 Sherwin Williams marketed another new paint that the company claimed would prevent mold growth. When asked by a reporter if this was really true, this writer's response was as follows: why should the company be allowed to put a new chemical on the market that had not been shown to be safe? Their track record of putting lead into paint supposedly to create a better product made such claims dubious at best. Indeed, the company marketed its mold paint as a "super paint…with sanitizing technology." Lead paint was also marketed as a paint that could "sanitize" homes, with disastrous conse-quences. Yet the company lobbied successfully to get its product into the market with an EPA registration (such registration does not evaluate human health issues).[46] Studies remain to be done on whether the new paint is truly healthy.

In 2016, together with Perry Gottesfeld, who with many others worked tirelessly to stop the production of new lead-based paint in other countries, some lead poisoning prevention professionals bought a few shares of Sherwin-Williams stock to attend the annual share-holders meeting in Cleveland.[47] Community members protested outside the meeting. They pleaded with the company to stop making new lead-based paint in other countries but were told that this was "part of our normal business operations that shareholders have no say over."[17] They and other major paint companies continued to market lead paint around the world, contaminating homes in other countries just as they did in the United States decades ago. Many nations have now banned the production of new residential lead paint.[48]

10.9 Nine lessons from lead paint poisoning prevention

One definition of the necessary steps to creating good policy says one must first define the problem, assemble the evidence, develop alternatives, select criteria, project the out-comes by understanding trade-offs, get policymakers to make decisions and implement them, and then tell the story.[49]

There were nine key inter-connected lessons in the lead paint poisoning prevention and healthy housing history, with important implications for other problems that may initially appear to be too big to solve.

- *Scientific research*: Research over the 1970s and 1980s (funded almost entirely by the federal government) established lead's toxicity at even low levels. It also identified well-intentioned yet harmful paint "removal" methods. Research on how exposures to hazards occurred helped to focus interventions correctly on the most important pathways and to ban dangerous paint removal methods. Research also identified the large numbers of children affected, the numbers of homes with hazards, and how best to remediate them.
- *Technical guidelines*: Guidelines translated this science in a way that could be used by practitioners to identify hazards and then remediate them in effective practical ways, that

is, putting the science into practice. This optimized solutions and at the same time contained costs to help preserve housing affordability. The HUD Guidelines for the Evaluation and Control of Lead-Based Paint Hazards in Housing in 1990 (for public housing only) and in 1995 and 2012 (for all housing) did not initially carry the force of law. But they eventually were incorporated into most federal, state, and local lead hazard control regulations, as well as private contracts and standards of care. International healthy housing guidelines from the World Health Organization appeared in 2018.

- *Legislation and regulations*: These guidelines and the supporting science were used to create, pass and implement legislation that defined hazards and how to control them. The legislation also assigned clear responsibilities to previously uncoordinated federal agencies and mandated the creation and promulgation of standards. The passage of Title X in 1992 resulted in reform of federal lead paint housing regulations in 1999 and related public health and environmental regulations.

- *Public education and advocacy*: The research, guidelines and regulations in turn were used to drive education through public campaigns to ensure citizens were informed so they could articulate what was needed. During the time period covered in this book, it was epitomized by the creation of the Alliance to End Childhood Lead Poisoning in 1990, the National Center for Healthy Housing in 1992, the campaign for a Lead Safe America in 1997, the United Parents Against Lead and other parent's groups, the National Safe and Healthy Housing Coalition in 2009 and the "Find It Fix It Fund It" campaign in 2016.

- *Strategic plans*: Scientifically sound, popular and feasible strategic plans devised at the highest levels of government, with specific budgets and a timeline for implementation were essential. Initially released only by a single agency one at a time [e.g., the Surgeon General in 1980, ATSDR in 1988, HUD in 1990, CDC in 1991, and EPA (also in 1991)], a truly government-wide plan was achieved through the President's Cabinet-level Task Force in 2000.

- *Market forces*: Private market forces were increasingly engaged through disclosure in 1996; competition for new lead paint XRF instruments in 1997; new standards for lead-contaminated dust and soil in 1999 and 2001; and for green energy-efficient healthy housing in 2012. It reflected a new broad-based consensus on the intersection of housing, health, and environment and a recognition that solutions could not be limited to government.

- *Subsidy and enforcement*: For low-income housing with the highest risks, subsidy and enforcement were essential because the private market failed to include the full cost of fixing housing hazards in the price of a home. Effective subsidies were achieved with the successful initiation in 1990 of the HUD lead hazard control grant program and its subsequent expansion in 1995, despite some initial growing pains in earlier years and attempted political interference in later years. Enforcement resulted in thousands of privately funded abated housing units beginning in 1997 and increasingly occurred in healthy housing codes starting in 2012 in the US.

- *Evaluation*: Scientific and programmatic evaluation of large-scale nationwide hazard elimination programs confirmed that they were working as intended. The methods identified by the science, articulated in guidelines, enacted into legislation, funded by the government, supported by the public, embodied in strategic plans, and included in markets, subsidies, and enforcement were all proven. Evaluation also refined approaches based on new technology and other innovations. This was achieved through the Congressionally mandated study of the effectiveness of the HUD lead hazard control grant program (led by the National Center for Healthy Housing), and other studies at Johns Hopkins University, the Kennedy Krieger Research Institute, the University of Rochester, the University of Cincinnati, and many others. Together, these produced confidence that the new methods were working.

- *True prevention*: Perhaps the most important lesson of all from these years was that there were extremely high costs to limiting responses to only after children and others had already been harmed. A reactive approach simply did not work, was more expensive, resulted in endless litigation, and caused needless injury and suffering. Succimer and other chelating drugs did not demonstrate effectiveness for the vast majority of lead-poisoned children (except for those with extremely high lead exposures). The reactive approach was increasingly replaced by new efforts to inspect and remediate houses, not limited to testing children's blood. The medical model that treated illness mostly after it had already appeared came to be seen as an abject failure in the case of lead paint. So too were environmental cleanup laws that applied only after pollution had already occurred and after dangerous building materials and products like lead paint had entered commerce. So too were housing laws and codes that responded to violations after they occurred instead of proactively preventing them. Increasingly, the reactive approach was replaced by true prevention and embodied in federal and local law, policies and practices.

10.10 Ending a policy paralysis paradox

Ill-informed attacks on science, government, and citizen's groups all increased during the 2010 decade. Decision-making based on facts that informed policy debates, a bedrock of democracy, became more challenging. Yet, the successes on lead paint showed the power of scientifically informed policy.

That progress sometimes came at a high price for those who had the courage to act, speak up, find solutions, and then implement them. Some were wrongly accused of terrible crimes, suffered from groundless attacks on their professional and personal integrity, were bankrupted, forced out of their jobs, or dismissed. In many other cases, local hardworking lead poisoning prevention and healthy homes professionals were dismissed from positions for acting on the evidence and explaining the facts. Yet because of those facts, most were vindicated.

The untold story of lead poisoning prevention over the past five decades is full of heroes: thousands of committed parents, housing and health and environmental professionals, scientists, doctors, advocates, parents, lawyers, and many others struggling to solve the problem. This book is the first to name some of them (see the Appendix).

A "policy paralysis paradox" gripped the nation from the early 1970s to the late 1980s: The lead paint problem was either too big to solve (there were too many homes to treat, and it would cost too much) or, paradoxically, it did not exist at all (because new lead paint had already been prohibited). Both proved false. Making the problem appear too big to solve and denial of its existence both resulted in the same thing—paralysis. The lead paint experience from 1970 to 2022 proved the paradox can be solved and paralysis overcome.

Lead poisoning's legacy can be one behind us, not in front of us.[50] If these same lessons are applied to other large seemingly intractable problems, they too can be solved—the

Covid-19 pandemic and climate change are examples. Both were either dismissed as non-existent or too big to solve, denying both the facts and the solutions.

President Franklin Roosevelt, in dedicating the National Archives Building in 1941 at another time of crisis in the world, remarked on the power of learning and acting:

> A Nation must believe in three things: It must believe in the past. It must believe in the future. It must, above all, believe in the capacity of its own people to learn from the past so that they can gain in judgment in creating their own future.[51]

Lead poisoning is ugly. It can rob us of our most precious gift—our future and the children who inherit it. Yet the story of lead paint poisoning prevention is one of hope. It is full of people:

- who cared;
- who had the right skills and expertise;
- who took chances;
- who empowered citizens to act and tell their stories;
- who listened carefully to those stories;
- who established enlightened, practical and scientifically validated policies;
- who succeeded in fighting for the resources to make progress;
- who put policy into practice; and
- who were in key positions of power at the right time.

To be fully triumphant means that no child's future will be limited because of lead exposure. It means that mold, pesticides, pests, accessibility, cost, and other housing problems will not cause emergency room visits or hospitalizations for asthma or other diseases and injuries. Elders and persons with disabilities will be able to stay in their homes and communities through better housing designs. No one will die in a house fire or because of carbon monoxide during a power outage because alarms were not working or were absent. New building products will be tested for safety before they come to market. Homes will support good health and be a source of happiness and pride, not a cause of injury or disease.

The course is clear and well mapped. It starts in the home. More than a building, a home is that special place in our hearts, that manifestation of who we are as a people and as individuals at the same time.

Together, we can end this strange policy paralysis paradox by using science to confront and solve problems that are too big *not* to solve.

I end with a child's observation: At the March for Science in 2017, a youngster's sign proclaimed: "Think like a proton—be positive."

References

1. von Goethe JW. *The Maxims and Reflections of Goethe.* Translated by Bailey Saunders. New York, London: Macmillan and co.; 1906: p. 324. <https://archive.org/details/maximsreflection00goetrich>.
2. Dignam T, Kaufmann RB, LeStourgeon L, Brown MJ. Control of lead sources in the United States, 1970–2017: public health progress and current challenges to eliminating lead exposure. *J Public Health Manag Prot.* 2019;25 (1 suppl):S13–S22.
3. Centers for Disease Control and Prevention. Ten great public health achievements—United States, 2001–2010. *MMWR Morb Mortal Wkly Rep.* 2011;60:619–623.

4. Blood Lead Levels in Children Aged 1–5 Years—United States, 1999–2010. Morbidity and Mortality Weekly Report. Vol. 62, No. 13 April 5; 2013: p. 245. Reported by William Wheeler and Mary Jean Brown.

5. Whitehead LS, Buchanan SD. Childhood lead poisoning: a perpetual environmental justice issue? *J Public Health Manag Pract*. 2019;25(1 Supp):S115–S120.

6. U.S. Department of Housing and Urban Development, Office of Lead Hazard Control and Healthy Homes. American Healthy Homes Survey II. Lead Findings, Final Report, October 29; 2021.

7. U.S. Department of Housing and Urban Development. National Survey of Lead and Allergens in Housing Volume I: Analysis of Lead Hazards, Final Report, Revision 7.1, October 31; 2002.

8. Adams J. Argument in Defense of the Soldiers in the Boston Massacre Trials, December 1770. National Archives. <https://founders.archives.gov/documents/Adams/05-03-02-0001-0004-0016>.

9. Marchant J. The Human Cosmos. Civilization and the Stars. Penguin Random House; 2020. <https://www.penguinrandomhouse.com/books/646570/the-human-cosmos-by-jo-marchant/>.

10. World Health Organization. Lead Poisoning and Health, August 23; 2019. <https://www.who.int/news-room/fact-sheets/detail/lead-poisoning-and-health>.

11. Geltman PL, Smock L, Cochran J. Trends in elevated blood lead levels using 5 and 10 μg/dL levels of concern among refugee children resettled in Massachusetts, 1998–2015. *Public Health Rep*. 2019;. Available from: https://doi.org/10.1177/0033354919874078.

12. United Nations Environment Program. Global Alliance to End Lead Paint. Update on the Global Status of Legal Limits on Lead in Paint; December 2020. <https://wedocs.unep.org/bitstream/handle/20.500.11822/35105/GS-2020.pdf?sequence = 3>.

13. Federal Action Plan to Reduce Childhood Lead Exposures and Associated Health Impacts. President's Task Force on Environmental Health Risks and Safety Risks to Children; December 2018. <https://ptfceh.niehs.nih.gov/resources/lead_action_plan_508.pdf>.

14. Lead Exposure and Prevention Advisory Committee (LEPAC); Notice of Establishment. A Notice by the Centers for Disease Control and Prevention on 02/13/2018. 83 Federal Register 6177. Section 2203 of Public Law 114–322 (Water Infrastructure Improvements for the Nation Act) (Registry for Lead Exposure and Advisory Committee), and the Federal Advisory Committee Act of October 6; 1972.

15. National Center for Healthy Housing. New York State Childhood Lead Poisoning Primary Prevention Program (NYS CLPPPP); 2014. <https://nchh.org/tools-and-data/technical-assistance/nys-clpppp/>.

16. Introducing Find It Fix It Fund It. National Center for Healthy Housing. (and others). 2016. < https://nchh.org/build-the-movement/find-fix-fund/ >.

17. Jacobs DE. Lead poisoning: focusing on the fix. Invited editorial. *J Public Health Manag Pract*. 2016;22(4):326–330.

18. Katrina Korfmacher personal communication with David Jacobs 2016.

19. International Lead Poisoning Prevention Week of Action, October 21–27; 2018. <https://ipen.org/sites/default/files/documents/ipen-lead-week-of-action-2018-v1_4.pdf>.

20. McPhillips K. *Local Heroes*. Australia: Pluto Press; 2002. ISBN 1-86403-058-5.

21. Jack S. Reed Honored by National and Local Groups; 2014. <https://nchh.org/2014/03/senator-jack-reed-honored-by-national-and-local-groups/>.

22. S Res.199—106th Congress (1999–2000) A resolution designating the week 24, 1999, through October 30, 1999, and the week of October 22, 2000, through October 28, 2000, as "National Childhood Lead Poisoning Prevention Week." <https://www.congress.gov/bill/106th-congress/senate-resolution/199/text?r = 57&s = 1>.

23. The United States, the State of Utah, the State of Rhode Island and the Commonwealth of Massachusetts Executive Office of Workforce Development, Department of Labor Standards v. The Home Depot, U.S.A., Inc., Civil Action No. 1:20CV5112, December 17; 2020. <https://www.justice.gov/enrd/consent-decree/file/1349441/download>.

24. Global Alliance to End Lead Paint. World Health Organization and United Nations Environment Program. 2022. Available from: https://www.unep.org/explore-topics/chemicals-waste/what-we-do/emerging-issues/global-alliance-eliminate-lead-paint-1 >.

25. AkzoNobel Works With United Nations to Eliminate the Use of Lead Paint, by Kerry Pianoforte, Coatings World, February 26, 2014. Available from: https://www.coatingsworld.com/contents/view_online-exclusives/2014-02-26/akzonobel-works-with-united-nations-to-eliminate-the-use-of-lead-paint.

26. CDC Education Interventions for Children Affected by Lead; April 2015. <https://www.cdc.gov/nceh/lead/publications/Educational_Interventions_Children_Affected_by_Lead.pdf>.

27. Jacobs DE, Tobin M, Targos L, et al. Replacing windows reduces childhood lead exposure: results from a state-funded program. *J Public Health Manag & Pract.* 2016;22(5):482–491.

28. David Eggert Associated Press. Michigan wins OK to spend federal funds on lead abatement. <https://www.crainsdetroit.com/article/20161114/NEWS01/161119887/michigan-gets-federal-ok-to-spend-119-million-on-lead-abatement>.

29. Disclosure of Known Lead Based Paint and Lead-Based Paint Hazards. Final Rule. 24 CFR Part 35 Subpart A; and 40 CFR Part 745. 1996.

30. H.R.2840 – Lead Contamination Control Act Amendments of 1991. 102nd Congress (1991–1992). SEC. 2803. disclosure of indoor lead hazards. <https://www.congress.gov/bill/102nd-congress/house-bill/2840/text?r = 23&s = 3>.

31. Jacobs DE, Forst L. Occupational safety, health and healthy housing: a review of opportunities and challenges. *J Public Health Manag Prot.* 2017;23(6). e-36 to e-45.

32. EPA. 2001. Identification of Dangerous Levels of Lead; Final Rule. 40 CFR Part 745. *Fed. Reg.* 66(4)1206.

33. Jacobs DE, et al. HUD Guidelines for the Evaluation and Control of Lead-Based Paint Hazards in Housing, Department of Housing and Urban Development, Washington DC, HUD-1539; 1995.

34. Agency for Toxic Substances and Disease Registry. The Nature and Extent of Lead Poisoning in Children in the United States: A Report to Congress. DHHS document no. 99-2966. Atlanta: US Department of Health and Human Services, Public Health Service; 1988.

35. US Department of Housing and Urban Development. Comprehensive and Workable Plan for the Abatement of Lead-Based Paint in Privately Owned Housing. U.S. Department of Housing and Urban Development Release Date: December 1990. <https://www.huduser.gov/portal/publications/affhsg/comp_work_plan_1990.html>.

36. Preventing Lead Poisoning in Young Children. U.S. Department of Health and Human Services, Public Health Service, Centers for Disease Control, October 1; 1991.

37. Preventing Lead Exposure in Children: A Statement by The Centers for Disease Control. U.S. Department of Health and Human Services/Public Health Service/Centers for Disease Control; January 1985.

38. Lead-based Paint Hazard Reduction and Financing Task Force. Putting the Pieces Together: Controlling Lead Hazards in the Nation's Housing. Washington, D.C.: Lead-based Paint Hazard Reduction and Financing Task Force; 1995.

39. Jacobs, DE, Matte, TD, Moos, LV, Nilles, B, Rodman, J. President's Task Force on Children's Environmental Health Risks and Safety Risks (2000, March). Eliminating Childhood Lead Poisoning: A Federal Strategy. Washington, DC: U.S. Centers for Disease Control and Protection. <http://www.cdc.gov/nceh/lead/about/fedstrategy2000.pdf>.

40. Jacobs DE, Nevin R. Validation of a twenty-year forecast of U.S. childhood lead poisoning: updated prospects for 2010. *Environ Res.* 2006;102(3):352–364.

41. Nevin R, Jacobs D. Updated costs and benefits of lead paint remediation. Unpublished.

42. Pew Charitable Trusts and Robert Wood Johnson Foundation. 10 Policies to Prevent and Respond to Childhood Lead Exposure; 2017. <https://nchh.org/information-and-evidence/healthy-housing-policy/10-policies/>.

43. Value of Prevention Tool; 2019. <http://valueofleadprevention.org/>.

44. Draft Executive Order: Ensuring Healthy Homes: Eliminating Lead and Other Housing Hazards. National Center for Healthy Housing. Dec 14, 2020. Available from: https://nchh.org/resource/draft-executive-order_-ensuring-healthy-homes_eliminating-lead-and-other-housing-hazards-national-center-for-healthy-housing/.

45. Hanna-Attisha M, Lanphear B, Landrigan P. Lead poisoning in the 21st century: the silent epidemic continues. *Am J Public Health.* 2018;108(11). Available from: https://doi.org/10.2105/AJPH.2018.304725. Available from: 30303719.

46. Sherwin William Paint Company. SuperPaint Interior Latex with Sanitizing Technology. Accessed 2022. <https://www.sherwin-williams.com/homeowners/products/superpaint-sanitizing?gclid = Cj0KCQjwz7uRBhDRARIsAFqjull45oNNlNhy95-sZ7WPf9_q31ELHmn-vIjn2qQV7PaTIgw3QwaSINwaAnAtEALw_wcB>.

47. Dissell R, Zeltner B. The Plain Dealer (Cleveland). Protesters Bring Lead Paint Fight to Sherwin-Williams Shareholder Meeting, April 20; 2016. <https://www.cleveland.com/healthfit/2016/04/protesters_bring_lead_paint_fi.html>.

48. United Nations Environment Program. Update on the Global Status of Legal Limits on Lead in Paint; 2019. <https://www.unenvironment.org/resources/report/2019-update-global-status-legal-limits-lead-paint>.

49. Bardach E. *A Practical Guide for Policy Analysis*. 4th ed CQ Press; 2012.

50. Jacobs DE. Lead Poisoning in private and public housing: the legacy still before us. Invited editorial. *Am J Public Health*. 2019;109(6):830–832.

51. President Franklin Roosevelt. Dedication of the National Archives Building. Washington DC. 1941. As Quoted in Jill Lepore, The Trump Papers: What Will Happen to the President's Records When He Leaves the White House? New Yorker Magazine, November 23, 2020. P 22.

1

US government agencies involved in lead paint

Centers for Disease Control and Prevention
Department of Housing and Urban Development
Occupational Safety and Health Administration
Environmental Protection Agency
Consumer Product Safety Commission
National Institute for Occupational Safety and Health
National Institute for Environmental Health Sciences
Department of Defense
National Aeronautics and Space Administration
State Department
Agency for Toxic Substances and Disease Registry
National Institutes of Health
Food and Drug Administration
National Institute of Standards and Technology
Maternal and Child Health Bureau
Health Resources and Services Administration
Department of Commerce
US Marshalls Service
Commission on Civil Rights
Local State and Federal Courts
US Senate and US House of Representatives
Federal Housing Finance Agency
General Services Administration
Internal Revenue Service
Rural Housing Service
Farmers Home Administration

Department of Veterans Affairs
White House Domestic Policy Council
Department of the Treasury
White House Office of Science and Technology
Department of the Interior
National Park Service
Council on Environmental Quality (President's Office)
US Department of Energy
White House Office of Management and Budget
Department of Agriculture
Department of Justice
Department of Education
Department of Health and Human Services
Centers for Medicare and Medicaid Services
Department of Transportation
Office of the Surgeon General
Fannie Mae and Freddie Mac
Local, City, County, and State Governments
Tribal Governments
Local Public and Indian Housing Authorities

2

Honor role-leaders in lead paint poisoning prevention and healthy housing

Thousands of people played significant roles and made major contributions to lead paint poisoning prevention and healthy housing in the United States and around the world. This list does not include elected officials or reporters because there were so many. This list does not necessarily include everyone who played an important role in the years from 1970–2020. Errors of omission are solely those of the author. Alphabetical by Last Name.

Roberta	Aaronson	Mark	Allen
Susan	Aceti	Paul	Allwood
Camille	Acevedo	Gizelle	Alvarez
Gary	Adamkiewicz	Peter	Ambrose
John	Adgate	Tadesse	Amera
Doris	Adler	Robert	Amler
Sia	Afshari	Rod	Amlor
Ravi	Agarwal	Matt	Ammon
Carlos	Aguilar	Mary Ann	Amrich
Emily	Ahonen	Ben	Anderson
Vicki	Ainslie	Charmayne	Anderson
Judith	Akoto	Henry	Anderson
Yaser	Al Sharif	Jackson	Anderson
Chris	Alexander	William	Angell
Mary	Alexander	Magaly	Angeloni
Terry	Allan	Jose	Anibel Gonzalez

Joseph	Annest	Noreen	Beatley
Steven	Antholt	Nancy	Beaudet
Conrad	Arnolts	Joe	Beck
Jeanine	Arrighi	Jennifer	Becker
Kevin	Ashley	Andrew	Beer
Peter	Ashley	Victoria	Belfit
Cathy	Atkins	David	Bellinger
Bob	Axelrod	John	Belt
Eric	Axelrod	Emily	Benfer
Kenichi	Azuma	Julie	Bennett
Andrea	Bader	Doug	Benson
Bruce	Bailey	Rebecca	Bentley
Emma	Baker	Martha	Berger
Richard	Baker	Michael	Bergman
Dan	Bakston	Linda	Bergofsky
Ken	Balbi	Kofi	Berko
Scott	Balfour	Reva	Berman
William	Ballou	Tom	Bertrand
John	Balmes	Graham	Best
Clinton	Bamberger	Paul	Biedryzycki
Desmond	Bannon	Jeff	Bigler
Tracy	Barlow	Lucy	Billings
Donald	Barltrop	Sue	Binder
Suzanne	Barnes	Eula	Bingham
Patricia	Barnes	Helen	Binns
Eugene	Barros	Jenae	Bjelland
Abigail	Bartlett	Elizabeth	Blackburn
John	Bartlett	Pam	Blais
Lynn	Battle	Tony	Blakely
David	Batts	Christopher	Bland
Jason	Bawdin-Smith	James	Bland
Michael	Beard	Ethyl	Bledsoe
Cynthia	Bearer	Jan	Block

Linda	Block	Mary Jean	Brown
Chris	Bloom	Stephanie	Brown
Evelyne	Bloomer	Sylvester	Brown
Michael	Blumenfeld	Yolanda	Brown
Xavier	Bonnefoy	Jim	Brownlee
Robert	Bornschein	Shelly	Bruce
Lynn	Boulay	Arthur	Bryant
Dana	Bourland	Sharunda	Buchanan
Phillip	Bouton	Susan	Buchanan
Kenneth	Boxley	Cathe	Bullwinkle
Treesa	Boyce	Julia	Burgess
Asa	Bradman	David	Burgoon
Matthias	Braubach	Terry	Burke
Joseph	Braun	Tony	Burmistrz
Mario	Bravo	Mary	Burns
Sharif	Braxton	Gina	Bushong
Joseph	Breen	Beth	Butler
Terry	Brennan	Bruce	Buxton
Michael	Breu	Kenneth	Byk
Jill	Breysse	Salvatore	Cali
Patrick	Breysse	Carla	Campbell
Holly	Brightwell	Richard	Canfield
Laura	Brion	Doreen	Cantor
David	Broadbent	Joan	Carbone
Charlotte	Brody	Dave	Carey
Debra	Brody	David	Carey
Catherine	Brook	Joe	Carra
Marissa	Brooks	Eileen	Carroll
Barry	Brooks	Matt	Carroll
Larry	Brooks	Thomas	Carroll
Paul	Brophy	George	Caruso
Merrill	Brophy	Melanie	Carver
Michael	Brown	Daniel	Casiato

Bogdan	Catalin	Vin	Collucio
Gordon	Cavanaugh	Liz	Colon
Mathew	Chachere	Patrick	Connor
Lawrence	Chadzinski	Barbara	Conrad
Mark	Chamberlain	Susan	Conrath
Rufus	Chaney	Lorraine	Conroy
Megan	Charlop	Brion	Cook
Pam	Chase	Marjorie	Coons
Patrick	Chaulk	Beth	Cooper
Flora	Chavez	Don	Cooper
Ginger	Chew	Catherine	Copley
Elinor	Chisholm	Charles	Copley
J Julian	Chisolm	Chris	Corcoran
Danny	Chou	Maryrose	Corrigan
Gale	Christopher	Deborah	Cory-Schlechta
Brian	Christopher	Gian	Cossa
Gail	Christopher	David	Cox
Dan	Chute	Kimball	Credle
Henry	Cisneros	Samantha	Crisci
Kyle	Clark	Annemarie	Crocetti
Nancy	Clark	Sheila	Crowley
Scott	Clark	John	Cullen
Mike	Clarke	Susan	Cummins
Dale	Clarkson	Larry	Dale
Thomas	Clarkson	Leslyn	Daligadu
Joan	Cleary	Karen	Dannemiller
Robert	Clickner	Andrew	Dannenberg
Simone	Cohen	Dale	Darrow
Daniel	Cohn	Steve	Davis
Karen	Cohn	Dorr	Dearborn
Robert	Cole	Allen	Dearry
Curtissa	Coleman	Mark	Demyanek
Gwen	Collman	Karen	Dennis

Ray	Dennis	Peyton	Eggleston
Dave	Dennison	Myra	Eggleton
William	Derbyshire	Lindsay	Eilers
Stephanie	DeScisciolo	Joycelyn	Elders
Gary	DeWalt	Robert	Elias
Paul	Diegelman	Miranda	Engberg
Kim	Dietrich	Marilyn	Engle
Carolyn	DiGuissepi	Pierre	Erville
Sherry	Dixon	Amanda	Escobar-Gramigna
Maria	Doa	Brenda	Eskanzi
Denny	Dobbin	Chris	Estes
Ella	Dobson	Connie	Etheridge
Phillip	Dodge	Adrienne	Ettinger
Cushing	Dolbeare	Ruth	Etzel
Yolanda	Domneys	Crystal	Evans
Andrew	Doniger	Anne	Evens
Shaun	Donovan	Bill	Ewall
Carlos	Dora	Andrew	Faciano
Sam	Dorevitch	Rebecca	Fahey
Jeroen	Douwes	Anna	Falicov
Mary	Doyle	Henry	Falk
Brendan	Doyle	Mark	Farfel
Susan	Drodz	Doug	Farquhar
Sherrell	Duggan	Walter (Nick)	Farr
Deanna	Durica	Norman	Faye
Peter	Earle	Alina	Fernandez
John	Eastman	Deeohn	Ferris
Kara	Eastman	Jeff	Feuer
Fred	Eberle	Stephanie	Filep
Sharon	Eberle	Michael	Firestone
Fred	Eberle	Alex	Fischberg
Conrad	Egan	Barrett	Fischer
Krista	Egger	Ann	Fisher-Durrah

William	Fisk	Andy	Geer
Fidelma	Fitzpatrick	Sarah	Geiger
Russ	Flegal	Mia	Gelles
Karen	Florini	Neil	Gendell
Carol	Foley	Rita	Gergely
Brenda	Foos	David	Gibbons
Kim	Foreman	Chuck	Gilbert
Linda	Forst	Pamela	Gilbert
Marisa	Fountain	James	Gilford
Bruce	Fowler	Marcheta	Gillam
Susan	Fox	Kiel	Gilleade
Cheryl	Fox	Lisa	Gilmore
Paul	Francisco	Dennis	Glancey
Oscar	Franklin	Michael	Godfrey
Eileen	Franko	Art	Godi
Cynthia	French	Shara	Godiwalla
Neal	Freuden	Gayle	Goewey
Nicholas	Freudenberg	Melanie	Goldberg
Warren	Friedman	Ellis	Goldman
Howard	Frumkin	Jonathan	Goldman
Laura	Fudala	Lynn	Goldman
Kim	Fuelling	Michael	Goldschmidt
David	Fukuzawa	Barry	Goldstein
Cora	Fulmore	Gary	Goldstein
Esther	Gadomski	Ali	Golshiri
Richard	Gaeta	Henry	Gonzales
Joanna	Gaitens	Rick	Goodeman
Stan	Galik	Andrew	Goodman
Warren	Galke	Sarah	Goodwin
Steven	Galson	Jeanne	Gorman
Jon	Gant	Ed	Gorman
Elizabeth	Garland	Damien	Gossett
Lorne	Garrettson	Perry	Gottesfeld

Elise	Gould	David	Harris
George	Gould	Reginald	Harris
Elise	Gould	William	Harris
David	Governo	Bart	Harvey
Robert	Goyer	Michelle	Harvey
John	Graef	Katharine	Hastings
Jim	Graham	Dolline	Hatchett
Lester	Grant	Ulla	Haverinen
Rose	Green Colby	Jeff	Havlena
Stu	Greenberg	Jeff	Havlena
Peter	Grevatt	Steve	Hays
Karen	Griego	Roberta	Hazen-Aaronson
Blaine	Griffin	Ben	Hecht
Anne	Griffith	Peggy	Hegarty-Stack
Carl	Grimes	Teresa	Heinz
Hal	Grodzins	Sue	Heller
Scott	Grosse	Mark	Henshall
Joanne	Grote	Yianice	Hernandez
Erin	Guay	Jerry	Hershovitz
Brian	Gulson	Rachel	Herzig
Brian	Gumm	Akilah	Hill
Sue	Gunderson	Michael	Hill
Bill	Gutknecht	Ed	Hinsberger
Al	Guyant	John	Hiscox
Susan	Guyaux	Samuel	Hodges III
Paul	Haan	Daniel	Hoffman
Bruce	Haber	Laura	Holcomb
Dale	Hagen	Rebecca	Hollenbach
Barbara	Haley	Kelvin	Holloway
Michael	Hanley	David	Homa
Mona	Hanna-Attisha	Reed	Homan
Elizabeth	Hansen	Paromita	Hore
Sharon	Harper	Eric	Hornbuckle

Chris	Hornig	Mark	James
Christopher	Hornig	Matti	Jantunen
Susan	Horowitz	Judy	Jarrell
Vernon	Houk	Sandra	Jibrell
Jane	Houlihan	Rebecca	Jim
Robert	Houston	Maria	Joao-Freitas
Philippa	Howden-Chapman	Alan	Johanns
Leann	Howell	Nancy	Johnson
Kathryn	Hubicki	Earl	Johnston
Peter	Hubicki	Desiree	Jones
Breanna	Hudson	Chris	Jones
Rebecca	Hudson	Robert	Jones
Harry	Hudson	Ron	Jones
Melanie	Hudson	Dennis	Jordan
Claudia	Huff	Jim	Jordan
Merideth	Hunter	Tara	Jordan-Radosevich
Paul	Hunt	Emile	Jorgenson
Paul	Hunter	Kristen	Joyner
Toni	Hurley	David	Kahane
Patricia	Hynes	Fred	Kalvelage
Nancy	Ibrahim	Arnie	Katz
Gabriela	Illa	Bruce	Katz
Margie	Isaacson	Carolyn	Kawecki
Maura	Jackson	Michael	Keall
Maurci	Jackson	Jim	Keck
Richard	Jackson	Ernest	Kelly
Victoria	Jackson	Kevin	Kelly
Robin	Jacobs	Tom	Kelly
David	Jacobs	Walt	Kelly
Michael	Jacobson	Eula	Kemmer
Rachel	Jacobson	Dick	Kennedy
Laverne	Jacobs-Robinson	John	Kennedy
Phillip	Jalbert	Kevin	Kennedy

Woody	Kessel	Kathaleen	Lamb
Joan	Ketterman	Larry	Lance
Victor	Kimm	Phil	Landrigan
Jack	Kindsey	Bruce	Lanphear
Eva	King	James	LaRue
Alan	King	Tom	Laubenthal
Phil	King	Kathy	Lauckner
Gary	Kirkmire	Ellen	Lazar
Kate	Kirkwood	Tess	Lea
Linda	Kite	Barbara	Leczynski
Michael	Kitto	Catherine	Lee
Michael	Kleinhammer	Brian	Lee
Susan	Kleinhammer	Charles	Lee
Mardi	Klevs	Eileen	Lee
Rebecca	Kliman-Hudson	Peter	Lees
Susan	Klitzman	Chris	Lehane
Ruth	Klotz-Chamberlin	Neal	Leiffer
John	Knox	Jessica	Leighton
Arthur	Kobrine	Amber	Lenhart
Chris	Kochititzky	Ken	Leventhal
Bill	Kojola	Ronnie	Levin
Laura	Kolb	Barry	Levy
Krista	Kolis	Debra	Lewis
Steven	Koons	Jamal	Lewis
Katrina	Korfmacher	Linda	Lewis
Michael	Kosnett	Stan	Lewis
Paul	Kowalski	Roger	Lewis
Greg	Kreuger	Richard	Lewis
James	Krieger	Dirk	Leyes
Glenda	Kruger	Albert	Liabastre
Julie	Kruze	Carmen	Liang-Pelletier
Judith	Kurland	JoAnne	Liebeler
Judy	LaKind	Ben	Lim

Ruth	Lindberg	Kevin	Marchman
Ian von	Lindern	Stephen	Margolis
Jane	Lin-Fu	David	Marker
Bruce	Lippy	Kenneth	Markison
Jill	Litt	Gerald	Markowitz
Amy	Liu	Morri	Markowitz
Dennis	Livingston	Christie	Marra
Mariann	Lloyd-Smith	Darcy Scott	Martin
Paul	Locke	Mel	Martinez
Courtland	Lohff	Angela	Mathee
Eleanor	Long	Joanna	Matheson
Inessa	Lopez	Marquonda	Mathis
Jorge	Lopez	Terry	Matlaga
Ray	Lopez	Elizabeth	Matsui
Jennifer	Lowry	Thomas	Matte
Jessica	Lucas	Mark	Matulef
Joan Cook	Luckhardt	Bela	Matyas
Ramona	Ludolph	Betsy	Mazahn
Kert	MacAfee	Mary	McAuliffe
John	MacIsaac	Janet	McCabe
Ashley	Mack	Jim	McCabe
Karen	Mack	Ellie	McCann
Patrick	MacRoy	John	McCarthy
Sharon	Maeda	Colleen	McCauley
Armand	Magnelli	John	McCauley
Kathryn	Mahaffey	Keegan	McChesney
Matt	Mahoney	Dianne	McCloskey
Miles	Mahoney	Bob	McConnell
Jack	Malgeri	Jack	McConnell
Giridhar	Mallya	David	McCormick
Jane	Malone	Lindsay	McCormick
Barry	Mankowitz	Chelsea	McCracken
Adelle	Mansour	Anna	McCreery

John	McFee	Vernice	Miller-Travis
Mike	McGeehan	Deborah	Millete
Shanea	McGeehan	Marie Lynn	Miranda
Michael	McGreevy	Aaron	Mitchell
Gordon	McKay	Kate	Mitchell
John	McKay	Clifford	Mitchell
Maureen	McKee	Gillian	Mittelstaedt
Mary	McKnight	Betsy	Mokrzycki
Michael	McKnight	John	Monahan
Patricia	McLaine	Kim	Monroe
Amy	McLean-Salls	Beth	Moore
John	McPhee	Lin	Moos
Patrick	Meehan	John	Moran
Kris	Meek	Lidia	Morawska
John	Melone	Laura	Moreno-Davis
Shannon	Melton	Olivia	Morgan
Magaly	Mendez	Rebecca	Morley
John	Menkedick	Ron	Morony
William	Menrath	Jacqueline	Mosby
Ron	Menton	Heidi	Most
Naomi	Mermin	Amy	Mucha
Paul	Messenger	Jean	Mugulusi
Pamela	Meyer	Johnna	Murphy
Michael	Meyerstein	Amy	Murphy
Angela	Mickalide	Calvin	Murphy
Howard	Mielke	Paul	Mushak
Elise	Miller	Damen	Music
Kevin	Miller	Brenda and Damon	Music
Mark	Miller	Michelle	Naccarati-Chapkis
Pamela	Miller	Deborah	Nagin
Barbara	Miller	Charley	Naney
Jo	Miller	Kim	Nathan
Michelle	Miller	Martin	Nee

Herbert	Needleman	Jerome	Paulson
Andrew	Nelson	Nancy	Pavur
Tom	Neltner	Teresa	Payne
Raymond	Neutra	Devon	Paynes-Sturges
Rick	Nevin	Jordan	Peccia
Carolyn	Newton	Sue	Pechilio Polis
Harrison	Newton	John	Peduto
Mark	Nicas	Ron	Peik
Leslie	Nickels	Sharon	Pendleton
Bruce	Nilles	Nick	Peneff
Brian	Nix	Otis	Perry
Patricia	Nolan	Andrew	Persily
Gary	Noonan	John	Pesce
John	Nordigian	Kieran	Phelan
Sarah	Norman	Peter	Phibbs
Ruth Ann	Norton	Dan	Phillips
Jerome	Nriagu	Tom	Phillips
Michelle	Nusum-Smith	Bathsheba	Philpott
Elizabeth	O'Brien	Miriam	Phily
Karen	O'Connor	Janet	Phoenix
Tom	O'Hagan	Tyler	Pigman
Rob	O'Haver	Kristen	Pike
Eric	Oetjen	Eugene	Pinzer
Ken	Olden	Sergio	Piomelli
Eric	Olson	James	Pirkle
David	Ormandy	Elise	Pivnick
Chris	Papanicolopolous	Anna	Plankey
Sharon	Park	Mark	Pocras
Andrew	Parker	Stephanie	Pollack
Patrick	Parsons	Joseph	Ponessa
Jack	Paster	Elena	Popp
Claire	Patterson	Victor	Powell
Sharyle	Patton	Ernestine	Powell

Preethi	Pratap	Corwin	Rhyan
Brad	Prenney	Deborah	Rice
Amos	Pressler	Kennedy	Rice
Michelle	Price	Lee Ann	Richardson
Wesley	Priem	Sally	Richman
David	Pugh	Knut	Ringen
Tim	Pye	Valerie	Rios
Brenda	Quarles	Rich	Ritota
Bill	Quigley	Michael	Rizer
Eileen	Quinn	Mike	Rizer
Ruth	Quinn	Deborah	Roane
Rick	Rabin	Nathalie	Robbel
Michael	Rabinowitz	John	Roberts
Felicia	Rabito	Dwight	Robinson
Bill	Radosevich	Nancy	Robinson
Samina	Raja	Sarah	Robinson
David	Rall	Simone	Robinson
Saul	Ramirez	K. W. Jim	Rochow
Rey	Ramsey	Sandy	Roda
Maria	Rapuano	Kit	Rodkey
Janice	Ratley	Joanne	Rodman
Amanda	Reddy	Bill	Rodosevich
Jack	Reed	Nathalie	Roebbel
Routt	Reigart	Walter	Rogan
Rick	Reinhart	John	Rogers
Richard	Reinhart	Charles	Rohde
John	Rekus	Maria Del Carmen	Rojas
Lee	Reno	Anne	Romasco
Janet	Reno	Liseth	Romero-Martinez
Nic	Retsinas	Brian	Rooney
Rebecca	Rex	Ken	Rose
Brenda	Reyes	William	Rose
George	Rhodes	John	Rosen

Susan	Rosmarin	Lisa	Schmidtfrerick-Miller
David	Rosner	Jay	Schneider
Ananya	Roy	Michael	Schock
Ellen	Roznowski	Mary Sue	Schottenfels
Tamara	Rubin	Brad	Schultz
Ron	Rupp	Harold	Schultz
Mary Ann	Russ	Joel	Schwartz
Pamela	Russo	Steve	Schwartzberg
Steve	Rust	John	Schwemberger
Don	Ryan	Darcy	Scott
Mark	Sabath	Kiernan	Scott
John	Salisbury	Ralph	Scott
Aaron	Salkoski	Karen	Segura
Paul	Sambanis	Katherine	Seikel
Megan	Sandel	Song	Seo
William	Sanders III	Steve	Shabazz
Anthony	Santiago	Zakia	Shabazz
Bob	Santucci	Donna	Shalala
Ginny	Sardone	Michael	Shannon
David	Satcher	Joshua	Sharfstein
Barbara	Sattler	Denise	Sharify
Margaret	Sauser	Michael	Sharp
John	Scalera	Richard	Shaughnessy
Dan	Scannell	Madeleine	Shea
Eric	Schaeffer	Kevin	Sheehan
Sally	Schaeffer	Peggy	Shepard
Ted	Schettler	Dianne	Sheridan
Lois	Schiffer	Wendy	Shields
Joe	Schirmer	Lois	Shiffer
Paul	Schlect	Joe	Shuldiner
Thomas	Schlenker	John	Shumway
Tom	Schlenker	David	Shutz
Dick	Schmehl	Larry	Siegelman

Ellen	Silbergeld	Roy	Sterner
Jennifer	Silverman	Wes	Stewart
Lisa	Simer	Louis	Stokes
Jonnette	Simmons	Anna	Storkson
Peter	Simon	Wes	Straub
Jeffrey	Simpkins	Warren	Strauss
Jim	Simpson	Steve	Strebel
Ron	Sims	Cindy	Stroup
Moira	Singer	Bob	Stryker
Gary	Singer	Amber	Sturdivant
Tom	Sinks	Paul	Succop
Greg	Siwinski	Mary Ann	Suero
Damien	Slaughter	Marissa	Sukosky
Louise	Slaughter	Brian	Sullivan
John	Sly	Louis	Sullivan
Tony	Smelgus	Pam	Susie
Lisa	Smestad	Aaron	Sussell
Andrew	Smith	Dick	Svenson
Kirk	Smith	Daniel	Symonik
Tammy	Smith	Sarah	Szanton
John	Sobolewski	Tim	Takaro
Lin	Song	Marc	Talley
Adam	Spanier	Shirlee	Tan
Amy	Spanier	YuAnn	Tan
Carin	Speidel	M. L.	Tanner
Jack	Spengler	Loreen	Targos
Scott	Spenser	Eileen	Tarlau Senn
Tammy	Sproule	Cynthia	Taylor
Gary	Stafford	Sergio	Tejadilla
Matt	Stefanik	Lucy	Telfar-Barnard
Michael	Stegman	Erin	Thanik
Kara	Stein	Kirsten	Thayer
Brent	Stephens	Ed	Thomas

David	Thompson	Reghan	Walsh
Nicole	Thomson	Tom	Wangerin
Susan	Thornfeldt	Gail	Ward
Susan	Timm	Nathaniel	Washington
Steven	Tise	Lee	Wasserman
Laura	Titus	Darlene	Watford
Janet	Tobacman	Darleen	Watkins
Dick	Tobin	Ken	Watts
Ellen	Tohn	William	Weber
Brain	Toll	John	Weicker
Theodore	Toon	Steve	Weil
David	Topol	Anita	Weinberg
David	Topping	Jack	Weinberg
Paul	Torzillo	Max	Weintraub
Pat	Tracy	Robert	Weisberg
Leo	Trasande	Steve	Weitz
Ramona	Travato	Michael	Weitzman
Chris	Trent	Alison	Welch
David	Turcotte	John	Wells
Margery	Turner	Anne	Wengrovitz
Myrtle	Turner	Dean	Wenrich
David	Turpin	Peter	Werwath
Kathy	Tyler-Harris	Will	Wheeler
Bruce	Upshaw	Kenn	White
Susan	Valenti	Sylvia	White
Bob	Vanderslice	Ellen	Widess
Howard	Varner	Amanda	Wiles
Jumana	Vasi	Charles	Wilkins
Stephen	Vega	Denis	Williams
Tom	Vernon	Emily	Williams
Susan Marie	Viet	James	Williams
William	Villalona	Jessica	Williams
Bailus	Walker	Jennifer	Willis
James	Walsh	Jonathan	Wilson

Robert	Wiseberg	Walter	Wynn III
Courtney	Wisinski	Joseph	Wysocki
Bill	Wisner	Jim	Yannarelly
Nse	Witherspoon	Cassandra	Yelverton
Hill	Wohl	Stephanie	Yendell
Michael	Wojtowycz	Kimberly	Yolton
Muriel	Wolf	Elizabeth	Zeldin
Betty	Wolverton	Jing	Zhang
Hofer	Wong	Hina	Zia
Robert	Wright	John	Zilka
Rosalind	Wright	Amy	Zimmerman
Dan	Wuenschel	Mark	Zuluaga
Sarah	Wylie	Ralph	Zumwalde
Neil	Wilson		

Glossary

Some definitions are from the Guidelines for the Evaluation and Control of Lead-Based Paint Hazards in Housing, released by the US Department of Housing and Urban Development, 2012

Abatement A measure or set of measures designed to permanently eliminate lead-based paint hazards or lead-based paint. Abatement strategies include the removal of lead-based paint, enclosure, encapsulation, replacement of building components coated with lead-based paint, removal of lead-contaminated dust, and removal of lead-contaminated soil or overlaying of soil with a durable covering such as asphalt (grass and sod are considered interim control measures). All of these strategies require site preparation; occupant and worker protection; cleanup; waste disposal; post-abatement clearance testing; recordkeeping; and, if applicable, monitoring. See, also, Interim controls.

AHHS American Healthy Housing Survey, the third nationally representative survey to count the number of homes with lead paint and lead paint hazards, preceded by National Survey of Lead and Allergens in Housing and the 1990 National Lead Paint Survey.

Binder Solid ingredients in a coating that hold the pigment particles in suspension and bind them to the substrate. Binders used in paints and coatings include oil, alkyd, acrylic, latex, and epoxy. The nature and amount of binder determine many of the coating's performance properties—washability, toughness, adhesion, gloss, etc.

Building component Any element of a building that may be painted or have dust on its surface, for example, walls, stair treads, floors, railings, doors, windowsills, etc.

Certified The designation for contractors who have completed training and other requirements to allow them to safely undertake risk assessments, inspections, abatement, or renovation repair and painting. Risk assessors, inspectors, abatement contractors, and renovation contractors should be certified (and licensed, if applicable) by the appropriate local, state, or federal agency.

Certified renovator An individual who has successfully completed a renovator course accredited by EPA or an EPA-authorized state or tribal program.

Cleaning The process of using a HEPA vacuum and wet cleaning agents to remove leaded dust; the process includes the removal of bulk debris from the work area, usually in a three-step process: HEPA vacuuming, wet washing or mopping, and a final HEPA vacuuming.

Clearance examination Visual examination and collection of lead dust samples by an inspector or risk assessor, or, in some circumstances, a sampling technician, and analysis by an EPA-recognized laboratory upon completion of an abatement project, interim control intervention, maintenance, or renovation job that disturbs lead-based paint (or paint presumed to be lead-based.) For abatement projects, the clearance examination is performed to ensure that lead exposure levels do not exceed clearance standards established by the EPA at 40 CFR 745.227(e)(8)(viii); HUD's dust-lead standards for clearance after interim control projects are found at 24 CFR 35.1320(b)(2)(i).

Deciliter (dL) One-tenth of a liter.

Detection limit The minimum amount of a substance that can be reliably measured by a particular method.

Deteriorated paint Any paint coating on a damaged or deteriorated surface or fixture, or any interior or exterior lead-based paint that is peeling, chipping, blistering, flaking, worn, chalking, alligatoring, cracking, or otherwise becoming separated from the substrate.

EBL   Elevated blood lead level as defined by the Centers for Disease Control and Prevention. Local standards may differ. In 2012 the CDC revised its definition to use a "reference value" of the blood lead level at the 97.5th percentile of children aged 1–5 years old based on its National Health and Nutrition Examination Survey (NHANES). As of 2021, the reference level was 3.5 μg/dL.

Encapsulation   Any covering or coating that acts as a barrier between lead-based paint and the environment, the durability of which relies on adhesion and the integrity of the existing bonds between multiple layers of paint and between the paint and the substrate. See, also, Enclosure.

Enclosure   The use of rigid, durable construction materials that are mechanically fastened to the substrate to act as a barrier between the lead-based paint and the environment.

Engineering controls   Measures other than respiratory and other personal protection or administrative controls that are implemented at the worksite to contain, control, and/or otherwise reduce exposure to lead-contaminated dust and debris usually in the occupational health setting. The measures include process and product substitution, isolation, and ventilation. The term may be used in the occupational health setting in regard to preventing workers' exposure to lead; it can also be used in other lead hazard control settings, such as in regard to preventing residents' exposure.

High-efficiency particulate air (HEPA) filter   A filter capable of removing particles of 0.3 μm or larger from air at 99.97% or greater efficiency.

HEPA vacuum   A vacuum cleaner that has been designed with a HEPA filter as the last filtration stage. The vacuum cleaner must be designed so that all the air drawn into the machine is expelled through the HEPA filter with none of the air leaking past it. (Note that HUD's definition in its Lead Safe Housing Rule, with its slightly different wording, is substantively identical.)

In-place management   see "Interim controls".

Inspection (of paint)   A surface-by-surface investigation to determine the presence of lead-based paint (in some cases including dust and soil sampling) and a report of the results.

Inspector (more formally, lead-based paint inspector)   An individual who has successfully completed training from an accredited program and been licensed or certified by the appropriate federal state or local agency to:

1. perform inspections to determine and report the presence of lead-based paint on a surface-by-surface basis through on-site testing;
2. report the findings of such an inspection;
3. collect environmental samples for laboratory analysis if needed;
4. perform clearance testing; and optionally
5. document successful compliance with lead-based paint hazard control requirements or standards.

Interim controls   A set of measures designed to temporarily reduce human exposure or possible exposure to lead-based paint hazards. Such measures include, but are not limited to, specialized cleaning, repairs, maintenance, painting, temporary containment, and the establishment and operation of management and resident education programs. Monitoring, conducted by owners or their agents, and reevaluations, conducted by professionals, are integral elements of interim control. Interim controls include dust removal; paint film stabilization; treatment of friction and impact surfaces; installation of soil coverings, such as grass or sod; and land use controls. Interim controls that disturb painted surfaces are renovation activities under EPA's Renovation, Repair, and Painting Rule.

Interior windowsill   The portion of the horizontal window ledge that protrudes into the interior of the room, adjacent to the window sash when the window is closed; often called the window stool.

Lead-based paint   Any paint, varnish, shellac, or other coating that contains lead equal to or greater than 1.0 mg/cm^2 as measured by XRF or laboratory analysis, or 0.5% by weight (5000 mg/g, 5000 ppm, or 5000 mg/kg) as measured by laboratory analysis. (Local definitions may vary.) This definition applies only to existing paint, not new paint.

Lead-based paint hazard A condition in which exposure to lead from lead-contaminated dust, lead-contaminated soil, or deteriorated lead-based paint would have an adverse effect on human health (as established by the EPA at 40 CFR 745.65, under Title IV of the Toxic Substances Control Act as amended by Title X of the 1992 Housing and Community Development Act). Lead-based paint hazards include, for example, paint-lead hazards, dust-lead hazards, and soil-lead hazards.

Lead-based paint hazard control Activities intended to control and eliminate lead-based paint hazards, such as interim controls and abatement.

Lead carbonate A pigment used in some lead-based paints as a hiding agent; also known as white lead or white lead carbonate.

Lead-containing paint As defined by the Consumer Product Safety Commission, paint or other similar surface-coating materials for consumer use that contain lead or lead compounds and in which the lead content (calculated as lead metal) is in excess of 0.009% by weight of the total nonvolatile content of the paint or the weight of the dried paint film (see 16 CFR 1303.1(c)). This definition applies to new paint, not existing paint in housing. See also "Lead-based paint."

NHANES National Health and Nutrition Examination Survey (the first nationally representative study to count the number of lead-poisoned children and measure blood lead levels in the US population).

NLLAP requirements Requirements specified by the EPA National Lead Laboratory Accreditation Program (NLLAP), for accreditation for the lead analysis of paint, soil, and dust matrices by an EPA-recognized laboratory accreditation organization.

Paint-lead hazard Lead-based paint on a friction surface that is subject to abrasion and where a dust-lead hazard is present on the nearest horizontal surface underneath the friction surface (e.g., the windowsill, or floor); Damaged or otherwise deteriorated lead-based paint on an impact surface that is caused by impact from a related building component; a chewable lead-based painted surface on which there is evidence of teeth marks; or any other deteriorated lead-based paint in any residential building or child-occupied facility or on the exterior of any residential building or child-occupied facility.

Paint stabilization The process of wet scraping, priming, and repainting surfaces coated with deteriorated lead-based paint. Paint stabilization also includes eliminating the cause(s) of paint deterioration, occupant and worker protection, and cleanup and clearance.

Paint removal The removal of lead-based paint from surfaces; this may be an abatement strategy or it may occur as a part of a renovation project or interim controls.

Patch test A test method or procedure to assess the adhesion of an encapsulant coating to a substrate covered with a layer or layers of lead-based paint.

Primary prevention The process of preventing lead hazards from occurring and, when they do occur, controlling lead hazards to prevent exposure before a child is poisoned. See, also, Secondary prevention and Tertiary prevention.

Replacement A strategy of abatement that involves the removal of building components coated with lead-based paint (such as windows, doors, and trim) and the installation of new components free of lead-based paint.

Risk assessment An on-site investigation of a residential dwelling to determine the existence, nature, severity, and location of lead-based paint hazards. Risk assessments, which must be conducted by a certified risk assessor, include an investigation of the age, history, management, and maintenance of the dwelling, and the number of children under age 6 and women of childbearing age who are residents; a visual assessment; environmental sampling (i.e., collection of dust wipe samples, soil samples, and deteriorated paint samples); and preparation of a report identifying abatement and interim control options based on specific conditions.

Risk assessor A certified individual who has successfully completed lead-based paint hazard risk assessment training with an accredited training program and who has been certified to:

1. perform risk assessments;
2. identify acceptable abatement and interim control strategies for reducing identified lead-based paint hazards;
3. perform clearance testing and reevaluations; and
4. document the successful completion of lead-based paint hazard control activities.

Screening   The process of testing children to determine if they have elevated blood lead levels.

Secondary prevention   The process of identifying children who have elevated blood lead levels, and controlling or eliminating the sources and/or pathways of further exposure. See, also, Primary prevention and Tertiary prevention.

Soil-lead hazard   Bare soil on residential property that contains lead in excess of the standard established by the EPA under Title IV of the Toxic Substances Control Act as amended by Title X of the 1992 Housing and Community Development Act. EPA standards for soil-lead hazards, published at 40 CFR 745.65(c), as of 2020, are 400 μg/g in play areas and 1200 μg/g in the rest of the yard. Also called lead-contaminated soil.

Substrate   A surface on which paint, varnish, or other coating has been applied or may be applied. Examples of substrates include wood, plaster, metal, and drywall.

Substrate effect   The radiation returned to an XRF analyzer by the paint, substrate, or underlying material, in addition to the radiation returned by any lead present. This radiation, when counted as lead X-rays by an XRF analyzer contributes to substrate equivalent lead (bias). The inspector may have to compensate for this effect when using XRF analyzers. See, also, XRF analyzer.

Target housing   Any housing constructed before 1978—except dwellings that do not contain bedrooms or dwellings that are designated specifically for the elderly or persons with disabilities, unless a child younger than 6 resides or is expected to reside in the dwelling. In the case of jurisdictions that banned the sale or use of lead-based paint before 1978, the Secretary of HUD may designate an earlier date for defining target housing.

Tertiary prevention   Providing medical treatment to children with elevated blood lead levels to prevent more serious injury or death.

Trough   See Window trough.

White lead   A white pigment, usually lead carbonate. See, also, Lead carbonate.

Windowsill   See Interior windowsill.

Window stool   See Interior windowsill.

Window trough   For a typical double-hung window, the portion of the exterior windowsill between the interior windowsill (or stool) and the frame of the storm window. If there is no storm window, the window trough is the area that receives both the upper and lower window sashes when they are both lowered. (Sometimes inaccurately called a window "well.") See, also, Window well.

Window well   The space that provides exterior access and/or light to a window that is below grade, that is, below the level of the surrounding earth or pavement. See, also, Window trough.

XRF analyzer   An instrument that determines lead loading and concentration in milligrams per square centimeter (mg/cm^2) using the principle of X-ray fluorescence (XRF). In this book, the term XRF analyzer generally refers to portable instruments manufactured to analyze paint and does not refer to laboratory-grade units.

Index

Note: Page numbers followed by "*f*" and "*t*" refer to figures and tables, respectively.

A

Abatement, 62–63, 89, 123, 128, 161, 211, 251, 254, 297, 299–300, 305
  plan, 305
  techniques, 46
Advocates, 80, 129, 204–206, 307, 387, 391
Affordable housing, 52–53, 247–248
Agency for Toxic Substances and Disease Registry (ATSDR), 67
  report to Congress, 68–69
Alameda County, 202
  Lead Abatement Program, 193–194
Alliance to End Childhood Lead Poisoning, 85–87, 89, 126, 167, 189, 204, 293–294, 332
Alliance's Primary Prevention Strategies project, 85
American Association for Laboratory Accreditation, 166
American Industrial Hygiene Association, 165–166
American National Standards Institute (ANSI), 21
American Public Health Association, 365
Ammon, Matt, 117, 169, 170*f*, 172, 231, 340
ARCO, 300
Ashton v Pierce court case, 81, 97, 249, 277, 377

B

Baltimore City Circuit Court, 244
Baltimore lead paint abatement and repair and maintenance study, 254–260
Baltimore Repair and Maintenance study vacuum method (BRM), 133–135
Baltimore Sun, 203
Basic Principles of Healthful Housing (1937), 364
Belmont report, 254
Bench Book, 173
Best of intentions, 276–280
Binder, Sue, 89–90, 103–104
"Bioavailable" lead, 137–138
  in dust, 16
Blood lead
  measurements, 14
  screening, 36–37
  tests, 90
Blue Cross Foundation, 328

"Blue nozzle" vacuum method, 133
BRM vacuum sampling method, 133–135
Brown, Mary Jean, 231, 325, 378
Brown, Sherrod, 340–341
Browner, Carol, 167, 194, 196, 207, 318

C

Cabinet, 206–219
  2010 goal, 218–219
California court decision, 299–306
Californians for Safe and Affordable Housing, 307–308
Campaign for lead safe America, 194–200
Cavanaugh, Gordon, 51, 71–72, 76, 116
Chelation, 36–37
  limitations of chelation therapy, 39
Childhood Lead Action Project, 203
Childhood Lead Poisoning Prevention Programs, 305
Children's Health Protection Advisory Committee, 230
Children's Lead Education and Poisoning Prevention program, 192–193
Children's Rehabilitation Institute, 247
Chisolm, Julian, 11–12, 14, 39, 49, 117, 218, 245–246
Chronic disease, 353
Cisneros, Henry, 78, 84, 114–120, 126–127, 167, 199, 322, 331–332, 378, 387
City Homes, 53, 247–248
Class-action lawsuits, 293
Clean Air Act, 163, 354–356
Clean Water Act, 354–356
Clearance, 124, 126
Clearance testing, 74–75
Cleveland Drop Ventilation System, 314
Cleveland mold mystery, 313–319
Climate change, 366
Colon, Liz, 193, 299
Common Rule, 254
Communicable disease, 353
Community Development Block Grants, 119
Community Environmental Health Resource Center (CEHRC), 200–203
Community Lead Action and Information Meetings, 193–194

Community Lead Education and Reduction Corps (CLEARCorps), 216–217
  evaluation of environmental impacts, 217
Community Legal Services, 188
Community/communities, 188–194
  groups, 188, 191
    importance, 189–190
    and press, 203–206
Comprehensive and Workable Plan, 84–85
Comprehensive Improvement Assistance Program, 71
ConAgra, 300, 307–308
Consumer Product Safety Commission (CPSC), 23
Cost-benefit analysis, 156
County Lead Abatement Program, 193–194
Courts, 243–248
Cranston, Alan, 93
Cuomo, Andrew, 115–116, 122, 147, 155, 159–160, 172, 174, 194, 196, 198–200, 201f, 378
Cuyahoga River, 355–356, 355f

D
Decent safe sanitary housing, 44–45
Deleading, 48
Disability-adjusted life years (DALYs), 334
Disclosure
  enforcement cases, 169f
  pamphlet and warning language, 168f
Doc4Kids Project, 319
Donovan, Shaun, 114, 160, 322, 326, 378
DuPont, 300

E
"Emergency" abatements, 83
Emission standards, 129
Enforcement actions, 166–174
Enforcement Center, 172, 174
Enterprise Community Partners, 328
Enterprise Foundation, 248
Environmental Health Watch, 203
Environmental justice, 189, 191–192
  and community participation in research, 280–281
Environmental Justice Conference, 191–192
Environmental Lead Proficiency Analytical Testing Program (ELPAT Program), 166
Erythrocyte protoporphyrin test (EP test), 89
Ethics in housing intervention research, 270–276
Evaluation of the HUD Lead Hazard Control Grant Program, 141
Executive Order, 207

F
Farfel, Mark, 14, 45, 49f, 103–104, 117, 134f, 245–247, 282, 378

Farr, Nick, 53, 102, 121, 127, 143, 248, 320
FDA, 7–9
Federal Emergency Management Agency (FEMA), 364–365
Federal Housing Administration (FHA), 153
"Find It, Fix It, Fund It" campaign (FFF campaign), 379–383
Fitzpatrick, Fidelma, 293, 298, 308
Flint lead drinking water disaster, 230
Florini, Karen, 100, 127, 129, 161–162
Food and Drug Administration, 254
Fox, Cheryl, 92, 199
"Framework for Action for Lead-Safety in Private Housing", 87
Framingham Study, 207
Fresh Air Funds, 355
Friedman, Warren, 117, 143, 150, 319, 322

G
Georgia Tech, 62, 64
  forum, 66
  training course, 63, 65–66, 71–72, 117
Global Alliance to End Lead Paint, 87
Goldman, Ellis, 78, 118, 218
Goldman, Lynn, 124
Goldstein, Barry, 263, 278
Gonzales, Henry, 93
"Good Samaritan" precedent, 244
Green Building Council, 358
Green Community Standards, 328
"Green" building organizations, 65
Grimes decision, 260–270
  Appeals Court comparison to Nazi and Tuskegee research, 269–270
  facts, 267
  federal oversight, 268–269
  informed consent, 266–267
  institutional review boards, 265
  lead dust testing methods, 267–268
  new legal ground, 264
  parents vs. court of appeals, 264–265
  protecting children, 263
  special duty, 265–266
  therapeutic and nontherapeutic research, 261–263
Grow Chemical Corporation, 39

H
Health, 351–354
  research, 245–247
Healthy Home Collaborative, 202–203
Healthy Homes, 316–317, 321
  assembling evidence, 326–330

systematic review of housing interventions and
health, 329–330
Cleveland mold mystery, 313–319
frameworks at HUD, CDC, and EPA, 325*t*
healthy homes frameworks at HUD, CDC, and EPA,
325*t*
housing deficiencies and associated health outcomes,
323*t*
movement, 358
National Safe and Healthy Housing Coalition,
330–331
public housing, 339–341
Surgeon General's Call to Action on Healthy
Housing, 325–326
World Health Organization Healthy Homes
Movement, 331–339
Healthy Homes and Schools Bond Act of 2018,
306–307
Healthy Homes Inspection Manual, 321–322
Healthy homes movement, 320
Healthy housing, 319
Healthy housing consensus, 366–368
Healthy Housing Reference Manual, 321–322
Heinz, Teresa, 85–86
Helsinki Declaration, 253
Hershovitz, Jerry, 89–90, 319
Hierarchy of controls, 25
High-efficiency particulate absorbing vacuums (HEPA
vacuums), 74, 124
Historic preservation, 82–83, 124
HOME program, 121–122
Houk, Vernon, 40, 89–90
Housing, 353
codes, 321–322, 356
deficiencies and associated health outcomes, 323*t*
finance, 362
intervention
research, 245
systematic review of housing interventions and
health, 329–330
law, 50–52
reframing wealth, affordability, and equity, 359–366
Housing Act (1937), 44
Housing Act (1987), 69–70
Housing Authority of New Orleans (HANO), 75
Housing Authority Risk Retention Group (HARRG),
71, 165
elements, 72–73
Housing Choice Voucher program, 152
"Housing is Infrastructure Act", 358
Housing Intervention Research, 245, 262, 273, 354
Housing Quality Standards (HQS), 151
Howden-Chapman, Philippa, 336

Hurricane Katrina, 365–366

I
Idiopathic pulmonary hemorrhage, 315–316
IG audit, 222
In-place management. *See* Interim controls
Indoor air quality, 357
Informed consent, 266–267
Institute of Medicine, 245
Institutional review boards (IRB), 253, 265
Integrated Exposure Uptake Biokinetic model (IEUBK
model), 131, 135
mechanistic, 136
InterAction, 246
Interim containment protocols, 71
Interim controls, 93–94, 123–124, 144, 249
International Code Council (ICC), 49, 322
International Labor Organization, 35–36
International Lead Poisoning Prevention Week of
Action, 332
International Organization for Standardization and
International Electrochemical Commission (ISO/
IEC), 166
Intervention solution research, 27–28

J
Jackson, Maurci, 127–128, 196–197, 218
Jacobs, David, 14*f*, 26*f*, 47*f*, 121, 127, 134*f*, 170*f*, 224, 305,
378
Johns Hopkins University, 244–246

K
Katz, Bruce, 78, 92, 114, 116, 118–119, 122, 198–199
Kemp, Jack, 51, 78, 84, 114–115, 117
Kennedy Krieger Institute (KKI), 244, 283
and health research, 245–247
Korfmacher, Katrina, 365

L
Lanphear, Bruce, 103, 136, 227–228, 301, 319
LARES, 333
Large Analysis and Review of European Housing and
Health Status (LARES), 333
Lawsuits, 52–53, 293–294, 365
against lead paint industry, 291–295
Lead
advantages and disadvantages of lead hazard
control methods, 124, 125*t*
exposure, 377
lead paint industry interference, 215–217
water pipes, 17–18
Lead and Environmental Hazards Association, 121,
229–230

Lead Based Paint Poisoning Prevention Act, 377
Lead chromate, 137–138
Lead dust, 46–48, 53–55
    standards, 129–140
        floor dust lead standards, 140t
        health basis of federal dust lead standard, 139t
        "model" wars, 135–140
        remediation methods, 140–146
    testing methods, 267–268
Lead Elimination Action Drive, 193
Lead in soil (PbS), 53–54
Lead paint, 3, 243, 388–389
    ban, 16, 18–21
    chips, 12
    corrective actions, 67–69
    disclosure
        enforcement cases, 169f
        pamphlet and warning language, 168f
    funding, 24t
    grantees, 121
    hazard, 81, 97–98, 123
        control methods used by HUD grantees, 145t
    Housing Act (1987), 69–70
    housing codes fail to regulate, 49–50
    inadequate measurement methods, 13–17
    industries, 3–4
        interference, 215–217
    lawsuits against lead paint industry, 291–295
    leaders in lead paint poisoning prevention and
        healthy housing, 399–416
    levels by jurisdiction, 144f
    office, 116, 119
    poisoning prevention, 389–391
    Public Housing Guidelines (1990), 79–84
    reforming housing regulations, 98–99
    removal, 45–49, 246
    right to know, 99–101
    risk assessments, 71–75
    science, policy, and practice, 62–67
    soil, 11–12
    Stewart McKinney Amendments (1988), 69–70
    struggle to reform federal housing lead paint
        regulations, 147–162
        cost of lead hazard control in federally assisted
            housing, 158t
    testing, 162–165
    Title X of 1992 Housing and Community
        Development Act, 91–102
    toxicity research and intervention solution research,
        27–28
    US government agencies, 21–24
Lead Paint Abatement and Repair and Maintenance
    study in Baltimore. See "R&M" study

Lead poisoning, 5, 35–36, 115, 192–193, 209
    Alliance to End Childhood Lead Poisoning marshals
        political will, 85–87
    and courts, 243–248
    forecast, 388
    government to work, 384–385
    influence of industry, 388–389
    lawsuits and affordable housing, 52–53
    medical interpretation of child blood lead classes,
        38t
    medical model, 36
    pathway studies, 53–55
    procedures right and recruiting expertise, 385–386
    strategic plans, 386–388
    surveillance and population surveys, 41–43
    treatment vs. prevention, 36–40
"Lead Tech" conferences, 121
Lead Wars, 276
Lead-Based Paint Poisoning Prevention Act (1971), 15,
    17, 22–23, 36, 41, 91, 123, 188
Lead-free, 25–27
Lead-safe, 25–27
    Housing Rule, 149
"Leadership for Environmental and Energy Design"
    standards for homes, 358
Legal evidence, 248–252
Legal theories, 293
Leveraged private funding, 214
Local governments, 188, 191
Local jurisdiction lawsuits against lead paint industry,
    291–295
Low-level lead exposures, 27
Lucifer Curves, 211
Lynch, Patrick, 293, 295–296

M
Mahaffey, Kathryn, 8, 89, 136, 202, 315, 321
Mahoney, Miles, 51, 65, 70–72, 103–104, 127
Maintaining a Lead-Safe Home, 197
Martinez, Mel, 84, 114, 118, 162, 172, 199, 214–216, 219,
    221, 224, 378, 384
Maryland Appeals Court, 244–245, 248–249, 258–259,
    263–265, 294
    comparison to Nazi and Tuskegee research, 269–270
    legacy of Maryland Court of Appeals Grimes
        decision, 281–283
Massachusetts lead law, 87
Matte, Tom, 89–90, 136, 143, 207–208, 319
McConnell, Bob, 293
McConnell, Jack, 293
Medicaid
    children, 90
    waivers, 213

Michigan Childhood Lead Poisoning Prevention Program, 230
Mikulski, Barbara, 51, 76, 221
"Model" wars, 135–140
Morley, Rebecca, 104, 117, 320, 332, 381
Multifamily mortgage insurance, 155
Mushak, Paul, 67, 301, 303f, 378, 386

N

National Association of Home Builders, 206
National Center for Healthy Housing (NCHH), 87, 101, 204, 329, 365
National Center for Lead-Safe Housing, 102–105, 141, 206
National Children's Study, 207
National Health and Nutrition Examination Survey (NHANES), 41, 136, 174–175, 209
  findings, 42
  health surveys, 84
  homes, 43
  HUD's dust sampling in homes of, 139
  surveys, 42–43, 68
National Healthy Housing Standard, 322
National Institute for Environmental Health Sciences, 204
National Institute for Occupational Safety and Health, 165, 206
National Institute of Building Sciences (NIBS), 79
National Lead Assessment and Abatement Council, 63–64
National Lead Company, 296
National Lead Information Center, 192
National Lead Laboratory Accreditation Program (NLLAP), 165–166
National Low Income Housing Coalition, 159
National Paint and Coatings Association (NPCA), 295
National Safe and Healthy Housing Coalition, 204, 230, 330–331
National Safe and Healthy Housing Coalition Hill Days, 330–331
National Safety Council, 192
National Toxicology Program, 27
Needleman, Herbert, 276–277
New Orleans housing authority, 76
New York City Housing Authority (NYCHA), 75
  inspection reports, 75
New York State Healthy Neighborhoods Program, 328
Nilles, Bruce, 166–169, 170f, 207–208
NL industries, 296, 300
Noncommunicable disease, 353
Nontherapeutic research, 261–263

O

Occupational and Child Health, 35
Occupational health professionals, 62, 65, 91
Occupational Safety and Health Administration (OSHA), 39–40, 62, 357
  construction standard for lead, 206
  lead occupational health standards, 163
  lead regulation for construction workers, 65
Office of Human Subject Research Protection (OHRP), 269
Operation Lead Elimination Action Program (Operation LEAP), 215
Ormandy, David, 331–332, 334

P

Paint chip laboratory analysis, 16
Parents, 188–194, 264–265
Partial abatement, 249–250
Pathway studies, 24, 53–55
Peerless Paint and Varnish Company, 39
Phoenix, Janet, 87, 193, 280
Physicians for Social Reform, 188
Pierce, Samuel, 114, 220
Pilot study, 132–133
Poison Squad, The, 253
Policy paralysis paradox, 391–392
Political sabotage, 219–231
  attack and counterattack at HUD, 219–227
  attempted elimination of CDC lead program, 227–231
Polluter Pays principle, 354
Pollution, 354
  exclusion, 53
Population attributable fraction, 334
Postabatement dust lead cleaning and testing, 46
Preliminary lead dust standard, 74
President's Task Force, 208, 212
  on Children's Environmental Health, 207
  on Lead, 378
Prevention, 36–40
Primary prevention, 36–37
Principal Responsible Party (PRP), 130
Private housing, new congressional appropriations for, 98
Product liability law, 292
Public housing, 50–52, 339–341
  quality control system, 73
  risk assessment program, 74
Public Housing Guidelines (1990), 79–84
Public nuisance law, 293, 297–299
Pure Food and Drug Act (1906), 253

Q

Quality assurance system of dust wipe samples, 73
Quality Management Review, 221

R

"R&M" study, 244–245, 248–249, 254–256
Rall, David, 103–104, 204
Ramirez, Saul, 199
Re-Use Stores, 65
Recycling, 65
Reddy, Amanda, 104
Reed, Jack, 83, 89–90, 118, 193, 292, 332, 340, 381
Remediation methods, 140–146
Renovation, repair, and painting rule (RRP rule),
    101–102
"Repeat offender" houses, 83
Repulsive tasting paints, 39
Research ethics and protection research of study
    participants, 252–254
Rhode Island, 193
    court decision, 295–299
    public nuisance law, 297–299
Risk assessments, 71–75
Roebbel, Nathalie, 332, 336
Ryan, Don, 79, 85, 87, 91–92, 103–104, 127, 129, 156,
    224, 277, 280, 320

S

Sanitation movement, 364
Sauser, Margaret, 167, 194, 220, 378
Scientific evidence, 248–252
Screening programs, 36
Secondary prevention, 36–37
Securities and Exchange Commission (SEC), 6
Shabazz, Zakia, 192
Shalala, Donna, 23, 207
Shared Commons, 354–359
Sherwin Williams, 291, 295, 389
Shumway, John, 118, 166–167, 169, 170f, 172
Sick building syndrome, 330
Single-family mortgage insurance
    office, 153–154
    program, 155
Social determinants of health, 354, 367
Special duty, 265–266
Stewart McKinney Amendments (1988), 69–70
Sticky tape, 133
Stokes, Carl, 355–356
Stokes, Louis, 116, 313, 341f, 355–356
Strategic plans, 386–388
Sudden Infant Death Syndrome (SIDS), 314
Sullivan, Louis, 88, 387
Superfund program, 129–131

Surgeon General's Call to Action on Healthy Housing,
    325–326
Surveillance systems, 90
Systematic review of housing interventions and health,
    329–330

T

Task Force's creation, 207
Tax Credit, 212
Tertiary prevention, 36–37
Therapeutic research, 261–263
Title X of 1992 Housing and Community Development
    Act, 91–102
    confidence, 105
    new congressional appropriations for private
        housing, 98
    principal purposes, 97
    reaction to prevention, 94–96
    renovation, repair, and painting, 101–102
    workforce, 96
Title X task force, 126–129
Toxicity research, 27–28
Treatment *vs.* prevention, 36–40
Tuskegee Syphilis Study, 253

U

Ultraclean techniques, 14
United Church of Christ (UCC), 191
United Parents Against Lead (UPAL), 188
Universal screening, 89
University of Cincinnati, 143
Urgent Lead Paint Hazard Prevention Act, 93
US Centers for Disease Control and Prevention (CDC),
    10, 20–22, 36–37, 43, 62, 113, 188–189
    attempted elimination of CDC lead program,
        227–231
    blood lead surveillance grants, 190
    CDC Strategic Plan of 1991, 88–91
    Grand Rounds session, 351
    lead poisoning
        1971 statement, 37
        1978 statement, 37–38
        1985 statement, 40
        1991 statement, 40
        2012 statement, 40
        prevention funding, 229f
    Recommended Minimum Housing Standards, 364
    surveillance program, 230
US Conference of Catholic Bishops, 191
US Department of Housing and Urban Development
    (HUD), 15, 21–23, 27, 39, 43, 113, 187–188, 245,
        300, 313, 351, 373–374
    1995 rescission and bringing science to, 114–122

attack and counterattack at, 219–227
budget and staff, 85*t*
Comprehensive and Workable Plan, 84–85
evaluation study, 249–250
healthy homes report to Congress, 319–324
housing programs, 52
lead
   hazard control grant program, 122
   paint guidelines, 122–126
Lead-Safe Housing Rule, 148
Office of Lead Hazard Control, 198
   and Healthy Housing, 319
regulations, 44–45
scandal prompts Congress, 75–79
US Environmental Protection Agency (EPA), 7–8, 8*f*,
   11–12, 15, 27, 63, 113, 188–189, 245
US government agencies involved in lead paint,
   397–398

V
Vacuum methods, 133–135

W
Washington Press Club in 1995, 117
Waxman, Henry, 88, 99–100, 383
Weatherization programs, 212–213
Weicker, John, 84
Weitz, Steve, 78, 118, 143, 150
Welfare Rights Organization, 188
Wet washing, 124
Whitehouse, Sheldon, 193, 293, 295–296
White House objections, 88
White House Office of Management and Budget,
   121–122
White House Press conference, 199
Wilson, Jonathan, 104, 143
Wipe method, 133–135
World Health Organization (WHO), 332
   Declaration of Fourth Ministerial Conference on
     Environment and Health, 333
   Healthy Homes Movement, 331–339

X
X-ray fluorescence (XRF), 15–16, 44, 64
   companies, 164
   instruments, 82, 119
   lead paint analyzers, 378
   performance characteristics sheets, 164
   portable lead paint analyzers, 82

Y
Young Lords, 188